Nutrition, Epigenetic Mechanisms, and Human Disease

Nutrition, Epigenetic Mechanisms, and Human Disease

Edited by
Nilanjana Maulik and Gautam Maulik

CRC Press
Taylor & Francis Group
Boca Raton London New York

CRC Press is an imprint of the
Taylor & Francis Group, an **informa** business

CRC Press
Taylor & Francis Group
6000 Broken Sound Parkway NW, Suite 300
Boca Raton, FL 33487-2742

Printed in the United States of America on acid-free paper
10 9 8 7 6 5 4 3 2 1

International Standard Book Number: 978-1-4398-0479-7 (Hardback)

Library of Congress Cataloging-in-Publication Data

Nutrition, epigenetic mechanisms, and human disease / edited by Nilanjana Maulik, Gautam Maulik.
p. ; cm.
Includes bibliographical references and index.
ISBN 978-1-4398-0479-7 (hardcover : alkaline pager)
1. Nutrition--Genetic aspects. I. Maulik, Nilanjana, editor. II. Maulik, Gautam, editor.
[DNLM: 1. Nutritional Physiological Phenomena--genetics. 2. Disease Susceptibility. 3. Epigenesis, Genetic. QU 145]

QP144.G45N886 2011
612.3--dc22 2010044037

Visit the Taylor & Francis Web site at
http://www.taylorandfrancis.com

and the CRC Press Web site at
http://www.crcpress.com

This book is dedicated to our parents whose unfailing support, encouragement, and affection tided us over many difficulties.

Nilanjana Maulik, PhD, FAHA, FACN
Gautam Maulik, PhD

Contents

Preface.....ix
Acknowledgments.....xi
The Editors.....xiii
Contributors.....xv

Chapter 1 Nutritional Epigenetics and Disease Prevention: Are We There Yet?.....1
Mukesh Verma

Chapter 2 Aging by Epigenetics: Nutrition, An Epigenetic Key to Long Life.....13
Nilanjana Maulik and Gautam Maulik

Chapter 3 Folate and DNA Methylation.....31
Julie Crowell, Anna Ly, and Young-In Kim

Chapter 4 Dietary Components, Epigenetics, and Cancer.....77
Cindy D. Davis and Sharon A. Ross

Chapter 5 Dietary Factors, Histone Modifications, and Cancer Prevention.....109
Igor P. Pogribny, Sharon A. Ross, and Igor Koturbash

Chapter 6 Nutrition, Epigenetics, and Vascular Function.....125
M. Carey Satterfield, Jason R. McKnight, Xilong Li, and Guoyao Wu

Chapter 7 Role of Epigenetic Machinery and MicRNAs in Diet-Induced Hepatocarcinogenesis.....141
Kalpana Ghoshal and Tasneem Motiwala

Chapter 8 Epigenetic Mechanisms in Lung Inflammation and Chronic Airway Diseases and Intervention by Dietary Polyphenols.....185
Irfan Rahman

Chapter 9 Glycemic Memory and Epigenetic Changes 215

Andrew L. Siebel, Ana Z. Fernandez, and Assam El-Osta

Chapter 10 Maternal Nutrition, Intrauterine Development, and Disease Risks in the Offspring through Epigenetic Regulation of Gene Expression 229

Yuan-Xiang Pan, Rita Strakovsky, and Shasha Zheng

Chapter 11 Nutritional Epigenetics: Impact on Metabolic Syndrome 259

Lakshminarasimhan Pavithra, Sreenivas Chavali, and Samit Chattopadhyay

Chapter 12 Nutrition and the Emerging Epigenetic Paradigm: Lessons from Neurobehavioral Disorders 287

Axel Schumacher

Chapter 13 Interactions between Folate, Other B Vitamins, DNA Methylation, and Neurodevelopmental Disorders 317

Rebecca J. Schmidt and Janine M. LaSalle

Chapter 14 Dietary Factors and the Emerging Role of Epigenetics in Neurodegenerative Diseases 363

Lucia Migliore and Fabio Coppedè

Index 381

Preface

A uniquely tailored diet that corresponds to the demands of our genetic signature is becoming an emerging indispensable need, as nutrition research is shifting its focus from epidemiology and physiology to effects of nutrients at the molecular level. Nutrigenomics relates to the application of high-throughput genomic tools in nutrition research to unravel the influence of micro- and macronutrients as potent dietary signals regulating metabolic pathways (dietary signature) and unmask how susceptible genotypes predispose to diet-related diseases. Since the last decade, extensive research on nutrigenomics has unveiled numerous epigenetic mechanisms that are influenced by our dietary signature and are capable of modifying an individual's susceptibility to diet-related disorders. The primary objective of this volume is to illustrate how nutrition can influence epigenetic inheritance and the mechanisms that underlie the modification of metabolic imprint of an individual, so that our enriched understanding of nutrigenomics can be applied to master a tailored diet that can alleviate imprinted metabolic syndromes. Specifically, the focus of the book will be on three key areas: discussion of the basics of nutrigenomics and epigenetic regulation, types of nutrition influencing the genetic imprinting, and the role of nutrition in modulating an individual's predisposition to cancer.

Nutrigenomics aims at devising dietary-intervention strategies to alleviate diet-related diseases and to restore normal metabolic homeostasis of the body. Epigenetic mechanisms like DNA methylation and transposon insertion have been shown to play at the nexus between nutrition and the genetic signature of an individual. Chromatin remodeling across the genome mediated via epigenetic mechanisms and transient nutritional stimuli can wield persistent changes on the genomic profile that are likely to be passed on to the subsequent generations. Genomic imprinting refers to a unique type of epigenetic regulation whereby differential modification of the parental alleles at certain genetic loci in the parental germlines (imprinting control regions) takes place depending on whether the allele is passed on to the offspring through the male or female gamete. Genomic imprinting mechanisms have been shown to be influenced by maternal modifier genes (after fertilization) resulting in the removal of paternal imprints on sperm DNA as well as by the dietary signature. Human epidemiologic studies reveal that metabolic imprinting is affected by poor perinatal and neonatal nutrition as well as maternal nutritional imbalance, which might result in predispositions to adult obesity, cardiovascular disease, atherosclerosis, hypertension, cancer, and type 2 diabetes.

This book addresses a very complex scenario related to nutrition—epigenetic changes related to human health and diseases. The contents are highly relevant, focused, and very timely. Recently, the National Institutes of Health (NIH) has added "epigenetic" to its roadmap; therefore a book on nutrition and epigenetics is certainly in demand. It is written by world-recognized experts in the field of nutrition, epigenetic regulations and gene expression related to aging, various cancers, vascular function, lung inflammation, diabetes, metabolic syndrome, and neurodegenerative diseases. Selected topics from this field have been covered in some books, but no comprehensive text on epigenetics, nutrition, and human health and disease is available. This handbook includes 14 contributions from leading scientists. After Chapter 1, "Nutritional Epigenetics and Disease Prevention: Are We There Yet?," the book deals with various ongoing researches on nutrition-mediated regulation of epigenetic mechanisms and various disease scenarios. We are very sure this invaluable reference book is of interest to all health care–related professionals as well as nutritionists, biochemists, cancer biologists, pharmacologists, and mutagenesists.

This book is intended for biochemists, molecular biologists, cell biologists, biomedical researchers, and clinical researchers.

Nilanjana Maulik, PhD, FAHA, FACN
Gautam Maulik, PhD

Acknowledgments

We gratefully acknowledge all the contributing authors for their excellent thought-provoking contributions in spite of their busy schedules. We express our sincere thanks to Professor Debasis Bagchi, Department of Pharmacological and Pharmaceutical Sciences, University of Houston College of Pharmacy, Houston, Texas, for his constant encouragement and help from the time of inception of this work to its completion.

We are grateful to acknowledge our colleague Mahesh Thirunavukkarasu, PhD, for his cooperation and insurmountable help in checking the format for each chapter.

We would also like to extend our thanks to Debayon Paul, MS, and Ram Sudheer Adluri, PhD, for their assistance during the preparation of this book.

The Editors

Nilanjana Maulik, PhD, FAHA, FACN, FICA, is a professor of molecular cardiology and heads the Angiogenesis Laboratory at the Department of Surgery, University of Connecticut Medical Center, Farmington. Professor Maulik earned her PhD in biochemistry in December 1990 from Calcutta University, India. After completion of her PhD, Professor Maulik joined the Department of Surgery at University of Connecticut Medical Center as a research fellow, continued as a faculty member, and now serves as tenured professor. She is also a faculty member of the Cell Biology Graduate Program at the University of Connecticut Health Center. She is heavily involved in NIH-funded research and has delivered more than 100 invited lectures both nationally and internationally. Professor Maulik has organized several international conferences/symposia. She is also a member of several prestigious societies, such as FASEB, AHA, ISHR, American College of Nutrition (ACN), and International College of Angiology (ICA). She has been a member of the Myocardial Ischemia Metabolism (MIM) study section of the NIH for the last 6 years and of the NHLBI Program Project Review Committee. Professor Maulik serves as a special panel board member (NIH) and as a member of the Northern Connecticut Chapter of AHA grant review process. She has also served in several other study sections of the NIH such as CVB, ECS, and VSCB. She is on several editorial boards of major cardiovascular journals and is an associate editor of *Molecular Cellular Biochemistry* journal. Teaching is an integral part of her professional path. She is a recipient of several prestigious awards including the Faculty Recognition Award from the University of Connecticut Health Center. Recently she has been appointed as the Director of Health Sciences for the International Academy of Cardiovascular Sciences, Manitoba, Canada. Her research focuses on the molecular mechanism of myocardial angiogenesis in the infarcted heart, ischemia/reperfusion injury, apoptosis, epigenetic modifications, and the development of cardioprotective strategies, which include gene and stem cell therapy. She has published 189 original peer-reviewed articles and 35 book chapters.

Gautam Maulik, PhD, is an instructor at the Department of Surgery, Brigham and Women's Hospital, Harvard Medical School, Boston, Massachusetts. Dr. Maulik earned his master's degree in biochemistry in 1983 from Calcutta University, Calcutta, India. Immediately after earning his degree, Dr. Maulik enrolled in the doctoral program in the Department of Biochemistry at Jadavpur University, Calcutta, India. After completing his PhD, Dr. Maulik came to the United States in 1994 to continue his research on free radical-mediated oxidative stress. Dr. Maulik has since joined the Dana Farber Cancer Institute/Harvard Medical School where he has continued to produce outstanding research. His most profound accomplishment since joining Harvard Medical School has been the development of a sophisticated cancer detection method and anticancer drugs. Based on his recent work, new anticancer drugs have been developed and are currently being tested for lung cancer and other malignancies. Dr. Maulik has already identified some of the factors and molecular pathways that are involved in the pathogenesis of lung cancer. He has published more than 50 articles in esteemed scientific journals, including *Cancer Research, Oncogene, Clinical Cancer Research, Nucleic Acids Research,* and *PNAS,* and has presented his research at important scientific conferences throughout the world, including in the United States, Japan, and India. Dr. Maulik has also published three book chapters.

Contributors

Samit Chattopadhyay
National Centre for Cell Science
Pune, Maharashtra, India

Sreenivas Chavali
University of Gothenburg
Gothenburg, Sweden

Fabio Coppedè
University of Pisa
Pisa, Italy

Julie Crowell
University of Toronto
Toronto, Ontario, Canada

Cindy D. Davis
National Cancer Institute
Bethesda, Maryland

Assam El-Osta
Baker IDI Heart and Diabetes Institute
Melbourne, Victoria, Australia

Ana Z. Fernandez
Venezuelan Institute for Scientific Research
Caracas, Venezuela

Kalpana Ghoshal
The Ohio State University
Columbus, Ohio

Young-In Kim
University of Toronto
St. Michael's Hospital
Toronto, Ontario, Canada

Igor Koturbash
National Center for Toxicological Research
Jefferson, Arkansas

Janine M. LaSalle
University of California–Davis School of Medicine
Davis, California

Xilong Li
Texas A&M University
College Station, Texas

Anna Ly
University of Toronto
Toronto, Ontario, Canada

Guatam Maulik
Harvard Medical School
Boston, Massachusetts

Nilanjana Maulik
University of Connecticut Health Center
Farmington, Connecticut

Jason R. McKnight
Texas A&M University
College Station, Texas

Lucia Migliore
University of Pisa
Pisa, Italy

Tasneem Motiwala
The Ohio State University
Columbus, Ohio

Yuan-Xiang Pan
University of Illinois–Urbana-Champaign
Urbana, Illinois

Lakshminarasimhan Pavithra
National Centre for Cell Science
Pune, Maharashtra, India
University of Gothenburg
Gothenburg, Sweden

Igor P. Pogribny
National Center for Toxicological Research
Jefferson, Arkansas

Irfan Rahman
University of Rochester Medical Center
Rochester, New York

Sharon A. Ross
National Cancer Institute
Bethesda, Maryland

M. Carey Satterfield
Texas A&M University
College Station, Texas

Rebecca J. Schmidt
The M.I.N.D. Institute
University of California
Davis School of Medicine
Davis, California

Axel Schumacher
Centre for Addiction and Mental Health
Toronto, Canada

Andrew L. Siebel
Baker IDI Heart and Diabetes Institute
Melbourne, Victoria, Australia

Rita Strakovsky
University of Illinois–Urbana-Champaign
Urbana, Illinois

Mukesh Verma
National Cancer Institute
Bethesda, Maryland

Guoyao Wu
Texas A&M University
College Station, Texas

Shasha Zheng
University of Illinois–Urbana-Champaign
Urbana, Illinois

1 Nutritional Epigenetics and Disease Prevention
Are We There Yet?

Mukesh Verma

CONTENTS

1.1 Epigenetics Mechanism and Gene Regulation ... 1
1.2 Components of Epigenetic Machinery ... 2
1.3 NIH Epigenomics Roadmap ... 4
1.4 Nutrients and Their Contribution in Epigenetic Regulation of Different Diseases ... 7
1.5 Challenges and Opportunities in the Field, Future Directions, and Concluding Remarks ... 8
Acknowledgment ... 9
References ... 9

1.1 EPIGENETICS MECHANISM AND GENE REGULATION

Our genome contains information related to gene structure and function but when and how long that information is utilized is determined by our epigenome. Epigenetics includes alterations in gene expression that do not include a change in DNA sequence during growth, development, and disease states (Ballestar and Esteller 2008). For normal function of a cell or organ, epigenetic regulation is needed; however, this regulation is disturbed during disease initiation and progression. Thus our genome is the "hardware" and our epigenome is the "software" of the body. Genetic information is static whereas epigenetic information is dynamic and transient. In the body, all the cells have the same genome but each cell has a different epigenome. The phenotype of a cell is determined by its epigenome (Murrell et al. 2005). Environmental factors (e.g., exposure to radiation, infectious agents), nutrients, toxins, and disease states affect the epigenome resulting in altered gene expression (Verma 2003; Kumar and Verma 2009) (Figure 1.1). Epigenetic changes can be measured quantitatively and followed during the progression/regression and recurrence of a disease (Ganesan et al. 2009; Feinberg 2010; Verma et al., 2004; 2006). This chapter compiles existing knowledge regarding the application of epigenetics toward understanding the dynamic interrelationship between bioactive food components (and/or a combination thereof) and cancer prevention.

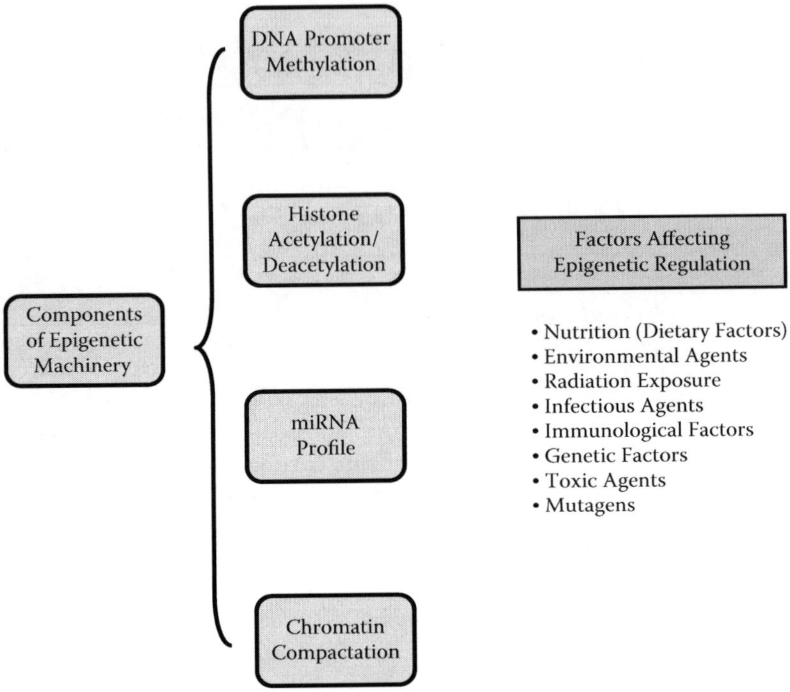

FIGURE 1.1 Components of epigenetics machinery and factors that influence gene regulation epigenetically. Factors mentioned here may work independently or in combination. A few factors affect DNA, whereas others affect proteins and nucleic acids simultaneously.

1.2 COMPONENTS OF EPIGENETIC MACHINERY

The major components of epigenetics are DNA methylation (methylation code), histone modification (histone code), chromatin compactation and relaxation, gene imprinting, and microRNA (miRNA) profile (Figure 1.1). Chromatin, which is composed of nucleosomes, is the key component of epigenetics. Nucleosomes are comprised of histone proteins arranged as octamers associated with 146 bp of DNA via its negatively charged phosphate backbone (Lustberg and Ramaswamy 2009; Mitsiades and Anderson 2009). The amino terminal part of histones protrudes out and becomes susceptible to enzymatic modifications, specifically at lysine residues, but also at other amino acids. More than 100 histone modifications of amino acids have been reported (Ganesan et al. 2009; Verma and Kumar 2009). The dimeric H3 and H4 form a tetramer, whereas H2A and H2B remain as dimmer.

The DNA in promoter regions of several genes contain CpG islands (regions rich in GC content), which are covalently modified due to methylation of cytosines at the 5′ position (Hitchins and Ward 2009; Laird 2010). This process is called hypermethylation. A number of tumor suppressor genes get inactivated due to hypermethylation of their promoters (Verma et al. 2006). On the other hand, a few genes, such as oncogenes, are methylated in their normal states and become hypomethylated in disease states, resulting in their activation (Ballestar and Esteller 2008; Laird 2010).

Throughout life, equilibrium is needed in the methylation state of the whole epigenome. Enzymes that are involved in methylation are called methyl transferases. These enzymes either initiate or maintain methylation. Proteins that bind to methylated DNA have also been identified and characterized and are referred to as methylated binding proteins (MBPs). The roles of other proteins, such as the polycomb group of proteins, have also been defined (Gieni and Hendzel 2009).

Repetitive regions, such as LINE and Alu, are hypermethylated in the normal state and are hypomethylated during growth and development. This process prevents chromosomal instability, translocation, and gene disruption caused by activation of transposons (Ballestar and Esteller 2008; Esteller 2008).

Quantitative measurements of methylation levels help in disease detection, progression, and follow-up to treatment. For example, hypermethylation of the glutathione gene (GSTP1) occurs only in prostate cancer and not in benign states (Bryzgunova et al. 2008). Thus human populations can be screened based on the methylation status of the GSTP1 gene to distinguish high-risk individuals. Furthermore, the methylation of cytosine can be reversed by chemicals, such as azacytidine and deoxycytidine, and inactive genes can be activated by chemical agents, both of which provide therapeutic potential (Ganesan et al. 2009). There are a few drugs that have been approved by the Food and Drug Administration (FDA) that have demonstrated promising results in clinical trials (Verma 2010).

MicroRNAs (miRNAs) are small noncoding RNAs with a length of 21–25 bp that possess the ability to suppress translation of a gene by binding to partially complementary messenger RNA (mRNA) (Ku and McManus 2008; Chen et al. 2009). For example, Let-7 and miR-15/miR-16 inactivate oncogenes RAS and BCL2, respectively (Esteller 2008; Garzon et al. 2009). Recent research indicates that selected miRNAs are tissue and disease specific. Cell development, differentiation, and death is affected by miRNAs (Sekine et al. 2009). miRNAs can also be used for disease detection and treatment follow-up (Chen et al. 2009). Technologies exist to perform epigenome-wide miRNA profiling to identify differentially expressed miRNAs in disease states.

The orientation and modulation of histones contribute to the heterochromatin and euchromatin states. Histone acetylation/deacetylation may result in turning off of the cell cycle regulatory genes, inactivation of tumor suppressor genes, and activation of oncogenes (Lane and Chabner 2009). Enzymes that mediate acetylation (acetyltransferases) and deacetylation (deacetylases) are well characterized in different cell types (Lane and Chabner 2009; Villagra et al. 2009). Histone modifications also include biotinylation, methylation, phosphorylation, sumoylation, and ubiquitination (Verma and Kumar 2009). Acetylation of H3-lysine has been observed at locations 9, 14, 18, and 23, whereas the lysine locations of that on H4 is at 5, 8, 12, and 16. The interaction of acetyl groups occurs at the epsilon amino group of lysine, resulting in histone neutralization. Other amino acids of histones that generally undergo alteration include arginine and serine (Ma et al. 2009; Marson 2009; Verma and Kumar 2009).

One additional type of gene regulation is gene imprinting, which is paternal or maternal allele-specific expression of a limited number of genes (50–80) (Jelinic and Shaw 2007; Vu et al. 2010). Without proper imprinting control, abnormal growth occurs. Examples of diseases regulated epigenetically via imprinting are

TABLE 1.1
Clinical Samples for Epigenetic Analysis

Sample	Suitable Analysis
Bronchoalveolar lavage	DNA isolation and methylation profiling
Buccal cells	DNA isolation and methylation profiling
Ductal lavage fluid	Proteomic analysis (histone and nonhistones) and methylation profiling
Cervical swab	Proteomic analysis (histone and nonhistones) and methylation profiling; methylation profile of infectious agents (HPV)
Duodenal fluid	Proteomic analysis (histone and nonhistones) and methylation profiling
Ejaculate	DNA isolation and methylation profiling (for example, GSTP1 methylation in prostate cancer)
Exfoliated cells	Proteomic analysis (histone and nonhistones) and methylation profiling
Nipple aspirate	Proteomic analysis (histone and nonhistones) and methylation profiling
Pleural lavage	DNA isolation and methylation profiling
Saliva	DNA isolation and methylation profiling, proteomic analysis, and miRNA profiling
Sputum	DNA isolation and methylation profiling, proteomic analysis, and miRNA profiling
Stool	DNA isolation and methylation profiling
Tissue	DNA isolation and methylation profiling, proteomic analysis, and miRNA profiling
Urine	DNA isolation and methylation profiling (for example, bladder cancer)

Beckwith-Wiedemann syndrome (BWS), Silver-Russell syndrome (SRS), and X-chromosome inactivation (Zhao et al., 2009). Methylation of DNA occurs in the imprinting loci called imprinting control regions (ICRs). Loss of imprinting (LOI) of IGF2 has been proposed in stem cell proliferation and cancer (Dammann et al. 2010; Timp et al. 2009). A variety of biospecimens can be utilized for epigenetic analysis (Table 1.1).

1.3 NIH EPIGENOMICS ROADMAP

The National Institutes of Health (NIH) has initiated the Epigenomics Roadmap Program (www.roadmapepigenomics.org), which is comprised of five major initiatives: Reference Epigenome Mapping Centers, Epigenomic Data Analysis and Coordinating Centers, Technology Development in Epigenetics, Discovery of Novel Epigenetic Marks, and Epigenomics of Human Health and Diseases. This program proposes that the origins of health and susceptibility to disease are the result of epigenetic regulation of the genetic information. Specifically, epigenetic mechanisms that control stem cell differentiation and organogenesis contribute to the biological response to environmental and other factors in the form of stimuli that contribute in disease development. To accomplish this, the Roadmap Epigenomics Program plans to develop standardized platforms, procedures, and reagents for epigenomics research; conduct demonstration projects to evaluate how epigenomes change;

develop new technologies for single-cell epigenomic analysis and in vivo imaging of epigenetic activity; and create a public data resource to accelerate the application of epigenomics approaches. This program will transform biomedical research by developing comprehensive reference epigenome maps, developing new technologies for comprehensive epigenomic analyses, and providing novel strategies for disease detection, diagnosis, treatment, and prognosis. Since several institutes are participating in this initiative, a number of diseases are covered in the roadmap. A few examples of epigenetic approaches, already funded by this program, in different diseases are described next.

In vascular epigenomics, Gary Gibbons's group (Morehouse School of Medicine) has proposed that the high prevalence of hypertensive vascular disease among African Americans is due to gene and environment interaction mediated via vascular epigenome. Hypertensive vasculature complications involve long-term changes in vessel function and structure and may contribute to diseases such as stroke. The underlying mechanisms are not completely understood. This group proposes that a group of genes are "vasculopathic" and the other group is "vasculoprotective." The methylation profile and histone modifications will be studied for whole genome and disease-specific methylation profile, and histone modifications will be identified. The participants in this proposal are age- and sex-matched controls and cases of African American origin. As a follow-up of this study, profiles will be developed from samples of cases undergoing treatment with different food habits and lifestyles. The data obtained will be available for public and will be an excellent resource to develop new prevention, intervention, and treatment strategies in vascular diseases.

Jessica Connelly of the University of Virginia is conducting research to test whether methylation plays a major role in endothelial cells and smooth muscle cells undergoing phenotypic switching during atherosclerosis initiation and progression. Normal and disease-affected tissue samples will be utilized in this case-control study to identify differentially expressed methylation profile. Contribution of genetic factors in epigenetic regulation of disease-related genes will also be accomplished. Another group, led by Yongmei Liu (Wake Forest University Health Sciences), has focused on investigating association of global methylation profile in circulating monocytes in relation to atherosclerosis and monocyte gene expression in the Multi-Ethnic Study of Atherosclerosis (MESA). More than 1500 samples of Caucasian, African American, and Hispanic origin will be analyzed by these investigators. After identifying disease-associated methylation marks, validation of these marks will be accomplished in more samples. Contribution of environmental, lifestyle, and dietary factors will also be evaluated.

A number of chronic conditions, such as cognitive decline and dementia, are developed during old age that impair older persons' ability to interact optimally in the community. Neuropathology of cognitive decline in old-age diseases, such as Alzheimer's disease, cerebrovascular disease, Lewy Body disease, accounts for only 20% of the cognitive behavior. David Benett (Rush University Medical Center) has proposed that there are other factors that may contribute to the remaining cognitive decline in old age. Life experiences (socioeconomic status, psychological distress, chronic non-neuronal diseases) are not related to known neuropathological process but contribute in disease development. Preliminary evidence of altered epigenetic marks in these diseases exists, but a systematic study has not been completed. David Benett is conducting epigenome-wide methylation analysis of brain tissues from participants in the Rush Memory and Aging Project and the Religious Orders Study. Furthermore, data from genomewide association studies (GWASs) of brain tissues will be used to establish correlation of epigenetics and genetics in cognitive diseases and brain disorders. Another group, led by Paul Coleman of Sun Health Research Institute, plans to utilize more than 500 samples of Alzheimer's disease from the Brain Bank at this institute to perform methylation profile and compare with genotyping data. It is our hope that the data will help in early diagnosis of the disease and in identifying new targets for treatment. Jonathan Mill of Kings College, London, is exploring epigenetic regulation in Alzheimer's disease using well-characterized postmortem Alzheimer's-disorder brains, and he will cover different regions of the brain in his analysis. Detailed clinical data on these patients is available before their death. Further functional analysis of potential genes will also be accomplished. Roel Ophoff of University of California at Los Angeles plans to investigate schizophrenia-associated epigenetic changes because mutations are rare in this disease and it seems logical to evaluate alternative mechanisms.

Type 2 diabetes mellitus (T2DM) is developed mostly in adults, but a few cases in younger ages have also been reported. Francine Einstein of the Albert Einstein College of Medicine is evaluating the epigenetic regulation in utero and its contribution to T2DM development during the lifetime. Stem cells will be utilized in this project. It is expected that understanding the contribution of intrauterine conditions to chronic adult disease may lead to novel epigenetic markers that may help in identifying high-risk individuals and populations. Evan Rosen of Beth Israel Deaconess Medical Center plans to study adipocyte methylation patterns to identify insulin resistance–associated epigenetic marks.

Autism is a neurological disorder with features like impaired social interaction and restricted and repetitive behavior, and it starts quite early in life. Margaret Fallin of Johns Hopkins University thinks that autism and related disorders have an epigenetic basis. Experiments are being conducted to test whether environment plays a major role in disease development and whether epigenetic regulation is predominant in that genetic regulation in autism. Samples from the Johns Hopkins National Children's study will be utilized in this study. The research will help us understand: are there regions of the epigenome susceptible to environment before and during pregnancy; and are there epigenomic regions that correlate with newborn and infant development phenotypes related to diabetes?

Glaucoma and age-related muscular degeneration and their regulation by epigenetic mechanisms will be studied by Shannath Merbs of Johns Hopkins University. About 4 million people are affected by these diseases in the United States. Samples from age- and sex-matched control cases will be analyzed for global methylation profiles. These samples were taken by laser capture microdissection so that the methylation profile can be obtained in retinal ganglia cells and in photoreceptor and retinal pigment epithelium cells.

Epigenetic regulation of bipolar disorders (BPDs)—such as the discordance of identical twins, significant fluctuations in disease initiation, progression, and development, and sex and paternal origin effects—will be studied by Art Petronis (Center for Addiction and Mental Health, Canada). The prefrontal cortex of individuals affected with BPD and schizophrenia will be utilized to discover disease-associated methylation profiles in various genes.

Asthma epigenetics will be studied by David Schwartz of the National Jewish Health Center by analyzing methylation profiles in T cells, airways epithelium, and mononuclear cells during disease development. Such studies will help in designing novel prevention and treatment strategies in asthma.

1.4 NUTRIENTS AND THEIR CONTRIBUTION IN EPIGENETIC REGULATION OF DIFFERENT DISEASES

Multiple factors interact with genes and contribute to phenotypes of disease development. Along with environmental factors, dietary components have a major role in both disease prevention and development (Coppedè 2009; Ross et al. 2008). The folate pathway has been studied as a candidate biochemical and metabolic pathway for colon cancer (Carr et al. 2008). This pathway has been conserved among species, indicating its significance (Johnson and Belshaw 2008). Genetic variants in relevant genes have shown associations with diseases such as cancer, heart disease, and neural tube defects. In colon cancer adenomas, dietary folic acid supplementation has a protective effect, whereas either no effects or adverse effects have been observed in relation to colon cancer recurrence. Genetic explanations alone cannot explain these observations; therefore attempts are being made to understand these associations by alternative mechanisms, such as epigenetics (Carr et al. 2008; Ulrich 2008).

Although reports indicate that nutrition plays a role in disease prevention, especially cancer, interactions among dietary bioactive food compounds and food combinations remain understudied. Colon cancer is one of the few areas of nutritional epigenetics that has been well studied (Johnson and Belshaw 2008). Folic acid is a well-known methyl donor, and several foods are fortified with folic acid. Folic acid one-carbon metabolism (FOCM) is an excellent example of a complex pathway with interconnected subpathways for folic acid and methionine metabolism, which in turn have their own feedback loops (Kim et al. 2009). Furthermore, folate biochemistry is well defined, and enzymes involved in the metabolism of folate, whether they exist in the cytoplasm or mitochondria, are well characterized (Ulrich 2008). Since methyl groups are the key component in CpG methylation, their levels influence gene expression. Alterations in homocysteine levels, DNA methylation, purine and thymidylate synthesis, and incorporation of uracil into DNA (misincorporation) occur simultaneously in the cells and contribute to DNA damage and repair pathway.

In metabolic syndromes, Plagemann et al. (2009) has proposed that overfeeding, by way of epigenetic factors, contributes to obesity, and subsequently to diabetes and cardiovascular diseases. Their conclusions are based on methylation profiling of the proopiomelanocortin gene, which encodes a polypeptide hormone precursor that undergoes extensive tissue-specific posttranslational modifications by an enzyme, prohormone convertase. The phenotypes included in the study were obesity, hyperleptinemia, hyperglycemia, hyperinsulinemia, and an increased insulin/glucose ratio. Histone demethylase has also been proposed to influence metabolic syndromes (Inagaki et al. 2009).

It is important to know the bioactive food components, their quality, and mode of interaction with the body in order to apply the effects of nutrients and their components to a healthy lifestyle. Genes controlling the synthesis and metabolism of bioactive food components are regulated genetically and epigenetically. A comprehensive understanding of genotype (genetics) and its relation with phenotype (epigenetics) is needed if nutrition is to be applied for disease intervention and prevention purposes (Milner 2008).

1.5 CHALLENGES AND OPPORTUNITIES IN THE FIELD, FUTURE DIRECTIONS, AND CONCLUDING REMARKS

Major challenges in the field of nutritional epigenetics are the large number of input variables, relatively few intermediate markers and measurements, dynamic nature of nutrients, and limited outcome measurements. One approach to address these problems could be the application of a systems biology approach where *in silico* models are developed based on biological information to test these models first in animals and then in human populations. Taking colon cancer as a prototype, existing databases, such as the Colon Cancer Family Registry Folate Study, should be utilized to develop models to understand dietary influences on epigenetics and disease development. As the next step, single-pathway approaches can be expanded to include a genomewide approach because technologies exist for measuring genomewide epigenetic changes (Feinberg 2010; Laird 2010). Combining observational studies with experimental studies may result in risk-prediction models with implications for identifying populations at high risk of developing diseases. Incorporating genomic information in epigenetic databases may also be useful in understanding the biology of the underlying disease, developing intervention targets, and ultimately improving health. There are a few bioactive food components that have activity with deacetylate histones, but this effect is general and not gene specific. Gene-specific inhibitors are needed to treat specific diseases.

Research questions for the future include the following:

- How do bioactive food components regulate epigenetic events in different diseases?
- How do bioactive food components alter epigenetic patterns and restore gene function?
- How do these components circumvent and compensate for pathways that are altered during the disease development?
- How can we make gene-specific epigenetic inhibitors?
- How can we measure the temporality in epigenetic profile caused by bioactive food components?

Epigenetics in general and nutrition epigenetics in particular have the potential to make a tremendous impact in disease prevention, control, and management. However, validation studies have not been completed to evaluate this potential in different diseases; therefore it would be premature to declare that nutrients can prevent or treat diseases. Once the human epigenome is completed and additional nutritional epigenetic studies have been conducted, it may be possible to achieve this goal.

ACKNOWLEDGMENT

We are thankful to Christine Kaefer and Britt Reid for reading the manuscript and providing their suggestions.

REFERENCES

Ballestar, E., and M. Esteller. 2008. Epigenetic gene regulation in cancer. *Adv Genet* 61:247–67.

Bryzgunova, O. E., E. S. Morozkin, S. V. Yarmoschuk, V. V. Vlassov, and P. P. Laktionov. 2008. Methylation-specific sequencing of GSTP1 gene promoter in circulating/extracellular DNA from blood and urine of healthy donors and prostate cancer patients. *Ann N Y Acad Sci* 1137:222–25.

Carr, D. F., G. Whiteley, A. Alfirevic, and M. Pirmohamed. 2008. Investigation of inter-individual variability of the one-carbon folate pathway: a bioinformatic and genetic review. *Pharmacogenomics J* 9:291–305.

Chen, Y., J. A. Gelfond, L. M. McManus, and P. K. Shireman. 2009. Reproducibility of quantitative RT-PCR array in miRNA expression profiling and comparison with microarray analysis. *BMC Genomics* 28:407–8.

Coppedè, F. 2009. The complex relationship between folate/homocysteine metabolism and risk of Down syndrome. *Mutat Res* 682:54–70.

Dammann, R. H., S. Kirsch, U. Schagdarsurengin, T. Dansranjavin, E. Gradhand, W. D. Schmitt, and S. Hauptmann. 2010. Frequent aberrant methylation of the imprinted IGF2/H19 locus and LINE1 hypomethylation in ovarian carcinoma. *Int J Oncol* 36:171–79.

Esteller, M. 2008. Epigenetics in cancer. *N Engl J Med* 358:1148–59.

Feinberg, A. P. 2010. Genome-scale approaches to the epigenetics of common human disease. *Virchows Arch* 456:13–21.

Ganesan, A., L. Nolan, S. J. Crabb, and G. Packham. 2009. Epigenetic therapy: histone acetylation, DNA methylation and anti-cancer drug discovery. *Curr Cancer Drug Targets* 9:963–81.

Garzon, R., G. A. Calin, and C. M. Croce. 2009. MicroRNAs in Cancer. *Annu Rev Med* 60:167–79.

Gieni, R. S., and M. J. Hendzel. 2009. Polycomb group protein gene silencing, non-coding RNA, stem cells, and cancer. *Biochem Cell Biol* 87:711–46.

Hitchins, M. P., and R. L. Ward. 2009. Favoritism in DNA methylation. *Cancer Prev Res (Phila Pa)* 2:847–49.

Inagaki, T., M. Tachibana, K. Magoori, H. Kudo, T. Tanaka, M. Okamura, M. Naito, T. Kodama, Y. Shinkai, and J. Sakai. 2009. Obesity and metabolic syndrome in histone demethylase JHDM2a-deficient mice. *Genes Cells* 14:991–1001.

Jelinic, P., and P. Shaw. 2007. Loss of imprinting and cancer. *J Pathol* 211:261–68.

Johnson, I. T., and N. J. Belshaw. 2008. Environment, diet and CpG island methylation: epigenetic signals in gastrointestinal neoplasia. *Food Chem Toxicol* 46:1346–59.

Kim, K.C., S. Friso, and S. W. Choi. 2009. DNA methylation, an epigenetic mechanism connecting folate to healthy embryonic development and aging. *J Nutr Biochem* 20:917–26.

Ku, G., and M. T. McManus. 2008. Behind the scenes of a small RNA gene-silencing pathway. *Hum Gene Ther* 19:17–26.

Kumar, D., and M. Verma. 2009. Methods in cancer epigenetics and epidemiology. *Methods Mol Biol* 471:273–88.

Laird, P. W. 2010. Principles and challenges of genome-wide DNA methylation analysis. *Nat Rev Genet* 1:191–203.

Lane, A. A., and B. A. Chabner. 2009. Histone deacetylase inhibitors in cancer therapy. *J Clin Oncol* 27:5459–68.

Lustberg, M. B., and B. Ramaswamy. 2009. Epigenetic targeting in breast cancer: therapeutic impact and future direction. *Drug News Perspect* 22:369–81

Ma, X., H. H. Ezzeldin, and R. B. Diasio. 2009. Histone deacetylase inhibitors: current status and overview of recent clinical trials. *Drugs* 69:1911–34.

Marson, C. M. 2009. Histone deacetylase inhibitors: design, structure-activity relationships and therapeutic implications for cancer. *Anticancer Agents Med Chem* 9:661–92.

Milner, J. A. 2008. Nutrition and cancer: essential elements for a roadmap. *Cancer Lett* 269:189–98.

Mitsiades, C. S., and K. C. Anderson. 2009. Epigenetic modulation in hematologic malignancies: challenges and progress. *J Natl Compr Canc Netw* 7:S1–12.

Murrell, A., V. K. Rakyan, and S. Beck. 2005. From genome to epigenome. *Hum Mol Genet* 15:R3–10.

Plagemann, A., T. Harder, M. Brunn, A. Harder, K. Roepke, M. Wittrock-Staar, T. Ziska, K. Schellong, E. Rodekamp, K. Melchior, and J. W. Dudenhausen. 2009. Hypothalamic proopiomelanocortin promoter methylation becomes altered by early overfeeding: an epigenetic model of obesity and the metabolic syndrome. *J Physiol* 587:4963–76.

Ross, S. A., J. Dwyer, A. Umar, J. Kagan, M. Verma, D. M. van Bemmel, and B. K. Dunn. 2008. Introduction: diet, epigenetic events and cancer prevention. *Nutr Rev* 66:S1–6.

Sekine, S., R. Ogawa , R. Ito, N. Hiraoka, M. T. McManus, Y. Kanai, and M. Hebrok. 2009. Disruption of Dicer1 induces dysregulated fetal gene expression and promotes hepatocarcinogenesis. *Gastroenterology* 136:2304–15.

Timp, W., A. Levchenko, and A. P. Feinberg. 2009. A new link between epigenetic progenitor lesions in cancer and the dynamics of signal transduction. *Cell Cycle* 8:383–90.

Ulrich, C. M. 2008. Folate and cancer prevention—where to next? Counterpoint. *Cancer Epidemiol Biomarkers Prev* 17:2226–30.

Verma, M. 2003. Viral genes and methylation. *Ann N Y Acad Sci* 983:170–80.

Verma, M. 2010. The human epigenome and cancer. In *Human genome epidemiology,* 2nd ed., ed. M. Khoury, S. Bedrosian, M. Gwinn, J. Higgins, J. Ioannidis, and J. L. Khoury. New York: Oxford University Press, 551–78.

Verma, M., and D. Kumar. 2009. In *Cancer epigenetics,* ed. T. Tollefsbol. New York: CRC Press, 347–57.

Verma, M., P. Maruvada, and S. Srivastava. 2004. Epigenetics and cancer. *Crit Rev Clin Lab Sci* 41:585–607.

Verma, M., D. Seminara, F. J. Arena, C. John, K. Iwamoto, and V. Hartmuller. (2006). Genetic and epigenetic biomarkers in cancer: improving diagnosis, risk assessment, and disease stratification. *Mol Diagn Ther* 10:1–15.

Villagra, A., E. M. Sotomayor, and E. Seto. 2009. Histone deacetylases and the immunological network: implications in cancer and inflammation. *Oncogene* 29:157–73.

Vu, T. H., A. H. Nguyen, and A. R. Hoffman. 2010. Loss of IGF2 imprinting is associated with abrogation of long-range intrachromosomal interactions in human cancer cells. *Hum Mol Genet* 19:901–19.

Zhao, R., J. F. DeCoteau, C. R. Geyer, M. Gao, H. Cui, and A. G. Casson. 2009. Loss of imprinting of the insulin-like growth factor II (IGF2) gene in esophageal normal and adenocarcinoma tissues. *Carcinogenesis* 30:2117–22.

2 Aging by Epigenetics
Nutrition, An Epigenetic Key to Long Life

Nilanjana Maulik and Gautam Maulik

CONTENTS

2.1 Introduction 13
  2.1.1 Epigenetics and Human Disease 13
  2.1.2 Epigenetics in Nutritional Science 15
  2.1.3 Shaping Life with Epigenetics 15
  2.1.4 List of Dietary Chemicals 16
2.2 Epigenetic Regulation of Aging 17
  2.2.1 Epigenetic Changes 17
  2.2.2 Gene Silencing or Gene Activation 20
2.3 The Emerging Role of Epigenetically Targeted Compounds in Aging 21
  2.3.1 Resveratrol: A Miracle Molecule 21
  2.3.2 Resveratrol and Longevity Genes: Epigenetic Therapy 22
2.4 Conclusion 25
References 25

2.1 INTRODUCTION

2.1.1 Epigenetics and Human Disease

The epigenetic regulation of the genome has evolved to bridge the gap between nature and nurture. Conrad Hal Waddington (1905–1975) was the person who coined the term "epigenetics" in 1942 while working with Honor B. Fell at the Strangeways Research Laboratory on cytonuclear interactions. Waddington's epigenetic landscape is a metaphor for how gene silencing or gene activation modulates development (Goldberg et al. 2007). The concept of Waddington's epigenetic landscape as described by his colleague Ralph Waldo Emerson is quite interesting:

> [A] small tortuous pass Winding through grassy shallows in and out, Two creeping miles of rushes, pads, and sponge … Northward the length of Follansbee we rowed, Under low mountains, whose unbroken ridge Ponderous with beechen forest sloped the shore. A pause and council: then, where near the head On the east a bay makes inward to the land between two rocky arms, we climb the bank. (Agassiz, 1869)

During the twenty-first century, epigenetics was redefined as "the study of heritable changes in gene expression that do not involve modifications in the DNA coding sequence." Epigenetic modifications principally include DNA methylation and a variety of histone modifications, of which the best characterized is acetylation. DNA hypermethylation and histone hypoacetylation are hallmarks of gene silencing, while DNA hypomethylation and acetylated histones promote active transcription. Metaphorically, Robertson described "epigenetics" at a cocktail party as:

> The general picture that emerges from all this is that the embryonic cell is rather like a room where a cocktail party is going on with the radio set with press button tuning in the center of it. The switching on of a particular battery of genes, controlling the synthesis, say, of nerve proteins, corresponds to pressing one particular button which brings in a programme precisely from one station. But you may succeed in getting this button pressed by jogging the elbow of somebody at the other side of the room, who stumbles against the next man, and so on down the line until somebody finally falls against the radio set, and this may be sufficient to click on whichever of the tuning buttons is most insecure. (Robertson 1977)

Abrupt disruption of DNA methylation and histone acetylation has been linked to a plethora of age-related disorders including cancer and autoimmune disorders, and understanding the mechanistic regulation of epigenome might afford the development of new therapies for treating these symptoms. DNA methylation helps to stabilize chromatin and hypomethylation, which can lead to genomic instability by predisposing to strand breakage and recombination within de-repressed repetitive sequences.

The relationship between epigenetics and cancer is far from clear, but tumor cells generally have comparatively low levels of DNA methylation. Methylation might switch off vital genes and contribute to the development of cancer. Several studies in humans as well as laboratory animals suggest a whole list of dietary chemicals from alcohol to zinc that might influence methylation and cancer susceptibility. For instance, a diet low in folic acid has actually been linked to excessive methylation of certain genes. Interestingly, global hypomethylation is seen in a number of cancers, including thyroid (Galusca et al. 2005), breast (Szyf et al. 2004), cervical (De Capoa et al. 2003), prostate (Florl et al. 2004), stomach (Kaneda et al. 2005), lung (Chalitchagorn et al. 2004), esophagus (Chalitchagorn et al. 2004), colorectal (Suter et al. 2004; Frigola et al. 2005), and liver (Chalitchagorn et al. 2004).

Aging of the immune system, or immunosenescence, is characterized by a decline in both T and B cell function. Several pieces of evidence suggest that epigenetic changes may be critical determinants of cellular senescence and organismal aging (Bandyopadhyay and Medrano 2003). It was observed that regions flanking the ITGAL (CD11a) promoter get demethylated in T cells from patients with active systemic lupus erythematosus (SLE). Demethylation of these sequences can contribute to increased ITGAL promoter activity, and thus could lead to increased LFA-1 expression (Lu et al. 2002). Again, LFA-1 overexpression is sufficient to cause T cell autoreactivity in vitro and a lupus-like disease in vivo (Yung et al. 1996). Another report suggested decreased ERK pathway signaling may be responsible for a decrease in DNA methyltransferase expression and DNA hypomethylation in lupus T cells (Deng et al. 2001).

2.1.2 Epigenetics in Nutritional Science

The new science of epigenetics explains how poor nutrition in the womb causes permanent genetic changes in offspring. "You are what you eat," as the old saying depicts. This might be true! But recent scientific research unveiled that "you are also what your mother ate during her pregnancy," which means that the genetic repertoire of an individual is a reflection of its mother's nutritional intake as well. Support for this claim stems from a research report by scientists from Utah in a series of elaborate experiments involving two groups of experimental rats. The first group included normal control rats, whereas the second group had the nutrients from their mother's placentas restricted in a way equivalent to preclampsia. Both groups of rats were examined right after birth and again after 21 days (preadolescent rats) for the amount of IGF-1 protein, which is known to play an indispensable role in the normal growth and development of rats and humans. Investigators found out that the lack of nutrients caused the gene responsible for IGF-1 to significantly reduce the amount of IGF-1 produced in the body before and after birth (Fu et al. 2009).

Again, diets deficient in methyl donor precursors (folate, methionine, and choline) have been consistently observed to induce DNA hypomethylation. We know that mammals cannot synthesize folate, choline, or methionine, yet dietary ingestion of these is essential for normal metabolic homeostasis. For example, restricting dietary folate intake diminishes S-adenosylmethionine (SAM) and increases plasma and cellular levels of homocysteine and S-adenosylhomocysteine (SAH) (Davis and Uthus 2003; Y. Kim 2004).

It is believed that the twentieth century was the golden age for nutritional research, an era when scientific discoveries related to nutrition in health and diseases flourished. The present century brings significant advances in biomedical and food technologies, opening the floodgates for crafted, prescription interventions. Though the complex regulation of the genome is not completely understood as yet, two crucial mechanisms that seem to be most prominent are the alterations in (1) chromatin-associated proteins, which act as scaffolds for the DNA during transcription (e.g., histones), and (2) degree of methylation of the nucleotide bases of DNA. Both these epigenetic regulation mechanisms of the genome are known to be associated with nutrient intake, and hence with diet and nutritional status of an individual. One well-known example of this kind of nutrient-mediated epigenetic regulation is the agouti mouse, where the coat color of mice and susceptibility to metabolic disorders can be meticulously controlled by the offering of dietary methyldonors (Cooney et al. 2002).

2.1.3 Shaping Life with Epigenetics

Epigenetic changes are known to occur generally during gestation, neonatal development, puberty, and old age. The field of genetics has become an integral part of the modern medicine in the last 50 years, since Watson and Crick first described the three-dimensional structure of the DNA double helix.

The loss of normal DNA methylation patterns is the best understood epigenetic cause behind a plethora of human diseases, based on a series of initial studies during the 1980s that focused on X-chromosome inactivation (Avner and Heard 2001), genomic imprinting (Verona et al. 2003), and cancer (Feinberg and Tycko 2004). Furthermore, DNA methylation involves the addition of a methyl group to cytosines of CpG (cytosine/guanine) pairs (Ehrlich and Wang 1981; Laird and Jaenisch 1994; Rodenhiser and Mann 2006) (Figure 2.1). DNA methylation pattern and histone modifications play a pleiotropic role in switching the genes on or off. Genetic imprinting broadly depends on these two phenomena. Genomic imprinting allows genes to remember whether they were inherited from the mother or father so that only the maternally or paternally inherited allele is expressed (Rodenhiser and Mann 2006) (Table 2.1). Genes carry the blueprints for the synthesis of the normal repertoire of proteins in cells. Every cell in the body has the same genetic information; what makes cells, tissues, and organs different is that different sets of genes are silenced or expressed. This can be likened to a complex musical score that remains lifeless without an orchestra of cells (players), and epigenotypes can be compared with the instruments that create varying notes.

2.1.4 List of Dietary Chemicals

Some evidence dating back to the 1930s shows how life span can at least be extended by reducing calorie intake. The U.S. National Institute on Aging is working to prove this hypothesis by investing in a long-term study called CALERIE (Comprehensive Assessment of Long-Term Effects of Reducing Intake of Energy) to investigate any direct link between calorie restriction, disease predisposition, and aging (calerie.dcri.duke.edu). Some biological events in our body (e.g., cell division, etc.) need a constant supply of new methyl groups, which can be provided directly from our food intake (methionine, betaine, choline, folic acid). However, additional components are needed (vitamin B_{12}, zinc, etc.) from food to transport the methyl groups within the body for epigenetic modification of the DNA. Healthy eating habits are intended to promote overall health while reducing the risk of developing nutrition-related diseases like cancer and cardiovascular pathophysiological complications. Scientific evidence points firmly toward the health benefits of a diet rich in fruits and vegetables. For instance, some polyphenolic constituents present in red wine have been shown to confer therapeutic benefits in the treatment of many neurodegenerative, metabolic, and heart diseases and even obesity. Leafy vegetables, peas, beans, sunflower seeds, fortified breads, cereals, etc. are rich sources of folic acid. In general, choline comes from eggs, lettuce, peanuts, and liver. In addition, garlic, nuts, kidney beans, tofu, fish, and chicken are the real source of methionine. In summary, we are in the process of realizing how specific molecules in our diet can influence epigenetic phenomena.

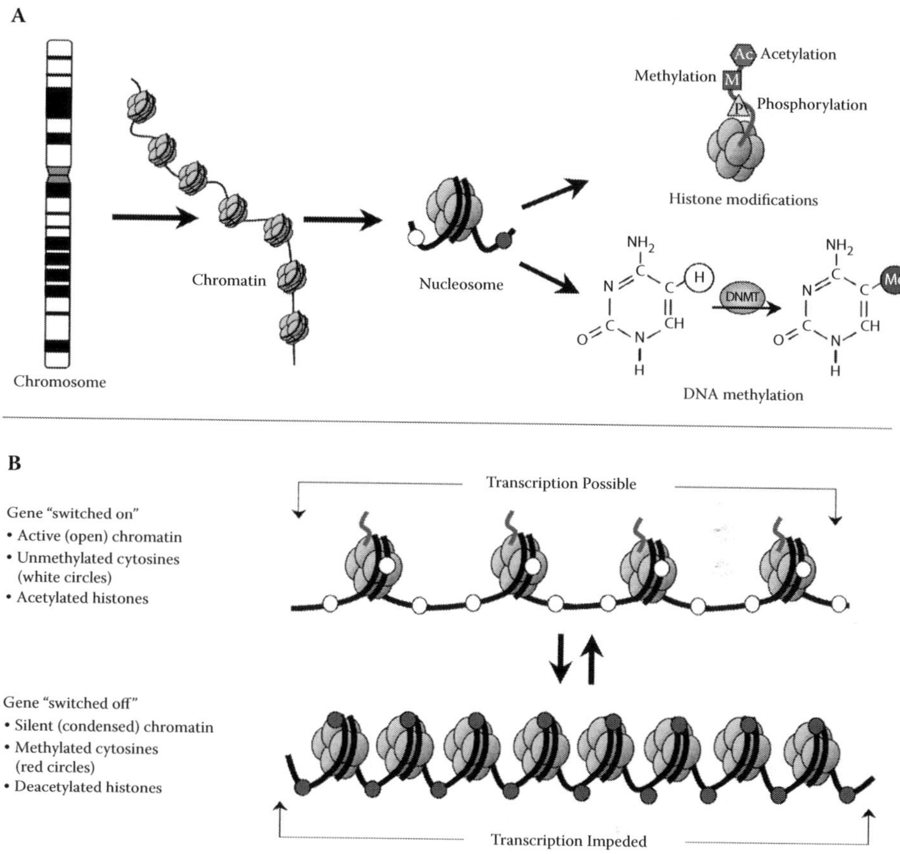

FIGURE 2.1 (Please see color insert following page 80.) A. Schematic of epigenetic modifications. Strands of DNA are wrapped around histone octamers, forming nucleosomes. These nucleosomes are organized into chromatin, the building block of a chromosome. Reversible and site-specific histone modifications occur at multiple sites through acetylation, methylation, and phosphorylation. DNA methylation occurs at the 5-position of cytosine residues in a reaction catalyzed by DNA methyltransferases (DNMTs). Together, these modifications provide a unique epigenetic signature that regulates chromatin organization and gene expression. B. Schematic of the reversible changes in chromatin organization that influence gene expression: genes are expressed (switched on) when the chromatin is open (active), and they are inactivated (switched off) when the chromatin is condensed (silent). White circles = unmethylated cytosines; red circles = methylated cytosines. (Adapted from Rodenhiser, D. and Mann, M., *Can Med Assoc J,* 174(4): 341, 2006. With permission.)

2.2 EPIGENETIC REGULATION OF AGING

2.2.1 Epigenetic Changes

Epigenetic modifications are the key regulators of developmental processes including differentiation, growth, and aging. An increase or decrease in the DNA methylation status of an individual might have a direct influence on the aging process. There

TABLE 2.1
Associations between Epigenetic Modifications and Human Diseases and Conditions

Disease/Condition	Gene	Biological Process
	Cancer	
Bladder	Multiple genes	Hypermethylation
Brain (glioma)	*RASSF1A*	Hypermethylation
Brain (glioblast)	*MGMT*	Hypermethylation
Breast	*BRCA1*	Hypermethylation
Breast	Multiple genes	Hypermethylation
Cervix	*p16*	Hypermethylation
Colon	Multiple genes	Hypermethylation
Colorectal	L1 repeats	Hypomethylation
Esophagus	*CDH1*	Hypermethylation
Head/neck	*p16, MGMT*	Hypermethylation
Kidney	*TIMP-3*	Hypermethylation
Leukemia	*p15*	Hypermethylation
Liver	Multiple genes	Hypermethylation
Lung	*p16, p73*	Hypermethylation
Lymphoma	*DAPK*	Hypermethylation
Myeloma	*DAPK*	Hypermethylation
Ovary	*BRCA1*	Hypermethylation
Ovary	*Sat2*	Hypomethylation
Pancreas	*APC*	Hypermethylation
Pancreas	Multiple genes	Hypomethylation
Prostate	*BRCA2*	Hypermethylation
Rhabdomyosarcoma	*PAX3*	Hypermethylation
Stomach	*Cyclin D2*	Hypomethylation
Thymus	*POMC*	Hypomethylation
Urothelial	Satellite DNA	Hypomethylation
Uterus	*hMLH1*	Hypermethylation
	Neurologic	
Schizophrenia	*RELN*	Hypermethylation
Bipolar disorder	*11p?*	Unknown
Memory formation	Multiple genes	Hypo-, hypermethylation
Lupus	Retroviral DNA	Hypomethylation
	Cardiovascular	
Atherosclerosis	Multiple genes	Hypo-, hypermethylation
Homocysteinemia	Multiple genes	Hypomethylation
Vascular endothelium	*eNOS*	Hypomethylation

TABLE 2.1
Associations between Epigenetic Modifications and Human Diseases and Conditions

Disease/Condition	Gene	Biological Process
	Imprinting and Pediatric Syndrome	
PWS or AS	15q11-q13	Imprinting
BWS	11p15	Imprinting
SRS	Chromosome 7	Imprinting
UDP14	14q23-q32	Imprinting
PHP, AHO, MAS	20q13.2	Imprinting
Rett syndrome	*MECP2*	Mutation
ICF syndrome	*DNMT3B*	Mutation
ATRX	*ATRX*	Chromatin structure
FraX	Triplet repeat	Silencing
FSHD	3.3 kb repeat	Chromatin structure
	Reproductive	
Ovarian teratoma	No paternal genome	Imprinting
CHM	No maternal genome	Imprinting
BiCHM	Maternal genome	Imprinting
Aging	Chromatin	Hypo-, hypermethylation

Source: Adapted from Rodenheiser, D. and Mann, M., *Can Med Assoc J,* 174(3), 341, 2006. With permission.

is evidence that age-dependent methylation changes are involved in the development of neurologic disorders, autoimmunity, and cancer in elderly people (Richardson 2003). The enzymes that add and remove acetyl groups, the histone acetyltransferases (HATs) and histone deacetylases (HDACs) respectively actually determine the steady-state level of histone acetylation (Berger 2007) and modulate the epigentic imprint. In mammals, there is an age-associated decline in total genomic DNA methylation (Romanov and Vanyushin 1981; Singhal et al. 1987; Wilson et al. 1987). This occurs mainly at repetitive DNA sequences, and so probably occurs predominantly in domains of constitutive heterochromatin. However, although genome-wide levels of methylation tend to decrease with age in mammals, site-specific hypermethylation of the DNA might increase (Issa et al. 1994, 1996; Ahuja et al. 1998; Yatabe et al. 2001; J. Kim et al. 2005). This again occurs at the CpG islands, some of which are in the promoter regions of genes. CpG islands are CG-rich sequences that are generally unmethylated, but can be regulated via methylation when required. Epigenetic effects occur throughout the life span of an individual. The silent information regulator (SIR) family of proteins in yeast or their homologs in higher mammals are involved in multiple cellular events including transcriptional silencing, chromatin remodeling, mitosis, and life span duration (Guarente 2000). Sir2-like enzymes catalyze a reaction in which the cleavage of NAD+ and histone and/or other protein deacetylation is coupled with the formation of O-acetyl-ADP-ribose. It was very recently found that in yeast, deletion of Sir2 shortens the life span, whereas an extra copy of this gene increases life span, demonstrating direct regulation of the Sir2 family in aging (Kaeberlein et al. 1999). Sirtuins, the homologs of yeast SIR2 family, belong to the atypical class III HDACs (Frye 2000). The mammalian sirtuin SIRT1 gene product encodes an NAD-dependent nuclear HDAC that closely resembles the yeast Sir2 protein (Frye 2000). Recently, SIRT1 was shown to control Bax-induced apoptosis by deacetylating Ku70, and by inhibiting Forkhead transcription factor-mediated cell death (Cohen et al. 2004; Motta et al. 2004). Several studies have demonstrated that human SIRT1 functions as a p53 deacetylase, which can impair the transcriptional activity of p53 (Vaziri et al. 2001), and prevent the cellular stress-induced senescence and apoptosis following DNA damage.

2.2.2 Gene Silencing or Gene Activation

One of the most widely studied and popular phenomena in aging research, known for over 70 years, is the effect of dietary restriction (underfeeding or malnutrition) in extending the life span of laboratory rodents and other species. When the gene expression profile is compared between diet-restricted and normally fed animals, a wide array of genes are found to be altered. The effects are significant, resulting in as much as a 50% increase in rodent longevity (Weindruch et al. 2002). The human genome encodes approximately 30,000 genes. Gene–environment interactions are thought to be mediated by epigenetic modifications of the genome. Alteration of gene expression mainly depends on gene–environment, gene–nutrition, gene–stress interactions that alter gene activities and lead to trigger cascades of cellular events to facilitate the adaptation of an individual cell to its environment. The term epigenetics was first used to describe gene environment interactions that lead to the manifestation of various phenotypes during development (L. Liu et al. 2008) (Figure 2.2). The key process in aging generally involves reduced expression of number of genes or gene silencing, which are extremely important in growth and function. Gene silencing is a complex mechanism, which mainly involves methylation of DNA, histone modification, and chromatin remodeling (Li 2002; Laird 2003; Roberts and Orkin 2004). Hypermethylation of the promoter mostly leads to silencing of the gene. Several groups have documented changes in the repertoire of expressed genes as a causal factor in aging. The genes related to aging are involved in cell cycle, apoptosis, detoxification, and cholesterol metabolism (Burzynski 2005). In general, the two biochemical processes (Burzynski 2003) that play a very important role in silencing of the genes are deacetylation of the histones and methylation of DNA. However, many additional DNA epigenetic regulatory mechanisms have been proposed in aging cells: (1) site-specific hypermethylation of promoter sequences and (2) genomewide hypomethylation. A landmark study involving microarray-based expression analysis of 11,000 genes in aging livers of mice by Cao et al. (2001) revealed that 46 known genes changed expression during aging (27-month-old vs. 7-month-old mice). It was found that 57% of genes were decreased and 43% increased in an age-dependent manner. Most of the increased genes were found to be associated with age-associated diseases. The tumor suppressor gene p53 plays a very important role in aging, apart from its pleiotropic role in maintaining normal cellular homeostasis. In humans, Sir2 inactivates p53 through

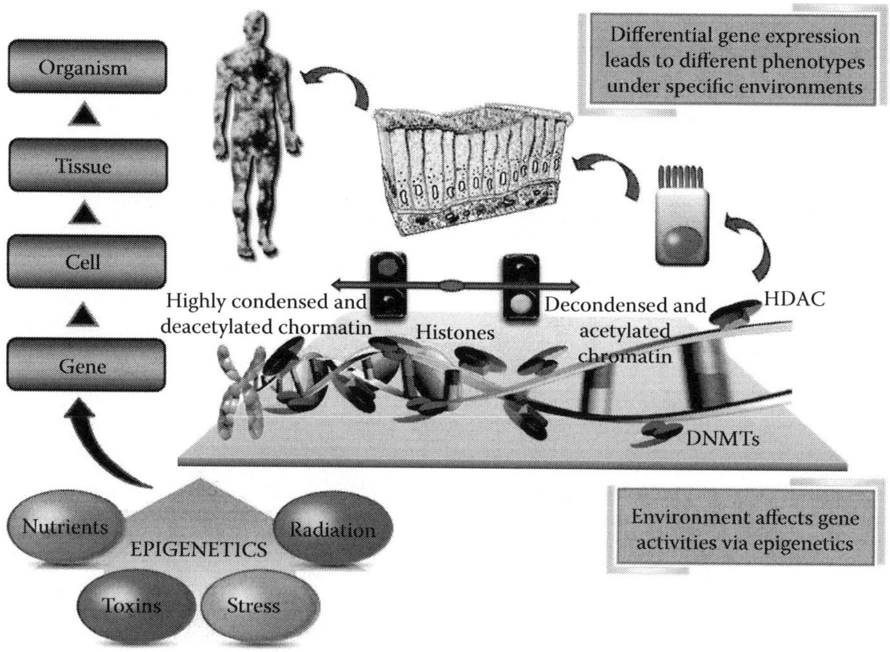

FIGURE 2.2 (Please see color insert.) A hierarchical view of gene-environment interactions during development. As depicted, environmental effects are integrated by epigenetic process including chromatin remodeling to either allow or inhibit gene expressions at the molecular level. Such effects will be manifested at the organismal level via ultimate functional output of the genome. (Adapted from Liu, L. et al., *Curr Issues Mol Biol,* 10(1–2): 25, 2008. With permission.)

deacetylation (Howitz et al. 2003). Among the silenced genes found during aging, 23% were found to be involved with the control of the cell cycle. The most prominent among them were the tumor suppressor gene Pten and Ifgbp1, which down-regulates IGF1. However, interestingly enough, the characteristics of growth and aging in humans are much different compared to that of the lower mammals.

2.3 THE EMERGING ROLE OF EPIGENETICALLY TARGETED COMPOUNDS IN AGING

2.3.1 Resveratrol: A Miracle Molecule

Almost 4500 years ago *Ayurveda,* the ancient medicinal book of the Hindus, described "darakchasava" (fermented juice of red grapes) as cardiotonic (Paul et al. 1999). Consequently, Jesus Christ described grape juice and red wine as a "gift of god," which was presumably used to purify body and soul. In 1940, resveratrol was first identified as the medicinal component of grapes, and was extracted from the dried roots of *Polygonum cuspidatum* (popularly known as Ko-jo-kon by the Japanese) and used to treat hyperlipidemic diseases (Vastano et al. 2000). In the modern era, resveratrol was rediscovered as an antiproliferative agent for cancer therapeutics. Antitumor activity of resveratrol was published in 1997 (Jang et al. 1997; Pezzuto 1997). The cardioprotective ability of resveratrol stemmed from the epidemiological studies indicating that mild to moderate alcohol consumption has been associated with a reduced incidence of morbidity and mortality from coronary heart disease (Renaud and De Lorgeril 1992). Rejuvenating resveratrol is a miracle molecule in cells that reduces the impact of free radicals that cause aging. "Rejuvenation" signifies restoration of youth and has been known for at least the last 200 years. The richest source of resveratrol is the roots of *Polygonum cuspidatum* (Ko-jo-kon), mainly cultivated in China and Japan. The skins of grapes contain about 50–100 mg resveratrol and are believed to contribute to the cardioprotective abilities of red wine, which contains about 0.2–7 mg resveratrol per liter of the wine. In addition to grapes, a large variety of fruits including mulberry, bilberry, lingonberry, sparkleberry, deerberry, partridgeberry, cranberry, blueberry, jackfruit, and peanut, as well as a wide variety of flowers and leaves including gnetum, white hellebore, corn lily, butterfly orchid tree, eucalyptus, spruce, poaceae, scots pine, and rheum, also contain resveratrol. Plants are known to synthesize resveratrol in response to environmental stress including water deprivation, ultraviolet (UV) irradiation, and especially fungal infection, and thus can be considered to be produced as part of the defense mechanism.

2.3.2 Resveratrol and Longevity Genes: Epigenetic Therapy

Over the last 20 years, many studies have described promising health benefits associated with wine consumption. Some studies suggest that red wine is more cardioprotective than white wine, hypothetically due to the enriched flavonoid antioxidants in red wine. Several experimental studies including ours (Penumathsa et al. 2008) supported the evidence that these beneficial effects are due to resveratrol, the polyphenolic compound present in red wine. Many studies have provided evidence that resveratrol affords antioxidant, anti-apoptotic effects apart from activation of longevity proteins (SIRT1). Green tea also not only confers powerful antioxidant effects but also helps to balance the normal DNA methylation status (Fang et al. 2003). Cruciferous vegetables such as broccoli, cauliflower, kale, and bok choy are powerful vegetables whose regular consumption might affect DNA methylation status, allowing tumor suppressor genes to function better. Grapes generally work via histone modulation. Histones are modified after translation by acetylation, methylation, phosphorylation, and ubiquitination. Resveratrol prevents high fat-accelerated aging by stimulating sirtuins (Sir2 proteins) such as SIRT1 activation (Pearson et al. 2008). There are at least seven members of Sir2 proteins such as HDACs that are believed to impart a "life preservation" effect. Activation of sirtuins by dietary polyphenols, such as resveratrol from red wine, has been found to increase life span by food restriction (Howitz et al. 2003). Resveratrol deacetylates histones reducing DNA transcription, which is thought to be the molecular mechanism of life prolongation and several other benefits. SIRT plays a very important role in the longevity response to dietary restriction. SIRT1 promotes longevity in part through epigenetic effects, including DNA methylation integrity, and it can be mimicked by specific components. The polyphenol resveratrol is known to be potential mimetic of dietary restriction. According to some reports (Baur et al. 2006; Lagouge et al. 2006), dietary resveratrol protected mice against diet-induced obesity and insulin resistance and was also found to induce other metabolic and physiological effects associated with longer life span. Again, resveratrol treatment was found to increase PGC-1α deacetylation in multiple tissues, which is consistent with SIRT1 activation, and deacetylation of PGC-1α in SIRT1$^{-/-}$ mouse embryonic fibroblasts. Resveratrol in very low dose partially mimics caloric restriction and retards various parameters related to aging in mice (Barger et al. 2008). Dietary resveratrol previously has been shown to extend life span in *Saccharomyces cerevisiae, Caenorhabditis elegans,* and *Drosophila* through a SIRT1-dependent mechanism (Howitz et al. 2003; Wood et al. 2004). According to some studies (Baur et al. 2006; Lagouge et al. 2006) feeding high levels of resveratrol to mice has been shown to be associated with increased SIRT1 activity determined by PGC-1α acetylation and the induction of its transcriptional targets resulting in extended life span as compared to the control animals. Most interestingly, SIRT1-mediated repression of p53 and NFκB might control life and death signals. However, these regulatory mechanisms are not clear as yet. SIRT1 interacts with the RelA/p65 subunit of NFκB and inhibits transcription by deacetylating RelA/p65 at lysine 310. Resveratrol treatment potentiates chromatin-associated SIRT1 protein on the cIAP-2 promoter region with a loss of NFκB-regulated gene expression (Yeung et al. 2004). Class 1 histone deacetylases (HDAC) regulate the transcriptional activity of NF-κB. It is shown that HDAC1, HDAC2, and HDAC3 deacetylate RelA/p65, resulting in increased IκBα association or loss of transactivation potential of the protein (Ashburner et al. 2001; Chen et al. 2001; Zhong et al. 2002).

Very recently, organosulfur compounds from garlic, such as diallyl disulfide, allyl mercaptan, and S-allylmercaptocysteine, as well as the isothiocyanates sulforaphane and 6-methylsulfinylhexyl isothiocyanate from several cruciferous vegetables, documented a capacity to alter histone acetylation and/or HDAC activity in vivo and in vitro. It is found that Class III HDACs (sirtuins) drew more attention after they were implicated in increasing life span and delay in aging-related diseases (Tissenbaum and Guarente 2001). A number of natural compounds found in the human diet can influence HDACs and the acetylation status of histones, as described in Table 2.2 (Delage and Dashwood 2008).

The future of epigenetic therapy seems to be very promising, particularly in treating life-threatening diseases such as some cancers and neurological disorders. Some drugs that inhibit the DNA methyltransferases, which add methyl groups on DNA, are now approved for clinical use in hospitals in the United States for the treatment of certain cancers (Yoo and Jones 2006; Issa 2007). Valproic acid, a histone deacetylase (HDAC) inhibitor, increases the effectiveness of antipsychotic medications in the treatment of schizophrenia and bipolar disorder (Citrome 2003). It is now widely accepted that histone modification and DNA methylation are significantly interrelated, almost working hand in hand to determine the extent of gene expression and to decide cell fate (Hashimshony et al. 2003). Resveratrol's excellent epigenetic properties can be used to design antiaging, anticancer, and heart-smart drugs in the future. Therefore the main goal of epigenetic therapy using various epigenetic drugs is to restore normal DNA methylation patterns and to prevent the cells from acquiring further methylation in DNA that could lead to silencing of genes crucial for normal cell function.

In several preclinical experimental models, genetic manipulation (Flurkey et al. 2001; Liang et al. 2003) and caloric restriction (Weindruch et al. 1986) have been

TABLE 2.2
Natural and/or Dietary Compounds Modulating Histone Acetylation and/or HDAC/HAT Activities

Dietary Components	Examples of Food/Plant Sources	References
S-allylmercaptocysteine	Garlic (*Allium sativum L.*)	(Lea et al. 2002)
6-methylsulfinylhexyl-isothiocyanate	Japanese horseradish (wasabi)	(Morimitsu et al. 2002)
Allyl mercaptan	Garlic (*Allium sativum L.*)	(Lea and Randolph 2001)
Anacardic acid	Cashew nut	(Balasubramanyam et al. 2003)
Butein	*Rhus verniciflua* (stems)	(Howitz et al. 2003; Porcu and Chiarugi 2005)
Butyrate	Dietary fiber fermentation	(Davie 2003)
Copper	Ubiquitous	(Kang et al. 2004; C. Lin et al. 2005)
Curcumin	*Curcuma longa* (turmeric roots)	(Balasubramanyam et al. 2004; H. Liu et al. 2005)
Diallyl disulfide (DADS)	Garlic (*Allium sativum L.*)	(Lea et al. 1999; Druesne et al. 2004; Marcu et al. 2006)
Dihydrocoumarin	*Melilotus officinalis* (sweet clover)	(Olaharski et al. 2005)
Fisetin	*Rhus toxicodendron* (leaves)	(Howitz et al. 2003; Porcu and Chiarugi 2005)
Garcinol	*Garcina indica* (fruit)	(Balasubramanyam et al. 2004)
Isoliquiritigenin	*Glycyrrhiza glabra* (licorice)	(Howitz et al. 2003; Porcu and Chiarugi 2005)
Luteolin	Sweet red pepper, celery, parsley	(Porcu and Chiarugi 2005)
Nickel	Ubiquitous	(Kang et al. 2003; Yan et al. 2003)
Piceatannol	Blueberries	(Porcu and Chiarugi 2005)
Psammapin A	Marine sponges	(C. Kim et al. 2007)
Quercetin	Apple, tea, onion, nuts, berries	(Howitz et al. 2003; Porcu and Chiarugi 2005)
Resveratrol	Red grapes, wines, eucalyptus, spruce	(Howitz et al. 2003; Porcu and Chiarugi 2005)
Sulforaphane	Broccoli, broccoli sprouts	(Myzak et al. 2004)
Theophylline	Black and green tea	(Ito et al. 2002; Cosio et al. 2004)

Source: Adapted from Delage, B. and Dashwood, R. H., *Annu Rev Nutr,* 28: 347–66, 2008. With permission.

shown to increase the life span as compared to their control littermates fed *ad libitum.* Diet-related studies have shown that the Mediterranean diet might decrease mortality and its associated susceptibility to cardiovascular disease and cancer and thereby increase longevity (Trichopoulou, Bamia et al. 2005; Trichopoulou, Orfanos et al. 2005). Most popular antiaging treatments involve human growth hormone (HGH) treatment (Rudman et al. 1990). Several animal studies supported the role of HGH in longevity (Flurkey et al. 2001; Liang et al. 2003; Al-Regaiey et al. 2005; Sun et al.

2005). However, some contradictory results also show that transgenic mice expressing HGH have a shorter life span than control mice (Bartke et al. 1999; Forster et al. 2003). In addition, several age-related studies revealed defects in stem cells that can limit proper organ maintenance and hence contribute to a shorter life span (Chambers et al. 2007). However, the possible role of stem cell aging in the determination of human aging is still far from being completely understood. Further mechanistic understanding of stem cell aging is required before it can be translated into human antiaging therapy. We all are trying to find the "magic bullet" that delays the natural aging process. A study by Khaw et al. (2008) documented that people who exercise regularly, eat a diet high in vitamin C, don't smoke, and consume moderate alcohol can add up to 14 years to their lives! Our near future awaits new therapies that could turn off "bad genes" and "turn on" the good ones to cure life-threatening diseases and extend longevity.

2.4 CONCLUSION

Though still in its infancy, nutritional epigenetics has revealed much about the complex interactions between diet and genes. It is very likely that nutrition affects gene expression and human health and disease through epigenetic mechanism. Great progress already has been made with folate metabolism, which affects DNA methylation status and gene silencing. Methylation of CpG islands increases with age and could therefore yield various chronic diseases in addition to cancer. Lifestyle changes such as exercise, controlled nutrition, and epigenetic drugs could bring about reversion of or slow down epigenetic modifications in patients with chronic diseases. Resveratrol has been documented as being involved in maintaining optimal health and longevity along with prevention or possible cure of chronic diseases such as atherosclerosis, diabetes, stroke, and various other cardiovascular diseases. However, further mechanistic research is needed to unravel the relation between diet, epigenetic events, and the predisposition to various cardiovascular diseases, aging, or cancer, in an attempt to exploit it for therapeutic purposes.

REFERENCES

Agassiz, L. 1869. De l'Espece et de la Classification en zoologie. *Balliere Paris* 375–91.

Ahuja, N., Q. Li, A. L. Mohan, S. B. Baylin, and J. P. Issa. 1998. Aging and DNA methylation in colorectal mucosa and cancer. *Cancer Res* 58 (23):5489–94.

Al-Regaiey, K. A., M. M. Masternak, M. Bonkowski, L. Sun, and A. Bartke. 2005. Long-lived growth hormone receptor knockout mice: interaction of reduced insulin-like growth factor i/insulin signaling and caloric restriction. *Endocrinology* 146 (2):851–60.

Ashburner, B. P., S. D. Westerheide, and A. S. Baldwin Jr. 2001. The p65 (RelA) subunit of NF-kappaB interacts with the histone deacetylase (HDAC) corepressors HDAC1 and HDAC2 to negatively regulate gene expression. *Mol Cell Biol* 21 (20):7065–77.

Avner, P., and E. Heard. 2001. X-chromosome inactivation: counting, choice and initiation. *Nat Rev Genet* 2(1):59–67.

Balasubramanyam, K., M. Altaf, R. Varier, V. Swaminathan, A. Ravindran, P. Sadhale, and T. Kundu. 2004. Polyisoprenylated benzophenone, garcinol, a natural histone acetyltransferase inhibitor, represses chromatin transcription and alters global gene expression. *J Biol Chem* 279 (32):33716–26.

Balasubramanyam, K., V. Swaminathan, A. Ranganathan, and T. Kundu. 2003. Small molecule modulators of histone acetyltransferase p300. *J Biol Chem* 278 (21):19134–40.

Balasubramanyam, K., R. Varier, M. Altaf, V. Swaminathan, N. Siddappa, U. Ranga, and T. Kundu. 2004. Curcumin, a novel p300/CREB-binding protein-specific inhibitor of acetyltransferase, represses the acetylation of histone/nonhistone proteins and histone acetyltransferase-dependent chromatin transcription. *J Biol Chem* 279 (49):51163–71.

Bandyopadhyay, D., and E. E. Medrano. 2003. The emerging role of epigenetics in cellular and organismal aging. *Exp Gerontol* 38 (11–12):1299–1307.

Barger, J. L., T. Kayo, J. M. Vann, E. B. Arias, J. Wang, T. A. Hacker, Y. Wang, D. Raederstorff, J. D. Morrow, C. Leeuwenburgh, D. B. Allison, K. W. Saupe, G. D. Cartee, R. Weindruch, and T. A. Prolla. 2008. A low dose of dietary resveratrol partially mimics caloric restriction and retards aging parameters in mice. *PLoS One* 3 (6):e2264.

Bartke, A., V. Chandrashekar, D. Turyn, R. W. Steger, L. Debeljuk, T. A. Winters, J. A. Mattison, N. A. Danilovich, W. Croson, D. R. Wernsing, and J. J. Kopchick. 1999. Effects of growth hormone overexpression and growth hormone resistance on neuroendocrine and reproductive functions in transgenic and knock-out mice. *Proc Soc Exp Biol Med* 222 (2):113–23.

Baur, J. A., K. J. Pearson, N. L. Price, H. A. Jamieson, C. Lerin, A. Kalra, V. V. Prabhu, J. S. Allard, G. Lopez-Lluch, K. Lewis, P. J. Pistell, S. Poosala, K. G. Becker, O. Boss, D. Gwinn, M. Wang, S. Ramaswamy, K. W. Fishbein, R. G. Spencer, E. G. Lakatta, D. Le Couteur, R. J. Shaw, P. Navas, P. Puigserver, D. K. Ingram, R. de Cabo, and D. A. Sinclair. 2006. Resveratrol improves health and survival of mice on a high-calorie diet. *Nature* 444 (7117):337–42.

Berger, S. 2007. The complex language of chromatin regulation during transcription. *Nature* 447 (7143):407–12.

Burzynski, S. 2003. Gene silencing-a new theory of aging. *Med Hypotheses* 60 (4):578–83.

Burzynski, S. 2005. Aging: gene silencing or gene activation? *Med Hypotheses* 64 (1):201–8.

Cao, S., J. Dhahbi, P. Mote, and S. Spindler. 2001. Genomic profiling of short- and long-term caloric restriction effects in the liver of aging mice. *P Natl Acad Sci U S A* 98 (19):10630.

Chalitchagorn, K., S. Shuangshoti, N. Hourpai, N. Kongruttanachok, P. Tangkijvanich, D. Thong-ngam, N. Voravud, V. Sriuranpong, and A. Mutirangura. 2004. Distinctive pattern of LINE-1 methylation level in normal tissues and the association with carcinogenesis. *Oncogene* 23 (54):8841–46.

Chambers, S., C. Shaw, C. Gatza, C. Fisk, L. Donehower, and M. Goodell. 2007. Aging hematopoietic stem cells decline in function and exhibit epigenetic dysregulation. *PLoS Biol* 5 (8):e201.

Chen, L., W. Fischle, E. Verdin, and W. Greene. 2001. Duration of nuclear NF-kappa B action regulated by reversible acetylation. *Science* 293 (5535):1653.

Citrome, L. 2003. Schizophrenia and valproate. *Psychopharmacol Bull* 37:74.

Cohen, H., S. Lavu, K. Bitterman, B. Hekking, T. Imahiyerobo, C. Miller, R. Frye, H. Ploegh, B. Kessler, and D. Sinclair. 2004. Acetylation of the C terminus of Ku70 by CBP and PCAF controls Bax-mediated apoptosis. *Mol Cell* 13 (5):627–38.

Cooney, C., A. Dave, and G. Wolff. 2002. Maternal methyl supplements in mice affect epigenetic variation and DNA methylation of offspring. *J Nutr* 132 (8):2393S.

Cosio, B., L. Tsaprouni, K. Ito, E. Jazrawi, I. Adcock, and P. Barnes. 2004. Theophylline restores histone deacetylase activity and steroid responses in COPD macrophages. *J Exp Med* 200 (5):689.

Davie, J. 2003. Inhibition of histone deacetylase activity by butyrate. *J Nutr* 133 (7):2485S.

Davis, C., and E. Uthus. 2003. Dietary folate and selenium affect dimethylhydrazine-induced aberrant crypt formation, global DNA methylation and one-carbon metabolism in rats. *J Nutr* 133 (9):2907.

De Capoa, A., A. Musolino, S. Della Rosa, P. Caiafa, L. Mariani, F. Del Nonno, A. Vocaturo, R. Donnorso, A. Niveleau, and C. Grappelli. 2003. DNA demethylation is directly related to tumour progression: evidence in normal, pre-malignant and malignant cells from uterine cervix samples. *Oncol Rep*10 (3):545.

Delage, B., and R. H. Dashwood. 2008. Dietary manipulation of histone structure and function. *Annu Rev Nutr* 28:347–66.

Deng, C., M. Kaplan, J. Yang, D. Ray, Z. Zhang, W. McCune, S. Hanash, and B. Richardson. 2001. Decreased ras-mitogen-activated protein kinase signaling may cause DNA hypomethylation in T lymphocytes from lupus patients. *Arthritis Rheum* 44 (2):397–407.

Druesne, N., A. Pagniez, C. Mayeur, M. Thomas, C. Cherbuy, P. Duee, P. Martel, and C. Chaumontet. 2004. Diallyl disulfide (DADS) increases histone acetylation and p21waf1/cip1 expression in human colon tumor cell lines. *Carcinogenesis* 25 (7):1227.

Ehrlich, M., and R. Wang. 1981. 5-Methylcytosine in eukaryotic DNA. *Science* 212 (4501):1350.

Fang, M., Y. Wang, N. Ai, Z. Hou, Y. Sun, H. Lu, W. Welshm and C. Yang, 2003. Tea polyphenol (-)-epigallocatechin-3-gallate inhibits DNA methyltransferase and reactivates methylation-silenced genes in cancer cell lines. *Cancer Res* 63 (22):7563.

Feinberg, A., and B. Tycko. 2004. The history of cancer epigenetics. *Nat Rev Cancer* 4 (2):143–53.

Florl, A., C. Steinhoff, M. Müller, H. Seifert, C. Hader, R. Engers, R. Ackermann, and W. Schulz. 2004. Coordinate hypermethylation at specific genes in prostate carcinoma precedes LINE-1 hypomethylation. *Brit J Cancer* 91 (5):985–94.

Flurkey, K., J. Papaconstantinou, R. Miller, and D. Harrison. 2001. Lifespan extension and delayed immune and collagen aging in mutant mice with defects in growth hormone production. *Proceedings Natl Acad Sci U S A* 98 (12):6736.

Forster, M., P. Morris, and R. Sohal. 2003. Genotype and age influence the effect of caloric intake on mortality in mice. *FASEB J:* 205331.

Frigola, J., X. Sole, M. Paz, V. Moreno, M. Esteller, G. Capella, and M. Peinado. 2005. Differential DNA hypermethylation and hypomethylation signatures in colorectal cancer. *Hum Mol Genet* 14 (2):319.

Frye, R. 2000. Phylogenetic classification of prokaryotic and eukaryotic Sir2-like proteins. *Biochem Bioph Res Co* 273 (2):793–98.

Fu, Q., X. Yu, C. Callaway, R. Lane, and R. McKnight. 2009. Epigenetics: intrauterine growth retardation (IUGR) modifies the histone code along the rat hepatic IGF-1 gene. *FASEB J* 23 (8):2438.

Galusca, B., J. Dumollard, S. Lassandre, A. Niveleau, J. Prades, B. Estour, and M. Peoc'h. 2005. Global DNA methylation evaluation: potential complementary marker in differential diagnosis of thyroid neoplasia. *Virchows Archiv* 447 (1):18–23.

Goldberg, A., C. Allis, and E. Bernstein. 2007. Epigenetics: a landscape takes shape. *Cell* 128(4):635–38.

Guarente, L. 2000. Sir2 links chromatin silencing, metabolism, and aging. *Gene Dev* 14(9):1021.

Hashimshony, T., J. Zhang, I. Keshet, M. Bustin, and H. Cedar. 2003. The role of DNA methylation in setting up chromatin structure during development. *Nat Genet* 34(2):187–92.

Howitz, K., K. Bitterman, H. Cohen, D. Lamming, S. Lavu, J. Wood, R. Zipkin, P. Chung, A. Kisielewski, and L. Zhang. 2003. Small molecule activators of sirtuins extend *Saccharomyces cerevisiae* lifespan. *Nature* 425 (6954):191–96.

Issa, J. 2007. DNA methylation as a therapeutic target in cancer. *Clin Cancer Res* 13 (6):1634.

Issa, J., Y. Ottaviano, P. Celano, S. Hamilton, N. Davidson, and S. Baylin. 1994. Methylation of the oestrogen receptor CpG island links ageing and neoplasia in human colon. *Nat Genet* 7 (4):536–40.

Issa, J., P. Vertino, C. Boehm, I. Newsham, and S. Baylin. 1996. Switch from monoallelic to biallelic human IGF2 promoter methylation during aging and carcinogenesis. *Proc Natl Acad Sci U S A* 93 (21):11757.

Ito, K., S. Lim, G. Caramori, B. Cosio, K. Chung, I. Adcock, and P. Barnes. 2002. A molecular mechanism of action of theophylline: induction of histone deacetylase activity to decrease inflammatory gene expression. *Proc Natl Acad Sci U S A* 99 (13): 921.

Jang, M., L. Cai, G. Udeani, K. Slowing, C. Thomas, C. Beecher, H. Fong, N. Farnsworth, A. Kinghorn, and R. Mehta. 1997. Cancer chemopreventive activity of resveratrol, a natural product derived from grapes. *Science* 275 (5297):218.

Kaeberlein, M., M. McVey, and L. Guarente. 1999. The SIR2/3/4 complex and SIR2 alone promote longevity in Saccharomyces cerevisiae by two different mechanisms. *Gene Dev* 13 (19):2570.

Kaneda, A., T. Tsukamoto, T. Takamura-Enya, N. Watanabe, M. Kaminishi, T. Sugimura, M. Tatematsu, and T. Ushijima. 2005. Frequent hypomethylation in multiple promoter CpG islands is associated with global hypomethylation, but not with frequent promoter hypermethylation. *Cancer Sci* 95 (1):58–64.

Kang, J., C. Lin, J. Chen, and Q. Liu. 2004. Copper induces histone hypoacetylation through directly inhibiting histone acetyltransferase activity. *Chem Biol Interact* 148(3):115–23.

Kang, J., Y. Zhang, J. Chen, H. Chen, C. Lin, Q. Wang, and Y. Ou. 2003. Nickel-induced histone hypoacetylation: the role of reactive oxygen species. *Toxicological Sciences* 74 (2):279.

Khaw, K., N. Wareham, S. Bingham, A. Welch, R. Luben, and N. Day. 2008. Combined impact of health behaviours and mortality in men and women: the EPIC-Norfolk prospective population study. *Obstet Gynecol Surv* 63 (6):376–77.

Kim, D., J. Shin, and H. Kwon. 2007. Psammaplin A is a natural prodrug that inhibits class I histone deacetylase. *Exp Mol Med* 39 (1):47.

Kim, J., S. Tavaré, and D. Shibata. 2005. Counting human somatic cell replications: methylation mirrors endometrial stem cell divisions. *Proceedings of the Natl Acad Sci U S A* 102 (49):17739.

Kim, Y. 2004. Folate and DNA methylation: a mechanistic link between folate deficiency and colorectal cancer? *Cancer Epidem Biomar* 13 (4):511.

Lagouge, M., C. Argmann, Z. Gerhart-Hines, H. Meziane, C. Lerin, F. Daussin, N. Messadeq, J. Milne, P. Lambert, and P. Elliott. 2006. Resveratrol improves mitochondrial function and protects against metabolic disease by activating SIRT1 and PGC-1α. *Cell* 127 (6):1109–22.

Laird, P. 2003. The power and the promise of DNA methylation markers. *Nat Rev Cancer* 3 (4):253–66.

Laird, P., and R. Jaenisch. 1994. DNA methylation and cancer. *Hum Mol Genet* 3:1487–96.

Lea, M., and V. Randolph. 2001. Induction of histone acetylation in rat liver and hepatoma by organosulfur compounds including diallyl disulfide. *Anticancer Res* 21 (4):2841–46.

Lea, M., V. Randolph, and M. Patel. 1999. Increased acetylation of histones induced by diallyl disulfide and structurally related molecules. *Int J Oncol* 15 (2):347.

Lea, M., M. Rasheed, V. Randolph, F. Khan, A. Shareef, and C. desBordes. 2002. Induction of histone acetylation and inhibition of growth of mouse erythroleukemia cells by S-allylmercaptocysteine. *Nutr Cancer* 43 (1):90–102.

Li, E. 2002. Chromatin modification and epigenetic reprogramming in mammalian development. *Nat Rev Genet* 3 (9):662–73.

Liang, H., E. Masoro, J. Nelson, R. Strong, C. McMahan, and A. Richardson. 2003. Genetic mouse models of extended lifespan. *Expe Geronto* 38 (11–12):1353–64.

Lin, C., J. Kang, and R. Zheng. 2005. Oxidative stress is involved in inhibition of copper on histone acetylation in cells. *Chem-Biol interact* 151 (3):167–76.

Liu, H., Y. Chen, G. Cui, and J. Zhou. 2005. Curcumin, a potent anti-tumor reagent, is a novel histone deacetylase inhibitor regulating B-NHL cell line Raji proliferation. *Acta Pharm Sinic* 26 (5):603–9.

Liu, L., Y. Li, and T. Tollefsbol. 2008. Gene-environment interactions and epigenetic basis of human diseases. *Curr Issues Mol Biol* 10 (1–2):25.

Lu, Q., M. Kaplan, D. Ray, S. Zacharek, D. Gutsch, and B. Richardson. 2002. Demethylation of ITGAL (CD11a) regulatory sequences in systemic lupus erythematosus. *Arthritis Rheum* 46 (5):1282–91.

Marcu, M., Y. Jung, S. Lee, E. Chung, M. Lee, J. Trepel, and L. Neckers. 2006. Curcumin is an inhibitor of p300 histone acetylatransferase. *Med Chem* 2 (2):169–74.

Morimitsu, Y., Y. Nakagawa, K. Hayashi, H. Fujii, T. Kumagai, Y. Nakamura, T. Osawa, F. Horio, K. Itoh, and K. Iida. 2002. A sulforaphane analogue that potently activates the Nrf2-dependent detoxification pathway. *J Biol Chem* 277 (5):3456.

Motta, M., N. Divecha, M. Lemieux, C. Kamel, D. Chen, W. Gu, Y. Bultsma, M. McBurney, and L. Guarente. 2004. Mammalian SIRT1 represses forkhead transcription factors. *Cell* 116(4): 551–63.

Myzak, M., P. Karplus, F. Chung, and R. Dashwood. 2004. A novel mechanism of chemoprotection by sulforaphane: inhibition of histone deacetylase. *Cancer Res* 64(16):5767.

Olaharski, A., J. Rine, B. Marshall, J. Babiarz, L. Zhang, E. Verdin, and M. Smith. 2005. The flavoring agent dihydrocoumarin reverses epigenetic silencing and inhibits sirtuin deacetylases. *PLoS Genet* 1 (6): e77.

Paul, B., I. Masih, J. Deopujari, and C. Charpentier. 1999. Occurrence of resveratrol and pterostilbene in age-old darakchasava, an ayurvedic medicine from India. *J Ethnopharmacol* 68 (1–3):71–76.

Pearson K.J., J.A. Baur, K.N. Lewis, L. Peshkin, N.L. Price, N. Labinsky, W.R. Swindell, D. Kamara, R.K. Minor, E. Perez, H.A. Jamieson, Y. Zhang, S.R. Dunn, K. Sharma, N. Pleshko, L.A. Woollett, A. Csiszar, Y. Ikeno, D. Le Couteur, P.J. Elliott, K.G. Becker, P. Navas, D.K. Ingram, N.S. Wolf, Z. Ungvari, D.A. Sinclair, R. de Cabo. 2008. Resveratrol delays age-related deterioration and mimics transcriptional aspects of dietary restriction without extending life span. *Cell Metab* 8(2):157–68.

Penumathsa S.V., S. Koneru, S.M. Samuel, G. Maulik, D. Bagchi, S.F. Yet, V.P. Menon, N. Maulik. 2008. Strategic targets to induce neovascularization by resveratrol in hypercholesterolemic rat myocardium: role of caveolin-1, endothelial nitric oxide synthase, hemeoxygenase-1, and vascular endothelial growth factor. *Free Radic Biol Med* 45(7):1027–34.

Pezzuto, J. 1997. Plant-derived anticancer agents. *Biochem Pharmacol* 53 (2):121–33.

Porcu, M., and A. Chiarugi. 2005. The emerging therapeutic potential of sirtuin-interacting drugs: from cell death to lifespan extension. *Trends Pharmacol Sci* 26 (2):94–103.

Renaud, S., and M. De Lorgeril. 1992. Wine, alcohol, platelets, and the French paradox for coronary heart disease. *The Lancet* 339 (8808):1523–26.

Richardson, B. 2003. Impact of aging on DNA methylation. *Ageing Res Rev* 2 (3):245–61.

Roberts, C., and S. Orkin. 2004. The SWI/SNF complex—chromatin and cancer. *Nat Rev Cancer* 4 (2):133–42.

Robertson, A. 1977. Conrad Hal Waddington. 8 November 1905–26 September 1975. *Biographical Memoirs of Fellows of the Royal Society* 23:575–622.

Rodenhiser, D., and M. Mann. 2006. Epigenetics and human disease: translating basic biology into clinical applications. *Can Med Assoc J* 174 (3):341.

Romanov, G., and B. Vanyushin. 1981. Methylation of reiterated sequences in mammalian DNAs: Effects of the tissue type, age, malignancy and hormonal induction. *BBA-Nucleic Acids Prot Syn* 653 (2):204–18.

Rudman, D., A. Feller, H. Nagraj, G. Gergans, P. Lalitha, A. Goldberg, R. Schlenker, L. Cohn, I. Rudman, and D. Mattson. 1990. Effects of human growth hormone in men over 60 years old. *New Engl J Med* 323 (1):1.

Singhal, R., L. Mays-Hoopes, and G. Eichhorn. 1987. DNA methylation in aging of mice. *Mech Ageing Dev* 41 (3):199–210.

Sun, L., K. Al-Regaiey, M. Masternak, J. Wang, and A. Bartke. 2005. Local expression of GH and IGF-1 in the hippocampus of GH-deficient long-lived mice. *Neurobiol Aging* 26 (6):929–37.

Suter, C., D. Martin, and R. Ward. 2004. Hypomethylation of L1 retrotransposons in colorectal cancer and adjacent normal tissue. *Int J Colorectal Dis* 19 (2):95–101.

Szyf, M., P. Pakneshan, and S. A. Rabbani. 2004. DNA methylation and breast cancer. *Biochem Pharmacol* 68 (6):1187–97.

Tissenbaum, H., and L. Guarente. 2001. Increased dosage of a sir-2 gene extends lifespan in *Caenorhabditis elegans*. *Nature* 410 (6825): 227–30.

Trichopoulou, A., C. Bamia, and D. Trichopoulos. 2005. Mediterranean diet and survival among patients with coronary heart disease in Greece. *Arch Int Med* 165 (8):929.

Trichopoulou, A., P. Orfanos, and T. Norat. 2005. Modified Mediterranean diet and survival: EPIC-elderly prospective cohort study. *Brit Med J* 330 (7498):991.

Vastano, B., Y. Chen, N. Zhu, C. Ho, Z. Zhou, and R. Rosen. 2000. Isolation and identification of stilbenes in two varieties of Polygonum cuspidatum. *J Agric Food Chem* 48(2):253–56.

Vaziri, H., S. Dessain, E. Eaton, S. Imai, R. Frye, T. Pandita, L. Guarente, and R. Weinberg. 2001. hSIR2SIRT1 functions as an NAD-dependent p53 deacetylase. *Cell* 107 (2):149–59.

Verona, R., M. Mann and M. Bartolomei 2003. Genomic imprinting: intricacies of epigenetic regulation in clusters. *Ann Rev Cell Dev Bi* 19 (1):237–59.

Weindruch, R., T. Kayo, C. Lee, and T. Prolla. 2002. Gene expression profiling of aging using DNA microarrays. *Mech Ageing Dev* 123 (2–3):177–93.

Weindruch, R., R. Walford, S. Fligiel, and D. Guthrie. 1986. The retardation of aging in mice by dietary restriction: longevity, cancer, immunity and lifetime energy intake. *J Nutr* 116 (4):641.

Wilson, V., R. Smith, S. Ma, and R. Cutler. 1987. Genomic 5-methyldeoxycytidine decreases with age. *J Biol Chem* 262 (21):9948.

Wood, J., B. Rogina, S. Lavu, K. Howitz, S. Helfand, M. Tatar, and D. Sinclair. 2004. Sirtuin activators mimic caloric restriction and delay ageing in metazoans. *Nature* 430 (7000):686–89.

Yan, Y., T. Kluz, P. Zhang, H. Chen, and M. Costa. 2003. Analysis of specific lysine histone H3 and H4 acetylation and methylation status in clones of cells with a gene silenced by nickel exposure. *Toxicol Appl Pharm* 190 (3):272–77.

Yatabe, Y., S. Tavaré, and D. Shibata. 2001. Investigating stem cells in human colon by using methylation patterns. *Proceedings of the National Academy of Sciences of the United States of America* 98 (19):10839.

Yeung, F., J. Hoberg, C. Ramsey, M. Keller, D. Jones, R. Frye, and M. Mayo. 2004. Modulation of NF- B-dependent transcription and cell survival by the SIRT1 deacetylase. *The EMBO Journal* 23(12): 2369.

Yoo, C. and P. Jones 2006. Epigenetic therapy of cancer: past, present and future. *Nat Rev Drug Discov* 5 (1):37–50.

Yung, R., D. Powers, K. Johnson, E. Amento, D. Carr, T. Laing, J. Yang, S. Chang, N. Hemati, and B. Richardson. 1996. Mechanisms of drug-induced lupus. II. T cells overexpressing lymphocyte function-associated antigen 1 become autoreactive and cause a lupuslike disease in syngeneic mice. *J Clin Invest* 97 (12):2866–71.

Zhong, H., M. May, E. Jimi, and S. Ghosh. 2002. The phosphorylation status of nuclear NF-[kappa] B determines its association with CBP/p300 or HDAC-1. *Mol Cell* 9 (3):625–36.

3 Folate and DNA Methylation

Julie Crowell, Anna Ly,* and Young-In Kim*

CONTENTS

3.1 Epigenetics 32
3.2 DNA Methylation 32
  3.2.1 DNA Methylation and Cancer 33
3.3 Folate and Cancer 36
3.4 Folate and DNA Methylation 38
  3.4.1 Effects of Folate Status on DNA Methylation in Animal Studies 38
    3.4.1.1 Effects of Methyl Group Deficiency and Supplementation on DNA Methylation in Rodents 38
    3.4.1.2 Effects of Isolated Folate Deficiency on DNA Methylation in Rodents 40
    3.4.1.3 Effects of Isolated Folate Supplementation on DNA Methylation in Rodents 45
    3.4.1.4 Interactions between Folate and Other Environmental Factors on DNA Methylation in Rodents 47
  3.4.2 Effects of Folate Deficiency and Supplementation on DNA Methylation in In Vitro Systems 48
  3.4.3 Effects of Folate Status on DNA Methylation in Human Studies 49
    3.4.3.1 Effect of Folate Deficiency and Supplementation on DNA Methylation in Human Clinical Trials 49
    3.4.3.2 Effects of Folate Status on Genomic DNA Methylation in Human Observational Studies 53
    3.4.3.3 Effects of Folate Status on Gene-Specific DNA Methylation in Human Observational Studies 56
    3.4.3.4 Summary of Effects of Folate on DNA Methylation in Human Studies 59
  3.4.4 Effects of Maternal Folate Intake during Pregnancy on DNA Methylation in the Offspring 60
  3.4.5 Effects of Folate and Aging on DNA Methylation 63
3.5 Conclusion 64
Acknowledgment 65
References 65

* These authors contributed equally to this chapter.

3.1 EPIGENETICS

The inheritance of information based on gene expression levels is known as epigenetics, as opposed to genetics, which refers to information transmitted on the basis of gene sequence (Esteller 2003). The field of epigenetics is therefore the study of modifications of DNA and DNA-binding proteins and histones that alter the structure of chromatin without altering the nucleotide sequence of DNA; some of these modifications may be associated with heritable changes in gene function (Egger et al. 2004). Silencing is a subset of epigenetics whereby gene expression and function are permanently lost. More recently, RNA interference has been emerging as an important mechanism in epigenetic silencing.

3.2 DNA METHYLATION

One of the most important epigenetic modifications in mammals is the methylation of cytosine located within the cytosine-guanine (CpG) dinucleotide sequences (Jones and Laird 1999; Jones and Baylin 2002). The pattern of methylation at cytosine residues in the CpG sequences is a heritable, tissue- and species-specific, postsynthetic modification of mammalian DNA (Jones and Laird 1999; Jones and Baylin 2002). Three to four percent of all cytosines in the human genome are methylated, and the resulting 5-methylcytosines make up 0.75–1% of all nucleotide bases in normal human DNA (Esteller 2003).

CpG sites are unevenly distributed in the mammalian genome; vast stretches of sequence (~99% of the genome) are deficient for CpGs and these are interspersed by CpG clusters called CpG islands. Generally, the CpG dinucleotide is greatly underrepresented throughout the mammalian genome (also termed CpG suppression) (Das and Singal 2004). The CpG dinucleotide should occur with a frequency of approximately 6%. However, the actual presence is only 5–10% of its predicted frequency (Das and Singal 2004). This CpG suppression may be related to the hypermutable capacity of 5-methylcytosine to deaminate spontaneously, which results in methylcytosine-to-thymine transition mutations (Das and Singal 2004). Seventy to eighty percent of all CpG sites in human DNA are normally methylated (Esteller 2003). However, this methylation occurs primarily in the bulk of the genome where CpG density is low, including exons, noncoding regions, and repeat DNA sites, and allows correct organization of chromatin in active and inactive states (Herman and Baylin 2003). Methylation of the CpG-depleted bulk of the genome facilitates transcriptional silencing of noncoding regions, which prevents the transcription of repeat DNA elements, inserted viral sequences, and transposons (Herman and Baylin 2003). Transposons are common and potentially mobile sequences of DNA that move from their usual location into a new region of the genome (Yoder et al. 1997). The human genome is littered with transposons and endogenous retroviruses acquired throughout the human history, and these parasitic sequences account for more than 35% of the human genome (Yoder et al. 1997). Parasitic DNA elements represent a significant threat to the structural integrity of the genome by promoting chromosome rearrangements or translocation or by directly disrupting genes (Robertson and Wolffe 2000). These parasitic sequences contain strong promoters that if integrated within a transcriptional unit could result in internal initiation (Robertson and Wolffe 2000). If integrated in the "antisense" orientation relative to the normal direction of transcription of the targeted genes, this could inhibit gene expression by transcriptional interference (Robertson and Wolffe 2000).

By contrast, about 1% of the genome consists of CpG-rich areas clustered in small stretches of DNA termed "CpG islands," which are defined as a 500-base pair window with a G:C content of at least 55% and an observed overexpected CpG frequency of at least 0.65 (Takai and Jones 2002). These motifs span the 5′ end of approximately half of the human genes including the promoter, untranslated region, and exon 1 (i.e., in and around the transcription start sites) (Takai and Jones 2002). Most CpG islands are unmethylated in normal cells, thereby allowing transcription, with the exception of CpG island on the inactive X chromosome in females and silenced alleles of imprinted genes (Robertson and Wolffe 2000). When methylated, CpG islands cause stable heritable transcriptional silencing. Transcriptional repression by CpG islands methylation is mediated by the transcriptional repressor, methyl-CpG binding proteins (MBDs), which binds methylated CpG islands and recruits a complex containing a transcriptional co-repressor and a histone deacetylase (HDAC) (Robertson and Wolffe 2000). Deacetylation of histones suppresses transcription by allowing tighter nucleosomal packaging and thus rendering an inactive chromatin conformation (Robertson and Wolffe 2000).

DNA methylation is a dynamic process between active methylation, mediated by CpG methyltransferases (DNMT13a, 3b) using S-adenosylmethionine (SAM) as the methyl donor, and removal of methyl groups from 5-methylcytosine residues by both passive and active mechanisms including demethylation by a purported demethylase (MBD2) (Li and Jaenisch 2000). DNA methylation patterns are reprogrammed during embryogenesis by genome-wide demethylation early in embryogenesis, which erases significant parts of the parental DNA methylation, followed by de novo methylation, which establishes a new DNA methylation pattern soon after implantation, with methylation limited to non-CpG island areas, except for the rare genes silenced in normal cells (Reik et al. 2001; Li and Jaenisch 2000). Maintenance methylase (DNMT1) uses hemimethylated sites to ensure DNA methylation patterns, whereas de novo methylases (DNMT3a, 3b) do not require preexisting methylation and establish a new DNA methylation pattern (Li and Jaenisch 2000).

3.2.1 DNA Methylation and Cancer

DNA methylation is an important epigenetic determinant in gene expression (an inverse relationship), in the maintenance of DNA integrity and stability, in chromatin modifications, and in the development of mutations (Jones and Laird 1999; Jones and Baylin 2002). As such, DNA methylation is mechanistically linked to the pathogenesis of several chronic diseases in humans, including cancer.

In contrast to methylated CpG sites in the CpG-poor bulk of the genome and unmethylated CpG islands in normal cells, cancer cells simultaneously harbor widespread loss of methylation in the CpG-depleted regions where most CpG dinucleotides should be methylated and gains in methylation of CpG islands in gene

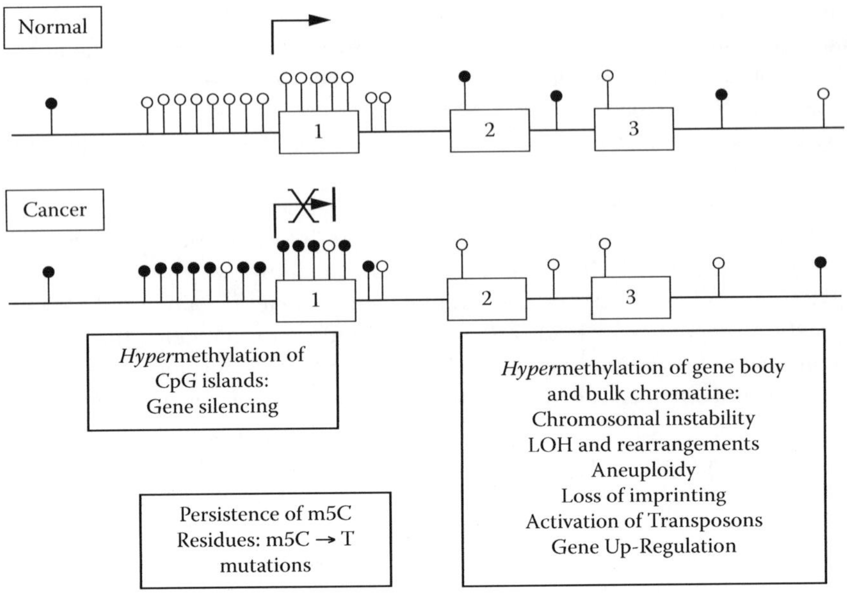

FIGURE 3.1 Distribution of CpG dinucleotides in the human genome and CpG methylation patterns in normal and tumor cells. In contrast to methylated CpG sites in the CpG-poor bulk of the genome and unmethylated CpG islands in normal cells, cancer cells simultaneously harbor widespread loss of methylation in the CpG-depleted regions where most CpG dinucleotides should be methylated and gains in methylation of CpG islands in gene promoter regions. Open circles represent unmethylated CpG sites whereas filled circles are methylated CpG sites. Boxes 1, 2, and 3 represent exons and the lines between exons are introns. X at the transcription start site represents transcriptional silencing.

promoter regions (Figure 3.1) (Jones and Laird 1999; Jones and Baylin 2002; Herman and Baylin 2003).

Global hypomethylation is an early and consistent event in carcinogenesis (Jones and Laird 1999; Jones and Baylin 2002; Herman and Baylin 2003). Global hypomethylation of the coding and noncoding regions and demethylation of repetitive DNA sequences contribute to the development of cancer through the following mechanisms: chromosomal instability, increased mutations, reactivation of intragenomic parasitic sequences that could be transcribed and moved to other sites where they could disrupt normal cellular genes, mitotic recombination leading to loss of heterozygosity and promotion of rearrangements, aneuploidy, loss of imprinting, and up-regulation of protooncogenes (Esteller 2003). However, animal studies have indicated that global DNA hypomethylation may promote or protect against tumor development in a site-specific manner. The combination of heterozygosity for a null mutation of the Dnmt1 gene and treatment with the DNMT inhibitor 5-aza-2′-deoxycytidine dramatically reduced the number of small intestinal polyps in *Apc*$^{\mathrm{Min/+}}$ mice (Laird et al. 1995). Similarly, mice deficient in one of the mismatch repair proteins, Mlh1, carrying the hypomorphic Dnmt1 allele had a lower incidence of small intestinal polyps; however, the incidence of lymphomas in these mice was increased (Trinh et al. 2002).

Mice carrying a hypomorphic Dnmt1 allele, which reduces Dnmt1 expression to 10% of wild-type levels and results in substantial genome-wide hypomethylaton in all tissues, developed aggressive T cell lymphomas that displayed a high frequency of chromosomal instability (Gaudet et al. 2003). Furthermore, genomic hypomethylation induced by a hypomorphic Dnmt1 allele in the $Nf1^{+/-}p53^{+/-}$ mice promoted the development of sarcomas at an earlier age compared with the $Nf1^{+/-}p53^{+/-}$ mice with normal levels of DNA methylation via increased chromosomal instability (Eden et al. 2003). Therefore genomic demethylation may protect against some cancers (e.g., intestinal tumors) but may promote chromosomal instability and increase the risk of cancer in other tissues (e.g., lymphoma, sarcoma).

Methylation at promoter CpG islands is an important mechanism of silencing transcription in carcinogenesis; the affected genes are silenced and their function is stably lost in a clonally propagated fashion (Esteller et al. 2001; Herman and Baylin 2003; Esteller 2003). That promoter CpG island methylation is essential for gene silencing is unequivocally established by several experiments including (1) loss of DNA methylation induced by homozygous deletion of the DNMT gene results in reexpression of previously silenced genes and (2) DNMT inhibition (pharmacologic or antisense) results in demethylation and reexpression of previously silenced genes (Esteller et al. 2001; Herman and Baylin 2003; Esteller 2003). The number of cancer-related genes silenced by promoter CpG island methylation equals or exceeds the number that are inactivated by mutation (Esteller et al. 2001; Herman and Baylin 2003; Esteller 2003). Many genes modified by promoter CpG methylation have classic tumor-suppressor function, and other genes play critical roles in cell cycle control, repair of DNA damage, apoptosis, differentiation, angiogenesis, metastasis, growth-factor response, drug resistance, and detoxification (Herman and Baylin 2003; Das and Singal 2004). Promoter CpG islands of over 60% of tumor suppressor and mismatch repair genes have been observed to be methylated in cancer (Herman and Baylin 2003; Das and Singal 2004).

Another means by which CpG methylation may contribute to carcinogenesis is the hypermutability of methylated cytosine. CpG dinucleotides within certain genes are not only the sites of DNA methylation but also mutational hot spots for human cancers (Zingg and Jones 1997). The majority of mutations observed in CpG sites are cytosine-to-thymine transitions mediated by the spontaneous deamination of 5-methylcytosine to thymine, by the enzymatic deamination of 5-methylcytosine to thymine by DNMT, and by the enzymatic deamination of unmethylated cytosine to uracil and subsequent methylation of uracil to thymine by DNMT (Zingg and Jones 1997). CpG sites have been shown to act as hot spots for germline mutations, contributing to 30% of all point mutations in the germline, and for acquired somatic mutations that lead to cancer (Robertson and Wolffe 2000). For example, methylated CpG sites in the *p53* tumor suppressor coding region contribute to as many as 50% of all inactivating mutations in colorectal cancer and to 25% of cancers in general (Robertson and Wolffe 2000).

Increased DNMT1, 3a, and 3b and decreased MBD2 expression and activity have been observed in many human cancers (Li and Jaenisch 2000). DNMT1 may promote tumorigenesis by its link to activation of the oncogenic ras signaling pathway; by increasing cellular proliferation by binding to proliferating cell nuclear antigen and by reducing cellular *p21*, a member of the cyclin-dependent kinase (CDK) inhibitor family that inhibits a wide range of cyclin-CDK complexes involved in G_1 and S phase progression; by inhibition of *p53*-dependent apoptosis; and by promoter CpG island methylation of tumor suppressor and mismatch repair genes (Li and Jaenisch 2000).

3.3 FOLATE AND CANCER

Folate is a water-soluble B vitamin that is naturally present in foods (e.g., green leafy vegetables, asparagus, broccoli, Brussels sprouts, citrus fruit, legumes, dry cereals, whole grain, yeast, lima beans, liver, and other organ meats). Folic acid is the synthetic form of this vitamin that is used commercially in supplements and in fortified foods.

Epidemiologic studies suggest an inverse association between folate status (assessed by dietary folate intake or by the measurement of blood folate levels) and the risk of several malignancies including cancer of the lungs, oropharynx, esophagus, stomach, colorectum, pancreas, cervix, ovary, prostate, and breast and the risk of neuroblastoma and leukemia (Y. I. Kim 1999, 2003, 2007, 2008). The precise nature and magnitude of the inverse relation between folate status and the risk of these malignancies, however, have not yet been clearly established (Y. I. Kim 1999, 2003, 2007, 2008).

The role of folate in carcinogenesis has been best studied for colorectal cancer. An accumulating body of evidence suggests that folate status is inversely related to the risk of sporadic and ulcerative colitis-associated colorectal cancer or its precursor adenoma (Y. I. Kim 1999, 2003, 2007, 2008). Although the results from epidemiologic and clinical studies are not uniformly consistent, the portfolio of evidence indicates ~20–40% reduction in the risk of colorectal cancer or adenoma in subjects with the highest folate status compared with those with the lowest status (Y. I. Kim 1999, 2003, 2007, 2008). The role of folate in colorectal carcinogenesis has been further strengthened by the observations that genetic polymorphisms in the folate metabolic pathway (e.g., the methylenetetrahydrofolate reductase [MTHFR] C677T polymorphism) modify colorectal cancer risk (Potter 2002; L. B. Bailey 2003; Y. I. Kim 2009).

Although there is no definitive evidence supporting the protective effect of folate supplementation on colorectal carcinogenesis from human experiments at present, several small intervention studies have demonstrated that folate supplementation can improve or reverse surrogate endpoint biomarkers of colorectal cancer (Cravo et al. 1994, 1998; Paspatis and Karamanolis 1994; Y. I. Kim et al. 2001; Khosraviani et al. 2002; Biasco et al. 1997; Lashner et al. 1999) and some epidemiologic studies have shown a beneficial effect of multivitamin supplements containing ≥400 μg folic acid on colorectal cancer risk and mortality (Giovannucci et al. 1995, 1998; Jacobs et al. 2001). However, in the recent Aspirin/Folate Polyp Prevention Study, folic acid supplementation was associated with a 67% increased risk of advanced lesions with a high malignant potential in men and women with a history of colorectal adenomas (Cole et al. 2007). In contrast, a recent trial has reported that folic acid supplementation had no beneficial effect on adenoma recurrence over three years among patients who previously had colorectal adenomas (Logan et al. 2008).

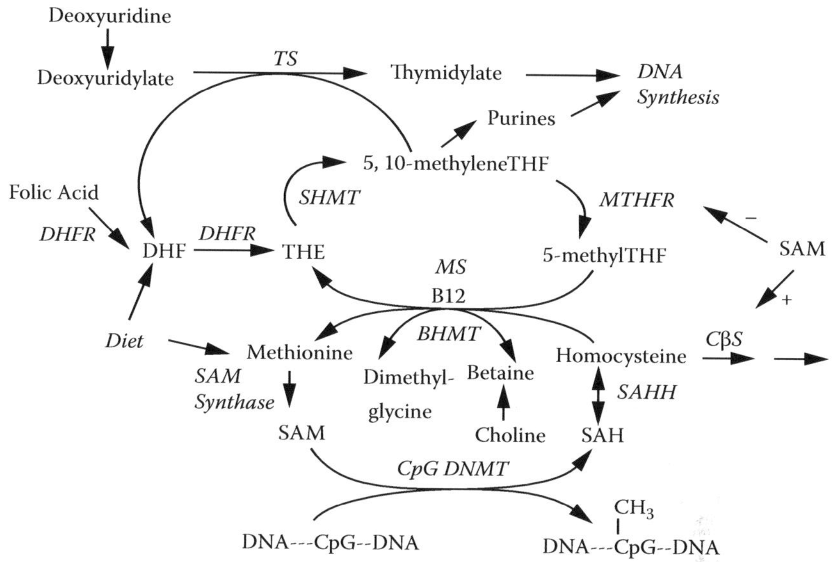

FIGURE 3.2 Simplified scheme of the role of 5,10-methylenetetrahydrofolate reductase (MTHFR) in folate metabolism and one-carbon transfer reactions involved in DNA synthesis and biological methylation reactions, including that of DNA. MTHFR catalyzes the irreversible conversion of 5,10-methylenetetrahydrofolate (5,10-methyleneTHF) to 5-methyltetrahydrofolate (5-methylTHF), and hence the MTHFR C677T polymorphism, which results in decreased MTHFR activity and increased thermolability of MTHFR, leads to lower levels of 5-methylTHF and an accumulation of 5,10-methyleneTHF. SAM is both an allosteric inhibitor of MTHFR and an activator of cystathionine β-synthase. B12, vitamin B-12; BHMT, betaine:homocysteine methyltransferase; CβS, cystathionine β-synthase; CH_3, methyl group; CpG, cytosine-guanine dinucleotide sequence; DHF, dihydrofolate; DHFR, dihydrofolate reductase; DNMT, DNA methyltransferase; MS, methionine synthase; SAH, S-adenosylhomocysteine; SAHH, S-adenosylhomocysteine hydrolase; SAM, S-adenosylmethionine; SHMT, serine hydroxymethyltransferase; THF, tetrahydrofolate; TS, thymidylate synthase.

The data from animal studies generally support a causal relationship between folate depletion and colorectal cancer risk and an inhibitory effect of modest levels of folate supplementation on colorectal carcinogenesis (Y. I. Kim 2003). However, animal studies have also shown that folate supplementation may increase colorectal risk and accelerate colorectal cancer progression if too much is given or if it is provided after neoplastic foci are established in the colorectum (Y. I. Kim 2003, 2004).

In addition, the role of folate in cancer risk is further complicated by common variants in critical genes involved in folate metabolism. Methylenetetrahydrofolate reductase (MTHFR) irreversibly catalyzes the conversion of 5,10-methylenetetrahydrofolate to 5-methyltetrahydrofolate, shuffling one-carbon units toward the methylation cycle at the expense of thymidylate and purine synthesis (Figure 3.2) (Friso and Choi 2002; Y. I. Kim 1999, 2000). The MTHFR C677T polymorphisms cause thermolability and reduced MTHFR activity, leading to lower levels of 5-methyltetrahydrofolate, an accumulation of 5,10-methylenetetrahydrofolate, increased plasma homocysteine levels (a sensitive inverse indicator of folate status), and changes in cellular composition of one-carbon folate derivatives (Figure 3.2) (Friso and Choi 2002; Y. I. Kim 1999, 2000). Another polymorphism in the MTHFR gene (A1298C) has also been recently associated with reduced enzyme activity (Castro et al. 2004; Friso et al. 2005). Several studies suggest that MTHFR polymorphisms in combination with a compromised folate status may modify the risk of colorectal adenomas or cancer (Ulrich et al. 1999; Levine et al. 2000; Marugame et al. 2003; Pufulete et al. 2003).

3.4 FOLATE AND DNA METHYLATION

The mechanisms by which folate deficiency enhances and supplementation suppresses colorectal carcinogenesis have not yet been clearly elucidated. However, several potential mechanisms relating to the disruption of the known biochemical function of folate have been proposed and investigated (Choi and Mason 2002; Y. I. Kim 1999; Duthie 1999; Ames 2001; Fenech 2001; Lamprecht and Lipkin 2003). Folate plays an important role in mediating the transfer of one-carbon moieties (Figure 3.2) (Wagner 1995). The substrate 5,10-methylenetetrahydrofolate, an intracellular coenzymatic form of folate, is required for conversion of deoxyuridylate to thymidylate and can be oxidized to 10-formyltetrahydrofolate for de novo purine synthesis (Figure 3.2) (Wagner 1995). Thus folate is an important factor in DNA synthesis, stability and integrity, and repair (Figure 3.2), aberrations of which have been implicated in colorectal carcinogenesis (Choi and Mason 2002; Y. I. Kim 1999; Duthie 1999; Ames 2001; Fenech 2001; Lamprecht and Lipkin 2003).

Folate, in the form of 5-methyltetrahydrofolate, is also involved in remethylation of homocysteine to methionine, which is a precursor of SAM, the primary methyl group donor for most biological methylation reactions including that of DNA (Figure 3.2) (Selhub and Miller 1992). After transfer of the methyl group, SAM is converted to S-adenosylhomocysteine (SAH), a potent inhibitor of most SAM-dependent methyltransferases (Figure 3.2) (Selhub and Miller 1992). Cravo and Mason first proposed that a mechanism by which folate deficiency enhances colorectal carcinogenesis might be through an induction of genomic DNA hypomethylation based on the biochemical function of folate in mediating one-carbon transfer and on evidence from animal experiments that demonstrated methyl group donor deficiency-induced DNA hypomethylation (Cravo et al. 1992). Genomic and site-specific DNA hypomethylation has been considered as a potential mechanism by which folate depletion enhances colorectal carcinogenesis (Choi and Mason 2002; Y. I. Kim 1999; Duthie 1999; Ames 2001; Fenech 2001).

3.4.1 Effects of Folate Status on DNA Methylation in Animal Studies

3.4.1.1 Effects of Methyl Group Deficiency and Supplementation on DNA Methylation in Rodents

Diets deficient in methyl group donors (choline, folate, methionine, and vitamin B_{12}) are associated with spontaneous and chemically induced development of hepatocellular carcinoma in rats (Newberne and Rogers 1986). Diets deficient in different combinations of methyl group donors have been consistently observed to induce genomic and proto-oncogene (*c-myc, c-fos, c-Ha-ras*) DNA hypomethylation and elevated steady-state levels of corresponding mRNAs (Zapisek et al. 1992; Wainfan et al. 1989; Wainfan and Poirier 1992; Dizik et al. 1991; Christman et al. 1993; Pogribny, Basnakian et al. 1995). Methyl group donor deficiency has also been shown to induce site-specific *p53* hypomethylation in rat liver (Pogribny, Poirier et al. 1995; Pogribny et al. 1997; Pogribny, Basnakian et al. 1995), although recent studies suggest that CpG methylation within the rat *p53* promoter region appears to be site-specific and varies throughout the carcinogenic process (Pogribny, Miller et al. 1997, 2000). Furthermore, in methyl-deficient rats, site-specific de novo methylation of the *p16* gene 5′ CpG island has been shown to precede tumor development, and with tumor progression the incidence and extent of de novo methylation increases (Pogribny and James 2002). Methyl group donor deficiency has also been shown to up-regulate Dnmt (Wainfan et al. 1988; Pogribny, Poirier et al. 1995; Pogribny et al. 1997; Wainfan et al. 1989; Wainfan and Poirier 1992).

Recent studies suggest that a prolonged diet deficient in methyl group donors can induce permanent changes in DNA methylation that cannot be reversed from a methyl-adequate diet (Pogribny et al. 2006, 2009). In one particular study, methyl group donor deficiency was shown to induce DNA damage and aberrant DNA methylation, resulting in a decrease in genomic DNA methylation and an increase in promoter CpG island methylation of the *Rassf1a* gene in rat liver (Pogribny et al. 2009). Interestingly, in rats fed this methyl-deficient diet for 9 weeks followed by a methyl-sufficient diet, DNA abnormalities were completely restored to the normal state, while feeding the methyl-deficient diet for 18 weeks followed by a methyl-sufficient diet repaired the DNA lesions but failed to restore the altered DNA methylation status to normal (Pogribny et al. 2009). A diet deficient in choline, methionine, and folate, which caused a 30% increase in DNA strand breaks, did not induce a significant degree of genomic DNA hypomethylation in rat colon, suggesting that the colorectum may be resistant to the hypomethylating effect of methyl group deficiency (Duthie, Narayanan, Brand et al. 2000). Interestingly, a recent study has shown that long-term administration of a methyl-deficient diet lead to genomic DNA hypermethylation in rat brain (Pogribny et al. 2008). Taken together, the results from these studies suggest that the effect of methyl group donor deficiency on DNA methylation is tissue-, site-, and gene-specific, and varies throughout the carcinogenesis process in liver.

Animal studies using viable yellow agouti (A^{vy}) mice have unequivocally demonstrated that maternal dietary methyl group supplementation with a modest amount of folic acid, vitamin B_{12}, choline, and betaine permanently alters the phenotype of the offspring via increased CpG methylation at the promoter CpG site of the *agouti* gene (Wolff et al. 1998; Cooney et al. 2002; Waterland and Jirtle 2003). Furthermore, Waterland and Jirtle (2003) have shown that the methylation status of the promoter CpG region of the *agouti* gene was highly correlated with the methylation status of the adjacent transposon gene (Waterland and Jirtle 2003). This indicates that there is a localized epigenetic instability in methylation that arises from an interaction between the transposon and its nearby genetic region and that genes that manifest a transposon region adjacent to a promoter region of DNA could be influenced by early life nutrition containing methyl group donors.

3.4.1.2 Effects of Isolated Folate Deficiency on DNA Methylation in Rodents

Although isolated folate deficiency has been shown to reduce SAM concentrations and SAM-to-SAH concentration ratios and increase SAH levels in rat liver (Y. I. Kim et al. 1994, 1995, 1997; Miller et al. 1994; Balaghi et al. 1993; Uthus et al. 2006; Choi et al. 2005), conflicting data exist for the effect of isolated folate deficiency on DNA methylation in rodent liver (Table 3.1). One study reported a significant 20% decrease in genomic DNA methylation associated with a severe degree of dietary folate deficiency of 4 weeks' duration in rat liver (Balaghi and Wagner 1993), while another study showed a paradoxical 56%, albeit nonsignificant, increase in genomic DNA methylation associated with the same severe folate deficient diet of 6 weeks' duration in rat liver (Y. I. Kim et al. 1997). A prolonged, moderate degree of dietary folate deficiency in weanling rats failed to induce significant genomic DNA hypomethylation, despite a decrease in hepatic SAM and an increase in both hepatic SAH and SAM-to-SAH concentration ratios (Uthus et al. 2006; Y. I. Kim et al. 1995). Similarly, in rats 6–7 weeks of age, a moderate degree of dietary folate deficiency for 24 weeks resulted in a significant decrease in hepatic SAM-to-SAH concentration ratios and genomic DNA methylation in rat liver was not altered (Duthie et al. 2010). In elder rats, however, dietary folate deficiency of a moderate degree for 8 and 20 weeks significantly increased hepatic SAH concentrations and in addition significantly decreased genomic DNA methylation in the liver compared with folate-supplemented rats (Choi et al. 2005).

In pregnant female rats, folate deficiency of a moderate degree and short duration (approximately 5 weeks) was shown to have no effect on maternal hepatic genomic DNA methylation (Maloney et al. 2007), while this same degree of folate deficiency in mice 6 weeks of age for 5 weeks' duration was shown to induce a significant 56% increase in genomic DNA methylation in the liver followed by the return of genomic DNA methylation value to that of the baseline by 8 weeks (Song et al. 2000).

A recent study examined the effects of both timing and duration of dietary folate intervention provided during the postweaning period on genomic DNA methylation in adult rat liver, and found that a moderate degree of folate deficiency provided at weaning and at 8 weeks of age and continued until 30 weeks of age failed to significantly modulate genomic DNA methylation in adult rat liver (Kotsopoulos et al. 2008). However, this same degree of dietary folate deficiency provided at weaning in rats and continued through early infancy and childhood for 5 weeks until puberty, followed by the control diet for 22 weeks, was shown to induce a significant 34–48% increase in genomic DNA methylation in adult rat liver compared with the control and folate-supplemented diet (Kotsopoulos et al. 2008).

Taken together, the results from these studies collectively suggest that dietary folate deficiency provided early on in life appears to induce genomic DNA hypermethylation in rodent liver, likely due to compensatory up-regulation of Dnmt and of the choline- and betaine-dependent transmethylation pathway, and if adequate levels of dietary folate and other methyl group donors are provided in adolescence and continued into adulthood, this pattern of genomic DNA hypermethylation is maintained. On the other hand, if continual dietary folate deficiency is imposed, this compensatory hypermethylation pattern will not be maintained. These studies also suggest that dietary folate may modulate DNA methylation more readily in the elderly compared to the young. Thus the results from these studies highlight the importance of timing of folate deficiency and subsequent supply of folate in establishing and maintaining the DNA methylation pattern in rodent liver.

An intriguing observation from one of these animal studies was that severe folate deficiency produced significant hypomethylation (by 40%) within mutation hot spot (exons 6–7), but not in exon 8, of the *p53* tumor suppressor gene despite a 56% increase in genomic DNA methylation in rat liver (Y. I. Kim et al. 1997). This observation raises a possibility that the effect of folate deficiency on DNA methylation may be site- and gene-specific and suggests that the changes in genomic and site-specific DNA methylation in response to folate deficiency may not be in the same direction.

Several animal studies have also examined the effect of isolated folate deficiency on DNA methylation in the colorectum, the primary target tissue that is particularly susceptible to the folate deficiency-induced carcinogeneic effect; however, this effect has not yet been clearly elucidated (Table 3.1). A moderate degree of dietary folate deficiency for 15–24 weeks in weanling rats did not significantly alter colonic SAH concentrations, and failed to induce significant genomic and *c-myc*-specific DNA hypomethylation in rat colon (Y. I. Kim et al. 1995). This same degree of folate deficiency for 10 weeks had no effect on genomic DNA methylation in rat colon, despite a significant increase in SAH concentrations and a significant decrease in SAM-to-SAH concentration ratios (Uthus et al. 2006). In rats 6–7 weeks of age, a moderate degree of dietary folate deficiency for 10 weeks was associated with a significant degree of DNA strand breaks (Duthie, Narayanan, Brand et al. 2000), while a longer duration of 24 weeks failed to significantly alter colonic SAM and SAH levels (Duthie et al. 2010) and in both cases, genomic DNA methylation in the colon was not altered (Duthie, Narayanan, Brand et al. 2000; Duthie et al. 2010). Furthermore, in mice 4 months of age, a moderate degree of dietary folate deficiency for 10 weeks did not modulate genomic and *Apc*-specific DNA methylation in mouse colon (Liu et al. 2007).

A moderate degree of folate deficiency for 20 weeks in conjunction with an alkylating colon carcinogen, dimethylhydrazine (DMH), did not significantly alter genomic DNA hypomethylation in rat colon (Y. I. Kim, Salomon et al. 1996). In two other animal studies using a similar degree of moderate folate deficiency in conjunction with azoxymethane (AOM), a metabolite of DMH, genomic DNA methylation in rat colon was not affected (Le Leu et al. 2000a, 2000b). However, these studies (Le Leu et al. 2000a, 2000b) were limited by the use of DMH or AOM, which can alter tissue SAM and SAH levels (Halline et al. 1988) and the extent of DNA methylation (Hepburn et al. 1991) independent of the effect of folate. Another study showed that moderate folate deficiency for 12 weeks in conjunction with DMH injection did significantly increase colonic SAH concentrations but did not change colonic SAM levels or the degree of genomic DNA methylation in rats (Davis and Uthus 2003).

TABLE 3.1
Summary of the Effect of Isolated Folate Deficiency on DNA Methylation in Rodents

Study (Reference)	Folate Deficiency	Age	Duration	Species	Organ/Tissue	DNA Methylation	Effect
Balaghi and Wagner 1993	Severe	3 weeks	4 weeks	Rat	Liver	Genomic	20% decrease ($p = 0.032$)
Y. I. Kim et al. 1997	Severe	3 weeks	6 weeks	Rat	Liver	Genomic	60% increase ($p = 0.1$)
						p53 (exons 6–7)	40% decrease ($p = 0.002$)
Y. I. Kim et al. 1995	Mild	3 weeks	15 and 24 weeks	Rat	Liver	Genomic	No effect
					Colon	Genomic	No effect
						c-myc	No effect
Uthus et al. 2006	Mild	3 weeks	10 weeks	Rat	Liver	Genomic	No effect
					Colon		No effect
Duthie et al. 2010	Mild	6–7 weeks	24 weeks	Rat	Liver	Genomic	No effect
					Colon		No effect
Choi et al. 2005	Mild	1 year	8 and 20 weeks	Rat	Liver	Genomic	Deficient group significantly < than supplemented group ($p < 0.05$)
Maloney et al. 2007	Mild	8–10 weeks	~5 weeks	Rat	Liver	Genomic	No effect
Song et al. 2000	Mild	3 weeks	8 weeks	Mouse	Liver	Genomic	No effect
		6 weeks	5 weeks				56% increase ($p < 0.05$)
Kotsopoulos et al. 2008	Mild	3 weeks	27 weeks	Rat	Liver	Genomic	No effect
		8 weeks	22 weeks				No effect
		3 weeks	5 weeks followed by control for 22 weeks				34–48% increase at 30 weeks of age ($p < 0.04$)
Duthie et al. 2000a	Mild	6–7 weeks	10 weeks	Rat	Colon	Genomic	No effect

Liu et al. 2007	Mild	4 months	10 weeks	Mouse	Colon	Genomic	No effect
						Apc (promoter and intron 1)	No effect
Kim et al. 1996	Mild + DMH	3 weeks	20 weeks	Rat	Colon	Genomic	No effect
Le Leu et al. 2000a	Mild + AOM	4 weeks	12 weeks	Rat	Liver	Genomic	No effect
					Colon		No effect
Le Leu et al. 2000b	Mild + AOM	3 weeks	26 weeks	Rat	Colon	Genomic	No effect
Davis and Uthus 2003	Mild + DMH	3 weeks	12 weeks	Rat	Liver	Genomic	No effect
					Colon		No effect
Sohn et al. 2003	Severe	3 weeks	5 weeks	Rat	Colon	Genomic	30% increase at week 3 ($p = 0.022$) and no effect at other time points
						p53 (promoter and exons 6–7)	Highly variable; decrease at CpG site 1 in exons 6–7 at week 5
Liu et al. 2008	Mild	4 months	10 weeks	Mouse	Colon	*p53* (exons 5–8)	No effect
Y. I. Kim et al. 1996	Mild + DMH	3 weeks	20 weeks	Rat	Colon	*p53* (exon 8)	25% decrease ($p = 0.038$)
Linhart et al. 2009	Severe	3 months	32 weeks	Mouse	Colon	Genomic	6% decrease ($p < 0.04$)
					Small intestine		No effect
					Spleen		No effect

One study investigated the time-dependent effects of dietary folate on genomic and *p53* (in the promoter region and exons 6–7) DNA methylation in rat colon, and how these changes are related to steady-state levels of *p53* transcript (Sohn et al. 2003). Despite a marked reduction in plasma and colonic folate concentrations, a large increase in plasma homocysteine, and a progressive decrease in colonic SAM-to-SAH ratio, a severe degree of folate deficiency did not induce significant genomic DNA hypomethylation in the colon (Sohn et al. 2003). Paradoxically, isolated folate deficiency significantly increased (by 30%) the extent of genomic DNA methylation in the colon at an intermediate time point (Sohn et al. 2003). The extent of *p53* methylation in the promoter and exons 6–7 was variable over time at each of the CpG sites examined, and no associations with time or dietary folate were observed at any CpG site except for a significant degree of hypomethylation with folate deficiency at the CpG site 1 in exons 6–7 at an extreme time point (Sohn et al. 2003). This change was not evident at any earlier time point; however, its significance is questionable. Dietary folate deprivation progressively decreased steady-state levels of *p53* transcript during the study period (Sohn et al. 2003). However, steady-state levels of *p53* mRNA did not significantly correlate with either genomic or *p53* methylation within the promoter region and exons 6–7 (Sohn et al. 2003). The observations from this study suggest that isolated folate deficiency, which significantly reduces steady-state levels of colonic *p53* mRNA, is not associated with a significant degree of genomic or *p53* DNA hypomethylation in rat colon. Furthermore, a recent study has shown that folate deficiency of a moderate degree for 10 weeks' duration in adult mice failed to induce a significant change in *p53*-specific methylation (exons 6–7); however, a nonsignificant, minor degree of colonic hypomethylation in exon 6 of the *p53* gene was observed, and this was significantly amplified by multiple B-vitamin depletion (Liu et al. 2008). Interestingly, significant *p53* hypomethylation in exon 8, but not in exons 6–7, was observed in the DMH-treated rat colon in conjunction with folate deficiency, although it remains unclear whether this was due to the DMH, the folate deficiency, or the combination of the two, and this was effectively overcome in a dose-dependent manner by increasing levels of dietary folate (Y. I. Kim, Pogribny et al. 1996). Taken together, the results from these studies suggest that isolated folate deficiency does not induce consistent and predictable changes in *p53* methylation in rat colon whereas it may produce *p53* hypomethylation in specific exons in rat liver and in rat colon in conjunction with alkylating agents. In contrast, dietary depletion of combined methyl donors predictably induces *p53* hypomethylation within exons 6–7 of the *p53* gene in rat liver (Sohn et al. 2003; Pogribny, Basnakian et al. 1995; Pogribny, Poirier et al. 1995; Pogribny et al. 1997). These observations suggest that *p53* methylation changes likely depend on the degree of methyl donor supply and consequent levels of methylation intermediates that are predictably and consistently achieved by combined methyl deficiency and not by isolated folate deficiency.

One critical factor that may explain the inability of folate deficiency to modulate genomic and site- and gene-specific DNA methylation in the colorectum appears to be SAM-to-SAH concentration ratios in the colorectum. Although isolated folate deficiency progressively decreased colonic mucosal SAM-to-SAH concentration ratios during 5 weeks of dietary intervention in the study by Sohn et al. (2003), only in the extreme deficient state, associated with 20% growth retardation and a 22-fold rise in plasma homocysteine, was there a significantly elevated level of colonic mucosal SAH and a significantly reduced colonic mucosal SAM-to-SAH concentration ratio compared with the control diet, indicating that modulation of SAM and SAH concentrations in the colonic mucosa is particularly resistant to the level of dietary folate. In contrast, folate deficiency, even to mild and moderate degrees, has been shown to modulate SAM and SAH in the brain (Ordonez and Wurtman 1974), kidney (Ordonez and Wurtman 1974), pancreas (Balaghi and Wagner 1992), and liver (Balaghi and Wagner 1993; Y. I. Kim et al. 1994, 1995, 1997; Miller et al. 1994) in rats. The reason for this tenacious resistance to altered SAM and SAH levels in the colorectum compared with other tissues is unclear at present. Intestinal microflora are capable of synthesizing folate, which has been shown to be taken up into colonic epithelial cells by the colonic folate carrier (Zimmerman 1990; Dudeja et al. 1997; Said et al. 2000), raising the possibility that compensatory mechanisms are available.

Despite the majority of these animal studies demonstrating a resistance to altered SAM and SAH levels in the colorectum, a significant increase in colonic SAH concentrations and a decrease in colonic SAM-to-SAH concentration ratios were observed in one of the aforementioned studies (Uthus et al. 2006), although the reason for this observation is unclear at present. Collectively, these data suggest that changes in SAM and SAH concentrations induced by folate deficiency may not be sufficient enough to modulate colonic genomic DNA methylation; however, there remains the possibility that alterations in gene-specific DNA methylation may have occurred.

The results from these studies suggest that a prolonged, moderate degree of folate deficiency does not significantly alter genomic DNA methylation in the colon; however, the study by Sohn et al. (2003) demonstrates that folate deficiency of a short duration and severe degree may induce genomic DNA hypermethylation in the colon (Sohn et al. 2003), possibly due to a compensatory up-regulation of Dnmt, consistent with observations in rat liver (Y. I. Kim et al. 1997; Song et al. 2000). Interestingly, a recent study reported that chronic, severe folate deficiency in adult mice induced significant genomic DNA hypomethylation in the colon, and a nonsignificant degree of genomic DNA hypomethylation in the small intestine and spleen (Linhart et al. 2009), suggesting that a prolonged and severe diet deficient in folate may induce genomic DNA hypomethylation. It is important to note, however, that the mice utilized in this study were of older age and therefore may be more vulnerable to changes in DNA methylation compared to the young.

Taken together, the results from studies illustrate the importance of timing, duration, and degree of folate deficiency, and further studies are warranted to determine the effect of folate deficiency on DNA methylation in the colon.

3.4.1.3 Effects of Isolated Folate Supplementation on DNA Methylation in Rodents

The effect of isolated folate supplementation on DNA methylation in rodent liver has not yet been clearly elucidated. A recent study has found that dietary folate supplementation at four times the basal dietary requirement of the rat (8 mg folic acid/kg) did not affect genomic DNA methylation in adult rat liver, regardless of timing or duration of supplementation (Kotsopoulos et al. 2008). Another study found that

TABLE 3.2
Summary of the Effect of Isolated Folate Supplementation on DNA Methylation in Rodents

Study (Reference)	Folate Supplementation	Age	Duration	Species	Organ	DNA Methylation	Effect
Kotsopoulos et al. 2008	8 mg folic acid/kg	3 weeks	27 weeks	Rat	Liver	Genomic	No effect
		3 weeks	5 weeks				No effect
		8 weeks	22 weeks				No effect
Achon et al. 2007	40 mg folic acid/kg	3 weeks	4 weeks	Rat	Liver	Genomic	No effect
Choi et al. 2005	18 μmol folate/kg	1 year	8 weeks	Rat	Liver	Genomic	Stepwise increase from folate-deplete, -replete, and -supplemented ($p_{trend} = 0.08$)
			20 weeks				Stepwise increase with increasing folate ($p_{trend} = 0.025$)
Kim et al. 1996	8 and 40 mg folic acid/kg + DMH	3 weeks	20 weeks	Rat	Colon	Genomic	No effect
Sohn et al. 2003	8 mg folic acid/kg	3 weeks	5 weeks	Rat	Colon	Genomic	No effect
						p 53	No effect
Choi et al. 2003	18 μmol folate/kg	3 weeks	8 weeks	Rat	Colon	Genomic	No effect
			20 weeks				No effect
		1 year	8 weeks				No effect
			20 weeks				No effect
Keyes et al. 2007	18 μmol folate/kg	4 months	20 weeks	Mouse	Colon	Genomic	No effect
						p16	No effect
		18 months				Genomic	Stepwise increase from folate-deplete, -replete, and -supplemented ($p_{trend} < 0.023$)
						p16	Stepwise increase from folate-deplete, -replete, and -supplemented ($p_{trend} = 0.009$)

dietary folate supplementation at 20 times the basal dietary requirement of the rat (40 mg folic acid/kg) provided for 4 weeks in weanling rats did not alter SAM and SAH, SAM-to-SAH concentration ratios, and genomic DNA methylation in the liver (Achon et al. 2007). Interestingly, in elder rats fed a folate-deficient (0 μmol folate/kg), replete (4.5 μmol folate/kg), or supplemented (18 μmol folate/kg) diet for 8 and 20 weeks, hepatic SAH levels decreased while genomic DNA methylation in liver increased incrementally with increasing levels of dietary folate, and folate-supplemented rats demonstrated significantly greater degree of genomic DNA methylation compared to folate-deplete rats at both 8 and 20 weeks (Choi et al. 2005).

In the colon, one study found that DMH administration in conjunction with folate supplementation of 8 mg folic acid/kg and 40 mg folic acid/kg for 20 weeks in weanling rats did not alter concentrations of SAM, SAH, SAM-to-SAH concentration ratios, and DNA methylation (Y. I. Kim, Salomon et al. 1996). As well, weanling rats fed a folate-supplemented diet (8 mg/kg folic acid) for 5 weeks showed no change in colonic concentrations of SAM, SAH, SAM-to-SAH concentration ratios, and colonic DNA methylation, and in addition, *p53*-specific DNA methylation was not altered (Sohn et al. 2003). Interestingly, dietary folate supplementation (18 μmol/kg) in both young and elder rats for 8 and 20 weeks resulted in a decrease in colonic SAH concentrations, although genomic DNA methylation in the colon was not altered (Choi et al. 2003). A recent study has also examined the effect of a folate-deficient, -replete, and -supplemented diet (0 μmol/kg, 4.5 μmol/kg, and 18 μmol/kg, respectively) in young (4 months) and old (18 months) mice for 20 weeks' duration, and reported that in old mice only, genomic and *p16*-specific DNA methylation increased in a manner that was directly related to dietary folate (Keyes et al. 2007), similar to findings in elder rat liver (Choi et al. 2005).

Taken together, the results from these studies suggest that in the young, liver and colon are generally resistant to changes in SAM and SAH levels from folate supplementation, and consequently, resistant to changes in DNA methylation. However, folate supplementation may modulate DNA methylation in the elderly. Further studies are warranted to investigate the effects of folate supplementation not only in colon and liver, but other tissues as well.

3.4.1.4 Interactions between Folate and Other Environmental Factors on DNA Methylation in Rodents

Although isolated folate has been shown to modify DNA methylation, several other animal studies highlight the importance of interactions between folate and other environmental factors and their synergistic effect on DNA methylation (Uthus et al. 2006; Davis and Uthus 2003; Liu et al. 2007, 2008). In rats fed a deficient or adequate diet in folate and selenium, plasma homocysteine and liver SAH levels were highest in folate-deficient rats fed a diet adequate in selenium, resulting in a significant interaction between dietary folate and selenium (Uthus et al. 2006), and a similar interaction was reported in a previous study (Davis and Uthus 2003). Multiple vitamin depletion (riboflavin, vitamin B_6, and vitamin B_{12}) has been shown to induce a 45% decrease in genomic DNA methylation of the colon, while isolated folate deficiency failed to induce significant colonic genomic DNA hypomethylation (Liu et al. 2007). Furthermore, moderate folate deficiency induced a minor degree of DNA hypomethylation of exon 6 of the *p53* gene, while combined vitamin deficiency significantly magnified these alterations in DNA methylation (Liu et al. 2008).

3.4.2 Effects of Folate Deficiency and Supplementation on DNA Methylation in In Vitro Systems

There are a limited number of in vitro studies examining the effect of folate deficiency and supplementation on DNA methylation. In one study by Duthie et al., normal human colonic epithelial cells were immortalized by SV40 T antigen and cultured in folate-deficient (<2.3 nmol/L) and control (9.1 μmol/L folic acid) medium for 14 days (Duthie, Narayanan, Blum et al. 2000). Folate deficiency led to genomic DNA hypomethylation, increased uracil misincorporation, and inhibition of DNA excision repair in colonic epithelial cells (Duthie, Narayanan, Blum et al. 2000). In a recent study using the human colonic adenocarcinoma cell line SW620, cells were grown in a folate-free medium or medium containing normal amounts of folic acid (3 μmol/L) for 7 and 14 days (Wasson et al. 2006). At day 7, genomic DNA hypomethylation was observed in cells grown in the folate-free medium compared with folate-replete cells, and a significant difference was seen between the two by day 14 (Wasson et al. 2006). Similarly, cells grown in the folate-free medium underwent a significant degree of DNA hypomethylation in the region of the *p53* gene, although folate depletion beyond day 7 did not further increase genomic DNA hypomethylation (Wasson et al. 2006). Interestingly, reintroducing folic acid into the cells returned both genomic and *p53* gene region-specific DNA methylation levels to normal, completely reversing the effects of folate deficiency (Wasson et al. 2006).

In another study using untransformed cell lines, folate deficiency (0.6 nmol/L in the medium) was shown to induce significant genomic DNA hypomethylation in both a mouse fibroblast cell line, NIH/3T3, and a Chinese hamster ovarian cell line, CHO-K1, by cell-specific mechanisms as indicated by cell-specific differential effects of folate deficiency on intracellular SAM, SAH, and DNMT (Stempak et al. 2005). In contrast, a similar experiment using two human colon adenocarcinoma cell lines, Caco2 and HCT116, has shown that that the extent and the direction of the changes of SAM and SAH in response to folate deficiency (0.6 nmol/L folic acid in the medium vs. 2.3 μmol/L in control medium) are cell specific, and that genomic-, site-, and gene-specific DNA methylation are not affected by the changes of SAM and SAH induced by folate depletion (Stempak et al. 2005). In a recent study using three cell lines derived from the normal human colon, HCEC, NCM460, NCM356, grown for 32–34 days in media containing 25, 50, 75, or 150 nmol/L folic acid, folate status failed to alter genomic DNA methylation in any of the cell lines (Crott et al. 2008).

In contrast, human nasopharyngeal carcinoma KB cells grown in folate-deplete (2–10 nmol/L folic acid) medium was associated with paradoxical hypermethylation in a 5′ CpG island (by 40%) and consequent down-regulation of the *H-cadherin* gene compared with cells grown in a folate-replete (2.0 μmol/L folic acid) medium (Jhaveri et al. 2001).

The results from these studies collectively suggest that the effects of folate deficiency on DNA methylation are site and gene specific. In addition, the direction of methylation changes may be cell, target organ, and stage of transformation specific, and may not be the same between genomic and gene- or site-specific DNA methylation. These conclusions are supported by prior observations that suggest that cancers from different organs and histologically different subtypes of cancer within a given organ exhibit distinct global and gene-specific methylation patterns (Virmani et al. 2002; Esteller et al. 2001). The major limitation of the in vitro system to study the effect of folate on DNA methylation is that the degree of folate deficiency and supplementation used in this system is not physiologically and clinically relevant and applicable to in vivo systems.

3.4.3 Effects of Folate Status on DNA Methylation in Human Studies

3.4.3.1 Effect of Folate Deficiency and Supplementation on DNA Methylation in Human Clinical Trials

There are some observations in humans suggesting that altered folate status can affect genomic DNA methylation. Folate depletion in healthy human volunteers in a metabolic unit setting has been observed to diminish genomic DNA methylation in leukocytes (Table 3.3) (Jacob et al. 1998; Rampersaud et al. 2000). Rampersaud et al. (2000) showed that lymphocyte genomic DNA methylation significantly decreased by 10% in response to moderate (118 μg folate/day) folate depletion for a period of 7 weeks in elderly women (60–85 years of age). No significant changes in leukocyte genomic DNA methylation were detected during the 7-week period of folate repletion with either 200 or 415 μg of folate/day (Rampersaud et al. 2000). Another study by Jacob et al. (1998) housed healthy, postmenopausal women (49–63 years of age) in a metabolic unit and fed them folate-deplete diets (56–111 μg folate/day) for 9 weeks. This resulted in a significant (by 120%) degree of lymphocyte genomic DNA hypomethylation, which was reversed during the 3-week period of folate supplementation (285–516 μg folate/day). However, an earlier study by the same group using healthy males (33–46 years of age) failed to show a change in in vivo methylation capacity (as measured by the ability to methylate orally administered nicotinamide as detected in the urine as methylated metabolites) in response to dietary folate and methyl group restriction (Table 3.3) (Jacob et al. 1995). In this study, male subjects housed in a metabolic unit were placed on a folate-deficient diet (25 μg folate/day) for 30 days (Jacob et al. 1995).

More recent controlled folate feeding studies include analysis of common MTHFR polymorphisms. Although daily folate restriction with 135 μg of dietary folate equivalents (DFEs) for 7 weeks followed by repletion with 400 or 800 μg DFEs for 7 weeks did not influence leukocyte genomic DNA methylation in a group of young and healthy women wild-type for MTHFR C677T (Axume et al. 2007a), another study by the same research group demonstrated a significant interaction between the timing of folate treatment and MTHFR polymorphism on DNA methylation (Axume et al. 2007b). Under the same controlled feeding conditions, Axume et al. observed that at the end of the 14-week period of folate treatment, leukocyte genomic DNA

TABLE 3.3
Summary of the Effect of Dietary Folate Deficiency on DNA Methylation in Humans

Study (Reference)	Subjects	Age	Dose	Duration	DNA Methylation	Effect
Rampersaud et al. 2000	Women	60–85 yrs	118 μg/day	7 weeks	Leukocytes Genomic	10% decrease ($p = 0.0012$)
Jacob et al. 1998	Women	49–63 yrs	56–111 μg/day	9 weeks	Lymphocytes Genomic	120% decrease ($p < 0.05$)
Jacob et al. 1999	Men		25 μg/day	30 days	Methylation capacity (not DNA)	No change
Axume et al. 2007a	Women	18–45yrs	135 μg/day	7 weeks	Leukocytes Genomic	No change
			400–800 μg/day	14 weeks		No change
Axume et al. 2007b	Women	18–45yrs	135 μg/day	7 weeks	Leukocytes Genomic	No change
			400–800 μg/day	14 weeks		4% decrease ($p < 0.05$)
Shelnutt et al. 2004	Women	20–30 yrs	115 μg/day	7 weeks	Leukocytes Genomic	5% decrease ($p = 0.08$)
			400 μg/day	14 weeks		8% increase ($p = 0.04$)
Quinlivan et al. 2004	Women	20–26 yrs	±115–120 μg/day	7 weeks	Monocytes, MdC enrichment (not DNA)	40% decrease ($p = 0.012$)

methylation was significantly lower in women with the 677TT genotype relative to women with the CC or CT genotype (Table 3.3). Similarly, Shelnutt et al. (2004) reported a trend toward a decrease in leukocyte genomic DNA methylation in women depleted of folate (115 μg DFE/day for 7 weeks). This was corrected by folate repletion of 400 μg DFE/day for 7 weeks but only in women with the MTHFR 677TT genotype. In addition, in a subgroup of women from this same population, in vivo analysis of genomic DNA methylation in monocytes as determined by methyldeoxycytidine enrichment following radiolabeled infusions of [$^{13}C_5$]methionine also indicated that folate-dependent intracellular one-carbon metabolism was suppressed after 7 weeks of folate restriction (115 to ±120 μg DFE/day) but this effect was independent of MTHFR genotype (Table 3.3) (Quinlivan et al. 2005).

In some human intervention studies, folate supplementation at 12.5–25 times the daily requirement for 3–12 months significantly increased the extent of colonic genomic DNA methylation in subjects with resected colorectal adenoma or cancer (Cravo et al. 1994, 1998; Y. I. Kim et al. 2001) whereas no such effect was observed in patients with chronic ulcerative colitis who were given folate supplementation at 12.5 times the daily requirement for 6 months (Table 3.4) (Y. I. Kim et al. 1995). Folate supplementation at three and five times the daily requirement, which was sufficient to improve and correct a marker of DNA damage, failed to modulate genomic DNA methylation in lymphocytes in healthy volunteers (Fenech et al. 1998; Basten et al. 2006) (Table 3.4). Similarly, global genomic methylation, measured by methylation of long interspersed nucleotide elements (LINE-1), in normal colonic mucosa derived from study participants from the Aspirin/Folate Polyp Prevention Study was not associated with folate treatment or circulating levels of folate and homocysteine (Table 3.4) (Figueiredo et al. 2009).

In another study, folate supplementation with 15 mg methyltetrahydrofolate a day for 8 weeks restored genomic DNA methylation in lymphocytes to normal levels in 32 men with uremia, hyperhomocysteinemia, and preexisting genomic DNA hypomethylation (Table 3.4) (Ingrosso et al. 2003). A physiological dose of folic acid (400 μg/day) for 10 weeks has been demonstrated to increase genomic DNA methylation in lymphocytes (by 31%; $p = 0.05$) and in colonic mucosa (by 25%; $p = 0.09$) compared with placebo in patients with colorectal adenomas (Table 3.4) (Pufulete, Al-Ghnaniem, Khushal et al. 2005). In a more recent study that investigated the combined effects of folic acid (12.5 times the daily requirement) and vitamin B_{12} (1.25 mg/day) supplementation for 6 months on promoter methylation of tumor suppressor and DNA repair genes frequently reported to be aberrantly methylated in colorectal cancer, a trend toward a 67% increase in promoter hypermethylation in rectal mucosal biopsies from patients with resected colorectal adenomas was reported in the intervention group compared with placebo, although this did not reach statistical significance (Table 3.4) (Van den Donk, Pellis et al. 2007).

The possibility of an inverse relationship between folate supplementation and DNA methylation status has recently been raised. Among reproductive-age women in China who were not previously exposed to folic acid, folic acid supplementation (100, 400, and 4000 μg/day) for 1 month significantly decreased genomic methylation by 13% ($p = 0.001$) and continued to remain significantly lower with further intervention (Table 3.4) (Quinlivan et al. 2008). An interesting observation in this study is that following a 3-month washout period after folic acid treatment, genomic methylation decreased even further (by 23%; $p < 0.0001$) relative to baseline values. This seemingly paradoxical effect of folic acid supplementation on global DNA methylation may be partly explained by the preferential shuttling of the flux of one-carbon units to the nucleotide synthesis pathway over the methionine cycle necessary for biological methylation reactions in response to folic acid supplementation. Although folic acid is an inhibitor of dihydrofolate reductase (DHFR) (S. W. Bailey and Ayling 2009), and an enzyme important in the maintenance of the intracellular folate pool, in certain situations it may up-regulate DHFR (Kamen et al. 1985) and this up-regulation may increase thymidylate synthase activity because the transcription of these genes is co-regulated by several transcription factors (Slansky et al. 1993; Sowers et al. 2003) (Figure 3.2). A mathematical modeling has indicated that this would increase thymidylate production, thereby increasing cellular proliferation, at the expense of biological methylation reactions (Nijhout et al. 2004).

TABLE 3.4
Summary of the Effect of Dietary Folate Supplementation on DNA Methylation in Human Intervention Trials

Study	Subjects	Dose	Duration	DNA Methylation	Effect
Cravo et al. 1994	Patients with CRC and adenoma	10 mg/day	6 months	Rectal mucosa Genomic	93% increase ($p < 0.002$)
Cravo et al. 1998	Patients with colonic adenoma	5 mg/day	3 months	Rectal mucosa Genomic	37% increase in patients with 1 adenoma ($p = 0.05$) and no change in those with >1 adenomas
Y. I. Kim et al. 2001	Patients with colonic adenoma	5 mg/day	6 months	Rectal mucosa Genomic	57% increase ($p = 0.001$)
			1 year		No change
Cravo et al. 1995	Patients with inflammatory bowel disease	5 mg/day	6 months	Rectal mucosa Genomic	No change
Fenech et al. 1998	Normal subjects	2 mg/day	12 weeks	Lymphocytes Genomic	No change
Basten et al. 2006	Normal subjects	1.2 mg/day	12 weeks	Lymphocytes Genomic	No change
Ingrosso et al. 2003	Uremic patients with hyperhomocysteinemia and preexisting DNA hypomethylation	15 mg/day 5-methyl THF	8 weeks	Lymphocytes Genomic	Restored to normal levels
Pufulete et al. 2005	Patients with colonic adenoma	400 μg/day	10 weeks	Lymphocytes Genomic	31% increase ($p = 0.05$)
				Rectal mucosa Genomic	25% increase ($p = 0.09$)
Figueiredo et al. 2009	Patients with colonic adenoma	1 mg/day	6–8 years	Colonic mucosa Genomic	No change
Van den Donk, Pellis et al. 2007	Patients with colorectal adenoma	5 mg/day	6 months	Rectal biopsies APC $p14^{ARF}$ $p16^{INK4A}$ hMLH1 O^6-MGMT RASSFIA	67% increase in probability of promoter hyper-methylation of all 6 genes ($p = 0.08$)

TABLE 3.4 (CONTINUED)
Summary of the Effect of Dietary Folate Supplementation on DNA Methylation in Human Intervention Trials

Study	Subjects	Dose	Duration	DNA Methylation	Effect
Quinlivan et al. 2008	Reproductive age Chinese women	100–4000 μg/day	1 month	Genomic	13% decrease ($p < 0.001$)
		100–4000 μg/day	3 months		No change
		100–4000 μg/day	6 months		No change
		0 μg/day	9 months		23% decrease ($p < 0.0001$)

3.4.3.2 Effects of Folate Status on Genomic DNA Methylation in Human Observational Studies

There are observational studies that have investigated the association between folate status and genomic DNA methylation. In human subjects with normal folate status, no significant correlations between genomic lymphocyte DNA methylation and red blood cell folate and plasma homocysteine concentrations were observed (Table 3.5) (Fenech et al. 1998; Narayanan et al. 2004). Studies evaluating the relationship between serum folate status and methylation-related intermediates have also failed to demonstrate a clear association (Table 3.5). In normal human subjects, significant differences in the SAM-to-SAH concentration ratio were not observed despite significantly higher serum SAM and SAH concentrations in study subjects with high serum folate levels (Hirsch et al. 2008). In an elderly population (50–70 years), serum and red blood cell folate were not associated with any of the aforementioned markers (Becker et al. 2003). However, in patients with colorectal adenocarcinomas, folate levels and SAM-to-SAH concentration ratios were lower in malignant tissue compared with normal-appearing adjacent colon mucosa ($p = 0.08$) (Alonso-Aperte et al. 2008).

Several other studies, on the other hand, have shown that global colonic DNA methylation is positively correlated with serum and red blood cell folate concentrations and negatively with plasma homocysteine concentrations in individuals with colonic adenomas and adenocarcinomas (Pufulete et al. 2003; Al-Ghnaniem et al. 2007) and in those without these lesions (Pufulete, Al-Ghnaniem, Rennie et al. 2005) (Table 3.5). A similar positive relationship between folate status and genomic DNA methylation has also been reported in peripheral blood mononuclear cells (Friso et al. 2002; Pilsner et al. 2007). In peripheral blood mononuclear cells, genomic DNA methylation was shown to directly correlate with folate status and inversely correlate with plasma homocysteine levels (Table 3.5) (Friso et al. 2002). Further analysis indicated that folate status interacted with MTHFR polymorphism to affect genomic methylation. MTHFR TT genotypes had a diminished level of

TABLE 3.5
Summary of the Effect of Folate Status on Genomic DNA Methylation in Human Observational Studies

Study	Subjects	Design	Folate Levels	DNA Methylation	Effect
Fenech et al. 1998	Normal subjects	Cross-sectional	RBC 364–440 nmol/l	Lymphocytes Genomic	No associations
Narayanan et al. 2004	Normal subjects	Cross-sectional	Plasma 15–21 nmol/l RBC 192–254 nmol/l	Lymphocytes Genomic	No associations
Hirsch et al. 2008	Men	Cross-sectional	Serum 21 vs. 63 nmol/l	Blood Genomic	No associations
Becker et al. 2003	Normal subjects	Cross-sectional	RBC 551 nmol/l	Blood Genomic	No associations
Alonso-Aperte et al. 2007	Patients with colon cancer	Cross-sectional	Colon 0.49 vs. 0.95 nmol/g	Neoplastic vs. Normal Colon Genomic	28% decrease ($p = 0.08$)
Pufulete et al. 2003	Patients with colon cancer or adenoma	Case-control	Serum 12 vs. 18 nmol/l	Colon mucosa Genomic	26% decrease ($p = 0.009$)
				Leukocytes Genomic	14% decrease ($p < 0.001$)
Pufulete, Al-Ghnaniem, Rennie, et al. 2005	Patients without colon cancer or adenoma	Cross-sectional	Serum 18.6 nmol/l (mean)	Colon mucosa Genomic	$r = -0.311$ ($p = 0.01$)
			RBC 648.1 nmol/l (median)		$r = 0.356$ ($p = 0.03$)
			Hcy 9.9 μmol/l (median)		$r = 0.256$ ($p = 0.04$)
Al-Ghnaniem et al. 2007	Patients with colon cancer or adenoma	Case-control	Serum 12.3 vs. 17.9 nmol/l	Colon mucosa Genomic	38% decrease ($p < 0.001$)
				Rectal mucosa Genomic	25% increase ($p = 0.09$)
Pilsner et al. 2007	Adults chronically exposed to arsenic	Cross-sectional	Plasma <9 vs. ≥9 nmol/l Plasma 8.6 nmol/l (mean)	Leukocytes Genomic	3% decrease ($p = 0.03$) $r = 0.12$ ($p \leq 0.05$)

TABLE 3.5 (CONTINUED)
Summary of the Effect of Folate Status on Genomic DNA Methylation in Human Observational Studies

Study	Subjects	Design	Folate Levels	DNA Methylation	Effect
Friso et al. 2002	Patients with and without coronary artery disease	Cross-sectional	Plasma <12 vs. ≥12 nmol/l	Lymphocytes Genomic	61% decrease ($p < 0.0001$) in subjects with MTHFR 677TT
			RBC <1.1 vs. ≥1.1 nmol/g Hb		65% decrease ($p < 0.0001$) in subjects with MTHFR 677TT
Stern et al. 2000	Healthy volunteers	Cross-sectional	RBC 2.3 nmol/g Hb (mean)	Leukocytes Genomic	r = –0.738 ($p = 0.02$) in subjects with MTHFR 677TT
Schernhammer et al. 2009	Patients with colon cancer	Prospective	Dietary intake ≥400 vs. <200 μg/day	Colon cancer Genomic	43% decreased risk of hypo-methylation ($p = 0.05$)
Lim et al. 2008	Women 50–79 yrs with colonic adenoma	Case-control	Pre-fortification: <317 vs. ≥317 μg/1000 kcal/day Post-fortification <413 vs. ≥413 DFE/day	Leukocytes Genomic	Methylation inversely associated with colonic adenoma risk, especially in subjects with low folate intake in pre- and post-fortification

genomic DNA methylation compared with those with the CC wild type (Friso et al. 2002). When analyzed according to folate status, however, only the TT subjects with low levels of folate accounted for the diminished genomic DNA methylation (Friso et al. 2002). In healthy human volunteers, leukocyte genomic DNA methylation was directly and significantly related to RBC folate concentrations in subjects with the MTHFR 677TT genotype (p-interaction < 0.02), but not in those with wild-type MTHFR (Table 3.5) (Stern et al. 2000). Furthermore, a strong trend toward diminished DNA methylation was also observed in subjects with the TT variant with lower plasma folate levels (p-interaction = 0.06). These studies are consistent with previous reports that the TT genotype is associated with impaired MTHFR enzyme activity and suggests that the cellular consequences of this impairment is mediated, in part, by a low folate status (Stern et al. 2000).

Additional evidence lends support to a direct relationship between folate status and global DNA methylation. In a combined analysis of colorectal cancers from participants from the Nurse's Health Study and the Health Professionals Follow-Up Study, the risk of genomic hypomethylation (determined by <55% LINE-1 methylation) was 43% lower in subjects with a high (≥400 μg) compared with a low (<200 μg) total daily folate intake (Table 3.5) (Schernhammer et al. 2009). In a study that stratified folate intake according to pre- and postfortification levels, the observed inverse association between leukocyte genomic DNA methylation and adenoma was stronger among subjects with low as compared to high total folate intake in either fortification period (Table 3.5) (Lim et al. 2008). One human study did report that serum and cervical tissue folate concentrations correlated inversely, albeit weakly, with cervical genomic DNA methylation (Fowler et al. 1998), however, such observations were not observed for red blood cell folate levels (Flatley et al. 2009).

3.4.3.3 Effects of Folate Status on Gene-Specific DNA Methylation in Human Observational Studies

Aberrant CpG island methylation is characteristic of tumor development, and a number of studies have demonstrated that dietary intake or blood levels of folate can modulate promoter CpG methylation in colorectal tumors. Specific promoter CpG islands are frequently and simultaneously methylated in sporadic colorectal cancer, leading to transcriptional silencing (Markowitz and Bertagnolli 2009; Kawakami et al. 2003; Curtin et al. 2007). This phenomenon, known as the CpG island methylator phenotype (CIMP) (Toyota et al. 1999; Issa 2004), accounts for approximately 15–30% of all colorectal cancers (Markowitz and Bertagnolli 2009; Kawakami et al. 2003). Although the panel of genes currently used as a marker of CIMP has not been clearly established, it often includes CpG targets found in the promoter regions of the cell cycle regulator *p16* and the mismatch repair gene *MLH1* (Toyota et al. 1999). Several epidemiological studies evaluated CIMP patterns or genes that can be found in the CIMP panel to provide further clarification of the role of folate in regional CpG island methylation.

In The Netherlands Cohort Study on Diet and Cancer, the prevalence of CpG island promoter hypermethylation was higher, albeit nonsignificantly, in colorectal cancers derived from patients with low folate/high alcohol intake compared with colorectal cancers from patients with high folate/low alcohol intake for each of the six tested genes (*APC, p14, p16, hMLH1, O^6-MGMT,* and *RASSF1A*) (Table 3.6) (Van Engeland et al. 2003). The number of colorectal cancers with at least one gene methylated was higher (84%) in the low-folate-intake/high-alcohol-intake group compared with the high-folate-intake/low-alcohol-intake group (70%; p = 0.085) (Van Engeland et al. 2003). A later follow-up analysis in a subcohort of this population did not report any effect of isolated dietary folate intake on risk of colorectal cancers specifically presenting with *MLH1* hypermethylation (Table 3.6) (de Vogel et al. 2008). However, further investigation in the same panel of genes revealed that folate intake interacted with MTHFR C677T polymorphism to influence CpG promoter methylation in colorectal adenomas (Table 3.6) (Van den Donk, Van Engeland et al. 2007). In this study, among individuals homozygous for this variant, the risk of promoter methylation was inversely related to dietary folate intake, but statistical significance was observed only for the O^6-*MGMT* DNA-methyltransferase gene (p-interaction = 0.02). The results from this suggest that higher folate intakes may increase methyltransferase expression and as a result, methylation activity, particularly in individuals with adenomas and reduced MTHFR enzyme activity (Van den Donk, Van Engeland et al. 2007). Similarly, Slattery et al. (2006) initially failed to identify a significant association between dietary folate and colon tumors showing CIMP (based on CpG island methylation of *p16, MLH1, and MINT-1, -2, and -3 loci*) (Table 3.6). In their follow-up study, however, subjects heterozygous or homozygous for the MTHFR A1298C genotype with low folate/low methionine/high alcohol intake had an over twofold greater risk (Odds Ratio [OR] = 2.1; 95% Confidence Interval [CI] 1.3–3.4, p-interaction = 0.03) of CIMP-positive tumors compared with subjects with the wild-type genotype and high folate/high methionine/low alcohol intake (Curtin et al. 2007). Al-Ghnaniem et al. (2007) also examined gene-specific methylation in biopsies of normal-appearing colorectal mucosa from subjects with and without colorectal neoplasia. In general, patients with neoplasia were reported to have lower serum folate and promoter CpG hypermethylation of the *ERα* and *MLH1* genes compared with disease-free patients. *ERα* methylation was also positively correlated with plasma homocysteine in all subjects but significant inverse correlations between promoter CpG methylation and folate status were not observed (Al-Ghnaniem et al. 2007).

In contrast, the prevalence of promoter methylation of *p16,* but not *hMLH1* or *hMSH2,* was higher in colorectal cancers from patients with high (>12.5 nmol/l) compared with low (<12.5 nmol/l) serum folate concentrations (Table 3.6; $p = 0.04$) (Mokarram et al. 2008). The risk of tumor promoter methylation was also 4.9-fold higher in patients with high circulating folate levels (OR = 4.9, 95% CI 1.4–17.7, $p =$ 0.01). This positive association was more apparent in males ($p = 0.02$) and the elderly (≥60 y) ($p = 0.03$) and was further modified by the MTHFR C677T polymorphism, reaching significance only in subjects heterozygous or homozygous for the MTHFR variant. A greater risk for CIMP positive colorectal tumors (OR = 2.96; 95% CI 1.24–7.08, $p < 0.05$), using promoter methylation of *CDKN2A, MLH1, CACNA1G, NEUROG1, RUNX3, SOCS1, IGF2, and CRABP1* as markers, has also been demonstrated in patients with high circulating levels of plasma folate (Table 3.6) (Van Guelpen et al. 2009). Furthermore, colorectal carcinomas with frequent promoter methylation have been shown to have higher tumor concentrations of different folate metabolites, including 5,10-methylenetetrahydrofolate and tetrahydrofolate (Table 3.6) (Kawakami et al. 2003).

The effect of folate status on gene-specific methylation patterns at other tissue sites has also been examined. In women with cervical dysplasia, although lower red blood cell folate and cervical promoter hypermethylation were both independently associated with increasing severity of cervical cancer, direct correlations

TABLE 3.6
Summary of the Effect of Folate Status on Gene-Specific DNA Methylation in Human Observational Studies

Study	Subjects	Design	Folate Levels	DNA Methylation	Effect
Van Engeland et al. 2003	Patients with colon cancer	Prospective	Dietary intake <215/high alcohol vs. ≥215/low alcohol μg/day	Colon cancer *APC-1A* *p14*ARF *p16*INK4A *hMLH1* *O*6*-MGMT* *RASSFIA*	Increased prevalence for all genes ($p > 0.05$)
de Vogel et al. 2008	Patients with colon cancer	Prospective	Dietary intake 142.4–163.2 vs 247–279.9 μg/day	Colon cancer *MLH1*	No associations
Van den Donk, Van Engeland, et al. 2007	Patients with CRC adenoma	Case-control	Dietary intake <183 vs. >212 μg/day	CRC adenoma *APC-1A* *p14*ARF *p16*INK4A *hMLH1* *O*6*-MGMT* *RASSFIA*	Increased with low folate and decreased with high folate in subjects with MTHFR 677TT
Slattery et al. 2006	Patients with colon cancer	Case-control	Dietary intake <135–152 (low), 135–201 (med), >180–201 (high) μg/1000kcal/day	Colon cancer *CIMP*	No associations
Curtin et al. 2007	Patients with colon cancer	Case-control	Dietary intake <135–152 (low), 135–201 (med), >180–201 (high) μg/1000kcal/day	Colon cancer *CIMP*	Increased in subjects with low folate/low methionine/high alcohol and MTFR 1228AC/CC
Al-Ghnaniem et al. 2007	Patients with colon cancer or adenoma	Case-control	Serum 12.3 vs.17.9 nmol/l	Colon mucosa *ERα* *MLH1*	No associations
Mokarram et al. 2008	Patients with colon cancer	Case-control	Serum >12.5 vs. <12.5 nmol/l	Colon cancer *p16*	Increased prevalence ($p = 0.04$)

TABLE 3.6 (CONTINUED)
Summary of the Effect of Folate Status on Gene-Specific DNA Methylation in Human Observational Studies

Study	Subjects	Design	Folate Levels	DNA Methylation	Effect
Van Guelpen et al. 2009	Patients with colon cancer	Case-control	Plasma ≥ 6.8 vs. <6.8 nmol/l	Colon cancer CIMP	~threefold increase ($p < 0.05$)
Kawakami et al. 2003	Patients with colon cancer	Cross-sectional	5,10-methylene-tetrahydrofolate 2.95 vs. 1.53 pmol/g tissue Tetrahydrofolate 3.71 vs. 1.99 pmol/g tissue	Colon cancer *hMLH* *p16* *TIMP3* *ARF* *MINT2* *DAPK* *APC*	Tumors with methylated promoters had higher tissue concentration of folate intermediates

between folate and methylation were not observed (Flatley et al. 2009). Similarly, no associations were found between dietary folate intake and promoter methylation of *E-cadherin, p16, and* $RAR\text{-}\beta_2$ in breast tumors (Tao et al. 2009). For the *ERα* gene, however, lower dietary folate intake was associated with a twofold increase risk (OR = 2.0; 95% CI 0.8–4.8), albeit nonsignificantly, for breast cancers with *ERα* promoter methylation. A greater risk of $p16^{INK4A}$ hypermethylation (OR = 2.3; 95% CI 1.1–4.8) was also observed in subjects with low compared with those with high dietary folate, which was further exacerbated in subjects with the MTHFR 677TT genotype (Kraunz et al. 2006). In addition, high folate intake has been shown to offer protection (OR = 0.84; 95% CI 0.72–0.99; $p = 0.04$) against methylation of several genes in subjects with a long history of smoking (Stidley et al. 2010). Whereas dietary folate intake derived from fruits was positively associated with a 34% increase in methylation frequency of the *MGMT* gene in esophageal squamous cell carcinoma (Wang et al. 2008) and tissue folate levels, measured as the sum of 5,10-methylenetetrahydrofolate and tethrahydrofolate, correlated positively with *LINE-1, CDH13*, and *RUNX3* methylation in non-small-cell lung cancer (Jin et al. 2009).

3.4.3.4 Summary of Effects of Folate on DNA Methylation in Human Studies

Although not entirely consistent, controlled folate depletion in a metabolic unit appears to reduce genomic DNA methylation in peripheral blood mononuclear cells. This observation is more apparent in elderly women and may interact with the MTHFR genotype in an as yet undefined manner. However, there are no conclusive data suggesting that folate deficiency of a physiologically and clinically relevant degree induces significant genomic DNA hypomethylation and/or site- and gene-specific aberrant DNA methylation in the colorectum. In contrast, folate supplementation, even at the modest supplemental levels, appears to be able to increase genomic DNA methylation in the colorectum in certain situations.

The majority of observational studies have described a direct relationship between dietary and blood levels of folate and genomic DNA methylation in both lymphocytes and colonic tissues such that a low folate status is associated with genomic hypomethylation. This positive association is more consistent in individuals with colorectal adenomas, adenocarcinomas, or previously resected neoplastic tumors as well as in those with a greater risk of health complications compared with normal subjects. In contrast, the direction and magnitude of effect due to dietary and blood folate concentrations on gene-specific methylation remain unclear. Some studies demonstrate a greater prevalence or risk of aberrant hypermethylation of certain genes involved in colorectal carcinogenesis in subjects with low folate while others have reported this in subjects with high folate. The discrepancies in identifying a clear association between folate status and gene-specific methylation may be explained in part by the different methods of stratifying folate levels for comparison and the use of different markers to evaluate folate status. Dietary intake and serum levels of folate may not necessarily be reflective of folate concentrations in the target organ. These studies are also complicated by the lack of consistency in the specific genes investigated and sampling of different tissues.

Moreover, the effect of folate on both genomic and gene-specific DNA methylation has been demonstrated to be dependent on MTHFR polymorphisms in some studies. Whether or not the MTHFR C677T and A1298C polymorphism in conjunction with marginal folate status affects DNA methylation in the colorectum in vivo needs to be further characterized. These studies emphasize the importance of taking into consideration interactions between folate status and critical genes in the folate and one-carbon metabolic pathway when investigating the effect of folate on DNA methylation.

3.4.4 Effects of Maternal Folate Intake during Pregnancy on DNA Methylation in the Offspring

Because epigenetic patterns are established in utero, it has been suggested that maternal dietary exposures during pregnancy may alter the intrauterine one-carbon precursor milieu and as a result, disrupt one-carbon metabolism in the developing offspring. While the potential functional ramifications of modifying DNA methylation patterns in the offspring have not been clearly elucidated, it has been implicated in the risk of several diseases in later life (Gluckman and Hanson 2004). At present, several preliminary studies have shown that maternal folic acid supplementation during pregnancy can modify the offspring's epigenome with subsequent changes in phenotype.

Studies using viable yellow agouti mice have unequivocally demonstrated that maternal dietary methyl group supplementation containing folic acid can permanently alter the phenotypic coat color of the offspring via increased CpG methylation in the promoter of the *agouti* gene (Waterland and Jirtle 2003; Wolff et al. 1998). Similarly, a methyl group-rich diet has been shown to significantly reduce the proportion of progeny with a kinked tail, and this was paralleled to increased CpG methylation in the promoter of the *axin fused* gene (Waterland et al. 2006). In the agouti mouse model, folate has also been shown to interact with other environmental exposures during the intrauterine period to modulate methylation patterns in the developing offspring. Bisphenol A, an estrogenic xenobiotic chemical used in the manufacturing of polycarbonate plastics and associated with higher body weight, increased risk of cancer, and other chronic health conditions (Maffini et al. 2006), has been shown to shift the coat color of agouti mice by decreasing CpG methylation of the *agouti* gene when provided in utero, and maternal methyl donor supplementation including folic acid successfully reversed the epigenetic and phenotypic effects of bisphenol A (Table 3.7) (Dolinoy et al. 2007). In rats, promoter methylation of the *Pparγ* and *Gr* genes have been observed to be significantly lower, by 20% ($p < 0.001$) and 22.8% ($p < 0.05$), respectively, in offspring from dams fed a protein-restricted diet compared with a control diet during pregnancy (Table 3.7) (Lillycrop et al. 2005). Concomitant increases in protein expression of the *Pparγ* and *Gr* genes were also reported, and maternal supplementation with folic acid prevented these changes (Lillycrop et al. 2005).

The effect of isolated folic acid supplementation in utero on epigenetic modulation in the offspring has been displayed in other animal studies. In mice heterozygous for the folate binding protein gene (Folbp1+/–), daily administration of folinic acid by gavage during the periconceptional period until day 15.5 of gestation resulted in a decrease in genomic DNA methylation in both the liver and brain of the offspring compared with controls (Table 3.7) (Finnell et al. 2002). In other rodent studies, folate deficiency and supplementation with other nutritional factors were not observed to affect liver genomic DNA methylation (Table 3.7) (Engeham et al. 2009; Maloney et al. 2007). However, in hyperhomocysteinmic rats, folic acid supplementation with 8 mg/kg during pregnancy resulted in significant positive correlations between placental genomic DNA methylation and folate levels in the placenta, plasma, and liver (Table 3.7) (J. M. Kim et al. 2009). Placental genomic DNA methylation was also inversely correlated with plasma homocysteine concentrations (J. M. Kim et al. 2009). In a study involving maternal periconceptional folate and vitamin B_{12} restriction, aberrant methylation patterns were observed in 4% of the 1400 CpG islands examined, and the adult male offspring displayed increased adiposity, insulin resistance, altered immune function, and high blood pressure (Table 3.7) (Sinclair et al. 2007).

Recent human studies provide further support of isolated folate supplementation in utero on epigenetic consequences in the offspring. A preliminary prospective study in the United Kingdom found an inverse correlation between cord plasma homocysteine concentrations and global methylation in cord lymphocyte samples in offspring of mothers taking daily folic acid supplements during pregnancy (Table 3.7) (Fryer et al. 2009). Although the results of this study are consistent with the biological functions of folate, significant associations between cord blood folate and genomic lymphocyte methylation were not reported. Interestingly, an association between genomic lymphocyte methylation and fetal birth weight was identified (Fryer et al. 2009). In another observational study conducted in The Netherlands, periconceptional maternal folic acid use of 400 μg/day significantly increased, by

TABLE 3.7
Summary of the Effect of Maternal Folate Intake during Pregnancy on DNA Methylation in the Offspring

Study	Species	Folate Level	Duration	DNA Methylation	Effect
Fryer et al. 2009	Humans	400 mg/day FA	Pregnancy	Cord blood Genomic	r = 0.364 ($p = 0.08$)
		Cord serum 15.8 μmol/l (mean)			r = 0.209 ($p > 0.05$)
		Cord Hcy 10.8 μmol/l (mean)			r = –0.688 ($p = 0.001$)
Steegers-Theunissen et al. 2009	Humans	400 μg/day FA	Peri-conceptional	Blood *IGF2*	4.5% higher ($p = 0.014$)
J. M. Kim et al. 2009	Rats	Plasma 30.1 nmol/l (mean)	Peri-conceptional and pregnancy	Placenta Genomic	r = 0.752 ($p = 0.0003$)
		Liver 11.4 nmol/g tissue (mean)		Placenta Genomic	r = 0.700 ($p = 0.0012$)
		Placenta 1.45 nmo/g tissue (mean)		Placenta Genomic	r = 0.819 ($p < 0.0001$)
Sinclair et al. 2007	Sheep	Vitamin B12 and folate deficient vs. control	Peri-conceptional	Liver 1400 CpG sites	4% of CpG sites had altered status ($p < 0.001$)
Lillycrop et al. 2005	Rats	Restricted protein with 1 vs. 5 mg/kg FA	Pregnancy	Liver *GR*	26% increase ($p < 0.01$)
				Liver *Pparα*	17% increase ($p < 0.001$)
				Liver *Pparγ*	No change
Dolinoy et al. 2005	Mice	BPA alone vs. BPA with: 4.3 mg/kg FA, 0.5 mg/kg B12, 5 mg/kg betaine, 8 g/kg choline	Peri-conceptional, pregnancy, and perinatal	Brain Liver Kidney Agouti	31% decrease ($p = 0.004$) and methyl groups reversed hypo-methylation
Engeham et al. 2009	Rats	1 vs. 5 mg FA	Pregnancy	Liver Genomic	No change
Pufulete, Al-Ghnaniem, Leather, et al. 2005	Patients with colonic adenoma	400 μg/day	10 weeks	Lymphocytes Genomic	31% increase ($p = 0.05$)

TABLE 3.7 (CONTINUED)
Summary of the Effect of Maternal Folate Intake during Pregnancy on DNA Methylation in the Offspring

Study	Species	Folate Level	Duration	DNA Methylation	Effect
				Rectal mucosa Genomic	25% increase ($p = 0.09$)
Finnell et al. 2002	MiceFolbp1$^{+/-}$	25 mg/kg/d folinic acid by gavage	Peri-conceptional and midgestation	Liver Genomic	~fourfold decrease ($p < 0.05$)
					~twofold decrease ($p < 0.05$)
Maloney et al. 2007	Rats	Folate deficient	Peri-conceptional and pregnancy	Liver Genomic	No change

4.5%, the methylation of *IGF2* gene in whole blood derived from the offspring at 17 months after delivery (Table 3.7) (Steegers-Theunissen et al. 2009). An independent inverse association was also observed between *IGF2* methylation and birth weight.

Although limited, several lines of evidence support the notion that DNA methylation patterns of the developing offspring can be significantly modulated by intrauterine exposure to varying levels of folic acid alone or in combination with other methyl donors or environmental factors. Furthermore, the changes in methylation status may be associated with permanent alterations in the phenotype of the offspring with potential health consequences in later adult life. Whether or not folate-mediated epigenetic changes in utero can affect the risk of colorectal cancer development in adulthood remains to be determined. Given the associations between folate and colorectal cancer risk, future studies are warranted to investigate whether intrauterine folic acid supplementation influences colorectal cancer risk via epigenetic mechanisms.

3.4.5 Effects of Folate and Aging on DNA Methylation

Aging has been shown to modify patterns of DNA methylation in a variety of species and tissues and is associated with genomic DNA hypomethylation and gene-specific promoter hypermethylation; however, this appears to be tissue specific (Vanyushin et al. 1973; Wilson et al. 1987; Fuke et al. 2004; Issa et al. 1994). Age-related methylation is a common event in human tissues and an important contributor to the promoter CpG island hypermethylation of several genes with consequent gene inactivation observed in colorectal and other cancers (Toyota and Issa 1999). Aging is a major determinant of colorectal cancer risk (Winawer et al. 1997). It is therefore possible that age-related DNA methylation of certain genes serves as a functional link between aging and colorectal cancer by providing selective advantage for normal colon cells through deregulating growth and differentiation (Toyota and Issa 1999). These cells may then be at higher susceptibility for acquiring genetic lesions such as mutations.

Although the causes of age-related methylation changes are largely unknown, it has been proposed that environmental exposures or modifier genes may play a role (Ahuja et al. 1998). In human studies, folate depletion has been shown to be associated with diminished genomic DNA methylation in lymphocytes (Jacob et al. 1998) and leukocytes (Rampersaud et al. 2000) in healthy, postmenopausal women. In animals, one study has shown that the aging colon is highly susceptible to folate depletion and consequent changes in SAM and SAH, compared with the young colon, in rats (Choi et al. 2003). Therefore, folate deficiency in the aging colon may predispose it to changes in SAM and SAH and consequent DNA methylation changes more readily than that in the young colon, and folate status and DNA methylation changes may serve as a functional link between aging and colorectal cancer.

A recent study has reported diminished genomic and increased *p16* promoter DNA methylation in the colon of aged mice compared with that of young mice (Keyes et al. 2007). Interestingly, both genomic and *p16* promoter DNA methylation increased in a manner that was directly related to dietary folate in old mice, whereas this pattern was not evident in young mice (Keyes et al. 2007). In elder rat liver, dietary folate over a wide range of intakes was also shown to modulate genomic DNA methylation, as genomic DNA methylation increased with increasing levels of dietary folate (Choi et al. 2005).

Taken together, the results from these studies suggest that the aging process can alter DNA methylation and folate may further modify DNA methylation in the colon and liver of the elderly. The precise mechanism of how folate and aging interact to modulate DNA methylation has not been clearly elucidated at present, and further studies are required to investigate the interaction between folate and aging on genomic and gene-specific DNA methylation.

3.5 CONCLUSION

Genetic changes in cancer are abrupt in onset, their effects are often all-or-nothing, the loss of function occurs at a fixed level, and they are not reversible in most cases. In contrast, epigenetic changes are gradual in onset and progressive; their effects are dose-dependent and are potentially reversible. These observations present new opportunities in cancer risk modification and prevention using dietary and lifestyle factors and potential chemopreventive drugs. In this regard, folate has been a focus of intense interest because of an inverse association between folate status and the risk of several malignancies and of its potential ability to modulate DNA methylation. The portfolio of evidence from animal, in vitro, and human studies collectively suggests that the effects of folate deficiency and supplementation on DNA methylation are highly complex, appear to depend on cell type, target organ, and stage of transformation, and are gene and site specific. These studies also suggest that changes in DNA methylation depend on the magnitude and duration of folate manipulations, on interactions with other methyl group donors and dietary factors, and on genetic variants in the folate metabolic and one-carbon transfer pathways. Although some similarities do exist, animal models differ in several important physiological aspects from humans including bioavailability, metabolism, and excretion of folate. Therefore any extrapolation of the observations from these models to human situations should be made very cautiously. Furthermore, animal models may produce variable results owing to species differences, different diet compositions, and variable dose, time, and duration of folate manipulations. Recent studies have demonstrated that the exposure to folate deficiency and supplementation in the intrauterine environment and early life and during the aging process may have profound effects on DNA methylation with significant functional ramifications. Although the jury is still out, the potential for folate to modulate DNA methylation and thus modify the risk of chronic diseases including cancer in humans remains provocative and is worthy of further studies.

ACKNOWLEDGMENT

The authors' own experimental work described in this chapter has been supported in part by operating grants from the Canadian Institutes for Health Research (Fund #14126) and Canadian Cancer Society Research Institute (Fund #17078).

REFERENCES

Achon, M., E. Alonso-Aperte, N. Ubeda, and G. Varela-Moreiras. 2007. Supranormal dietary folic acid supplementation: effects on methionine metabolism in weanling rats. *Br J Nutr* 98 (3):490–96.

Ahuja, N., Q. Li, A. L. Mohan, S. B. Baylin, and J. P. Issa. 1998. Aging and DNA methylation in colorectal mucosa and cancer. *Cancer Res.* 58 (23):5489-94.

Al-Ghnaniem, R., J. Peters, R. Foresti, N. Heaton, and M. Pufulete. 2007. Methylation of estrogen receptor alpha and mutL homolog 1 in normal colonic mucosa: association with folate and vitamin B-12 status in subjects with and without colorectal neoplasia. *Am J Clin Nutr* 86 (4):1064–72.

Alonso-Aperte, E., M. P. Gonzalez, R. Poo-Prieto, and G. Varela-Moreiras. 2008. Folate status and S-adenosylmethionine/S-adenosylhomocysteine ratio in colorectal adenocarcinoma in humans. *Eur J Clin Nutr* 62 (2):295–98.

Ames, B. N. 2001. DNA damage from micronutrient deficiencies is likely to be a major cause of cancer. *Mutat Res* 475 (1–2):7–20.

Axume, J., S. S. Smith, I. P. Pogribny, D. J. Moriarty, and M. A. Caudill. 2007a. Global leukocyte DNA methylation is similar in African American and Caucasian women under conditions of controlled folate intake. *Epigenetics* 2 (1):66–68.

Axume, J., S. S. Smith, I. P. Pogribny, D. J. Moriarty, and M. A. Caudill. 2007b. The MTHFR 677TT genotype and folate intake interact to lower global leukocyte DNA methylation in young Mexican American women. *Nutr Res* 27 (1):1365–17.

Bailey, L. B. 2003. Folate, methyl-related nutrients, alcohol, and the MTHFR 677C⊠T polymorphism affect cancer risk: intake recommendations. *J Nutr* 133 (11 Suppl 1):3748S–53S.

Bailey, S. W., and J. E. Ayling. 2009. The extremely slow and variable activity of dihydrofolate reductase in human liver and its implications for high folic acid intake. *Proc Natl Acad Sci U S A* 106 (36):15424–29.

Balaghi, M., D. W. Horne, and C. Wagner. 1993. Hepatic one-carbon metabolism in early folate deficiency in rats. *Biochem J* 291:145–49.

Balaghi, M., and C. Wagner. 1992. Methyl group metabolism in the pancreas of folate-deficient rats. *J Nutr* 122 (7):1391–96.

Balaghi, M., and C. Wagner. 1993. DNA methylation in folate deficiency: use of CpG methylase. *Biochem Biophys Res Commun* 193 (3):1184-90.

Basten, G. P., S. J. Duthie, L. Pirie, N. Vaughan, M. H. Hill, and H. J. Powers. 2006. Sensitivity of markers of DNA stability and DNA repair activity to folate supplementation in healthy volunteers. *Br J Cancer* 94 (12):1942–47.

Becker, A., Y. M. Smulders, T. Teerlink, E. A. Struys, K. de Meer, P. J. Kostense, C. Jakobs, J. M. Dekker, G. Nijpels, R. J. Heine, L. M. Bouter, and C. D. Stehouwer. 2003. S-adenosylhomocysteine and the ratio of S-adenosylmethionine to S-adenosylhomocysteine are not related to folate, cobalamin and vitamin B6 concentrations. *Eur J Clin Invest* 33 (1):17–25.

Biasco, G., U. Zannoni, G. M. Paganelli, R. Santucci, P. Gionchetti, G. Rivolta, R. Miniero, L. Pironi, C. Calabrese, G. Di Febo, and M. Miglioli. 1997. Folic acid supplementation and cell kinetics of rectal mucosa in patients with ulcerative colitis. *Cancer Epidemiol Biomarkers Prev* 6 (6):469–71.

Castro, R., I. Rivera, P. Ravasco, M. E. Camilo, C. Jakobs, H. J. Blom, and I. T. de Almeida. 2004. 5,10-methylenetetrahydrofolate reductase (MTHFR) 677C→T and 1298A→C mutations are associated with DNA hypomethylation. *J Med Genet* 41 (6):454–58.

Choi, S. W., and J. B. Mason. 2002. Folate status: effects on pathways of colorectal carcinogenesis. *J Nutr* 132 (8 Suppl):2413S–2418S.

Choi, S. W., S. Friso, G. G. Dolnikowski, P. J. Bagley, A. N. Edmondson, D. E. Smith, and J. B. Mason. 2003. Biochemical and molecular aberrations in the rat colon due to folate depletion are age-specific. *J Nutr* 133 (4):1206–12.

Choi, S. W., S. Friso, M. K. Keyes, and J. B. Mason. 2005. Folate supplementation increases genomic DNA methylation in the liver of elder rats. *Br J Nutr* 93 (1):31–35.

Christman, J. K., G. Sheikhnejad, M. Dizik, S. Abileah, and E. Wainfan. 1993. Reversibility of changes in nucleic acid methylation and gene expression induced in rat liver by severe dietary methyl deficiency. *Carcinogenesis* 14 (4):551–57.

Cole, B. F., J. A. Baron, R. S. Sandler, R. W. Haile, D. J. Ahnen, R. S. Bresalier, G. E. McKeown-Eyssen, R. W. Summers, R. I. Rothstein, C. A. Burke, D. C. Snover, T. R. Church, J. I. Allen, D. J. Robertson, G. J. Beck, J. H. Bond, T. Byers, J. S. Mandel, L. A. Mott, L. H. Pearson, E. L. Barry, J. R. Rees, N. Marcon, F. Saibil, P. M. Ueland, and E. R. Greenberg. 2007. Folic acid for the prevention of colorectal adenomas: a randomized clinical trial. *JAMA* 297 (21):2351–59.

Cooney, C. A., A. A. Dave, and G. L. Wolff. 2002. Maternal methyl supplements in mice affect epigenetic variation and DNA methylation of offspring. *J Nutr* 132 (8 Suppl):2393S–2400S.

Cravo, M., P. Fidalgo, A. D. Pereira, A. Gouveia-Oliveira, P. Chaves, J. Selhub, J. B. Mason, F. C. Mira, and C. N. Leitao. 1994. DNA methylation as an intermediate biomarker in colorectal cancer: modulation by folic acid supplementation. *Eur J Cancer Prev* 3 (6):473–79.

Cravo, M. L., J. B. Mason, Y. Dayal, M. Hutchinson, D. Smith, J. Selhub, and I. H. Rosenberg. 1992. Folate deficiency enhances the development of colonic neoplasia in dimethylhydrazine-treated rats. *Cancer Res* 52 (18):5002–6.

Cravo, M. L., A. G. Pinto, P. Chaves, J. A. Cruz, P. Lage, C. Nobre Leitao, and F. Costa Mira. 1998. Effect of folate supplementation on DNA methylation of rectal mucosa in patients with colonic adenomas: correlation with nutrient intake. *Clin Nutr* 17 (2):45–49.

Crott, J. W., Z. Liu, M. K. Keyes, S. W. Choi, H. Jang, M.P . Moyer, and J. B. Mason. 2008. Moderate folate depletion modulates the expression of selected genes involved in cell cycle, intracellular signaling and folate uptake in human colonic epithelial cell lines. *J Nutr Biochem* 19 (5):328–35.

Curtin, K., M. L. Slattery, C. M. Ulrich, J. Bigler, T. R. Levin, R. K. Wolff, H. Albertsen, J. D. Potter, and W. S. Samowitz. 2007. Genetic polymorphisms in one-carbon metabolism: associations with CpG island methylator phenotype (CIMP) in colon cancer and the modifying effects of diet. *Carcinogenesis* 28 (8):1672–79.

Das, P. M., and R. Singal. 2004. DNA methylation and cancer. *J Clin Oncol* 22 (22):4632–42.

Davis, C. D., and E. O. Uthus. 2003. Dietary folate and selenium affect dimethylhydrazine-induced aberrant crypt formation, global DNA methylation and one-carbon metabolism in rats. *J Nutr* 133 (9):2907–14.

de Vogel, S., B. W. Bongaerts, K. A. Wouters, A. D. Kester, L. J. Schouten, A. F. de Goeij, A. P. de Bruine, R. A. Goldbohm, P. A. van den Brandt, M. van Engeland, and M. P. Weijenberg. 2008. Associations of dietary methyl donor intake with MLH1 promoter hypermethylation and related molecular phenotypes in sporadic colorectal cancer. *Carcinogenesis* 29 (9):1765–73.

Dizik, M., J. K. Christman, and E. Wainfan. 1991. Alterations in expression and methylation of specific genes in livers of rats fed a cancer promoting methyl-deficient diet. *Carcinogenesis* 12 (7):1307–12.

Dolinoy, D. C., D. Huang, and R. L. Jirtle. 2007. Maternal nutrient supplementation counteracts bisphenol A-induced DNA hypomethylation in early development. *Proc Natl Acad Sci U S A* 104 (32):13056–61.

Dudeja, P. K., S. A. Torania, and H. M. Said. 1997. Evidence for the existence of a carrier-mediated folate uptake mechanism in human colonic luminal membranes. *Am J Physiol* 272:G1408–15.

Duthie, S. J. 1999. Folic acid deficiency and cancer: mechanisms of DNA instability. *Br Med Bull* 55 (3):578–92.

Duthie, S. J., G. Grant, L. P. Pirie, A. J. Watson, and G. P. Margison. 2010. Folate deficiency alters hepatic and colon MGMT and OGG-1 DNA repair protein expression in rats but has no effect on genome-wide DNA methylation. *Cancer Prev Res* 3 (1):92–100.

Duthie, S. J., S. Narayanan, S. Blum, L. Pirie, and G. M. Brand. 2000. Folate deficiency in vitro induces uracil misincorporation and DNA hypomethylation and inhibits DNA excision repair in immortalized normal human colon epithelial cells. *Nutr Cancer* 37 (2):245–51.

Duthie, S. J., S. Narayanan, G. M. Brand, and G. Grant. 2000. DNA stability and genomic methylation status in colonocytes isolated from methyl-donor-deficient rats. *Eur J Nutr* 39 (3):106–11.

Eden, A., F. Gaudet, A. Waghmare, and R. Jaenisch. 2003. Chromosomal instability and tumors promoted by DNA hypomethylation. *Science* 300 (5618):455.

Egger, G., G. Liang, A. Aparicio, and P. A. Jones. 2004. Epigenetics in human disease and prospects for epigenetic therapy. *Nature* 429 (6990):457–63.

Engeham, S. F., A. Haase, and S.C. Langley-Evans. 2009. Supplementation of a maternal low-protein diet in rat pregnancy with folic acid ameliorates programming effects upon feeding behaviour in the absence of disturbances to the methionine-homocysteine cycle. *Br J Nutr* 1–12.

Esteller, M. 2003. Relevance of DNA methylation in the management of cancer. *Lancet Oncol* 4 (6):351–58.

Esteller, M., P. G. Corn, S. B. Baylin, and J. G. Herman. 2001. A gene hypermethylation profile of human cancer. *Cancer Res* 61 (8):3225–29.

Fenech, M. 2001. The role of folic acid and Vitamin B12 in genomic stability of human cells. *Mutat Res* 475 (1–2):57–67.

Fenech, M., C. Aitken, and J. Rinaldi. 1998. Folate, vitamin B12, homocysteine status and DNA damage in young Australian adults. *Carcinogenesis* 19 (7):1163–71.

Figueiredo, J. C., M. V. Grau, K. Wallace, A. J. Levine, L. Shen, R. Hamdan, X. Chen, R. S. Bresalier, G. McKeown-Eyssen, R. W. Haile, J. A. Baron, and J. P. Issa. 2009. Global DNA hypomethylation (LINE-1) in the normal colon and lifestyle characteristics and dietary and genetic factors. *Cancer Epidemiol Biomarkers Prev* 18 (4):1041–49.

Finnell, R. H., O. Spiegelstein, B. Wlodarczyk, A. Triplett, I. P. Pogribny, S. Melnyk, and J. S. James. 2002. DNA methylation in Folbp1 knockout mice supplemented with folic acid during gestation. *J Nutr* 132 (8 Suppl):2457S–2461S.

Flatley, J. E., K. McNeir, L. Balasubramani, J. Tidy, E. L. Stuart, T. A Young, and H.J. Powers. 2009. Folate status and aberrant DNA methylation are associated with HPV infection and cervical pathogenesis. *Cancer Epidemiol Biomarkers Prev* 18 (10):2782–89.

Fowler, B. M., A. R. Giuliano, C. Piyathilake, M. Nour, and K. Hatch. 1998. Hypomethylation in cervical tissue: is there a correlation with folate status? *Cancer Epidemiol Biomarkers Prev* 7 (10):901–6.

Friso, S., and S. W. Choi. 2002. Gene-nutrient interactions and DNA methylation. *J Nutr* 132 (8 Suppl):2382S–2387S.

Friso, S., S. W. Choi, D. Girelli, J. B. Mason, G. G. Dolnikowski, P. J. Bagley, O. Olivieri, P. F. Jacques, I. H. Rosenberg, R. Corrocher, and J. Selhub. 2002. A common mutation in the 5,10-methylenetetrahydrofolate reductase gene affects genomic DNA methylation through an interaction with folate status. *Proc Natl Acad Sci U S A* 99 (8):5606–11.

Friso, S., D. Girelli, E. Trabetti, O. Olivieri, P. Guarini, P. F. Pignatti, R. Corrocher, and S. W. Choi. 2005. The MTHFR 1298A>C polymorphism and genomic DNA methylation in human lymphocytes. *Cancer Epidemiol Biomarkers Prev* 14 (4):938–43.

Fryer, A. A., T. M. Nafee, K. M. Ismail, W. D. Carroll, R. D. Emes, and W. E. Farrell. 2009. LINE-1 DNA methylation is inversely correlated with cord plasma homocysteine in man: a preliminary study. *Epigenetics* 4 (6):394–98.

Fuke, C., M. Shimabukuro, A. Petronis, J. Sugimoto, T. Oda, K. Miura, T. Miyazaki, C. Ogura, Y. Okazaki, and Y. Jinno. 2004. Age related changes in 5-methylcytosine content in human peripheral leukocytes and placentas: an HPLC-based study. *Ann Hum Genet* 68 (3):196–204.

Gaudet, F., J. G. Hodgson, A. Eden, L. Jackson-Grusby, J. Dausman, J. W. Gray, H. Leonhardt, and R. Jaenisch. 2003. Induction of tumors in mice by genomic hypomethylation. *Science* 300 (5618):489–92.

Giovannucci, E., E. B. Rimm, A. Ascherio, M. J. Stampfer, G. A. Colditz, and W. C. Willett. 1995. Alcohol, low-methionine—low-folate diets, and risk of colon cancer in men. *J Natl Cancer Inst* 87 (4):265–73.

Giovannucci, E., M. J. Stampfer, G. A. Colditz, D. J. Hunter, C. Fuchs, B. A. Rosner, F. E. Speizer, and W. C. Willett. 1998. Multivitamin use, folate, and colon cancer in women in the Nurses' Health Study. *Ann Intern Med* 129 (7):517–24.

Gluckman, P. D., and M. A. Hanson. 2004. Developmental origins of disease paradigm: a mechanistic and evolutionary perspective. *Pediatr Res* 56 (3):311–17.

Halline, A. G., P. K. Dudeja, and T. A. Brasitus. 1988. 1,2-Dimethylhydrazine-induced premalignant alterations in the S-adenosylmethionine/S-adenosylhomocysteine ratio and membrane lipid lateral diffusion of the rat distal colon. *Biochim Biophys Acta* 944 (1):101–7.

Hepburn, P. A., G. P. Margison, and M. J. Tisdale. 1991. Enzymatic methylation of cytosine in DNA is prevented by adjacent O6-methylguanine residues. *J Biol Chem* 266 (13):7985–87.

Herman, J. G., and S. B. Baylin. 2003. Gene silencing in cancer in association with promoter hypermethylation. *N Engl J Med* 349 (21):2042–54.

Hirsch, S., A. M. Ronco, C. Guerrero-Bosagna, M. P. de la Maza, L. Leiva, G. Barrera, M. Llanos, M. A. Alliende, F. Silva, and D. Bunout. 2008. Methylation status in healthy subjects with normal and high serum folate concentration. *Nutrition* 24 (11–12):1103–9.

Ingrosso, D., A. Cimmino, A. F. Perna, L. Masella, N. G. De Santo, M. L. De Bonis, M. Vacca, M. D'Esposito, M. D'Urso, P. Galletti, and V. Zappia. 2003. Folate treatment and unbalanced methylation and changes of allelic expression induced by hyperhomocysteinaemia in patients with uraemia. *Lancet* 361 (9370):1693–99.

Issa, J. P. 2004. CpG island methylator phenotype in cancer. *Nat Rev Cancer* 4 (12):988–93.

Issa, J. P., Y. L. Ottaviano, P. Celano, S. R. Hamilton, N. E. Davidson, and S. B. Baylin. 1994. Methylation of the oestrogen receptor CpG island links ageing and neoplasia in human colon. *Nat Genet* 7 (4):536–40.

Jacob, R. A., D. M. Gretz, P. C. Taylor, S. J. James, I. P. Pogribny, B. J. Miller, S. M. Henning, and M. E. Swendseid. 1998. Moderate folate depletion increases plasma homocysteine and decreases lymphocyte DNA methylation in postmenopausal women. *J Nutr* 128 (7):1204–12.

Jacob, R. A., F. S. Pianalto, S. M. Henning, J. Z. Zhang, and M. E. Swendseid. 1995. In vivo methylation capacity is not impaired in healthy men during short-term dietary folate and methyl group restriction. *J Nutr* 125 (6):1495–1502.

Jacobs, E. J., C. J. Connell, A. V. Patel, A. Chao, C. Rodriguez, J. Seymour, M. L. McCullough, E. E. Calle, and M. J. Thun. 2001. Multivitamin use and colon cancer mortality in the Cancer Prevention Study II cohort (United States). *Cancer Cause Control* 12 (10):927–34.

Jhaveri, M. S., C. Wagner, and J. B. Trepel. 2001. Impact of extracellular folate levels on global gene expression. *Mol Pharmacol* 60 (6):1288–95.

Jin, M., K. Kawakami, Y. Fukui, S. Tsukioka, M. Oda, G. Watanabe, T. Takechi, T. Oka, and T. Minamoto. 2009. Different histological types of non-small cell lung cancer have distinct folate and DNA methylation levels. *Cancer Sci* 100 (12): 2325–30.

Jones, P. A., and S. B. Baylin. 2002. The fundamental role of epigenetic events in cancer. *Nat Rev Genet* 3 (6):415–28.

Jones, P. A., and P. W. Laird. 1999. Cancer epigenetics comes of age. *Nat Genet* 21 (2):163–67.

Kamen, B. A., P. A. Nylen, V. M. Whitehead, H. T. Abelson, B. J. Dolnick, and D. W. Peterson. 1985. Lack of dihydrofolate reductase in human tumor and leukemia cells in vivo. *Cancer Drug Deliv* 2 (2):133–38.

Kawakami, K., A. Ruszkiewicz, G. Bennett, J. Moore, G. Watanabe, and B. Iacopetta. 2003. The folate pool in colorectal cancers is associated with DNA hypermethylation and with a polymorphism in methylenetetrahydrofolate reductase. *Clin Cancer Res* 9 (16 Pt 1):5860–65.

Keyes, M. K., H. Jang, J. B. Mason, Z. Liu, J. W. Crott, D. E. Smith, S. Friso, and S.W. Choi. 2007. Older age and dietary folate are determinants of genomic and p16-specific DNA methylation in mouse colon. *J Nutr* 137 (7):1713–17.

Khosraviani, K., H. P. Weir, P. Hamilton, J. Moorehead, and K. Williamson. 2002. Effect of folate supplementation on mucosal cell proliferation in high risk patients for colon cancer. *Gut* 51 (2):195–99.

Kim, J. M., K. Hong, J. H. Lee, S. Lee, and N. Chang. 2009. Effect of folate deficiency on placental DNA methylation in hyperhomocysteinemic rats. *J Nutr Biochem* 20 (3):172–76.

Kim, Y. I. 1999. Folate and carcinogenesis: Evidence, mechanisms, and implications. *J Nutr Biochem* 10:66–88.

———. 2000. Methylenetetrahydrofolate reductase polymorphisms, folate, and cancer risk: a paradigm of gene-nutrient interactions in carcinogenesis. *Nutr Rev* 58 (7):205–9.

———. 2003. Role of folate in colon cancer development and progression. *J Nutr* 133:3731S–3739S.

———. 2004. Will mandatory folic acid fortification prevent or promote cancer? *Am J Clin Nutr* 80 (5):1123–28.

———. 2007. Folate and colorectal cancer: an evidence-based critical review. *Mol Nutr Food Res* 51 (3):267–92.

———. 2008. Folic acid supplementation and cancer risk: point. *Cancer Epidemiol Biomarkers Prev* 17 (9):2220–25.

———. 2009. Role of the MTHFR polymorphisms in cancer risk modification and treatment. *Future Oncol* 5 (4):523–42.

Kim, Y. I., H. W. Baik, K. Fawaz, T. Knox, Y. M. Lee, R. Norton, E. Libby, and J. B. Mason. 2001. Effects of folate supplementation on two provisional molecular markers of colon cancer: a prospective, randomized trial. *Am J Gastroenterol* 96 (1):184–95.

Kim, Y. I., J. W. Miller, K. A. da Costa, M. Nadeau, D. Smith, J. Selhub, S. H. Zeisel, and J. B. Mason. 1994. Severe folate deficiency causes secondary depletion of choline and phosphocholine in rat liver. *J Nutr* 124 (11):2197–2203.

Kim, Y. I., J. K. Christman, J. C. Fleet, M. L. Cravo, R. N. Salomon, D. Smith, J. Ordovas, J. Selhub, and J. B. Mason. 1995. Moderate folate deficiency does not cause global hypomethylation of hepatic and colonic DNA or c-myc-specific hypomethylation of colonic DNA in rats. *Am J Clin Nutr* 61 (5):1083–90.

Kim, Y. I., I. P. Pogribny, A. G. Basnakian, J. W. Miller, J. Selhub, S. J. James, and J. B. Mason. 1997. Folate deficiency in rats induces DNA strand breaks and hypomethylation within the p53 tumor suppressor gene. *Am J Clin Nutr* 65 (1):46–52.

Kim, Y. I., I. P. Pogribny, R. N. Salomon, S. W. Choi, D. E. Smith, S. J. James, and J. B. Mason. 1996. Exon-specific DNA hypomethylation of the p53 gene of rat colon induced by dimethylhydrazine. Modulation by dietary folate. *Am J Pathol* 149 (4):1129–37.

Kim, Y. I., R. N. Salomon, F. Graeme-Cook, S. W. Choi, D. E. Smith, G. E. Dallal, and J. B. Mason. 1996. Dietary folate protects against the development of macroscopic colonic neoplasia in a dose responsive manner in rats. *Gut* 39 (5):732–40.

Kotsopoulos, J., K. J. Sohn, and Y. I. Kim. 2008. Postweaning dietary folate deficiency provided through childhood to puberty permanently increases genomic DNA methylation in adult rat liver. *J Nutr* 138 (4):703–9.

Kraunz, K. S., D. Hsiung, M. D. McClean, M. Liu, J. Osanyingbemi, H. H. Nelson, and K. T. Kelsey. 2006. Dietary folate is associated with p16(INK4A) methylation in head and neck squamous cell carcinoma. *Int J Cancer* 119 (7):1553–57.

Laird, P. W., L. Jackson-Grusby, A. Fazeli, S. L. Dickinson, W. E. Jung, E. Li, R. A. Weinberg, and R. Jaenisch. 1995. Suppression of intestinal neoplasia by DNA hypomethylation. *Cell* 81 (2):197–205.

Lamprecht, S. A., and M. Lipkin. 2003. Chemoprevention of colon cancer by calcium, vitamin D and folate: molecular mechanisms. *Nat Rev Cancer* 3 (8):601–14.

Lashner, B. A., B. D. Shapiro, A. Husain, and J. R. Goldblum. 1999. Evaluation of the usefulness of testing for p53 mutations in colorectal cancer surveillance for ulcerative colitis. *Am J Gastroenterol* 94 (2):456–62.

Le Leu, R. K., G. P. Young, and G. H. McIntosh. 2000a. Folate deficiency diminishes the occurrence of aberrant crypt foci in the rat colon but does not alter global DNA methylation status. *J Gastroenterol Hepatol* 15 (10):1158–64.

Le Leu, R. K., G. P. Young, and G. H. McIntosh. 2000b. Folate deficiency reduces the development of colorectal cancer in rats. *Carcinogenesis* 21 (12):2261–65.

Levine, A. J., K. D. Siegmund, C. M. Ervin, A. Diep, E. R. Lee, H. D. Frankl, and R. W. Haile. 2000. The methylenetetrahydrofolate reductase 677C→T polymorphism and distal colorectal adenoma risk. *Cancer Epidemiol Biomarkers Prev* 9 (7):657–63.

Li, E., and R. Jaenisch. 2000. DNA methylation and methyltransferases. In *DNA alterations in cancer: genetic and epigenetic changes,* ed. M. Ehrlich. Natick, MA: Eaton.

Lillycrop, K. A., E. S. Phillips, A. A. Jackson, M. A. Hanson, and G. C. Burdge. 2005. Dietary protein restriction of pregnant rats induces and folic acid supplementation prevents epigenetic modification of hepatic gene expression in the offspring. *J Nutr* 135 (6):1382–86.

Lim, U., A. Flood, S. W. Choi, D. Albanes, A. J. Cross, A. Schatzkin, R. Sinha, H. A. Katki, B. Cash, P. Schoenfeld, and R. Stolzenberg-Solomon. 2008. Genomic methylation of leukocyte DNA in relation to colorectal adenoma among asymptomatic women. *Gastroenterology* 134 (1):47–55.

Linhart, H. G., A. M. Troen, G. W. Bell, E. Cantu, W. H. Chao, E. Moran, E. Steine, T. He, and R. Jaenisch. 2009. Folate deficiency induces genomic uracil misincorporation and hypomethylation but does not increase DNA point mutations. *Gastroenterology* 136 (1):227–35.

Liu, Z., S. W. Choi, J. W. Crott, M. K. Keyes, H. Jang, D. E. Smith, M. Kim, P. W. Laird, R. Bronson, and J. B. Mason. 2007. Mild depletion of dietary folate combined with other B vitamins alters multiple components of the Wnt pathway in mouse colon. *J Nutr* 137 (12):2701–8.

Liu, Z., S. W. Choi, J. W. Crott, D. E. Smith, and J. B. Mason. 2008. Multiple B-vitamin inadequacy amplifies alterations induced by folate depletion in p53 expression and its downstream effector MDM2. *Int J Cancer* 123 (3):519–25.

Logan, R. F., M. J. Grainge, V. C. Shepherd, N. C. Armitage, and K. R. Muir. 2008. Aspirin and folic acid for the prevention of recurrent colorectal adenomas. *Gastroenterology* 134 (1):29–38.

Maffini, M. V., B. S. Rubin, C. Sonnenschein, and A. M. Soto. 2006. Endocrine disruptors and reproductive health: the case of bisphenol-A. *Mol Cell Endocrinol* 254–55:179–86.

Maloney, C. A., S. M. Hay, and W. D. Rees. 2007. Folate deficiency during pregnancy impacts on methyl metabolism without affecting global DNA methylation in the rat fetus. *Br J Nutr* 97 (6):1090–98.

Markowitz, S. D., and M. M. Bertagnolli. 2009. Molecular origins of cancer: Molecular basis of colorectal cancer. *N Engl J Med* 361 (25):2449–60.

Marugame, T., E. Tsuji, C. Kiyohara, H. Eguchi, T. Oda, K. Shinchi, and S. Kono. 2003. Relation of plasma folate and methylenetetrahydrofolate reductase C677T polymorphism to colorectal adenomas. *Int J Epidemiol* 32 (1):64–66.

Miller, J. W., M. R. Nadeau, J. Smith, D. Smith, and J. Selhub. 1994. Folate-deficiency-induced homocysteinaemia in rats: disruption of S-adenosylmethionine's co-ordinate regulation of homocysteine metabolism. *Biochem J* 298:415–19.

Mokarram, P., F. Naghibalhossaini, M. Saberi Firoozi, S. V. Hosseini, A. Izadpanah, H. Salahi, S. A. Malek-Hosseini, A. Talei, and M. Mojallal. 2008. Methylenetetrahydrofolate reductase C677T genotype affects promoter methylation of tumor-specific genes in sporadic colorectal cancer through an interaction with folate/vitamin B12 status. *World J Gastroenterol* 14 (23):3662–71.

Narayanan, S., J. McConnell, J. Little, L. Sharp, C. J. Piyathilake, H. Powers, G. Basten, and S. J. Duthie. 2004. Associations between two common variants C677T and A1298C in the methylenetetrahydrofolate reductase gene and measures of folate metabolism and DNA stability (strand breaks, misincorporated uracil, and DNA methylation status) in human lymphocytes in vivo. *Cancer Epidemiol Biomarkers Prev* 13 (9):1436–43.

Newberne, P. M., and A. E. Rogers. 1986. Labile methyl groups and the promotion of cancer. *Annu Rev Nutr* 6:407–32.

Nijhout, H. F., M. C. Reed, P. Budu, and C. M. Ulrich. 2004. A mathematical model of the folate cycle: new insights into folate homeostasis. *J Biol Chem* 279 (53):55008–16.

Ordonez, L. A., and R. J. Wurtman. 1974. Folic acid deficiency and methyl group metabolism in rat brain: effects of L-dopa. *Arch Biochem Biophys* 160 (2):372–76.

Paspatis, G. A., and D. G. Karamanolis. 1994. Folate supplementation and adenomatous colonic polyps. *Dis Colon Rectum* 37 (12):1340–41.

Pilsner, J. R., X. Liu, H. Ahsan, V. Ilievski, V. Slavkovich, D. Levy, P. Factor-Litvak, J. H. Graziano, and M. V. Gamble. 2007. Genomic methylation of peripheral blood leukocyte DNA: influences of arsenic and folate in Bangladeshi adults. *Am J Clin Nutr* 86 (4):1179–86.

Pogribny, I. P., A. G. Basnakian, B. J. Miller, N. G. Lopatina, L. A. Poirier, and S. J. James. 1995. Breaks in genomic DNA and within the p53 gene are associated with hypomethylation in livers of folate/methyl-deficient rats. *Cancer Res* 55 (9):1894–1901.

Pogribny, I. P., and S. J. James. 2002. De novo methylation of the p16INK4A gene in early preneoplastic liver and tumors induced by folate/methyl deficiency in rats. *Cancer Lett* 187 (1–2):69–75.

Pogribny, I. P., A. R. Karpf, S. R. James, S. Melnyk, T. Han, and V. P. Tryndyak. 2008. Epigenetic alterations in the brains of Fisher 344 rats induced by long-term administration of folate/methyl-deficient diet. *Brain Res* 1237:25–34.

Pogribny, I. P., B. J. Miller, and S. J. James. 1997. Alterations in hepatic p53 gene methylation patterns during tumor progression with folate/methyl deficiency in the rat. *Cancer Lett* 115 (1):31–38.

Pogribny, I. P., M. Pogribna, J. K. Christman, and S. J. James. 2000. Single-site methylation within the p53 promoter region reduces gene expression in a reporter gene construct: possible in vivo relevance during tumorigenesis. *Cancer Res* 60 (3):588–94.

Pogribny, I. P., L. A. Poirier, and S. J. James. 1995. Differential sensitivity to loss of cytosine methyl groups within the hepatic p53 gene of folate/methyl deficient rats. *Carcinogenesis* 16 (11):2863–67.

Pogribny, I. P., S. A. Ross, C. Wise, M. Pogribna, E. A. Jones, V. P. Tryndyak, S. J. James, Y. P. Dragan, and L. A. Poirier. 2006. Irreversible global DNA hypomethylation as a key step in hepatocarcinogenesis induced by dietary methyl deficiency. *Mutat Res* 593 (1–2):80–87.

Pogribny, I. P., S. I. Shpyleva, L. Muskhelishvili, T. V. Bagnyukova, S. J. James, and F. A. Beland. 2009. Role of DNA damage and alterations in cytosine DNA methylation in rat liver carcinogenesis induced by a methyl-deficient diet. *Mutat Res* 669 (1–2):56–62.

Potter, J. D. 2002. Methyl supply, methyl metabolizing enzymes and colorectal neoplasia. *J Nutr* 132 (8 Suppl):2410S–2412S.

Pufulete, M., R. Al-Ghnaniem, A. Khushal, P. Appleby, N. Harris, S. Gout, P. W. Emery, and T. A. Sanders. 2005. Effect of folic acid supplementation on genomic DNA methylation in patients with colorectal adenoma. *Gut* 54 (5):648–53.

Pufulete, M., R. Al-Ghnaniem, A. J. Leather, P. Appleby, S. Gout, C. Terry, P. W. Emery, and T. A. Sanders. 2003. Folate status, genomic DNA hypomethylation, and risk of colorectal adenoma and cancer: a case control study. *Gastroenterology* 124 (5):1240–48.

Pufulete, M., R. Al-Ghnaniem, J. A. Rennie, P. Appleby, N. Harris, S. Gout, P. W. Emery, and T. A. Sanders. 2005. Influence of folate status on genomic DNA methylation in colonic mucosa of subjects without colorectal adenoma or cancer. *Br J Cancer* 92 (5):838–42.

Quinlivan, E. P., K. Crider, R. J. Berry, L. Hao, Z. Li, J. H. Zhu, D. Maneval, T.P. Young, and L. B. Bailey. 2008. Global DNA methylation changes in response to chronic consumption and withdrawal of low, moderate, and high folic acid doses. *Faseb J* 22:689.7.

Quinlivan, E. P., S. R. Davis, K. P. Shelnutt, G. N. Henderson, H. Ghandour, B. Shane, J. Selhub, L. B. Bailey, P. W. Stacpoole, and J. F. Gregory, 3rd. 2005. Methylenetetrahydrofolate reductase 677C→T polymorphism and folate status affect one-carbon incorporation into human DNA deoxynucleosides. *J Nutr* 135 (3):389–96.

Rampersaud, G. C., G. P. Kauwell, A. D. Hutson, J. J. Cerda, and L. B. Bailey. 2000. Genomic DNA methylation decreases in response to moderate folate depletion in elderly women. *Am J Clin Nutr* 72 (4):998–1003.

Reik, W., W. Dean, and J. Walter. 2001. Epigenetic reprogramming in mammalian development. *Science* 293 (5532):1089–93.

Robertson, K. D., and A. P. Wolffe. 2000. DNA methylation in health and disease. *Nat Rev Genet* 1 (1):11–19.

Said, H. M., N. Chatterjee, R. U. Haq, V. S. Subramanian, A. Ortiz, L. H. Matherly, F. M. Sirotnak, C. Halsted, and S. A. Rubin. 2000. Adaptive regulation of intestinal folate uptake: effect of dietary folate deficiency. *Am J Physiol Cell Physiol* 279 (6):C1889–95.

Schernhammer, E. S., E. Giovannucci, T. Kawasaki, B. Rosner, C. Fuchs, and S. Ogino. 2009. Dietary folate, alcohol, and B vitamins in relation to LINE-1 hypomethylation in colon cancer. *Gut* (e-pub).

Selhub, J., and J. W. Miller. 1992. The pathogenesis of homocysteinemia: interruption of the coordinate regulation by S-adenosylmethionine of the remethylation and transsulfuration of homocysteine. *Am J Clin Nutr* 55 (1):131–38.

Shelnutt, K. P., G. P. Kauwell, J. F. Gregory III, D. R. Maneval, E. P. Quinlivan, D. W. Theriaque, G. N. Henderson, and L. B. Bailey. 2004. Methylenetetrahydrofolate reductase 677C→T polymorphism affects DNA methylation in response to controlled folate intake in young women. *J Nutr Biochem* 15 (9):554–60.

Sinclair, K. D., C. Allegrucci, R. Singh, D. S. Gardner, S. Sebastian, J. Bispham, A. Thurston, J. F. Huntley, W. D. Rees, C. A. Maloney, R. G. Lea, J. Craigon, T. G. McEvoy, and L. E. Young. 2007. DNA methylation, insulin resistance, and blood pressure in offspring determined by maternal periconceptional B vitamin and methionine status. *Proc Natl Acad Sci U S A* 104 (49):19351–56.

Slansky, J. E., Y. Li, W. G. Kaelin, and P. J. Farnham. 1993. A protein synthesis-dependent increase in E2F1 mRNA correlates with growth regulation of the dihydrofolate reductase promoter. *Mol Cell Biol* 13 (3):1610–18.

Slattery, M. L., K. Curtin, C. Sweeny, T. R. Levin, J. D. Potter, R. K. Wolff, H. Albertsen, and W. S. Samowitz. 2006. Diet and lifestype factor associations with CpG island methylator phenotype and BRAF mutations in colon cancer. *Int J Cancer* 120:656–63.

Sohn, K. J., J. M. Stempak, S. Reid, S. Shirwadkar, J. B. Mason, and Y. I. Kim. 2003. The effect of dietary folate on genomic and p53-specific DNA methylation in rat colon. *Carcinogenesis* 24 (1):81–90.

Song, J., K. J. Sohn, A. Medline, C. Ash, S. Gallinger, and Y. I. Kim. 2000. Chemopreventive effects of dietary folate on intestinal polyps in Apc+/-Msh2-/- mice. *Cancer Res* 60 (12):3191–99.

Sowers, R., J. Toguchida, J. Qin, P. A. Meyers, J. H. Healey, A. Huvos, D. Banerjee, J. R. Bertino, and R. Gorlick. 2003. mRNA expression levels of E2F transcription factors correlate with dihydrofolate reductase, reduced folate carrier, and thymidylate synthase mRNA expression in osteosarcoma. *Mol Cancer Ther* 2 (6):535–41.

Steegers-Theunissen, R. P., S. A. Obermann-Borst, D. Kremer, J. Lindemans, C. Siebel, E. A. Steegers, P. E. Slagboom, and B. T. Heijmans. 2009. Periconceptional maternal folic acid use of 400 microg per day is related to increased methylation of the IGF2 gene in the very young child. *PLoS One* 4 (11):e7845.

Stempak, J. M., K. J. Sohn, E. P. Chiang, B. Shane, and Y. I. Kim. 2005. Cell and stage of transformation-specific effects of folate deficiency on methionine cycle intermediates and DNA methylation in an in vitro model. *Carcinogenesis* 26 (5):981–90.

Stern, L. L., J. B. Mason, J. Selhub, and S. W. Choi. 2000. Genomic DNA hypomethylation, a characteristic of most cancers, is present in peripheral leukocytes of individuals who are homozygous for the C677T polymorphism in the methylenetetrahydrofolate reductase gene. *Cancer Epidemiol Biomarkers Prev* 9 (8):849–53.

Stidley, C. A., M. A. Picchi, S. Leng, R. Willink, R. E. Crowell, K. G. Flores, H. Kang, T. Byers, F. D. Gilliland, and S. A. Belinsky. 2010. Multivitamins, folate, and green vegetables protect against gene promoter methylation in the aerodigestive tract of smokers. *Cancer Res* 70 (2):568–74.

Takai, D., and P. A. Jones. 2002. Comprehensive analysis of CpG islands in human chromosomes 21 and 22. *Proc Natl Acad Sci U S A* 99 (6):3740–45.

Tao, M. H., P. G. Shields, J. Nie, C. Marian, C. B. Ambrosone, S. E. McCann, M. Platek, S. S. Krishnan, B. Xie, S. B. Edge, J. Winston, D. Vito, M. Trevisan, and J. L. Freudenheim. 2009. DNA promoter methylation in breast tumors: no association with genetic polymorphisms in MTHFR and MTR. *Cancer Epidemiol Biomarkers Prev* 18 (3):998–1002.

Toyota, M., N. Ahuja, M. Ohe-Toyota, J. G. Herman, S. B. Baylin, and J. P. Issa. 1999. CpG island methylator phenotype in colorectal cancer. *Proc Natl Acad Sci U S A* 96 (15):8681–86.

Toyota, M., and J.P. Issa. 1999. CpG island methylator phenotypes in aging and cancer. *Semin Cancer Biol* 9 (5):349–57.

Trinh, B. N., T. I. Long, A. E. Nickel, D. Shibata, and P. W. Laird. 2002. DNA methyltransferase deficiency modifies cancer susceptibility in mice lacking DNA mismatch repair. *Mol Cell Biol* 22 (9):2906–17.

Ulrich, C. M., E. Kampman, J. Bigler, S. M. Schwartz, C. Chen, R. Bostick, L. Fosdick, S. A. Beresford, Y. Yasui, and J. D. Potter. 1999. Colorectal adenomas and the C677T MTHFR polymorphism: evidence for gene-environment interaction? *Cancer Epidemiol Biomarkers Prev* 8 (8):659–68.

Uthus, E.O., S.A. Ross, and C.D. Davis. 2006. Differential effects of dietary selenium (se) and folate on methyl metabolism in liver and colon of rats. *Biol Trace Elem Res* 109 (3):201–14.

van den Donk, M., L. Pellis, J. W. Crott, M. van Engeland, P. Friederich, F. M. Nagengast, J. D. van Bergeijk, S. Y. de Boer, J. B. Mason, F. J. Kok, J. Keijer, and E. Kampman. 2007. Folic acid and vitamin B-12 supplementation does not favorably influence uracil incorporation and promoter methylation in rectal mucosa DNA of subjects with previous colorectal adenomas. *J Nutr* 137 (9):2114–20.

van den Donk, M., M. van Engeland, L. Pellis, B. J. Witteman, F. J. Kok, J. Keijer, and E. Kampman. 2007. Dietary folate intake in combination with MTHFR C677T genotype and promoter methylation of tumor suppressor and DNA repair genes in sporadic colorectal adenomas. *Cancer Epidemiol Biomarkers Prev* 16 (2):327–33.

van Engeland, M., M. P. Weijenberg, G. M. Roemen, M. Brink, A. P. de Bruine, R. A. Goldbohm, P. A. van den Brandt, S. B. Baylin, A. F. de Goeij, and J. G. Herman. 2003. Effects of dietary folate and alcohol intake on promoter methylation in sporadic colorectal cancer: The Netherlands cohort study on diet and cancer. *Cancer Res* 63 (12):3133–37.

van Guelpen, B., A. M. Dahlin, J. Hultdin, V. Eklof, J. Ingegerd, M.L. Henriksson, I. Cullman, G. Hallmans, and R. Palmqvist. 2009. One-carbon metabolism and CpG island methylator phenotype status in incident colorectal cancer: a nested case-referent study. *Cancer Cause Control* (e-pub).

Vanyushin, B. F., A. L. Mazin, V. K. Vasilyev, and A. N. Belozersky. 1973. The content of 5-methylcytosine in animal DNA: the species and tissue specificity. *Biochim Biophys Acta* 299 (3):397–403.

Virmani, A. K., J. A. Tsou, K. D. Siegmund, L. Y. Shen, T. I. Long, P. W. Laird, A. F. Gazdar, and I. A. Laird-Offringa. 2002. Hierarchical clustering of lung cancer cell lines using DNA methylation markers. *Cancer Epidemiol Biomarkers Prev* 11 (3):291–97.

Wagner, C. 1995. Biochemical role of folate in cellular metabolism. In *Folate in health and disease,* ed. L. B. Bailey. New York: Marcel Dekker.

Wainfan, E., M. Dizik, M. Stender, and J. K. Christman. 1989. Rapid appearance of hypomethylated DNA in livers of rats fed cancer-promoting, methyl-deficient diets. *Cancer Res* 49 (15):4094–97.

Wainfan, E., M. Kilkenny, and M. Dizik. 1988. Comparison of methyltransferase activities of pair-fed rats given adequate or methyl-deficient diets. *Carcinogenesis* 9 (5):861–63.

Wainfan, E., and L. A. Poirier. 1992. Methyl groups in carcinogenesis: effects on DNA methylation and gene expression. *Cancer Res* 52 (7 Suppl):2071s–2077s.

Wang, J., A. J. Sasco, C. Fu, H. Xue, G. Guo, Z. Hua, Q. Zhou, Q. Jiang, and B. Xu. 2008. Aberrant DNA methylation of P16, MGMT, and hMLH1 genes in combination with MTHFR C677T genetic polymorphism in esophageal squamous cell carcinoma. *Cancer Epidemiol Biomarkers Prev* 17 (1):118–25.

Wasson, G. R., A. P. McGlynn, H. McNulty, S. L. O'Reilly, V. J. McKelvey-Martin, G. McKerr, J. J. Strain, J. M. Scott, and C. S. Downes. 2006. Global DNA and p53 region-specific hypomethylation in human colonic cells is induced by folate depletion and reversed by folate supplementation. *J Nutr* 136 (11):2748–53.

Waterland, R. A., D. C. Dolinoy, J. R. Lin, C. A. Smith, X. Shi, and K. G. Tahiliani. 2006. Maternal methyl supplements increase offspring DNA methylation at Axin Fused. *Genesis* 44 (9):401–6.

Waterland, R. A., and R. L. Jirtle. 2003. Transposable elements: targets for early nutritional effects on epigenetic gene regulation. *Mol Cell Biol* 23 (15):5293–5300.

Wilson, V. L., R. A. Smith, S. Ma, and R. G. Cutler. 1987. Genomic 5-methyldeoxycytidine decreases with age. *J Biol Chem* 262:9948–51.

Winawer, S. J., R. H. Fletcher, L. Miller, F. Godlee, M. H. Stolar, C. D. Mulrow, S. H. Woolf, S. N. Glick, T. G. Ganiats, J. H. Bond, L. Rosen, J. G. Zapka, S. J. Olsen, F. M. Giardiello, J. E. Sisk, R. van Antwerp, C. Brown-Davis, D. A. Marciniak, and R. J. Mayer. 1997. Colorectal cancer screening: clinical guidelines and rationale. *Gastroenterology* 112 (2):594–642.

Wolff, G. L., R. L. Kodell, S. R. Moore, and C. A. Cooney. 1998. Maternal epigenetics and methyl supplements affect *agouti* gene expression in Avy/a mice. *Faseb J* 12 (11):949–57.

Yoder, J. A., C. P. Walsh, and T. H. Bestor. 1997. Cytosine methylation and the ecology of intragenomic parasites. *Trends Genet* 13 (8):335–40.

Zapisek, W. F., G. M. Cronin, B. D. Lyn-Cook, and L. A. Poirier. 1992. The onset of oncogene hypomethylation in the livers of rats fed methyl-deficient, amino acid-defined diets. *Carcinogenesis* 13 (10):1869–72.

Zimmerman, J. 1990. Folic acid transport in organ-cultured mucosa of human intestine. Evidence for distinct carriers. *Gastroenterology* 99 (4):964–72.

Zingg, J. M., and P. A. Jones. 1997. Genetic and epigenetic aspects of DNA methylation on genome expression, evolution, mutation and carcinogenesis. *Carcinogenesis* 18 (5):869–82.

4 Dietary Components, Epigenetics, and Cancer

Cindy D. Davis and Sharon A. Ross

CONTENTS

4.1 Introduction 77
4.2 Epigenetics and Cancer 79
4.3 Diet and DNA Methylation: Timing of Exposure 84
4.4 Histone Modification by Bioactive Food Components and Diet Composition 90
4.5 Dietary Modulation of Polycomb Repressive Complexes 95
4.6 Small, Noncoding RNA, Epigenetics, and Dietary Factors 96
4.7 Conclusions 99
References 99

4.1 INTRODUCTION

Cancer is a leading cause of death in the United States. "Cancer" is a general term that represents more than 100 diseases, each with its own etiology. Cancer risk is influenced by both genetic and environmental factors, including dietary habits. While each type of cancer has unique characteristics, they share one common feature: unregulated cell division. All cancers begin when a single cell acquires multiple genetic changes and loses control of its normal growth and replication processes (Hanahan and Weinberg 2000). The cancer process, which can occur over decades, includes fundamental yet diverse, wide-ranging cellular processes that can be influenced by diet, such as carcinogen bioactivation, cellular differentiation, DNA repair, cellular proliferation/signaling, and apoptosis (Davis and Milner 2007). These cellular processes are altered via deregulation of key genes, resulting in an altered cellular phenotype (Wiseman 2008). Such anomalous gene expression may result from genetic disruption, i.e., mutation, or from epigenetic modulation by silencing genes that should be active or activating genes that should be silent. Diet and bioactive food factors may directly influence both processes.

Several lines of evidence, including epidemiological and preclinical studies, suggest that the increased intake of certain bioactive food components may modulate cancer risk and tumor behavior (Table 4.1). However, the specific molecular mechanisms for these observations and the quantities needed (as well as issues of frequency, duration, and timing of exposure in the life span) to bring about the cancer-protective effect remain largely unanswered.

TABLE 4.1
Dietary Components Linked with Cancer Prevention through Epigenetic Modifications

Dietary Component	Food Sources
Allyl mercaptan	Garlic and other *Allium* vegetables
Betaine	Wheat germ, spinach, and shellfish
Biotin	Yeast, egg yolk, organ meats, and grains
Butyrate	Fermentation of dietary fiber
Choline	Egg yolks, organ meats, and wheat germ
Copper	Seafood, nuts, legumes, and organ meats
Curcumin	Turmeric
Diallyl disulfide	Garlic and other *Allium* vegetables
Epigallocatechin 3-gallate	Green tea
Folate	Green leafy vegetables, dried beans and peas, and enriched breads and cereals
Genistein	Soybeans
Isothiocyanates	Cruciferous vegetables
Lunasin	Soybeans
Resveratrol	Grapes, wine, and peanuts
Selenium	Seafood, meat, and whole grains
Vitamin A	Fat-containing and fortified dairy products, liver, and provitamin carotenoids in fruits and vegetables
Vitamin B_{12}	Animal products
Zinc	Meats, seafood, and whole grains

Many studies provide intriguing evidence that part of the cancer-inhibiting properties associated with several dietary components may relate to modulation of epigenetic processes. These include DNA methylation of the cytosine phosphate guanine dinucleotide (CpG) islands in promoters as well as other regions of the genome, chromatin remodeling and higher-order chromatin structural alterations, posttranslational ATP-dependent modifications of histone tail domains including methylation, acetylation, ubiquitination, and phosphorylation, and regulation through noncoding RNAs (Figure 4.1). We have previously delineated (Ross 2003) at least four different mechanisms in which nutrients may modify DNA methylation. The first is that nutrients may influence the supply of methyl groups for the formation of S-adenosylmethionine (SAM), which is the universal methyl donor. The second mechanism is that nutrients may modify utilization of methyl groups by processes including altered DNA methyltransferase activity. A third possible mechanism may relate to DNA demethylation activity. Finally, the DNA methylation patterns may influence the response to dietary component(s). Interestingly, such interactions may apply similarly to the way in which diet impacts histone methylation marks and processes. Moreover, diet can also affect other histone tail modifications. The interrelationship between dietary components, epigenetics, and cancer will be further explored in this review by providing additional examples and highlighting areas for further research.

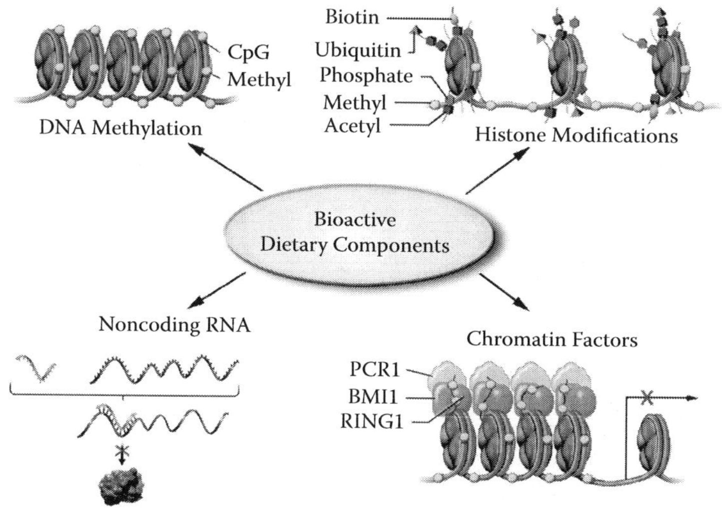

FIGURE 4.1 (Please see color insert following page 80.) Evidence suggests that bioactive food components can modify several epigenetic mechanisms, including DNA methylation, histone modifications, noncoding RNA, and chromatin factors such as the polycomb repressive complex 1.

4.2 EPIGENETICS AND CANCER

The cellular epigenetic apparatus consists of chromatin, which contains a histone protein-based structure around which DNA is wrapped, histone posttranslational modifications, and covalent modifications of a methyl group to cytosines residing at the dinucleotide sequence CG in DNA (McGowan et al. 2008). These modifications determine the accessibility of the transcriptional machinery to the genome and therefore influence gene expression: genes are inactivated (switched off) when the chromatin is closed (heterochromatin), and they are expressed (switched on) when the chromatin is open (euchromatin) (Rodenhiser and Mann 2006). Recently, small, noncoding RNAs have also been shown to provide an additional level of epigenetic regulation in the nucleus (D. H. Kim et al. 2008).

The most widely studied epigenetic modification in humans involves the covalent addition of a methyl group (CH_3) to the 5′ position of a cytosine that precedes a guanosine in the DNA sequence (the CpG dinucleotide) (Esteller 2005). DNA methylation does not act in isolation, but interacts with histone modifications and chromatin-remodeling complexes to influence chromatin structure and gene regulation (Tost 2009). This reaction is catalyzed by DNA methyltransferases (DNMTs), and SAM serves as the universal methyl donor. Three DNMTs—DNMT1, DNMT3A, and DNMT3B—have been extensively studied in developmental processes and in cancer. DNMT1 has been referred to as a "maintenance" methyltransferase, because it has a preference for a hemimethylated substrate and is involved in copying DNA methylation patterns during cellular replication (Razin and Riggs 1980). In contrast, DNMT3A and DNMT3B are responsible for de novo methylation.

Emerging evidence suggests that the targeting of DNMT toward DNA may be more complex in that these enzymes appear to target specific genes with the assistance of sequence-specific factors (Brenner et al. 2005). DNA methylation is thought to be removed passively by blocking methylation of newly synthesized DNA during DNA replication (Ooi and Bestor 2008). However, the presence of an active DNA demethylase has long been speculated in mammalian cells, and recent experimental findings suggest that the methyl DNA binding protein MBD2 and de novo DNA methyltransferases DNMT3A and DNMT3B possess DNA demethylase activity in mammalian cells (Ooi and Bestor 2008; J. K. Kim et al. 2009). These observations need to be confirmed and further examined to determine whether active mammalian demethylation plays a role in the regulation of DNA methylation.

Regions rich in CpG dinucleotides, termed CpG islands, often occur in the promoter regions of genes (Esteller 2007). These regions are usually unmethylated, which is associated with the ability of CpG-island-containing genes to be transcribed in the presence of the required transcriptional activators. Conversely, methylation at these critical sites inhibits the binding of transcription factors to their recognition elements, recruits methylated DNA binding proteins such as MeCP2 and MBD2 to the gene, and activates chromatin modification enzymes such as histone deacetylases (HDACs), which in turn introduce histone modifications, resulting in chromatin silencing (Li 2002). The transcriptional silencing of tumor suppressor genes by CpG-island-promoter hypermethylation is at least as common as DNA mutations as a mechanism for inactivation of classical tumor suppressor genes in human cancer (Jones and Baylin 2002; Tsou et al. 2002). Furthermore, a number of candidate tumor suppressor genes not commonly inactivated by mutation are transcriptionally silenced by this mechanism (Jones and Baylin 2002). The aberrant methylation of genes that suppresses tumorigenesis appears to occur early in tumor development and increases progressively, eventually leading to the malignant phenotype (Fearon and Vogelstein 1990; Y. Kim and Mason 1995). In addition to region-specific hypermethylation, widespread global DNA hypomethylation (Ehrlich 2002) and increased DNA methyltransferase (Kautiainen and Jones 1986) activity are common characteristics of tumor cells. Thus aberrant DNA methylation in cancer cells may result in inappropriate under- and overexpression of specific genes, which may, in turn, promote malignant transformation and progression. Importantly, DNA methylation changes are thought to be inherited mitotically in somatic cells, providing a potential mechanism whereby environmental factors, including dietary exposures, can have long-term effects on gene expression (Wolffe 1994).

In addition to the specific sequence of bases within DNA, the way the genetic material is packaged into chromatin can influence gene expression. The fundamental repeating unit of chromatin is the nucleosome. A single nucleosomal core particle is composed of a fragment of DNA (146 bp) wrapped around a histone octomer formed by four histone partners, an H3-H4 tetramer and two H2A-H2B dimers. Each successive core is separated by a DNA linker of approximately 60 bp associated with a single molecule of histone H1. Changes to the structure of chromatin influence gene expression: genes are inactivated (switched off) when the chromatin is closed (heterochromatin), and they are expressed (switched on) when chromatin is open (euchromatin) (Kornberg and Lorch 1999; Rodenhiser and Mann 2006).

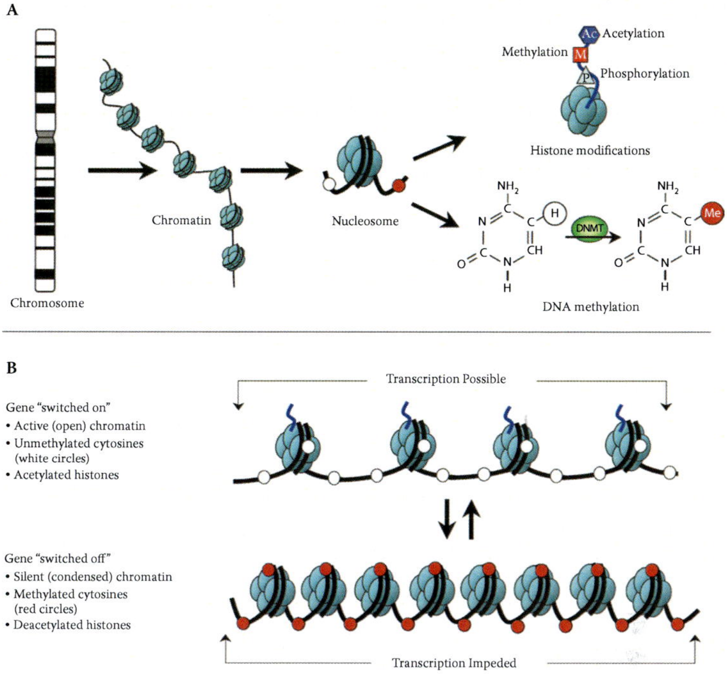

FIGURE 2.1 A. Schematic of epigenetic modifications. Strands of DNA are wrapped around histone octamers, forming nucleosomes. These nucleosomes are organized into chromatin, the building block of a chromosome. Reversible and site-specific histone modifications occur at multiple sites through acetylation, methylation, and phosphorylation. DNA methylation occurs at the 5-position of cytosine residues in a reaction catalyzed by DNA methyltransferases (DNMTs). Together, these modifications provide a unique epigenetic signature that regulates chromatin organization and gene expression. B. Schematic of the reversible changes in chromatin organization that influence gene expression: genes are expressed (switched on) when the chromatin is open (active), and they are inactivated (switched off) when the chromatin is condensed (silent). White circles = unmethylated cytosines; red circles = methylated cytosines. (Adapted from Rodenhiser, D. and Mann, M., *Can Med Assoc J,* 174(4): 341, 2006. With permission.)

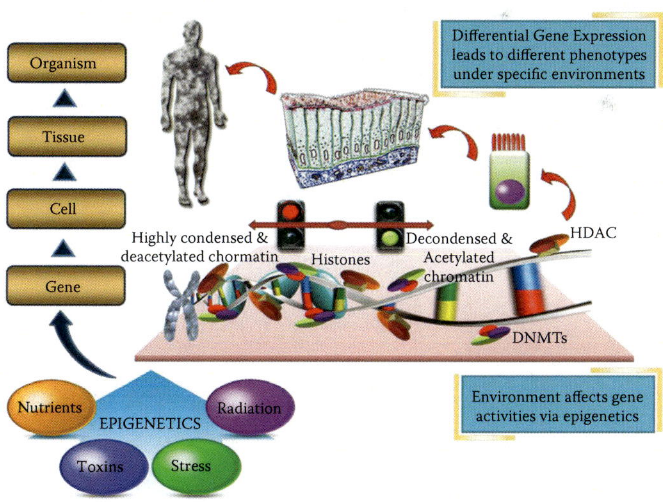

FIGURE 2.2 A hierarchical view of gene-environment interactions during development. As depicted, environmental effects are integrated by epigenetic process including chromatin remodeling to either allow or inhibit gene expressions at the molecular level. Such effects will be manifested at the organismal level via ultimate functional output of the genome. (Adapted from Liu, L. et al., *Curr Issues Mol Biol,* 10(1–2): 25, 2008. With permission.)

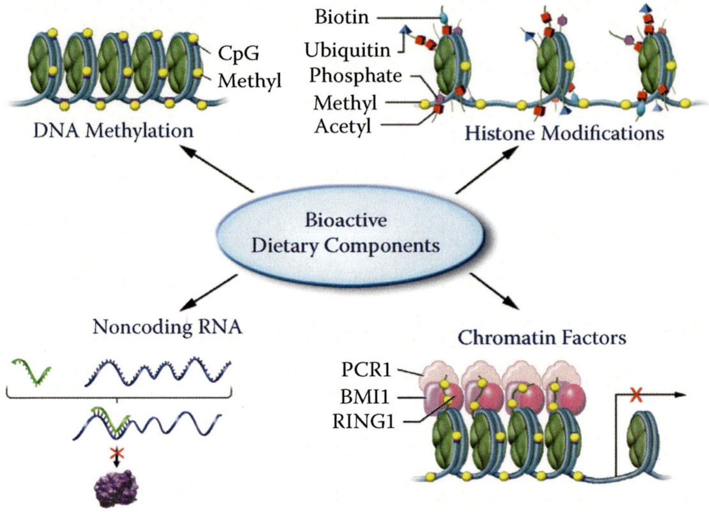

FIGURE 4.1 Evidence suggests that bioactive food components can modify several epigenetic mechanisms, including DNA methylation, histone modifications, noncoding RNA, and chromatin factors such as the polycomb repressive complex 1.

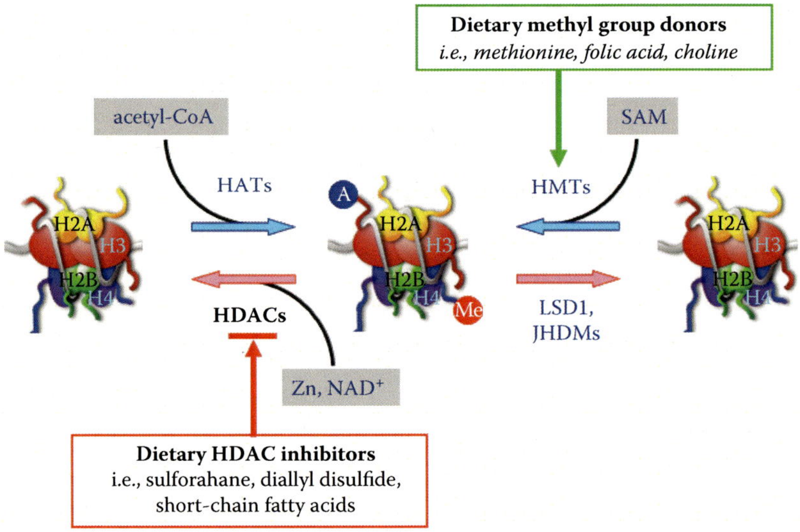

FIGURE 5.1 Schematic model showing a dynamic state of histone lysine acetylation and methylation. The dysbalance between acetylation and deacetylation processes in cancer cells resulted in inappropriate histone lysine deacetylation during tumorigenesis. Dietary HDAC inhibitors restore the normal pattern of histone lysine acetylation and prevent and/or halt tumorigenesis. Likewise, aberrant histone lysine methylation may be corrected by adequate dietary supplementation of methyl group donors and normalization of HMT functioning.

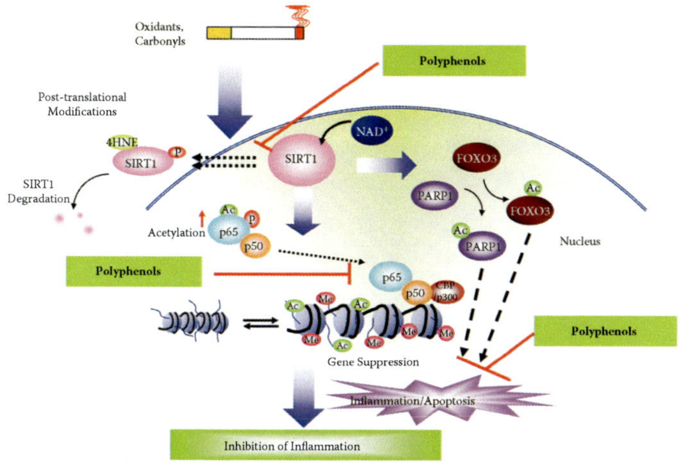

FIGURE 8.2 SIRT1 and regulation of lung inflammation by polyphenols. Cigarette smoke increased abnormal human lung inflammation by degradation of SIRT1, a class III HDAC. SIRT1 is modified by phosphorylation, carbonylation, and aldehyde adducts formation, leading to export of modified SIRT1 in the cytosol and subsequent degradation. SIRT1 suppresses transcription factor NF-κB by deacetylation of RelA/p65 at lys 310 residue. Acetylation of RelA/p65 by cigarette smoke induces the expression of proinflammatory mediators. Decreased SIRT1 leads to increased histone acetylation, which causes increased chromatin modifications and release of proinflammatory mediators. CBP/p300 functions as intrinsic histone acetyltransferases (HATs). Polyphenol, such as resveratrol, can restore the activity of SIRT1, leading to repression of proinflammatory response. SIRT1 regulates FOXO transcription factor acetylation/deacetylation, which is involved in DNA repair, cell cycle, oxidative stress, and apoptosis. Acetylation of FOXO is associated with inflammatory gene induction, stress response genes, and cellular apoptosis. Under the resting condition, inflammatory-related genes are controlled by SIRT1 deacetylase. However, when NF-κB and FOXO3 are acetylated by CBP/p300, transcription of pro-inflammatory and apoptotic/stress genes are up-regulated. Polyphenols inhibit the acetylation of FOXO3 via activation of SIRT1 by increasing NAD^+ or inhibiting PARP1. SIRT1: sirtuin1; P: phosphate; Ac: acetylation; 4HNE: 4-hydroxy-2-nonenal; CBP; CREB (cAMP response element binding protein)-binding protein; PARP: poly (ADP-ribose) polymerase; FOXO: drosophila forkhead transcription factor.

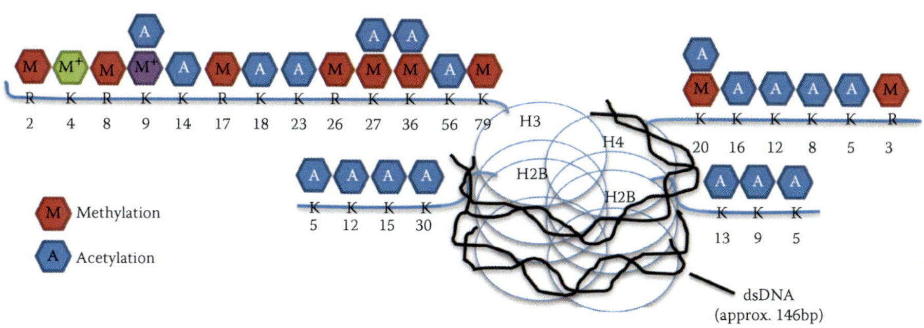

FIGURE 9.3 Nucleosome structure with known histone modifications on specific amino acid residues. A single nucleosome is composed of approximately 146 bp of double-stranded DNA (dsDNA) wrapped around an octamer core of histone proteins from which histone N-terminal tails protrude. Specific amino acids (R: arginine; K: lysine) are subjected to different posttranslational modifications. Here only acetylation (A: blue) and methylation (M: red) marks are included, the most relevant to diabetes and hyperglycemic response being: H3K4 methylation (active mark: green +) and H3K9 methylation (repressive mark: purple –).

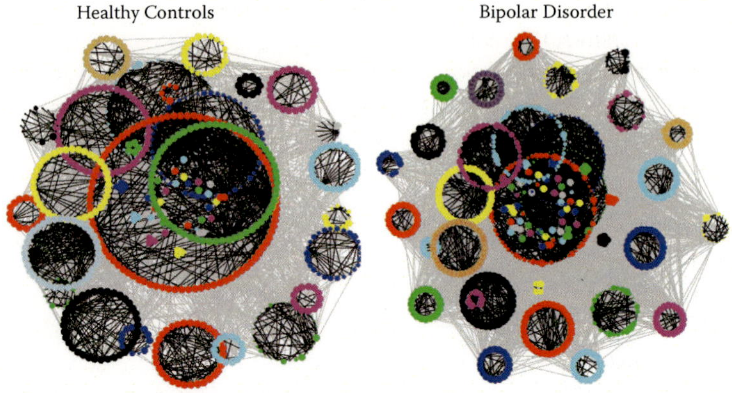

FIGURE 12.2 Decreased germline epigenetic modularity in major psychosis. Partial-correlation network analysis of microarray data illustrating significant connections between nodes, demonstrating strong hierarchical modularity for male healthy controls and BD samples. (Data from Mill, J. et al., *Am J Hum Genet,* 141B, 4, 421–5, 2008). Although modularity is apparent in both sample groups, it is significantly lower in the BD group, resulting in increased between-module interference, reflected by the higher number of interconnections between specific genomic regions in the psychosis group. A similar pattern of modularity was seen in the comparison of methylation between brain SCZ samples and unaffected controls.

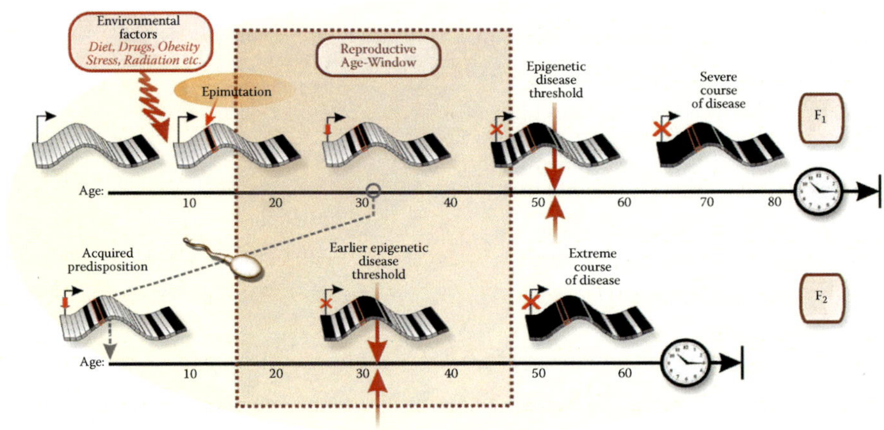

FIGURE 12.4 Anticipation and heritable epigenetic variation. Shown is an example of a gene promoter, which is primarily unmethylated (white boxes) and active in early life, but may become aberrantly methylated (black boxes) during aging and therefore increasingly inactivated. The relatively high frequency of de novo epimutations suggests that epigenetic alterations accumulate during aging. In order to have a deleterious effect, the number of de novo methylation events must cross a certain threshold. Depending on the overall effect of the series of pre- and postnatal impacts on the preepimutation, only some predisposed individuals will reach the "threshold" of epigenetic deregulation that causes the phenotypic changes that meet the diagnostic criteria for a clinical disorder. Some individuals may already be epigenetically predisposed at birth, due to transgenerational epigenetic effects. For example, if epimutations occur in the germline and are not cleared following fertilization, meiotically heritable epigenetic modifications may spread from the inherited epimutation. As a result, the symptoms of the disorder become apparent at an earlier age as it is passed on to the next generation, potentially accompanied with an increase of severity of symptoms or even premature death. Deleterious epigenetic drift occurring after the reproductive phase is relatively neutral to selection, because their bearers have already transmitted their genes (and potentially epigenetic information) to the next generation.

The core histones undergo a wide range of posttranslational modifications, most of which are reversible, that cause structural changes in the chromatin. These modifications, which usually occur at the amino acids that constitute the N-terminal tails of histones, can either facilitate or hinder the association of DNA repair proteins and transcription factors with chromatin and include methylation (Jenuwein 2001), phosphorylation (Oki et al. 2007), acetylation (Wade et al. 1997), sumoylation (Shiio and Eisenman 2003), ubiquitination (Shilatifard 2006), and biotinylation (Kothapalli et al. 2005). These modifications confer functional properties. For example, histone acetylation neutralizes the positive charge on histones and disrupts the electrostatic interactions between DNA and histone proteins; this promotes chromatin unfolding, which has been associated with gene expression (Zhang and Dent 2005). In contrast, deacetylation and condensation generally suppress transcription (Shukla et al. 2008). The specific pattern of histone modifications has been proposed to form a "histone code," which explains, in part, the sections of the genome to be expressed at a given point in time in a given cell type (Jenuwein and Allis 2001). It has been hypothesized that similar to a genetic mutation, a change in the posttranslational modification(s) of histone tails around a regulatory region of a gene can silence an active gene, resulting in "loss of function," or activate a silent gene, leading to "gain of function." Such modifications may also enhance or impair the extent of gene expression in the absence of complete gene silencing or activation. In fact, aberrant histone posttranslational modifications have been associated with cancer. For example, a common hallmark of human tumor cells is the loss of monoacetylation and trimethylation of histone H4 (Fraga, Ballestar, Villar-Garea et al. 2005).

Enzymes that add or remove histone modifications affect a wide range of DNA-based events including transcription, replication, recombination, and repair, as well as chromosome condensation and nuclear organization (Hake et al. 2004). The acetylation state of histone tails is controlled by the antagonistic action of two enzyme families: histone acetyltransferases (HATs), which transfer an acetyl group from acetyl CoA to an epsilon-amino group of lysine residues of histones, and histone deacetylases (HDACs), which catalyze the hydrolysis of these acetamides. HDACs can be divided into three groups based on sequence homology, enzymology, and biological properties. Type I HDACs are nuclear-localized; type II HDACs shuttle between the cytoplasm and the nucleus; and type III HDACs, which are often referred to as the SIR2 family, encompass those HDACs with homology to yeast Sir2. Type III HDACs require nicotinamide adenine dinucleotide (NAD) to function.

Histone modifications can alter gene expression and modify cancer risk (Gayther et al. 2000; Peters et al. 2001; Fischle et al. 2003; P. Zhu et al. 2004; Bannister and Kouzarides 2005; Fraga, Ballestar, Villar-Garea et al. 2005; Gibbons 2005; Marks and Dokmanovic 2005; Atsumi et al. 2006). Abnormal activities of both HATs and HDACs have been linked to the pathogenesis of cancer. Although inactivating mutations of HATs and overexpression of HDACs have been described (Mahlknecht and Hoelzer 2000), the best functional link between HDACs and cancer progression comes from models of acute promyelocytic leukemias, where chromosomal translocations result in chimeric proteins that alter transcriptional events and thereby interfere with normal cell growth, differentiation, and apoptosis (R. J. Lin et al. 1998). Chromosomal translocations between retinoic acid receptor (RAR) and promyelocytic leukemia zinc finger or promyleocytic leukemia protein lead to inappropriate recruitment of co-repressors and HDACs, and abolish the ability of RAR to mediate myelocytic differentiation (Atsumi et al. 2006). It is interesting to note that aberrant targeting of HDACs has been associated with transcriptional silencing of tumor suppressor genes, including *p21,* which encodes a cyclin-dependent kinase inhibitor that blocks cell cycle progression from G_1 into S phase (Gibbons 2005). The expression of p21 has been found to be reduced in many different tumors, allowing uncontrolled cell division. Interestingly, HDAC inhibitors have been shown to reactivate p21 expression, thereby inhibiting tumor cell proliferation (Gibbons 2005). Furthermore, the HDAC inhibitor-induced expression of p21 correlates with an increase in the acetylation of histones associated with the *p21* promoter region. These discoveries have led to the development of HDAC inhibitors as chemotherapeutic agents in clinical trials (Rosato and Grant 2003). Recent studies implicating histone posttranslational modifications with cell identity, including stem cell identity and characteristics such as pluripotency, suggest that HATs and histone acetylation in conjunction with other chromatin modifications may regulate stem cell pluripotency, and deregulation of such histone marking may lead to tumorigenesis (Shukla et al. 2008).

Several proteins or protein complexes have recently been recognized for their ability to regulate chromatin structure and dynamics. For example, ATP-dependent chromatin-remodeling factors define accessibility to the transcription machinery via altering the position of nucleosomes around the transcription start site (Varga-Weisz and Becker 2006). The SWItch/Sucrose NonFermentable (SWI/SNF) complex is a chromatin-remodeling complex that uses the energy produced from ATP hydrolysis to modify chromatin structure and therefore regulate gene expression; recent observations link an aberrant SWI/SNF complex to cancer (Medina and Sanchez-Cespedes 2008). Emerging evidence suggests that both gene activation and gene repressive complexes participate in ATP-dependent chromatin-remodeling activities (Xue et al. 1998). In fact, depending on the context, the SWI/SNF complex can be involved in either transcriptional activation or repression. Another ATP-dependent chromatin-remodeling complex, a DNA helicase/ATPase-containing complex termed nucleosome remodeling and deacetylase corepressor complex (or NuRD), which represses transcription through chromatin remodeling (Xue et al. 1998), has recently been found to direct aberrant gene repression and transmission of epigenetic repressive marks in acute promyelocytic leukemia (Morey et al. 2008).

The polycomb group (PcG) of proteins, which contains at least two distinct complexes, PcG complex 1 and 2, function as transcriptional repressors that silence specific sets of genes through chromatin modification. These proteins may also contribute to the pathogenesis of cancer (Sparmann and Van Lohuizen 2006). PcG complex 2, composed of several factors/proteins including histone methylase activity, is first recruited to silence chromatin with concomitant methylation of histone H3 at lysine 27 (K27me3). This is followed by PcG complex 1 recruitment through recognition of this histone mark, which then triggers ubiquitination of histone H2A and/or inhibits chromatin remodeling to maintain the silenced state of the locus (Takihara 2008). Knockout mice lacking PcG have provided biological evidence that these chromatin repressive complexes are essential for sustaining stem cell activity. Furthermore, enrichment of polycomb repressive complexes has been correlated with cancer progression and prognosis, in addition to cancer stem cell activity. Similar to developmental genes in embryonic stem cells, genes hypermethylated in cancer cells adopt a bivalent pattern of histone marks and low-level expression when treated with DNA methylation inhibitors (McGarvey et al. 2008). Moreover, recent findings showing that PcG target genes in normal cells are more likely to acquire aberrant promoter hypermethylation in cancers (Widschwendter et al. 2007), along with evidence of interactions between DNMT1/DNMT3B and several PcG complex subunits (EZH2 and BMI1) (Vire et al. 2006) are beginning to provide some insight as to how this process is occurring. Additional research is needed to better understand how and in what context these and other epigenetic mechanisms and regulatory factors may interact to regulate chromatin structure, dynamics, and gene expression. Moreover, what is the relationship between these chromatin regulatory factors and cancer development and how can this knowledge be exploited for cancer prevention?

The role of small, noncoding RNA molecules in the regulation of gene expression is an emerging area of research. Noncoding RNAs, such as microRNA (miRNA or MiR), have been shown to modulate posttranscriptional silencing through the targeted degradation of mRNAs. There is considerable scientific interest in studying deregulation of these small RNAs in various diseases, including cancer (Fabbri et al. 2008). Recent evidence suggests that miRNAs may also transcriptionally silence gene expression in the nucleus (D. H. Kim et al. 2008). In this study, investigators performed a bioinformatic search for miRNA target sites proximal to known gene transcription start sites in the human genome. One conserved miRNA, *miR-320,* that was identified is encoded within the promoter region of the cell cycle gene POLR3D in the antisense orientation. Evidence for a cis-regulatory role for *miR-320* in transcriptional silencing of POLR3D expression was provided. Interestingly, using chromatin immunoprecipitation (ChIP) assays, *miR-320* was suggested to direct the association of RNA interference (RNAi) protein Argonaute-1 (AGO1), polycomb group component EZH2, and trimethylation of histone H3 lysine 27 (H3K27me3) to the POLR3D promoter. These results support the existence of an epigenetic mechanism for miRNA-directed transcriptional gene silencing (TGS) in mammalian cells. These investigators hypothesized that misregulation of endogenous miRNAs that target gene promoters may potentially play a role in the aberrant epigenetic silencing of cancer-related genes. Moreover, epigenetic modifications have recently been found to be induced and directed by other small RNA molecules in human cells (Hawkins and Morris 2008). These small RNAs are thought to act like the exogenous small inhibitory RNAs (siRNAs) in gene inactivation. In fact, this endogenous small RNA-mediated transcriptional gene silencing was shown to be correlated with changes in chromatin structure (including modulation of histone marks and DNA methylation) at specific sites in promoter regions (Hawkins and Morris 2008).

Accumulating evidence is revealing an important role of miRNAs in the pathogenesis of all types of human cancer. This is characterized by abnormal levels of expression for mature and/or precursor miRNA transcripts in comparison to the corresponding normal tissues (Bartel 2004). miRNAs have been shown to regulate cell proliferation, differentiation, and apoptosis, all of which are dysregulated during carcinogenesis (Barbarotto et al. 2008). These data suggest that disturbance in miRNA expression and function may contribute to the initiation and maintenance of tumors (Calin and Croce 2006).

miRNA microarrays and real-time quantitative reverse transcription polymerase chain reaction have been used to identify and compare the miRNA expression profiles in normal cells and tissues with those in tumors such as lung, colorectal, breast, thyroid cancers, glioblastomas, and lymphomas (J. Q. Yin et al. 2008). These types of studies demonstrate that there are distinct miRNA expression patterns associated with various tumor types and that cancer samples appear to have miRNA expression profiles that are distinct from normal tissues (reviewed in J. Q. Yin et al. 2008). For instance, the expression of miR-126, miR-143, and miR-145 were significantly decreased in >80% of the tumor samples compared with the corresponding normal tissue, whereas miR-21 was found to be overexpressed in 80% of the tumors (J. Q. Yin et al. 2008). Interestingly, when Lu and coworkers (2005) published a large miRNA expression profile study of >200 miRNAs in primary tumors, cancer cell lines, and normal tissues using a novel bead-based technology, they found that unsupervised hierarchal clustering of the miRNA profiles was able to determine the histologic origin of highly undifferentiated cancer samples with a significantly higher success rate than profiles obtained by measuring 16,000 protein-coding mRNAs (12 out of 17 correct versus 1 out of 17 correct, respectively) (Lu et al. 2005). While these studies suggest that miRNA profiling may be a useful tool for molecular diagnosis and prognosis, additional research is needed to better understand the role of miRs in cancer development and prevention.

Bioactive dietary components have been shown to impact each of the epigenetic mechanisms described previously—DNA methylation, histone posttranslational modifications, chromatin remodeling and other chromatin factors, and noncoding RNA expression. The impact of dietary components on these processes and their role in cancer development and prevention are highlighted in the sections that follow. Ultimately, it will be important to understand how these and other newly identified epigenetic components interact to regulate gene expression and influence disease risk.

4.3 DIET AND DNA METHYLATION: TIMING OF EXPOSURE

Recent evidence from preclinical models suggests that prenatal and early postnatal diet may alter DNA methylation, which may impact the risk of developing disease, such as cancer, later in life (Wolff et al. 1998; Waterland and Jirtle 2003; Cropley et al. 2006; Dolinoy et al. 2006; Waterland, Dolinoy et al. 2006; Waterland, Lin et al. 2006). For example, maternal nutrient restriction had long-term programming effects on fetal baboon development (Unterberger et al. 2009). These effects were organ specific and gestational age specific. Moreover, diet-induced epigenetic alterations might also be inherited transgenerationally, thereby potentially affecting the health of future generations. In this regard, it was recently shown that individuals who were prenatally exposed to famine during the Dutch Hunger Winter in 1944–1945 had, 6 decades later, less DNA methylation of the imprinted insulin-like growth factor II (*IGF2*) and increased methylation of *IL10, LEP, ABCA1, GNASAS,* and *MEGF* compared with their unexposed, same-sex siblings (Heijmans et al. 2008; Tobi et al. 2009). Additionally, a significant interaction with gender was observed for a number of these genes (*INSIGF, LEP, and GNSAS*). These data indicate that persistent changes in DNA methylation may be a common consequence of prenatal famine exposure and that these changes depend on the gender of the exposed individual as well as the gestational timing of the exposure. The results of these studies support the fetal basis or developmental origins of the adult-onset disease hypothesis. This intriguing hypothesis implies that an organism can adapt to environmental signals in early life, but that these adaptations may also increase the risk of developing chronic diseases, including cancer, later in life when there is a disparity between the perceived environment and that which is encountered in adulthood.

There is also evidence to suggest that environmental exposures after birth can influence phenotype later in life through epigenetic changes. Monozygotic twins have been shown to be epigenetically indistinguishable during the early years of life, while older monozygotic twins exhibited remarkable differences in their content and genomic distribution of DNA methylation and histone acetylation, affecting their gene expression portrait (Fraga, Ballestar, Paz et al. 2005). Another investigation found that obesity-discordant monozygotic twins had several changes in the transcription profiles of adipose tissue between the twins (Pietilainen et al. 2008). These results suggest that the effects of acquired human obesity, which is independent of genetic factors, may be related to epigenetic modulation of the genome.

Some of the best evidence for the impact of diet on DNA methylation comes from studies utilizing the yellow agouti (A^{vy}) mouse model. In this model, an endogenous retrovirus-like transposon sequence is inserted close to the gene coding for the agouti protein (Duhl et al. 1994). Normally, a cryptic promoter within the retrotransposon is silenced by methylation allowing normal tissue-specific and regulated agouti expression. However, if this site is hypomethylated, the promoter is active and drives constitutive ectopic expression of the agouti gene. Agouti expression results in a yellow coat color and obesity, as well as increased susceptibility to other chronic diseases, including cancer. Dietary supplementation with a combination of folic acid, vitamin B_{12}, choline, betaine, and zinc to yellow agouti dams in utero has been shown to lead to changes in DNA methylation as well as profound effects on phenotype of the offspring (Wolff et al. 1998; Cooney et al. 2002; Waterland and Jirtle 2003; Cropley et al. 2006; Dolinoy et al. 2006). In initial experiments this supplementation regime to maternal diets was associated with a change in coat color from a yellow to an agouti or pseudo-agouti coat in the offspring (Waterland and Jirtle 2003). This phenotype change is typically associated with a lower risk of cancer, diabetes, and obesity, and with prolonged life in this model (Cooney et al. 2002). Furthermore, representative yellow mice displayed more hypomethylated long terminal repeats 5′ of the *agouti* gene compared to the mice with the agouti coat color (Cooney et al. 2002).

Another group of investigators verified these findings when pregnant dams were fed a diet supplemented with folic acid, vitamin B_{12}, choline, and betaine, but not zinc, and found that coat color changes were directly associated with alterations in DNA methylation and there was a distribution shift toward increased CpG methylation at the A^{vy} locus with methyl supplementation (Cropley et al. 2006). Moreover, the coat color phenotype and A^{vy} methylation relationship persisted into adulthood as evidenced by a comparison of tail DNA at 21 days and liver DNA at 100 days. These studies clearly demonstrate that maternal methyl donor supplementation during gestation can alter the phenotype of the offspring phenotype by methylation changes in the epigenome. It is not yet evident, however, which of the dietary constituents are necessary or sufficient for the DNA methylation and phenotypic change.

Similar alterations in coat color and DNA methylation were induced in offspring through maternal supplementation with genistein, the major phytoestrogen in soy, at amounts comparable to those a human might receive through a high-soy diet (250 mg/kg diet) (Dolinoy et al. 2006). Furthermore, the offspring exposed to genistein in utero and had DNA hypermethylation at the A^{vy} locus appeared to be protected against obesity in adulthood, indicating that maternal dietary supplementation is associated with not only altered fetal methylation patterns but also methylation-dependent susceptibility to disease. The mechanism of how genistein affects methylation and epigenetic pathways has yet to be determined as these investigators did not find an association between genistein supplementation and one-carbon metabolism. Changes in other epigenetic marks or regulation of specific nuclear transcription factors are potential sites of action of genistein that should be explored further.

The agouti mouse has also been utilized as an environmental biosensor to evaluate the effects of maternal exposure to bisphenol A (BPA), a xenobiotic chemical used in the manufacturing of polycarbonate plastic, on the fetal epigenome (Dolinoy et al. 2007). In utero or neonatal exposure to BPA was associated with higher body weight, increased breast and prostate cancer, and altered reproductive function. Maternal exposure to 50 mg of BPA/kg body weight shifted the coat color distribution of A^{vy} mouse offspring toward yellow by decreasing CpG methylation in an intracisternal-A particle (IAP) retrotransposon upstream of the *agouti* gene. Moreover, maternal supplementation with either dietary methyl donors or the phytoestrogen genistein rescued the BPA-exposed animals by shifting the coat color distribution toward the control animals and by negating the DNA hypomethylating effect of BPA. These studies present convincing evidence that maternal exposure to a xenobiotic chemical such as BPA during pregnancy in rodents can alter the phenotype of their offspring by stably altering the epigenome, an effect that can be counteracted by maternal dietary supplements.

The agouti model has also recently been utilized to examine whether diet has transgenerational effects on phenotype. Interestingly, passing the A^{vy} allele through three maternal generations resulted in amplification of obesity in the offspring of the mice fed a standard diet in each successive generation (Waterland et al. 2008). In contrast, by the third generation, offspring of methyl-supplemented animals had a significant decrease in body weight relative to the unsupplemented group. These results suggest a preventive effect on transgenerational amplification of obesity in adulthood. The transgenerational effect, however, was not explained by progressive A^{vy} methylation in the methyl-supplemented group, and the authors speculate that rather than occurring at A^{vy} locus, the transgenerational effect of methyl supplementation on body weight among isogenic A^{vy}/a mice may involve epigenetic mechanisms operating at other loci. This study is extremely interesting because it examines the effect of maternal obesity among three generations of genetically identical mice that had an inclination to overeat due to the A^{vy} allele. Furthermore, the authors hypothesize that the supplements investigated affected body weight by interfering with the area of the brain that regulates appetite.

Studies using another murine metastable epiallele, *axin fused* (*Axin*^{Fu}), found epigenetic plasticity to maternal diet similar to the agouti mouse model (Waterland et al. 2006). *Axin*^{Fu} is a mutation in mice that causes kinked, fused tails and other developmental abnormalities including axial duplications. Female mice supplemented with methyl donors and factors (including choline, betaine, folic acid, vitamin B_{12}, methionine, and zinc) before and during pregnancy were found to have offspring with an increase in DNA methylation at the *Axin*^{Fu} locus and reduced incidence of tail kinking (Waterland et al. 2006). The hypermethylation was tail specific, suggesting a midgestation effect. The investigators speculate that the results indicate a stochastic establishment of an epigenotype at metastable epialleles. The epigenotype can be modified by dietary supplementation, and such influences are likely not limited to early embryonic development.

It is important to clarify how observations in the agouti and *axin fused* models may relate to DNA methylation, phenotype, and disease risk in humans. Although the A^{vy} locus and *Axin*^{Fu} locus—retrovirus-like transposon sequences—are not found in the human genome, there is the possibility that metastable epialleles ("metastable" refers to the labile nature of the epigenetic state of these alleles; "epiallele" defines their potential to maintain epigenetic marks transgenerationally [Rakyan et al. 2002]) associated with other transposable elements could similarly be influenced by methylation or another epigenetic regulating process via in utero exposure to dietary factors. An intriguing hypothesis is that transposable elements in the mammalian genome may cause considerable phenotypic variability, making each individual mammal a "compound epigenetic mosaic" (Whitelaw and Martin 2001). Whether or not such an epigenetic mosaic exists in humans, whether it can be modulated by early diet, whether there is a critical time period for exposure, and how such phenotypes alter susceptibility to chronic disease in adulthood require further study. Additionally, other regions of the genome that may be susceptible to epigenetic variation need to be identified and characterized in human tissues.

Folate is the nutrient that has been most extensively studied in relation to dietary modification of DNA methylation and cancer risk. The portfolio of evidence from animal, human, and in vitro studies suggest that the effects of folate deficiency and supplementation on DNA methylation are gene and site specific, and appear to depend on cell type, target organ, stage of transformation, and the timing and duration of folate depletion/supplementation (Y. I. Kim 2005). Folate deficiency beginning at weaning and continued throughout puberty followed by control diet in adulthood until 30 weeks of age (when animals were sacrificed) induced a significant 34–49% increase in genomic DNA methylation in adult rat liver compared with control and folate supplemented diets provided in the same manner (Kotsopoulos et al. 2008). The authors hypothesized that a compensatory up-regulation of DNMT and of choline and betaine-dependent transmethylation pathways occurred in response to folate deficiency during the postweaning period. This was thought to result in genomic DNA hypermethylation in the liver, and this pattern was maintained in the presence of adequate folate when the animals were fed the control diet at puberty through adulthood. In contrast, dietary folate deficiency or supplementation imposed continuously from weaning to adulthood or from puberty to adulthood did not significantly affect genomic DNA methylation in adult rat liver. Similarly, rats fed a methyl-deficient diet for 9 weeks experienced DNA damage (as evidenced by up-regulation of base excision DNA repair genes and accumulation of strand breaks in DNA), pronounced global loss of DNA methylation, and hypermethylation of CpG islands in the rat livers (Pogribny et al. 2009). These abnormalities were completely restored in the livers of rat exposed to methyl-deficiency for 9 weeks after removal of the methyl-deficient diet and feeding of a methyl-sufficient diet. However, when rats were fed a methyl-deficient diet for 18 weeks and then given a methyl-sufficient diet, only DNA lesions were repaired. These data suggest that early folate nutrition during postnatal development can impact epigenetic programming, which can have a permanent effect in adulthood. More studies on the long-term functional consequences of such dietary epigenetic programming are encouraged.

Folate supplementation may also modify DNA methylation in adult humans (Ingrosso et al. 2003). Although global DNA methylation was reduced in patients with uremia and hyperhomocysteinemia, both DNA hypomethylation and hyperhomocysteinemia were reversed by administration of folate (15 mg oral methyltetrahydrofolate/day for 8 weeks). Furthermore, the DNA hypomethylation was linked to defects in the expression of genes controlled by methylation. To study this gene regulation, the pattern of allelic expression for the normally imprinted *H19* gene in peripheral mononuclear cells in seven of the dialysis patients that were heterozygous for *H19* RsaI restriction fragment length polymorphisms (RFLP) was examined. The RFLP analysis showed a shift from monoallelic to biallelic expression of *H19* (expression of both *H19* T and *H19* C alleles in the heterozygotes could be identified) when plasma total homocysteine concentration were between 39 μmol/L and 62 μmol/L. In contrast, in the three patients with high total homocysteine concentrations (<62 μmol/L) reverse transcriptase-PCR analysis showed a shift back to monoallelic expression of *H19* after folate treatment. These data suggest that treatment of hyperhomocysteinemia with folate corrects DNA hypomethylation and provides a mechanism for the relevant changes in gene expression. Further work is needed to determine other genes and diseases that might be regulated in a similar fashion.

In addition to silencing inappropriately activated genes (e.g., imprinted genes) by DNA methylation, dietary components have also been shown to reactivate inappropriately silenced genes (e.g., tumor suppressor genes) by preventing demethylation or reversing hypermethylation-induced inactivation of key tumor suppression or DNA repair genes. For example, many different dietary components have been shown to modulate DNA methylation by interfering with DNMT activity; this has been shown to be an effective approach for cancer prevention. Epigallocatechin 3-gallate (EGCG) (5–50 microM) from green tea and genistein (2–20 μmol/L) from soybean have been found to restore methylation patterns and gene expression of tumor suppressor genes in neoplastic cells in culture (M. Z. Fang et al. 2003, 2005; M. Fang et al. 2007). Both dietary components have been shown to inhibit DNA methyltransferase activity by binding to the enzyme that was associated with demethylation of CpG islands in the gene promoters and the reactivation of methylation-silenced genes such as *p16INK4a, retinoic acid receptor beta, O6-methylguanine methyltransferase, human mutL homolog 1,* and *glutathione S-transferase-pi* (M. Z. Fang et al. 2003, 2005; M. Fang et al. 2007). Changes in the expression of these genes have been associated with growth inhibition in human esophageal, colon, prostate, and mammary cancer cell lines. Rats fed selenium- or folate-deficient diets had significantly reduced liver and colon DNMT activity; however, the mechanism(s) for their inhibitory effect is not known (Davis and Uthus 2003). Cadmium is another active inhibitor of DNMT as has been shown in both cultured rat liver cells and animals (Takiguchi et al. 2003) In addition, both cadmium and zinc inhibited DNMT activity in the nuclear extracts from rats fed either a methyl-deficient or a control diet (Poirier and Vlasova 2002). The inhibitory effect of cadmium and zinc may be caused by binding of these metals to the cysteine residue of the active center of DNMT.

The Annurca apple, a variety of southern Italy, is rich in cancer-protective polyphenols. For example, populations in southern Italy who consume these apples have lower incidences of colorectal cancer than elsewhere in the Western world. Annurca polyphenol extract (APE) was evaluated for cancer-protective effects in RKO, SW48, and SW480 human colorectal cancer cells (Fini et al. 2007). Because some sporadic colorectal cancers display the CpG island methylator (CIMP) phenotype, DNA methylation of selected tumor suppressor genes was also evaluated in these cells after treatment with APE and compared with the synthetic demethylating agent 5-aza-2′deoxycytidine (5-aza-2dC). APE treatment (polyphenol dose comparable to that from dietary consumption of one apple) decreased cell viability and enhanced apoptosis in the RKO and SW48 cell lines, both in vitro models for CIMP. A similar dose of APE reduced DNA methylation in the promoters of *hMLH1, p14(ARF)*, and *p16(INK4a)* genes with subsequent restoration of normal mRNA expression in RKO cells. These effects were qualitatively comparable with those obtained with 5-aza-2dC. In addition, a significant reduction in expression of DNMT-1 and -3b proteins without changes in mRNA was also observed after treatment with APE. Thus APE appears to decrease gene-specific DNA methylation through the inhibition of DNMT proteins.

The putative mechanism of action for selenium in cancer prevention has been an active area of research since the observation that selenium supplementation decreased prostate cancer incidence in the Nutritional Prevention of Cancer trial (Clark et al. 1996). Recently, selenium has been shown to induce promoter DNA demethylation and gene reexpression in LNCaP prostate cancer cells, suggesting that epigenetic modifications may be a possible mechanism for cancer prevention (Xiang et al. 2008). In particular, selenite treatment (at physiologically achievable levels, 1.5 μmol/L) caused partial promoter DNA demethylation and reexpression of the *pi-class glutathione-S-transferase* (*GSTP1*) in a dose- and time-dependent manner. Selenite treatment decreased mRNA levels of DNMTs 1 and 3A and protein levels of DNMT1. Additionally, selenite treatment caused partial promoter demethylation and reexpression of the tumor suppressor adenomatous polyposis coli and cellular stress response 1. Both of these genes have been shown to be hypermethylated in human prostate cancer cells. Interestingly, selenite also decreased histone deacetylase activity and increased levels of acetylated lysine 9 on histone H3 (H3K9), but decreased levels of methylated H3K9. More specifically, selenite treatment influenced epigenetic marks associated with the *GSTP1* promoter, including reduced levels of DNMT1 and methylated H3K9, but increased levels of acetylated H3K9. These epigenetic marks are correlated with gene activation. The relationship between particular histone modifications and DNA methylation in gene reactivation may be gene specific and requires additional study, preferably in an in vivo setting.

Although the interactions between bioactive food components and DNA methylation are among the earliest studies of the relationship between diet and epigenetics in cancer prevention, there continues to be a growing body of literature encompassing more and more dietary factors that may impact DNA methylation at various times of vulnerability, including during cancer development and prevention. How these single observations will be united to reflect the complexity of human dietary patterns as well as how the many epigenetic mechanisms and marks will be integrated to provide an understanding of specific and global gene regulation are the hopeful outcomes of future endeavors.

4.4 HISTONE MODIFICATION BY BIOACTIVE FOOD COMPONENTS AND DIET COMPOSITION

Several dietary factors, including butyrate (formed in the colon from the fermentation of dietary fiber), diallyl disulfide (present in garlic and other *Allium* vegetables), and isothiocyanates (found in cruciferous vegetables) have been found to inhibit HDAC enzymes (Myzak and Dashwood 2006) and to enhance histone acetylation (Myzak and Dashwood 2006), and have been associated with cancer prevention in various preclinical and clinical studies. Moreover, these dietary components have all been shown to inhibit cell proliferation and stimulate apoptosis in a manner analogous to other nondietary HDAC inhibitors.

Butyrate has been reported to inhibit HDAC activity and increase histone acetylation in a number of cell lines (Myzak and Dashwood 2006). These butyrate-induced alterations in histone enzymes and marks have been associated with several processes that are important for inhibiting carcinogenesis, including cellular differentiation, cell cycle arrest, apoptosis, and inhibition of invasion and metastasis. Butyrate has also been found to alter transcriptional regulation in a manner comparable to other HDAC inhibitors, such as trichostatin A (Mariadason et al. 2000). Similar to trichostatin A, butyrate caused down-regulation of c-myc mRNA expression, which likely accounts for its effect on reducing cell proliferation (Bernhard et al. 1999). Furthermore, both trichostatin A and butyrate increase histone H3 and H4 acetylation within the *CDKN1A* promoter (using ChIP experiments), which regulates the p21 protein, in colorectal cancer cells in culture (J. Y. Fang et al. 2004). Because only 2% of genes are thought to be regulated by butyrate, specificity of HDAC inhibition and consequent transcriptional regulation may be due to sequences such as butyrate response elements in the promoter region of certain genes (Davie 2003). In addition to increasing global histone acetylation, butyrate may induce the deacetylation of certain histones that are close to transcription start sites, which is accompanied by decreased RNA levels at nearby genes (Rada-Iglesias et al. 2007). The epigenetic influence of butyrate now extends beyond chromatin remodeling to include DNA methylation (Spurling et al. 2008). (Epi)genome-wide analysis (of both genetic sequence and epigenetic marks) could assist in uncovering this specificity.

The active garlic constituent diallyl disulfide (DADS) has been shown to induce cell cycle arrest in the G_2/M phase in both HT-29 and Caco-2 human colon cancer cell lines. These antiproliferative effects were correlated with increased CDKN1A mRNA and p21 protein levels, as well as increased H4 and/or H3 acetylation within the *CDKN1A* promoter (Druesne et al. 2004; Druesne-Pecollo et al. 2006). The findings suggest that histone hyperacetylation of the promoter region may account for the cell cycle arrest induced by DADS. Histone acetylation changes have been observed in rat liver and transplanted Morris hepatoma 7777 cells with DADS treatment (200 mg/kg body weight), demonstrating that altered acetylated histone status is achievable in vivo (Lea and Randolph 2001). Furthermore, using a nontumorigenic animal model, investigators found that DADS (200 mg/kg) treatment increased histone H4 and H3 acetylation in isolated colonocytes from rats (Druesne-Pecollo et al. 2007). Moreover, these experiments revealed the ability of DADS to modulate the expression of a few genes (e.g., glutathione S-transferase B class, mitogen-activated protein kinase 3, inhibitor of DNA binding 1) in colonocytes in vivo. More research that compares the effects of garlic constituents on normal and cancer cells is advocated in order to decipher the biological effects. Physiologic concentrations (low micromolar range) of *S*-allyl-mercaptocysteine (SAC), another organosulfur compound found in garlic, have also been reported to induce growth arrest in various cell lines and increase the levels of acetylated histones H3 and H4 (Lea et al. 2002). When several garlic organosulfur compounds were screened, allyl mercaptan (AM) was found to be much more potent than DADS or SAC (Nian et al. 2008). Molecular modeling, structure activity, and enzyme kinetics studies with purified HDAC8 provided evidence for a competitive mechanism (K_1 = 24 μM AM). In human colon cancer cells, AM induced the accumulation of acetylated histones and enhanced the binding of Sp3 and p53 transcription factors to the *p21WAF1* gene promoter. There was a corresponding increase in p21 mRNA and protein expression, resulting in cell cycle arrest and growth inhibition. These data suggest that multiple organosulfur compounds present in garlic serve as HDAC inhibitors. Humans normally consume foods rather than the isolated components. Therefore future work is needed to determine whether the various compounds in garlic would have additive, synergistic or antagonistic effects.

Sulforaphane (SFN) (3-15 μM) has been shown to inhibit HDAC activity and in parallel increase acetylated histones in several cell systems, including human embryonic kidney 293 cells, HCT116 human colorectal cancer cells, and various prostate epithelial cells lines (BPH-1, LnCaP, and PC-3) (Myzak and Dashwood 2006). This increased histone acetylation induced by sulforaphane was linked with increased apoptosis and cell cycle arrest at the G_2/M phase (Myzak and Dashwood 2006). Moreover, these observations were associated with increased acetylated histone H4 in the *p21* promoter and concomitant increased p21 protein expression (Myzak et al. 2004). Furthermore, sulforaphane dose-dependently increased the amount of acetylated histone H4 associated with the *p21* promoter (Myzak Hardin et al. 2006). Using the *Apc*^{min} mouse model, sulforaphane (443 mg/kg diet) suppressed tumor development and increased acetylated histones in gastrointestinal polyps, including acetylated histones specifically associated with the promoter region of the *p21* and *bax* genes (Myzak, Dashwood et al. 2006). Most remarkable is the finding that in healthy human subjects ($N = 3$), a single ingestion of 68 g (1 cup) of broccoli sprouts rich in SFN inhibited HDAC activity in circulating peripheral blood mononuclear cells 3–6 h after consumption, with simultaneous induction of histone H3 and H4 acetylation (Myzak et al. 2007). The biological consequences of reduced HDAC activity and enhanced histone acetylation in normal compared to cancer cells requires further study. Although SFN has been shown to selectively induce apoptosis and growth inhibition in cancer cells but not in normal cells, additional research is warranted to determine beneficial versus harmful responses to bioactive dietary components during vulnerable periods.

Another isothiocyanate present in cruciferous vegetables, namely phenethyl isothiocyanate (PEITC, 1 μM), was also found to inhibit the level and activity of HDACs in both androgen-dependent and -independent prostate cancer cells, and induce selective histone acetylation and methylation for chromatin unfolding (L. G. Wang et al. 2007). Chromatin immunoprecipitation revealed that the *p21* gene was associated with PEITC-induced hyperactylated histones (L. G. Wang et al. 2008). As a result, the chromatin unfolding permitted the transcriptional activation of the *p21* gene. PEITC also significantly reduced the expression of c-Myc, which represses *p21*. These results demonstrate that isothiocyanates can inhibit HDAC activity in vitro and in vivo and suggest that this inhibition might contribute to the cancer-preventive effects of cruciferous vegetables.

Other components of vegetables may also exert cancer prevention via HDAC inhibition. *Momordica charantia,* known as bitter melon, is a plant that grows in tropical areas worldwide and is eaten as a vegetable. A protein present in bitter melon seeds (MCP30) induces apoptosis in prostatic intraepithelial neoplasia and prostate cancer cell lines but has no effect on normal prostate cells (Xiong et al. 2009). Mechanistically, MCP30 inhibits HDAC activity and promotes acetylation of histones H3 and H4.

As mentioned previously, the mechanism of action of selenium in cancer prevention has been an active area of research. Of particular interest is why selenium-enriched yeast was protective against prostate cancer in the Nutritional Prevention of Cancer trial yet selenomethionine was not protective in the SELECT trial (Lippman et al. 2009). Very recent observations suggest that this may be related to selenium metabolites influencing HDAC expression (J. I. Lee et al. 2009; Nian et al. 2009). While the majority of selenium in selenium-enriched yeast is selenomethionine, other selenium metabolites, such as Se-methyl-selenocysteine (MSC), are also present. Colon cancer cells treated with MSC but not selenomethionine had increased acetylation of histone H3 and dose-dependent inhibition of HDAC activity. Many mammalian cells, such as prostate and colon, contain the enzyme glutamine transaminase K, which can metabolize MSC but not selenomethionine to its α-keto acid metabolite. The α-keto acid metabolite of MSC is β-methylselenopyruvate (MSP). MSP is a competitive inhibitor of HDAC8, and computational modeling supported a mechanism involving reversible interaction with the zinc atom at the active site. Moreover, MSP increased the activity of a *P21WAF1* luciferase reporter in HT29 cells, except when the Sp1/Sp3 sites were eliminated, and p21 mRNA and protein levels were markedly elevated. These results suggest that MSC, and possibly other organoselenium compounds, can generate α-keto acid metabolites, which are HDAC inhibitors with the potential to modulate histone status and chromatin remodeling, leading to derepression of silenced tumor suppressor genes.

A recent report revealed that genistein (at both 10 and 25 μmol/L) induced the expression of the tumor suppressor genes *p21* and *p16* (INK4a) with a concomitant decrease in cyclins in prostate cancer cells (Majid et al. 2008). These investigators found that genistein increased acetylation of histones H3, H4, and H3 lysine 4 (H3K4) at the *p21* and *p16* transcription start sites with concomitant increased expression of histone acetyltransferases. Interestingly, DNA methylation analysis revealed the absence of *p21* promoter methylation prior to genistein exposure. Furthermore, these same investigators found that genistein (50 μM) activated expression of several aberrantly silenced tumor suppressor genes that have unmethylated promoters such as PTEN, CYLD, p53, and FOXO3a in prostate cancer cells (Kikuno et al. 2008). Instead of turning on tumor suppressor genes through promoter demethylation, these investigators found that genistein influenced remodeling of the heterochromatic domains at promoters by reducing/modulating histone H3-Lysine 9 (H3K9) methylation and deacetylation. These findings suggest that genistein may be protective against cancers with various epigenetic profiles. Furthermore, the relationship between genistein, histone, and DNA methylation modifications in gene reactivation, which may be gene specific, is not entirely clear and requires further investigation.

The story is further complicated for soy consumption because of the activity of other constituents found in soy that may modify chromatin in a somewhat opposing fashion. For example, lunasin, a unique 43-amino acid soybean peptide that has cancer prevention properties, has been found to inhibit acetylation of core histones in mammalian cells and selectively kill cells that are in the process of transformation (e.g., E1A-transfected mouse fibroblast NIH 3T3 cells), but does not affect the growth rate of normal and established cancer cell lines at 10 μM concentrations (Lam et al. 2003). An epigenetic mechanism of action has been proposed whereby lunasin selectively kills cells being transformed or newly transformed cells by binding to deacetylated core histones exposed by the transformation event, thereby disrupting the dynamics of histone acetylation-deacetylation. These results point to the importance of understanding the timing of cellular vulnerability to epigenetic modulation. Recently, investigators observed the histone H3- and H4-acetylation inhibitory properties of lunasin from different Korean soybean varieties used for various food purposes (Jeong et al. 2007). They found that various amounts of lunasin are found in the soybean varieties (4.40–70.49 ng of lunasin per μg of protein), and that the amount was correlated with the extent of inhibition of core histone acetylation. Furthermore, the blood from rats fed lunasin-enriched soy protein, but not the blood from control fed rats, was found to inhibit histone acetylation activity. Delineating the epigenetic activity of various soy constituents and products in different cellular contexts is another potential area for future research.

Recently, in primates maternal diet was shown to alter histone acetylation and gene expression profiles in the developing offspring (Aagaard-Tillery et al. 2008). In this study, chronic consumption of a maternal high-fat diet (35% fat versus 13% fat for control animals) resulted in a threefold increase in fetal liver triglycerides and histologic correlates of fatty liver disease, which was accompanied by hyperacetylation of fetal hepatic tissue at histone 3 lysine 14 (H3K14) and decreased HDAC1 mRNA, protein, and activity. Gene expression changes were also observed, including increased glutamic pyruvate transaminase (alanine aminotransferase) 2 (*GPT2*), *DNAJA2* (a heat shock protein 70 co-chaperone), and *Rdh12* (an all trans and 9-cis retinol dehydrogenase responsive to oxidative stress) in fetal hepatic tissue from maternal caloric-dense diet animals when compared with control. Furthermore, the gene *Npas2,* a peripheral circadian regulator, was significantly down-modulated in the offspring of animals fed the high-fat diet. Definitive conclusions regarding the role of H3K14 acetylation with respect to the observed altered gene expression requires additional study. These results, however, suggest that a caloric-dense maternal diet leading to obesity epigenetically alters fetal chromatin structure in primates via covalent modifications of histones and hence offers a molecular basis to the fetal origins of adult onset of disease hypothesis.

Caloric restriction has been shown to decrease cancer susceptibility in many different animal models. These effects may be mediated by increasing the activity of type III HDACs. In yeast and worms, caloric restriction increases longevity by modulating the expression of the NAD-dependent HDAC SIR2 (Vaquero et al. 2006). Furthermore, the glucose analog 2-deoxyglucose, which blocks glucose metabolism and results in decreased energy metabolism, decreased carcinogen-induced rat mammary tumor incidence and multiplicity and prolonged tumor latency (Z. Zhu et al. 2005). The expression of Sirt-1, a mammalian homolog of SIR2, was induced in a dose- and time-dependent manner by treatment with 2-deoxyglucose in these animals, suggesting that the sirtuin family of genes may be involved in the cancer-protective effects of caloric restriction in mammals (Z. Zhu et al. 2005). Caloric restriction has also been shown to increase carbonylation of histones (Sharma et al. 2006). However, the functional significance of this observation is not known. A number of plant polyphenols have been shown to activate Sirt1 including quercetin and resveratrol (Allard et al. 2009). Resveratrol, a polyphenolic compound found in grapes, wine, and peanuts, has been shown to increase SIR2 as well as to mimic the caloric restriction pathways for longevity in both *Caenorhabditis elegans* and *Drosophila melanogaster* models (Wood et al. 2004). A recent report indicates that resveratrol treatment produces beneficial effects similar to the effects of caloric restriction in mice (Lagouge et al. 2006). These data suggest that dietary components can either stimulate or inhibit various HDAC enzymes.

Histone acetyltransferases (HAT) are another important molecular target for dietary components. Both curcumin (20 μM) and copper (50 μM) have been shown to induce histone hypoacetylation by inhibiting HAT activity in vitro (Kang et al. 2005; C. Lin et al. 2005). Exposure of human leukemia (HL60) cells to copper resulted in cell proliferation arrest and a concentration- and time-dependent decrease of histone acetylation (C. Lin et al. 2005). The histone acetylation was suppressed by exogenous hydrogen peroxide and enhanced by superoxide dismutase, catalase, or the combination of both, indicating that copper at least partially inhibits histone acetylation through triggering of oxidative stress (C. Lin et al. 2005). Similar results for curcumin were observed in human hepatoma Hep3B cells (Kang et al. 2005). Future studies are warranted to determine whether similar effects would be observed in vivo.

The majority of the evidence for dietary-induced histone posttranslational modifications concerns the effects of dietary HDAC inhibitors and the effects of diet on histone acetylation, much of which has been described in the preceding sections. Additional histone posttranslational modifications and their enzymatic partners have been shown to be influenced by dietary factors that have also been implicated in cancer prevention pathways. These include interactions between a methyl-deficient diet and histone methylation in early hepatocellular carcinogenesis (Pogribny, Ross, Wise et al. 2006) and the nutrient biotin and histone biotinylation in repression of transposable elements in cancer cells (Chew et al. 2008). Biotinylation of histone 4 (at lysine 8 and 12) has been associated with heterochromatin structures, gene silencing, mitotic condensation of chromatin, and DNA repair. Like many of the histone modifications, biotinylation appears to be a reversible process, although debiotinylases have not been characterized. Histone biotinylation depends on the exogenous biotin supply, and it has been hypothesized that some effects of biotin deficiency can be attributed to abnormal chromatin structures (Hassan and Zempleni 2006). These reports provide evidence for the impact of dietary factors on histone modification and in determining chromatin structures, including whether the chromatin is in the open (euchromatin, active) or closed (heterochromatin, inactive) state. Research on the identification and characterization of dietary triggers of histone modifications and associated affects such as gene silencing or activation is emerging. Some of these efforts will likely examine specificity of bioactive food factors for particular histone-modifying enzymes and perhaps will utilize epigenomic approaches to map the numerous histone posttranslational marks in normal and cancer cells following dietary exposure(s).

4.5 DIETARY MODULATION OF POLYCOMB REPRESSIVE COMPLEXES

Histone modifications triggered by polycomb repressive complex signaling are thought to be important during embryonic stem (ES) cell differentiation. For example, PcG complex 2 binding mediates trimethylation of lysine 27 (K27me3) on histone H3, but this histone mark is lost on developmental genes that are transcriptionally induced during ES cell differentiation. The active vitamin A constituent retinoic acid (RA) is involved in differentiation of ES cells as well as differentiation of various cancer cells in culture. Interestingly, a decrease in the H3K27me3 mark was recently observed in as little as 3 days after differentiation of mouse ES cells induced by RA (1 μM) treatment (E. R. Lee et al. 2007). The enzyme histone K27 methyltransferase EZH2, which mediates the K27me3 mark, also decreased with RA treatment. A loss of EZH2 binding and H3K27me3 was observed locally on PcG complex 2 target genes induced after 3 days of RA, including the gene *nestin.* In contrast, direct RA-responsive genes that are rapidly induced, such as *Hoxa1,* showed a loss of EZH2 binding and K27me3 within only a few hours of RA treatment. These observations suggest that there are likely temporal stages of derepression of polycomb complex target genes during early differentiation and also emphasize the complexity of the histone code in regulating gene transcription as increased histone acetylation was found to override this H3K27me3 repressive mark to induce gene transcription in some genes.

After the PcG complex 2 binds and increases H3K27me3 in a specific gene region, the second polycomb repressive complex 1 (PcG complex 1), which contains the protein Bmi-1, binds to the K27me3 in histone H3, and catalyzes the ubiquitinylation of histone H2A. This cooperation between the two PcG complexes is what leads to silencing of gene expression. PcG complex 1, including Bmi-1, appears to remain attached to the chromatin after these events are completed. Bmi-1 is overexpressed in some human cancers, including colorectal cancer (J. H. Kim et al. 2004) and human non-small-cell lung cancer (Vonlanthen et al. 2001), as well as markedly elevated in epidermal squamous cell carcinoma cells (K. Lee et al. 2008).

The polyphenol EGCG reduces skin cancer cell survival. The impact of EGCG on the PcG complex 1 chromatin factor in cultured squamous cell carcinoma cells was recently examined to determine the involvement of Bmi-1 in the activity of EGCG (Balasubramanian et al. 2008). EGCG (40 μM) was found to suppress Bmi-1 levels and reduce Bmi-1 phosphorylation, resulting in displacement of the Bmi-1 polycomb protein complex from chromatin and reducing survival of transformed cells. These observations provide additional evidence for the role of dietary components in reducing cancer cell survival by altering epigenetic control of gene expression. The importance of the polycomb repressive complexes in the development of cancer is currently an active research enterprise. Future research is needed to clarify the role of dietary regulation of PcGs (primarily focusing on cancer stem cells and perhaps adult stem cells) during the carcinogenic process.

4.6 SMALL, NONCODING RNA, EPIGENETICS, AND DIETARY FACTORS

The methyl-deficient model of spontaneous hepatocarcinogenesis (HCC) in rodents is unique in that dietary inadequacy rather than exposure to chemical carcinogens or viral agents can cause tumor formation (Pogribny et al. 2007). Specifically, deficiency of the major dietary sources of methyl groups and cofactors (methionine, choline, folic acid, and vitamin B_{12}) is sufficient to induce liver tumor formation in male rats and certain mouse strains (Newberne 1986; Wainfan and Poirier 1992; Christman et al. 1993; Poirier 1994; Denda et al. 2002). The methyl deficiency induced in these animals has been associated with several defects, including increased genomewide and gene-specific hypomethylation (Wainfan and Poirier 1992; Christman et al. 1993; Pogribny et al. 2004). Recent studies examining the early stages of hepatocarcinogenesis induced by methyl deficiency in rats found significant alterations in other aspects of the epigenetic machinery, including aberrant expression of DNA methyltransferases and methyl CpG binding proteins (Ghoshal et al. 2006), defects in histone methyltransferase protein expression and histone posttranslational modifications (Pogribny, Ross, Wise et al. 2006), and changes in the expression of microRNAs (miRNA) (Kutay et al. 2006; Tryndyak et al. 2009). The aberrant epigenetic alterations imposed by this diet have been hypothesized to be the primary mechanism responsible for malignant transformation of rat liver cells (Pogribny, Ross, Tryndyak et al. 2006; Wainfan and Poirier 1992; Christman et al. 1993; Ghoshal et al. 2006; Pogribny, Ross, Wise et al. 2006), but which of the epigenetic defects is responsible initially for transformation has not been determined. In this regard, development of methyl-deficient induced HCC was shown to be characterized by prominent early changes in the expression of miRNA genes that are involved in the regulation of apoptosis, cell proliferation, cell-to-cell connection, and epithelial-mesenchymal transition (Tryndyak et al. 2009). These changes include inhibition of the expression of *miR-34a, miR-127, miR-200b,* and *miR-16a* with corresponding changes in the levels of E2F3, NOTCH1, BCL6, ZFHX1B, and BCL2, proteins that are targeted by these miRNAs. The significance of the disruption of miRNAs expression in HCC was confirmed by the persistence of these miRNA alterations in the livers of methyl-deficient rats re-fed a methyl-adequate diet. These investigators hypothesized that the early occurrence of alterations in miRNA expression and their persistence during the entire process of hepatocarcinogenesis indicated that the dysregulation of microRNA expression is likely to be an important contributing factor in the development of HCC. Whether the inhibition of expression of these specific miRNAs in this HCC model is the earliest trigger(s) toward fixing the neoplastic state requires further study. However, it is interesting to note that the sequence of pathological and molecular events in the methyl-deficient model of HCC is remarkably similar to the development of HCC associated with viral hepatitis B and C infections, alcohol exposure, and metabolic liver diseases (Powell et al. 2005).

The incidence of nonalcoholic fatty liver disease is increasing dramatically, which can lead to an increase in the prevalence of nonalcoholic steatohepatitis (NASH) and associated complications such as HCC (Torres and Harrison 2008). Mice fed a choline-deficient diet had up-regulation of miR-155, miR-221/222, and miR-21 and down-regulation of the most abundant liver-specific miRNA, miR-122 (B. Wang et al. 2009). Western blot analysis showed reduced expression of hepatic phosphatase and tensin homolog (PTEN) and CCAAT/enhancer binding protein beta (C/EBPbeta), respective targets of miR-21 and miR-155 at early stages of HCC, long before preneoplastic transformation, implicating their role in the initiation of tumorigenesis. Similarly, long-term exposure to a high-fat diet in genetically susceptible (C57BL/6J) mice leads to NASH and a high rate of spontaneous HCC (Hill-Baskin et al. 2009). miRNA profiles revealed increased expression of miR-31, miR-146, and miR-182 and decreased expression of miR-191. A switch from a high-fat to a low-fat diet reversed these changes. While these studies demonstrate that dietary factors can modulate miRNA expression and cancer risk, it is not clear why different miRNA targets were observed between the two studies. Clearly additional research is needed.

Curcumin, derived from the rhizome of *Curcuma longa,* is a naturally occurring flavanoid that stimulates apoptosis and has recently been shown to alter miRNA expression profiles in human BxPC-3 pancreatic cancer cells (Sun et al. 2008). In this study, 11 miRNAs were significantly up-regulated and 18 miRNAs were significantly down-regulated after 72 h of treatment with 10 μmol/L curcumin in these cells. For example, curcumin up-regulated miRNA-22 and down-regulated miRNA-199a*, and these findings were confirmed by RT-PCR analysis. Furthermore, the expression of two computationally predicted targets for miRNA-22 (because miRNA-22 function is unknown), SP1 transcription factor (SP1) and estrogen receptor 1 (ESR1), were investigated in the pancreatic cancer cells. Up-regulation of miRNA-22 expression by either treatment with 10 μmol/L curcumin or transfections with synthetic miRNA-22 mimics, reduced the expression of its target genes SP1 and ESR1, while experiments using miRNA-22 antisense-enhanced SP1 and ESR1 expression. These findings suggest that alterations of miRNA expression by curcumin may be an important mediator of its cancer-protective effects in pancreatic cancer cells.

One of the most important phytochemicals in cruciferous vegetables is indole-3-carbinol (I3C). Considerable evidence has shown that I3C inhibits experimentally induced carcinogenesis in a number of rodent models at different sites, including lung, through induction of phase I and phase II enzymes, inhibition of proliferation and induction of apoptosis in tumor cells, and modulation of estrogen metabolism (Bradlow 2008). Mice treated with the chemical carcinogen vinyl-carbamate and given I3C in the diet have decreased expression of miR-21, miR-130a, miR-146b, and miR-377 compared to mice treated with the carcinogen only (Melkamu et al. 2010). All of these miRNAs have been shown to be aberrantly expressed in human tumors. Moreover, miR-21 is the most frequently up-regulated oncomir in solid tumors (Guil and Esteller 2009). Increased levels of miR-21 have been reported in cancer of the lung, glial cells, breast, liver, and pancreas. Increased expression of miR-21 leads to increased cell proliferation, reduced apoptosis, and enhanced tumor growth and invasion by down-regulating the expression of tumor genes such as PTEN, p53, and TGF-beta (Melkamu et al. 2010). Whether or not I3C would inhibit miR-21 expression in other types of tumors is not currently known. However, these data suggest that miRNAs and their target genes are promising biomarkers for the cancer-protective effects of dietary components.

The preceding evidence suggests that nutrient deficiency as well as bioactive food component supplementation may modify the expression of miRNAs and that this modulation has consequences for carcinogenesis. Whether the affected miRNA are acting in an epigenetic fashion through miRNA-directed transcriptional gene silencing (Melkamu et al. 2010) as well as by modulating posttranscriptional silencing (i.e., the targeted degradation of mRNAs) has not yet been delineated. In fact, the role that microRNAs themselves can have as chromatin modifiers is only beginning to be understood (Guil and Esteller 2009). Recent evidence suggests that this regulation can involve a direct or indirect repression of DNA and histone-modifying enzymes, as well as chromatin-remodeling factors. It is also interesting to note that microRNA genes are also epigenetically modified in cancer cells (Guil and Esteller 2009). Evidence suggests that miRNA genes are subject to hypermethylation and hypomethylation in a tumor- and tissue-specific manner. Further characterization of the downstream mRNA targets for these miRNAs will shed light on the functional consequences of their altered epigenetic regulation and how this contributes to human tumorigenesis. Whether dietary factors also influence DNA methylation near or within miRNA genes is another research topic for consideration. In addition to miRNA, other noncoding RNAs will likely to be impacted by dietary modulation as well as have activity in epigenetic pathways and in cancer. Investigators have only begun to understand how small, noncoding RNAs act in gene regulation and disease.

4.7 CONCLUSIONS

Recent evidence suggests that many different bioactive food components exert cancer-protective effects through modulation of epigenetic mechanisms, such as DNA methylation of CpG islands in promoters and other regions of the genome, chromatin-silencing complexes, posttranslational modifications of histone tail domains, and regulation of noncoding RNAs. Moreover, dietary alteration of epigenetic events has been associated with modulation of several cellular processes associated with carcinogenesis, including differentiation, inflammation, apoptosis, cell cycle control/proliferation, carcinogen metabolism, and angiogenesis, among others.

In the near future, epigenomic approaches are likely to assist in characterizing genomewide epigenetic marks that are targets for dietary regulation. The ability to characterize reference epigenomes (be they profiles of DNA methylation or histone modifications) will greatly impact the ability to determine, on a global level, how diet impacts differential epigenetic effects on normal versus cancer tissue and in different tissues. This information will also provide the tools to elucidate epigenetic changes resulting from dietary exposures during critical periods of prenatal and postnatal development, adolescence, and senescence, as well as investigate the potential impact of diet on transgenerational transmission of epigenetic changes. In addition, links between genetics and epigenetics may provide additional insights about transcriptional regulation during carcinogenesis and how dietary factors participate in these interactions. Moreover, the identification and characterization of novel epigenetic marks and mechanisms with the capacity to differentially silence and activate gene expression are likely to surface over the next few years.

Although the cancer epigenetic field has advanced in the last decade, much remains to be revealed especially with respect to potential modification by bioactive dietary components. Issues remain about the quantity of dietary components needed to bring about a biological effect, the timing of exposure, and other variables (chemical form, duration of exposure) that can influence the response. Importantly, for the future of nutrigenomics and personalized nutrition, epigenetic marks may be useful as biomarkers of cancer prevention, early disease, or nutritional status, as well as function as potential molecular targets that are modulated by dietary interventions.

REFERENCES

Aagaard-Tillery, K. M., K. Grove, J. Bishop, X. Ke, Q. Fu, R. McKnight, and R. H. Lane. 2008. Developmental origins of disease and determinants of chromatin structure: maternal diet modifies the primate fetal epigenome. *J Mol Endocrinol* 41 (2):91–102.

Allard, J. S., E. Perez, S. Zou, and R. de Cabo. 2009. Dietary activators of Sirt1. *Mol Cell Endocrinol* 299 (1):58–63.

Atsumi, A., A. Tomita, H. Kiyoi, and T. Naoe. 2006. Histone deacetylase 3 (HDAC3) is recruited to target promoters by PML-RARalpha as a component of the N-CoR co-repressor complex to repress transcription in vivo. *Biochem Biophys Res Commun* 345 (4):1471–80.

Balasubramanian, S., K. Lee et al. 2008. The Bmi-1 polycomb group gene in skin cancer: regulation of function by (-)-epigallocatechin-3-gallate. *Nutr Rev* 66 (Suppl 1):S65–68.

Bannister, A. J., and T. Kouzarides. 2005. Reversing histone methylation. *Nature* 436 (7054):1103–6.

Barbarotto, E., T. D. Schmittgen, and G. A. Calen. 2008. MicroRNAs and cancer: profile, profile, profile. *Int J Cancer* 122 (5):969–77.

Bartel, D. P. 2004. MicroRNAs: genomics, biogenesis, mechanism, and function. *Cell* 116 (2):281–97.

Bernhard, D., M. J. Ausserlechner, M. Tonko, M. Löffler, B. L. Hartmann, A. Csordas, and R. Kofler. (1999). Apoptosis induced by the histone deacetylase inhibitor sodium butyrate in human leukemic lymphoblasts. *FASEB J* 13 (14):1991–2001.

Bradlow, H. L. 2008. Review. Indole-3-carbinol as a chemoprotective agent in breast and prostate cancer. *In Vivo* 22 (4):441–45.

Brenner, C., R. Deplus, C. Didelot, A. Loriot, E. Viré, C. De Smet, A. Gutierrez, D. Danovi, D. Bernard, T. Boon, P. G. Pelicci, B. Amati, T. Kouaarides, Y. de Launoit, L. Di Croce, and F. Fuks. 2005. Myc represses transcription through recruitment of DNA methyltransferase corepressor. *EMBO J* 24 (2):336–46.

Calin, G. A., and C. M. Croce. 2006. MicroRNA signatures in human cancers. *Nat Rev Cancer* 6 (11): 857-66.

Chew, Y. C., J. T. West, S. J. Kratzer, A. M. Ilvarsonn, J. C. Eissenberg, B. J. Dave, D. Klinkebiel, J. K. Christman, and J. Zempleni. 2008. Biotinylation of histones represses transposable elements in human and mouse cells and cell lines and in *Drosophila* melanogaster. *J Nutr* 138 (12):2316–22.

Christman, J. K., G. Sheikhnejad, M. Dizik, S. Abileah, and E. Wainfan. 1993. Reversibility of changes in nucleic acid methylation and gene expression induced in rat liver by severe dietary methyl deficiency. *Carcinogenesis* 14 (4): 551–57.

Clark, L. C., G. F. Combs Jr., B. W. Turnbull, E. H. Slate, D. K. Chalker, J. Chow, L. S. Davis, R. A. Glover, G. F. Graham, E. G. Gross, A. Krongrad, J. L. Lesher Jr., H. K. Park, B. B. Sanders Jr., C. L. Smith, and J. R. Taylor. 1996. Effects of selenium supplementation for cancer prevention in patients with carcinoma of the skin. A randomized controlled trial. Nutritional Prevention of Cancer Study Group. *JAMA* 276 (24):1957–63.

Cooney, C. A., A. A. Dave, and G. L. Wolff. 2002. Maternal methyl supplements in mice affect epigenetic variation and DNA methylation of offspring. *J Nutr* 132 (8 Suppl): 2393S–2400S.

Cropley, J. E., C. M. Suter, K. B. Beckman, and D. I. K. Martin. 2006. Germ-line epigenetic modification of the murine A vy allele by nutritional supplementation. *Proc Natl Acad Sci U S A* 103 (46):17308–12.

Davie, J. R. 2003. Inhibition of histone deacetylase activity by butyrate. *J Nutr* 133 (7 Suppl):2485S–2493S.

Davis, C. D., and J. A. Milner. 2007. Molecular targets for nutritional preemption of cancer. *Curr Cancer Drug Targets* 7 (5):410–15.

Davis, C. D., and E. O. Uthus. 2003. Dietary folate and selenium affect dimethylhydrazine-induced aberrant crypt formation, global DNA methylation and one-carbon metabolism in rats. *J Nutr* 133 (9): 2907–14.

Denda, A., W. Kitayama, H. Kishida, N. Murata, M. Tsutsumi, T. Tsujiuchi, D. Nakae, and Y. Konishi. 2002. Development of hepatocellular adenomas and carcinomas associated with fibrosis in C57BL/6J male mice given a choline-deficient, L-amino acid-defined diet. *Jpn J Cancer Res* 93 (2):125–32.

Dolinoy, D. C., D. Huang, and R. L. Jirtle. (2007). Maternal nutrient supplementation counteracts bisphenol A-induced DNA hypomethylation in early development. *Proc Natl Acad Sci U S A* 104 (32):13056–61.

Dolinoy, D. C., J. R. Weidman, R. A. Waterland, and R. L. Jirtle. 2006. Maternal genistein alters coat color and protects Avy mouse offspring from obesity by modifying the fetal epigenome. *Environ Health Perspect* 114 (4):567–72.

Druesne, N., A. Pagniez, C. Mayeur, M. Thomas, C. Cherbuy, P.-H. Duée, P. Martel, and C. Chaumontet. 2004. Diallyl disulfide (DADS) increases histone acetylation and p21(waf1/cip1) expression in human colon tumor cell lines. *Carcinogenesis* 25 (7):1227–36.

Druesne-Pecollo, N., C. Chaumontet, A. Pagniez, P. Vaugelade, A. Bruneau, M. Thomas, C. Cherbuy, P. H. Duée, and P. Martel. 2007. In vivo treatment by diallyl disulfide increases histone acetylation in rat colonocytes. *Biochem Biophys Res Commun* 354 (1):140–47.

Druesne-Pecollo, N., A. Pagniez, M. Thomas, C. Cherbuy, P.-H. Duée, P. Martel, and C. Chaumontet. 2006. Diallyl disulfide increases CDKN1A promoter-associated histone acetylation in human colon tumor cell lines. *J Agric Food Chem* 54 (20):7503–7.

Duhl, D. M., H. Vrieling, K. A. Miller, G. L. Wolff, and G. S. Barsh. 1994. Neomorphic agouti mutations in obese yellow mice. *Nat Genet* 8 (1):59–65.

Ehrlich, M. 2002. DNA methylation in cancer: too much, but also too little. *Oncogene* 21 (35):5400–13.

Esteller, M. 2005. Aberrant DNA methylation as a cancer-inducing mechanism. *Annu Rev Pharmacol Toxicol* 45:629–56.

Esteller, M. 2007. Cancer epigenomics: DNA methylomes and histone-modification maps. *Nat Rev Genet* 8 (4):286–98.

Fabbri, M., C. M. Croce, and G. A. Galin. 2008. MicroRNAs. *Cancer J* 14 (1):1–6.

Fang, J. Y., Y. X. Chen, J. Lu, R. Lu, L. Yang, H. Y. Zhu, W. Q. Gu, and L. G. Lu. 2004. Epigenetic modification regulates both expression of tumor-associated genes and cell cycle progressing in human colon cancer cell lines: Colo-320 and SW1116. *Cell Res* 14 (3):217–26.

Fang, M., D. Chen, and C. S. Yang. 2007. Dietary polyphenols may affect DNA methylation. *J Nutr* 137 (1 Suppl):223S–228S.

Fang, M. Z., D. Chen, Y. Sun, Z. Jin, J. K. Christman, and C. S. Yang. 2005. Reversal of hypermethylation and reactivation of p16INK4a, RARbeta, and MGMT genes by genistein and other isoflavones from soy. *Clin Cancer Res* 11 (19 Pt 1):7033–41.

Fang, M. Z., Y. Wang, N. Ai, Z. Hou, Y. Sun, H. Lu, W. Welsh, and C. S. Yang. 2003. Tea polyphenol (-)-epigallocatechin-3-gallate inhibits DNA methyltransferase and reactivates methylation-silenced genes in cancer cell lines. *Cancer Res* 63 (22):7563–70.

Fearon, E. R., and B. Vogelstein 1990. A genetic model for colorectal tumorigenesis. *Cell* 61 (5):759–67.

Fini, L., M. Selgrad, V. Fogliano, G. Graziani, M. Romano, E. Hotchkiss, Y. A. Daoud, E. B. De Vol, C. R. Bland, and L. Ricciardiello. 2007. Annurca apple polyphenols have potent demethylating activity and can reactivate silenced tumor suppressor genes in colorectal cancer cells. *J Nutr* 137 (12):2622–28.

Fischle, W., Y. Wang, and C. D. Allis. 2003. Histone and chromatin cross-talk. *Curr Opin Cell Biol* 15 (2):172–83.

Fraga, M. F., E. Ballestar, M. F. Paz, S. Ropero, F. Setien, M. L. Ballestar, D. Heine-Suñer, J. D. Cigudosa, M. Urioste, J. Benitez, M. Boix-Chornet, A. Sanchez-Aguilera, C. Ling, E. Carlsson, P. Poulsen, A. Vaag, Z. Stephan, T. D. Spector, Y. Z. Wu, C. Plass, and M. Esteller. 2005. Epigenetic differences arise during the lifetime of monozygotic twins. *Proc Natl Acad Sci U S A* 102 (30):10604–9.

Fraga, M. F., E. Ballestar, A. Villar-Garea, M. Boix-Chornet, J. Espada, G. Schotta, T. Bonaldi, C. Haydon, S. Ropero, K. Petrie, N. G. Iyer, A. Pérez-Rosado, E. Calvo, J. A. Lopez, A. Cano, M. J. Calasanz, D. Colomer, M. A. Piris, N. Ahn, A. Imhof, C. Caldas, T. Jenuwein, and M. Esteller. 2005. Loss of acetylation at Lys16 and trimethylation at Lys20 of histone H4 is a common hallmark of human cancer. *Nat Genet* 37 (4):391–400.

Gayther, S. A., S. J. Batley, L. Linger, A. Bannister, K. Thorpe, S. F. Chin, Y. Daigo, P. Russell, A. Wilson, H. M. Sowter, J. D. Delhanty, B. A. Ponder, T. Kouzarides, and C. Caldas. 2000. Mutations truncating the EP300 acetylase in human cancers. *Nat Genet* 24 (3):300–303.

Ghoshal, K., X. Li, J. Datta, S. Bai, I. Pogribny, M. Pogribny, Y. Huang, D. Young, and S. T. Jacob. 2006. A folate- and methyl-deficient diet alters the expression of DNA methyltransferases and methyl CpG binding proteins involved in epigenetic gene silencing in livers of F344 rats. *J Nutr* 136 (6):1522–27.

Gibbons, R. J. 2005. Histone modifying and chromatin remodelling enzymes in cancer and dysplastic syndromes. *Hum Mol Genet* 14 Spec No 1:R85–92.

Guil, S., and M. Esteller 2009. DNA methylomes, histone codes and miRNAs: tying it all together. *Int J Biochem Cell Biol* 41 (1):87–95.

Hake, S. B., A. Xiao, and C. D. Allis. 2004. Linking the epigenetic "language" of covalent histone modifications to cancer. *Br J Cancer* 90 (4):761–69.

Hanahan, D., and R. A. Weinberg 2000. The hallmarks of cancer. *Cell* 100 (1):57–70.

Hassan, Y. I. and J. Zempleni 2006. Epigenetic regulation of chromatin structure and gene function by biotin. *J Nutr* 136 (7):1763–65.

Hawkins, P. G., and K. V. Morris. 2008. RNA and transcriptional modulation of gene expression. *Cell Cycle* 7 (5):602–7.

Heijmans, B. T., E. W. Tobi, A. D. Stein, H. Putter, G. J. Glauw, E. S. Susser, P. E. Slagboom, and L. H. Lumey. 2008. Persistent epigenetic differences associated with prenatal exposure to famine in humans. *Proc Natl Acad Sci U S A* 105 (44):17046–49.

Hill-Baskin, A. E., M. M. Markiewski, D. A. Buchner, H. Shao, D. DeSantis, G. Hsiao, S. Subramaniam, N. A. Berger, C. Croniger, J. D. Lambris, and J. H. Nadeau. 2009. Diet-induced hepatocellular carcinoma in genetically predisposed mice. *Hum Mol Genet* 18 (16):2975–88.

Ingrosso, D., A. Cimmino, A. F. Perna, L. Masella, N. G. De Santo, M. L. De Bonis, M. Vacca, M. D'Esposito, M. D'Urso, P. Galletti, and V. Zappia. 2003. Folate treatment and unbalanced methylation and changes of allelic expression induced by hyperhomocysteinaemia in patients with uraemia. *Lancet* 361 (9370):1693–99.

Jenuwein, T. 2001. Re-SET-ting heterochromatin by histone methyltransferases. *Trends Cell Biol* 11 (6):266–73.

Jenuwein, T., and C. D. Allis. 2001. Translating the histone code. *Science* 293 (5532): 1074–80.

Jeong, H. J., J. B. Jeong, D. S. Kim, and B. O. Delumen. 2007. Inhibition of core histone acetylation by the cancer preventive peptide lunasin. *J Agric Food Chem* 55 (3):632–37.

Jones, P. A., and S. B. Baylin. 2002. The fundamental role of epigenetic events in cancer. *Nat Rev Genet* 3 (6):415–28.

Kang, J., J. Chen, Y. Shi, J. Jia, and Y. Zhang. 2005. Curcumin-induced histone hypoacetylation: the role of reactive oxygen species. *Biochem Pharmacol* 69 (8):1205–13.

Kautiainen, T. L., and P. A. Jones. 1986. DNA methyltransferase levels in tumorigenic and nontumorigenic cells in culture. *J Biol Chem* 261 (4):1594–98.

Kikuno, N., H. Shiina, S. Urakami, K. Kawamoto, H. Hirata, Y. Tanaka, S. Majid, M. Igawa, and R. Dahiya. 2008. Genistein mediated histone acetylation and demethylation activates tumor suppressor genes in prostate cancer cells. *Int J Cancer* 123 (3):552–60.

Kim, D. H., P. Saetrom, O. Snøve, and J. J. Rossi. 2008. MicroRNA-directed transcriptional gene silencing in mammalian cells. *Proc Natl Acad Sci U S A* 105 (42):16230–35.

Kim, J. H., S. Y. Yoon, C. N. Kim, J. H. Joo, S. K. Moon, I. S. Choe, Y. K. Choe, and J. W. Kim. 2004. The Bmi-1 oncoprotein is overexpressed in human colorectal cancer and correlates with the reduced p16INK4a/p14ARF proteins. *Cancer Lett* 203 (2):217–24.

Kim, J. K., M. Samaranayake, and S. Pradhan. 2009. Epigenetic mechanisms in mammals. *Cell Mol Life Sci* 66 (4):596–612.

Kim, Y., and J. B. Mason. 1995. Folate, epithelial dysplasia and colon cancer. *Proc Assoc Am Physicians* 107 (2):218–27.

Kim, Y. I. 2005. Nutritional epigenetics: impact of folate deficiency on DNA methylation and colon cancer susceptibility. *J Nutr* 135 (11):2703–9.

Kornberg, R. D., and Y. Lorch. 1999. Twenty-five years of the nucleosome, fundamental particle of the eukaryote chromosome. *Cell* 98 (3):285–94.

Kothapalli, N., G. Camporeale, A. Kueh, Y. C. Chew, A. M. Oommen, J. B. Griffin, and J. Zempleni. 2005. Biological functions of biotinylated histones. *J Nutr Biochem* 16 (7):446–48.

Kotsopoulos, J., K. J. Sohn, and Y. I. Kim. 2008. Postweaning dietary folate deficiency provided through childhood to puberty permanently increases genomic DNA methylation in adult rat liver. *J Nutr* 138 (4):703–9.

Kutay, H., S. Bai, J. Datta, T. Motiwala, I. Pogribny, W. Frankel, S. T. Jacob, and K. Ghosahl. 2006. Downregulation of miR-122 in the rodent and human hepatocellular carcinomas. *J Cell Biochem* 99 (3):671–78.

Lagouge, M., C. Argmann, Z. Gerhart-Hines, H. Meziane, C. Lerin, F. Daussin, N. Messadeq, J. Milne, P. Lambert, B. Geny, M. Laasko, P. Puigserver, and J. Auwerx. 2006. Resveratrol improves mitochondrial function and protects against metabolic disease by activating SIRT1 and PGC-1alpha. *Cell* 127 (6):1109–22.

Lam, Y., A. Galvez, and B. O. de Lumen. 2003. Lunasin suppresses E1A-mediated transformation of mammalian cells but does not inhibit growth of immortalized and established cancer cell lines. *Nutr Cancer* 47 (1):88–94.

Lea, M. A., and V. M. Randolph 2001. Induction of histone acetylation in rat liver and hepatoma by organosulfur compounds including diallyl disulfide. *Anticancer Res* 21 (4A):2841–45.

Lea, M. A., M. Rasheed, V. M. Randolph, F. Khan, A. Shareef, and C. desBordes. 2002. Induction of histone acetylation and inhibition of growth of mouse erythroleukemia cells by S-allylmercaptocysteine. *Nutr Cancer* 43 (1):90–102.

Lee, E. R., F. E. Murdoch, and M. K. Fritsch. 2007. High histone acetylation and decreased polycomb repressive complex 2 member levels regulate gene specific transcriptional changes during early embryonic stem cell differentiation induced by retinoic acid. *Stem Cells* 25 (9):2191–99.

Lee, J. I., H. Nian, A. J. Cooper, R. Sinha, J. Dai, W. H. Bisson, R. H. Dashwood, and J. T. Pinto. 2009. Alpha-keto acid metabolites of naturally occurring organoselenium compounds as inhibitors of histone deacetylase in human prostate cancer cells. *Cancer Prev Res (Phila Pa)* 2 (7):683–93.

Lee, K., G. Adhikary, S. Balasubramanian, R. Gopalakrishnan, T. McCormick, G. P. Dimri, R. L. Eckert, and E. A. Rorke. 2008. Expression of Bmi-1 in epidermis enhances cell survival by altering cell cycle regulatory protein expression and inhibiting apoptosis. *J Invest Dermatol* 128 (1):9–17.

Li, E. 2002. Chromatin modification and epigenetic reprogramming in mammalian development. *Nat Rev Genet* 3 (9):662–73.

Lin, C., J. Kang, and R. Zheng. 2005. Oxidative stress is involved in inhibition of copper on histone acetylation in cells. *Chem Biol Interact* 151 (3):167–76.

Lin, R. J., L. Nagy, S. Inoue, W. Shao, W. H. Miller Jr., and R. M. Evans. 1998. Role of the histone deacetylase complex in acute promyelocytic leukaemia. *Nature* 391 (6669):811–14.

Lippman, S. M., E. A. Klein, P. J. Goodman, M. S. Lucia, I. M. Thompson, L. G. Ford, H. L. Parnes, L. M. Minasian, J. M. Gaziano, J. A. Hartline, J. K. Parsons, J. D. Bearden III, E. D. Crawford, G. E. Goodman, J. Claudio, E. Winquist, E. D. Cook, D. D. Karp,

P. Walther, M. M. Lieber, A. R. Kristal, A. K. Darke, K. B. Arnold, P. Ganz, R. M. Santella, D. Albanes, P. R. Taylor, J. L. Probstfield, T. J. Jagpal, J. J. Crowley, F. L. Meyskens Jr., L. H. Baker, and C. A. Coltman Jr. 2009. Effect of selenium and vitamin E on risk of prostate cancer and other cancers: the Selenium and Vitamin E Cancer Prevention Trial (SELECT). *JAMA* 301 (1):39–51.

Lu, J., G. Getz, E. A. Miska, E. Alvarez-Saavedra, J. Lamb, D. Peck, A. Sweet-Cordero, B. L. Ebert, R. H. Mark, A. A. Ferrando, J. R. Downing, T. Jacks, H. R. Horvitz, and T. R. Golub. 2005. MicroRNA expression profiles classify human cancers. *Nature* 435 (7043):834–38.

Mahlknecht, U., and D. Hoelzer. 2000. Histone acetylation modifiers in the pathogenesis of malignant disease. *Mol Med* 6 (8):623–44.

Majid, S., N. Kikuno, J. Nelles, E. Noonan, Y. Tanaka, K. Kawamoto, H. Hirata, L. C. Li, H. Zhao, S. T. Okino, R. F. Place, D. Pookot, and R. Dahiya. 2008. Genistein induces the p21WAF1/CIP1 and p16INK4a tumor suppressor genes in prostate cancer cells by epigenetic mechanisms involving active chromatin modification. *Cancer Res* 68 (8):2736–44.

Mariadason, J. M., G. A. Corner, and L. H. Augenlicht. 2000. Genetic reprogramming in pathways of colonic cell maturation induced by short chain fatty acids: comparison with trichostatin A, sulindac, and curcumin and implications for chemoprevention of colon cancer. *Cancer Res* 60 (16):4561–72.

Marks, P. A., and M. Dokmanovic. 2005. Histone deacetylase inhibitors: discovery and development as anticancer agents. *Expert Opin Investig Drugs* 14 (12):1497–1511.

McGarvey, K. M., L. van Neste, L. Cope, J. E. Ohm, J. G. Herman, W. V. Criekinge, K. E. Schuebel, and S. B. Baylin. 2008. Defining a chromatin pattern that characterizes DNA-hypermethylated genes in colon cancer cells. *Cancer Res* 68 (14):5753–59.

McGowan, P. O., M. J. Meaney, and M. Szyf. 2008. Diet and the epigenetic (re)programming of phenotypic differences in behavior. *Brain Res* 1237:12–24.

Medina, P. P., and M. Sanchez-Cespedes 2008. Involvement of the chromatin-remodeling factor BRG1/SMARCA4 in human cancer. *Epigenetics* 3 (2):64–68.

Melkamu, T., X. Zhang, J. Tan, Y. Zeng, and F. Kassie. 2010. Alteration of microRNA expression in vinyl carbamate-induced mouse lung tumors and modulation by the chemopreventive agent indole-3-carbinol. *Carcinogenesis* 31 (2):252–58.

Morey, L., C. Brenner, F. Fazi, R. Villa, A. Gutierrez, M. Buschbeck, C. Nevvi, S. Minucci, F. Fuks, and L. Di Croce. 2008. MBD3, a component of the NuRD complex, facilitates chromatin alteration and deposition of epigenetic marks. *Mol Cell Biol* 28 (19):5912–23.

Myzak, M. C., and R. H. Dashwood. 2006. Histone deacetylases as targets for dietary cancer preventive agents: lessons learned with butyrate, diallyl disulfide, and sulforaphane. *Curr Drug Targets* 7 (4):443–52.

Myzak, M. C., W. M. Dashwood, G. A. Orner, E. Ho, and R. H. Dashwood. 2006. Sulforaphane inhibits histone deacetylase in vivo and suppresses tumorigenesis in Apc-minus mice. *FASEB J* 20 (3):506–8.

Myzak, M. C., K. Hardin, R. Wang, R. H. Dashwood, and E. Ho. 2006. Sulforaphane inhibits histone deacetylase activity in BPH-1, LnCaP and PC-3 prostate epithelial cells. *Carcinogenesis* 27 (4):811–19.

Myzak, M. C., P. A. Karplus, F.-L. Chung, and R. H. Dashwood. 2004. A novel mechanism of chemoprotection by sulforaphane: inhibition of histone deacetylase. *Cancer Res* 64 (16):5767–74.

Myzak, M. C., P. Tong, W.-M. Dashwood, R. H. Dashwood, and E. Ho. 2007. Sulforaphane retards the growth of human PC-3 xenografts and inhibits HDAC activity in human subjects. *Exp Biol Med (Maywood)* 232 (2):227–34.

Newberne, P. M. 1986. Lipotropic factors and oncogenesis. *Adv Exp Med Biol* 206:223–51.

Nian, H., W. H. Bisson, W.-M. Dashwood, J. T. Pinto, and R. H. Dashwood. 2009. Alpha-keto acid metabolites of organoselenium compounds inhibit histone deacetylase activity in human colon cancer cells. *Carcinogenesis* 30 (8):1416–23.

Nian, H., B. Delage, J. T. Pinto, and R. H. Dashwood. 2008. Allyl mercaptan, a garlic-derived organosulfur compound, inhibits histone deacetylase and enhances Sp3 binding on the P21WAF1 promoter. *Carcinogenesis* 29 (9):1816–24.

Oki, M., H. Aihara, T. Ito. 2007. Role of histone phosphorylation in chromatin dynamics and its implications in diseases. *Subcell Biochem* 41:319–36.

Ooi, S. K., and T. H. Bestor 2008. The colorful history of active DNA demethylation. *Cell* 133 (7):1145–48.

Peters, A. H., D. O'Carroll, H. Scherthan, K. Mechtler, S. Sauer, C. Schöfer, K. Weipoltshammer, M. Pagani, M. Lachner, A. Kohlmaier, S. Opravil, M. Doyle, M. Sibilia, and T. Jenuwein. 2001. Loss of the Suv39h histone methyltransferases impairs mammalian heterochromatin and genome stability. *Cell* 107 (3):323–37.

Pietiläinen, K. H., J. Naukkarinen, A. Rissanen, J. Saharinen, P. Ellonen, H. Keränen, A. Suomalainen, A. Götz, T. Suortti, H. Yki-Järvinen, M. Oresic, J. Kaprio, and L. Peltonen. 2008. Global transcript profiles of fat in monozygotic twins discordant for BMI: pathways behind acquired obesity. *PLoS Med* 5 (3):e51.

Pogribny, I. P., S. J. James, S. Jernigan, and M. Progribna. 2004. Genomic hypomethylation is specific for preneoplastic liver in folate/methyl deficient rats and does not occur in non-target tissues. *Mutat Res* 548 (1–2):53–59.

Pogribny, I. P., S. A. Ross, V. P. Tryndyak, M. Progribna, L. A. Poirier, and T. V. Karpinets. 2006. Histone H3 lysine 9 and H4 lysine 20 trimethylation and the expression of Suv4-20h2 and Suv-39h1 histone methyltransferases in hepatocarcinogenesis induced by methyl deficiency in rats. *Carcinogenesis* 27 (6):1180–86.

Pogribny, I. P., S. A. Ross, C. Wise, M. Progribna, E. A. Jones, V. P. Tryndyak, S. J. James, Y. P. Dragan, and L. A. Poirier. 2006. Irreversible global DNA hypomethylation as a key step in hepatocarcinogenesis induced by dietary methyl deficiency. *Mutat Res* 593 (1–2):80–87.

Pogribny, I. P., S. I. Shpyleva, L. Muskhelishvili, T. V. Bagnyokova, S. J. James, and F. A. Beland. 2009. Role of DNA damage and alterations in cytosine DNA methylation in rat liver carcinogenesis induced by a methyl-deficient diet. *Mutat Res* 669 (1–2):56–62.

Pogribny, I. P., V. P. Tryndyak, L. Muskhelishvili, I. Rusyn, and S. A. Ross. 2007. Methyl deficiency, alterations in global histone modifications, and carcinogenesis. *J Nutr* 137 (1 Suppl):216S–222S.

Poirier, L. A. 1994. Methyl group deficiency in hepatocarcinogenesis. *Drug Metab Rev* 26 (1–2):185–99.

Poirier, L. A., and T. I. Vlasova. 2002. The prospective role of abnormal methyl metabolism in cadmium toxicity. *Environ Health Perspect* 110 (Suppl 5):793–95.

Powell, C. L., O. Kosyk, B. U. Bradford, J. S. Parker, E. K. Lobenhofer, A. Denda, F. Uematsu, D. Nakae, and I. Rusyn. 2005. Temporal correlation of pathology and DNA damage with gene expression in a choline-deficient model of rat liver injury. *Hepatology* 42 (5):1137–47.

Rada-Iglesias, A., S. Enroth, A. Ameur, C. M. Koch, G. K. Clelland, P. Respuela-Alonso, S. Wilcox, O. M. Dovey, P. D. Ellis, C. F. Langford, I. Dunham, J. Komorowski, and C. Wadelius. 2007. Butyrate mediates decrease of histone acetylation centered on transcription start sites and down-regulation of associated genes. *Genome Res* 17 (6):708–19.

Rakyan, V. K., M. E. Blewitt, R. Druker, J. I. Preis, and E. Whitelaw. 2002. Metastable epialleles in mammals. *Trends Genet* 18 (7):348–51.

Razin, A., and A. D. Riggs 1980. DNA methylation and gene function. *Science* 210 (4470):604–10.

Rodenhiser, D. and M. Mann 2006. Epigenetics and human disease: translating basic biology into clinical applications. *CMAJ* 174 (3):341–48.

Rosato, R. R., and S. Grant. 2003. Histone deacetylase inhibitors in cancer therapy. *Cancer Biol Ther* 2 (1):30–37.

Ross, S. A. 2003. Diet and DNA methylation interactions in cancer prevention. *Ann N Y Acad Sci* 983:197–207.

Sharma, R., A. Nakamura, H. Nakamoto, and S. Goto. 2006. Carbonyl modification in rat liver histones: decrease with age and increase by dietary restriction. *Free Radic Biol Med* 40 (7):1179–84.

Shiio, Y., and R. N. Eisenman. 2003. Histone sumoylation is associated with transcriptional repression. *Proc Natl Acad Sci U S A* 100 (23):13225–30.

Shilatifard, A. 2006. Chromatin modifications by methylation and ubiquitination: implications in the regulation of gene expression. *Annu Rev Biochem* 75:243–69.

Shukla, V., T. Vaissiere, and Z. Herceg. 2008. Histone acetylation and chromatin signature in stem cell identity and cancer. *Mutat Res* 637 (1–2):1–15.

Sparmann, A., and M. van Lohuizen. 2006. Polycomb silencers control cell fate, development and cancer. *Nat Rev Cancer* 6 (11):846–56.

Spurling, C. C., J. A. Suhl, N. Boucher, C. E. Nelson, D. W. Rosenberg, and C. Giardina. 2008. The short chain fatty acid butyrate induces promoter demethylation and reactivation of RARbeta2 in colon cancer cells. *Nutr Cancer* 60 (5):692–702.

Sun, M., Z. Estrov, Y. Ji, K. R. Coombes, D. H. Harris, and R. Kurzrock. 2008. Curcumin (diferuloylmethane) alters the expression profiles of microRNAs in human pancreatic cancer cells. *Mol Cancer Ther* 7 (3):464–73.

Takiguchi, M., W. E. Achanzar, W. Qu, G. Li, and M. P. Waalkes. 2003. Effect of cadmium on DBA-(cytosine-5) methyltransferase activity and DNA methylation status during cadmium-induced cellular transformation. *Exp Cell Res* 286 (2):355–65.

Takihara, Y. 2008. Role of polycomb-group genes in sustaining activities of normal and malignant stem cells. *Int J Hematol* 87 (1):25–34.

Tobi, E. W., L. H. Lumey, R. P. Talens, D. Kremer, H. Putter, A. D. Stein, P. E. Slagboom, and B. T. Heijmans. 2009. DNA methylation differences after exposure to prenatal famine are common and timing- and sex-specific. *Hum Mol Genet* 18 (21):4046–53.

Torres, D. M., and S. A. Harrison. 2008. Diagnosis and therapy of nonalcoholic steatohepatitis. *Gastroenterology* 134 (6):1682–98.

Tost, J. 2009. DNA methylation: an introduction to the biology and the disease-associated changes of a promising biomarker. *Methods Mol Biol* 507:3–20.

Tryndyak, V. P., S. A. Ross, F. A. Beland, and I. P. Pogribny. 2009. Down-regulation of the microRNAs miR-34a, miR-127, and miR-200b in rat liver during hepatocarcinogenesis induced by a methyl-deficient diet. *Mol Carcinog* 48 (6):479–87.

Tsou, J. A., J. A. Hagen, C. L. Carpenter, and I. A. Laird-Offringa. 2002. DNA methylation analysis: a powerful new tool for lung cancer diagnosis. *Oncogene* 21 (35):5450–61.

Unterberger, A., M. Szyf, P. W. Nathanielsz, and L. A. Cox. 2009. Organ and gestational age effects of maternal nutrient restriction on global methylation in fetal baboons. *J Med Primatol* 38 (4):219–27.

Vaquero, A., M. B. Scher, D. H. Lee, A. Sutton, H. L. Cheng, F. W. Alt, L. Serrano, R. Sternglanz, and D. Renberg. 2006. SirT2 is a histone deacetylase with preference for histone H4 Lys 16 during mitosis. *Genes Dev* 20 (10):1256–61.

Varga-Weisz, P. D., and P. B. Becker. 2006. Regulation of higher-order chromatin structures by nucleosome-remodelling factors. *Curr Opin Genet Dev* 16 (2):151–56.

Vire, E., C. Brenner, R. Deplus, L. Blanchon, M. Fraga, C. Didelot, L. Morey, A. van Eynde, D. Bernard, A.-M. Vanderwinden, M. Bollen, M. Esteller, L. Di Croce, Y. de Launoit, and F. Fuks. 2006. The polycomb group protein EZH2 directly controls DNA methylation. *Nature* 439 (7078):871–74.

Vonlanthen, S. J. Heighway, H. J. Altermatt, M. Gugger, A. Kappeler, M. M. Borner, M. van Lohuizen, and D. C. Betticher. 2001. The bmi-1 oncoprotein is differentially expressed in non-small cell lung cancer and correlates with INK4A-ARF locus expression. *Br J Cancer* 84 (10):1372–76.

Wade, P. A., D. Pruss, and A. P. Wolffe. 1997. Histone acetylation: chromatin in action. *Trends Biochem Sci* 22 (4):128–32.

Wainfan, E., and L. A. Poirier. 1992. Methyl groups in carcinogenesis: effects on DNA methylation and gene expression. *Cancer Res* 52 (7 Suppl):2071s–2077s.

Wang, B., S. Majumder, G. Nuovo, H. Kutay, S. Volinia, T. Patel, T. D. Schmittgen, C. Croce, K. Ghoshal, and S. T. Jacob. 2009. Role of microRNA-155 at early stages of hepatocarcinogenesis induced by choline-deficient and amino acid-defined diet in C57BL/6 mice. *Hepatology* 50 (4):1152–61.

Wang, L. G., A. Beklemisheva, X. M. Liu, A. C. Ferrari, J. Feng, and J. W. Chiao. 2007. Dual action on promoter demethylation and chromatin by an isothiocyanate restored GSTP1 silenced in prostate cancer. *Mol Carcinog* 46 (1):24–31.

Wang, L. G., X. M. Liu, Y. Fang, W. Dai, F. B. Chiao, G. M. Puccio, J. Feng, D. Liu, and J. W. Chiao. 2008. De-repression of the p21 promoter in prostate cancer cells by an isothiocyanate via inhibition of HDACs and c-Myc. *Int J Oncol* 33 (2):375–80.

Waterland, R. A., D. C. Dolinoy, J. R. Lin, C. A. Smith, X. Chi, and K. J. Tahiliani. 2006. Maternal methyl supplements increase offspring DNA methylation at Axin Fused. *Genesis* 44 (9):401–6.

Waterland, R. A., and R. L. Jirtle 2003. Transposable elements: targets for early nutritional effects on epigenetic gene regulation. *Mol Cell Biol* 23 (15):5293–5300.

Waterland, R. A., J. R. Lin, C. A. Smith, and R. L. Jirtle. 2006. Post-weaning diet affects genomic imprinting at the insulin-like growth factor 2 (Igf2) locus. *Hum Mol Genet* 15 (5):705–16.

Waterland, R. A., M. Travisano, K. G. Tahiliani, M. T. Rached, and S. Mirza. 2008. Methyl donor supplementation prevents transgenerational amplification of obesity. *Int J Obes (Lond)* 32 (9):1373–79.

Whitelaw, E., and D. I. Martin. 2001. Retrotransposons as epigenetic mediators of phenotypic variation in mammals. *Nat Genet* 27 (4):361–65.

Widschwendter, M., H. Fiegl, D. Egel, E. Mueller-Holzner, G. Spizzo, C. Marth, D. J. Weisenberger, M. Campan, J. Young, I. Jacobs, and P. W. Laird. 2007. Epigenetic stem cell signature in cancer. *Nat Genet* 39 (2):157–58.

Wiseman, M. 2008. The second World Cancer Research Fund/American Institute for Cancer Research expert report. Food, nutrition, physical activity, and the prevention of cancer: a global perspective. *Proc Nutr Soc* 67 (3):253–56.

Wolff, G. L., R. L. Kodell, S. R. Moore, and C. A. Cooney. 1998. Maternal epigenetics and methyl supplements affect agouti gene expression in Avy/a mice. *FASEB J* 12 (11):949–57.

Wolffe, A. P. 1994. Inheritance of chromatin states. *Dev Genet* 15 (6):463–70.

Wood, J. G., B. Rogina, S. Lavu, K. Howitz, S. L. Helfand, M. Tatar, and D. Sinclair. 2004. Sirtuin activators mimic caloric restriction and delay ageing in metazoans. *Nature* 430 (7000):686–89.

Xiang, N., R. Zhao, G. Song, and W. Zhong. 2008. Selenite reactivates silenced genes by modifying DNA methylation and histones in prostate cancer cells. *Carcinogenesis* 29 (11):2175–81.

Xiong, S. D., K. Yu, X. H. Liu, L. H. Yin, A. Kirschenbaum, S. Yao, G. Narla, A. DiFeo, J. B. Wu, Y. Yuan, S. M. Ho, Y. W. Lam, and A. C. Levine. 2009. Ribosome-inactivating proteins isolated from dietary bitter melon induce apoptosis and inhibit histone deacetylase-1 selectively in premalignant and malignant prostate cancer cells. *Int J Cancer* 125 (4):774–82.

Xue, Y., J. Wong, G. T. Moreno, M. K. Young, J. Côté, and W. Wang. 1998. NURD, a novel complex with both ATP-dependent chromatin-remodeling and histone deacetylase activities. *Mol Cell* 2 (6):851–61.

Yin, J. Q., R. C. Zhao, and K. V. Morris. 2008. Profiling microRNA expression with microarrays. *Trends Biotechnol* 26 (2):70–76.

Zhang, K., and S. Y. Dent 2005. Histone modifying enzymes and cancer: going beyond histones. *J Cell Biochem* 96 (6):1137-48.

Zhu, P., E. Martin, J. Mengwasser, P. Schlag, K. P. Janssen, and M. Gottlicher. 2004. Induction of HDAC2 expression upon loss of APC in colorectal tumorigenesis. *Cancer Cell* 5 (5): 455–63.

Zhu, Z., W. Jiang, J. N. McGinley, and H. J. Thompson. 2005. 2-Deoxyglucose as an energy restriction mimetic agent: effects on mammary carcinogenesis and on mammary tumor cell growth in vitro. *Cancer Res* 65 (15): 7023–30.

5 Dietary Factors, Histone Modifications, and Cancer Prevention

Igor P. Pogribny, Sharon A. Ross, and Igor Koturbash

CONTENTS

5.1 Introduction 109
5.2 Histone Modifications and Cancer 111
  5.2.1 Histone Lysine Acetylation and Cancer 111
  5.2.2 Histone Lysine Methylation and Cancer 112
5.3 Histone Modifications and Tumorigenesis 112
5.4 Dietary Modifications of Histone Lysine Acetylation and Methylation Patterns and Cancer Prevention 114
  5.4.1 Dietary HDAC Inhibitors and Cancer Prevention 114
  5.4.2 Dietary Methyl Group Donors and Cancer Prevention 115
  5.4.3 Diet and Other Histone Modifications and Cancer Prevention 116
5.5 Conclusions and Remarks 117
References 117

5.1 INTRODUCTION

Classically, the development of cancer in humans has been viewed as a progressive multistep process of transformation of normal cells into malignant cells driven by genetic alterations (Hanahan and Weinberg 2000; Vogelstein and Kinzler 2004). However, a wealth of data in the past decade indicates that an accurate epigenomic landscape is critical for normal cell functioning, with evidence of extensive epigenetic distortion in tumor cells having largely changed the view on cancer as being a solely genetic disease (Jones and Baylin 2007). Currently, cancer is recognized as a disease provoked by a combination of genetic and epigenetic aberrations, and both of these components cooperate and complement each other at every stage of tumor development (Jones and Baylin 2007). The remarkable feature of epigenetic abnormalities, unlike genetic alterations, is their potential reversibility. This capability of reversal has resulted in the emergence of a novel innovative epigenetic approach for the treatment of cancer (Brown and Strathdee 2002; Cortez and Jones 2008; Ellis et al. 2009). However, the timely correction of epigenetic alterations is a promising avenue not only for cancer treatment but also for prevention of the development of cancer (Kopelovich et al. 2003; Davis and Ross 2007; Yang and Seto 2007).

Epigenetics is defined as heritable changes in gene expression associated with modifications of DNA or chromatin proteins that are not due to any alteration in the DNA sequence (Egger et al. 2004; Feinberg 2008). The most common epigenetic modifications include methylation of cytosine bases in DNA, posttranslational modifications of histone proteins, and their location along the DNA (Sharma et al. 2010). DNA methylation, a well-known primary epigenetic regulator of gene expression, arises from the addition of a methyl group from the universal methyl donor, S-adenosyl-L-methionine (SAM), to the fifth carbon atom in the cytosine pyridine ring, resulting in the formation of 5-methylcytosine in DNA (Chiang et al. 1996). This reaction is catalyzed by DNA methyltransferases (DNMTs) (Goll and Bestor 2005; Li and Bird 2007). In eukaryotes, this stable post-synthetic epigenetic mark is found exclusively at cytosine residues at CpG sequences (Bird and Wolffe 1999).

DNA methylation is initiated and established by means of the de novo DNA methyltransferases DNMT3A and DNMT3B (Goll and Bestor 2005; Li and Bird 2007), whose expression is coordinated by DNMT3L (Gowher et al. 2005), Lsh (lymphoid-specific helicase) (Zhu et al. 2006), microRNAs (Fabbri et al. 2007), and piRNAs (Aravin et al. 2008). During DNA replication, DNA methylation is maintained by a complex cooperative interplay of the maintenance methyltransferase DNMT1 with the de novo DNA methyltransferases DNMT3A and DNMT3B (Liang et al. 2002; El-Osta 2003), methyl-CpG-binding protein 2 (MeCP2) (Li and Bird 2007; Kimura and Shiota 2003), histone-modifying enzymes (Kim et al. 2009; J. Wang et al. 2009), and the UHRF1 (ubiquitinlike, containing PHD and RING finger domains 1) protein (Arita et al. 2008; Avvakumov et al. 2008; Hashimoto et al. 2008).

A second, well-studied epigenetic mechanism involves a variety of posttranslational covalent modifications of histones (H2A, H2B, H3, and H4). Histones are evolutionary conserved proteins that have a globular carboxy-terminal domain critical to nucleosome formation and a flexible amino-terminal tail that protrudes from the nucleosome core and contacts adjacent nucleosomes in a higher-order structure. The amino-terminal tails of histones are subject of at least eight different classes of modifications, including acetylation, methylation, phosphorylation, ubiquitination, sumoylation, biotinylation, and ADP-ribosylation (Jenuwein and Allis 2001; Kouzarides 2007; Lennartsson and Ekwall 2009). These histone modifications function either by influencing chromatin packaging or by recruiting and/or occluding other protein complexes (Neff and Armstrong 2009; Sharma et al. 2010).

DNA methylation and modifications of histones are essential for normal development and the maintenance of cellular homeostasis and functions in adult organisms, particularly for X chromosome inactivation in females (Singer-Sam and Riggs 1993; Brinkman et al. 2006), genomic imprinting (Razin and Cedar 1994; Ideraabdullah et al. 2008), silencing of repetitive DNA elements (Yoder et al. 1997; Martens et al. 2005), regulation of chromatin structure (Bernstein et al. 2007; Miranda and Jones 2007), and proper expression of genetic information (Razin and Riggs 1980; Berger 2007). The precise epigenetic status is balanced in normal cells; however, a number of exogenous and endogenous factors may disrupt this balance and compromise the stability of the epigenome, leading to the development of a wide range of human pathologies, including cancer.

5.2 HISTONE MODIFICATIONS AND CANCER

Although a number of histone modifications have been described, acetylation and methylation of histones are the two main modifications, which disturbances have been well established in cancer cells.

5.2.1 Histone Lysine Acetylation and Cancer

The association between histone lysine acetylation and gene transcriptional activation was first reported in 1964 (Allfrey et al. 1964). In contrast, histone lysine deacetylation correlates with gene silencing (Iizuka and Smith 2003; Ropero and Esteller 2007). Deacetylated lysines are positively charged and react strongly with the negatively charged DNA. This leads to dense chromatin condensation at gene promoters and transcriptional gene silencing by limiting the accessibility to transcription machinery (Grønbaek et al. 2008; Ropero and Esteller 2007). On the other hand, acetylation of the lysines neutralizes this charge, which generates the open chromatin structure and activation of gene transcription. The accurate balance between histone lysine acetylation and deacetylation in normal cells is maintained by the opposing activities of histone acetyltransferases (HATs or KATs) and histone deacetylases (HDACs). Three main families of HATs, including the MOZ/YBF2/SAS2/TIP60 (MYST) family, the GCN5 *N*-acetyltransferase (GNAT), and the p300/CBT family, catalyze the transfer of acetyl groups from acetyl-CoA to the lysine ε-amino group, resulting in acetylated lysine and CoA (Smith and Denu 2009). HDACs catalyze the removal of acetyl groups from the acetyl-lysine residues within histones. There are four classes of HDACs. Class I consists of HDAC1-3 and HDAC8 that are localized in the nucleus, whereas class II HDACs, including HDAC4-7, HDAC9, and HDAC10, shuttle between nuclear and cytoplasmic compartments. Class III, which consists of SIR2 (Silent Information Regulator 2) family members, or sirtuins (SIRT1-7), can be found in the cytoplasm, nucleus, and mitochondria (Michan and Sinclair 2007), while HDAC11, which represents the most recently described class IV HDACs, is primarily localized in the nucleus. Class I, II, and IV HDACs share homology in both sequence and structure and utilize an active-site metal-dependent (Zn) catalytic mechanism, whereas class III HDACs utilizes a distinct nicotinamide adenine nucleotide (NAD^+)-dependent catalytic mechanism (Smith and Denu 2009).

The results of numerous studies have convincingly established that the balance between histone lysine acetylation and deacetylation in cancer cells is extensively altered, resulting in profound loss of histone lysine acetylation. Deacetylation of lysine residues, especially lysine 9 on histone H3 (H3K9) and lysine 16 on histone H4, has been detected in various human tumors including leukemia, colon, lung, and breast cancers (Fraga et al. 2005; Kondo and Issa 2004; Elsheikh et al. 2009). It is widely believed that the abnormal state of histone lysine acetylation in cancer cells is associated with altered activity of HATs and/or HDACs. Indeed, down-regulation of HAT and up-regulation of various classes of HDACs have been reported in both hematological and solid tumors (Ropero and Esteller 2007; Mariadason 2008; Ellis et al. 2009).

5.2.2 Histone Lysine Methylation and Cancer

In addition to widespread abnormal histone lysine acetylation, cancer cells also display profound changes in histone methylation patterns. Histones can be methylated on either their lysine or arginine residues. Lysine methyltransferases (HMTs or KMTs) catalyze mono-, di-, and trimethylation by transferring a methyl group from SAM to the lysine ε-amino group. Until recently, histone lysine methylation has been considered as a permanent modification. However, the discovery and characterization of a lysine specific demethylase LSD1/KDM1 in 2004 (Shi et al. 2004) and family of jumonji histone demethylases (JHDMs) in 2006 (Klose et al. 2006; Tsukada et al. 2006) has changed this view.

Unlike histone lysine acetylation, methylation of lysine residues leads to either transcriptional activation or transcriptional repression depending upon which residue is modified and the extent of methylation. For instance, trimethylation of lysine 4 on histone H3 (H3K4) and lysine 36 on histone H3 (H3K36) generally correlates with transcriptional activity. In contrast, demethylation of H3K4 is required for gene silencing. Trimethylation of H3K9 and H3K27 are hallmarks of transcriptional repression. Likewise, trimethylation of H4K20 is an integral part of heterochromatin mediated silencing, whereas monomethylation of H4K20 correlates with ongoing transcription (Jenuwein 2006; Hublitz et al. 2009).

The results of numerous studies have clearly established that cancer cells are characterized by two different types of alterations in histone lysine methylation patterns: loss of global histone H3K9, H3K27, and H4K20 trimethylation and increase of these marks at gene promoter regions. Substantially reduced trimethylation of histone H3K9, H3K27, and H4K20 has been documented in many types of human tumors, including colon, breast, ovarian, pancreatic, and lung cancers (Kondo and Issa 2004; Fraga et al. 2005; Wei et al. 2008; Van den Broeck et al. 2008). In contrast, increased trimethylation of histone H3K9 and H3K27 is associated with aberrant gene silencing in various forms of cancer (Snowden et al. 2002; Schlesinger et al. 2007; Kondo, Shen, Cheng et al. 2008). It is widely believed that alterations in histone lysine methylation patterns are associated with dysregulation of HMTs responsible for these marks. For instance, down-regulation of G9a, RIZ1, and Suv4-20h2 HMTs is tightly correlated with the loss of global histone H3K9 and H4K20 trimethylation in cancer (Kondo, Shen, Ahmed et al. 2008; Zhou et al. 2008; Van den Broeck et al. 2008).

5.3 HISTONE MODIFICATIONS AND TUMORIGENESIS

The development of cancer is a complex multifactorial process characterized by many biologically significant and interdependent alterations. One of these changes is epigenetic dysregulation. It is clear that cancer, by itself, can trigger epigenetic alterations that reflect the transformed state of neoplastic cells; however, the distortion of the cellular epigenetic status in normal cells can also have an impact on the predisposition to precancer-specific pathological states and cancer development. This leads to the suggestion that epigenetic alterations, including aberrant histone modifications, not only are important features of cancer cells, but also play a major role in the initiation and propagation of cancer. The epigenetic model of cancer initiation suggests that epigenetic alterations that occur in stem, progenitor, or differentiated cells are the earliest events in cancer initiation that may predispose to mutational events, give rise to cancer stem cells, and contribute to tumor heterogeneity (Jaffe 2003; Feinberg 2004; Karpinets and Foy 2005; Feinberg et al. 2006; Shukla et al. 2008). The results of recent studies demonstrating the early emergence of epigenetically reprogrammed cells, which have epigenetic alterations that are similar to those found in malignant cells, provided strong experimental support to this hypothesis. Specifically, using different models of liver carcinogenesis induced by genotoxic or nongenotoxic carcinogens, a marked decrease in the trimethylation of H3K9 and, especially, H4K20 in the preneoplastic livers has been demonstrated (Tryndyak et al. 2006; Pogribny, Tryndyak, Woods et al. 2007). Similar changes were also detected during estrogen-induced mammary gland carcinogenesis (Kovalchuk et al. 2007) and lung tumorigenesis (Van den Broeck et al. 2008). More importantly, changes in the normal histone modification pattern precede the formation of preneoplastic lesions, thereby indicating the significance of epigenetic events in the induction of oncogenic pathways during the early stages of carcinogenesis (Kovalchuk et al. 2007; Van den Broeck et al. 2008; Starlard-Davenport et al. 2010; Pogribny et al. 2010).

Emerging evidence suggests a crucial role of histone H3K9 and H4K20 trimethylation in the maintenance of genomic stability (Schotta et al. 2004; Kourmouli et al. 2004). One of the primary functions of H3K9, H3K27, and H4K20 trimethylation is the formation of constitutive heterochromatin. These marks are concentrated at centric, pericentric, and telomeric chromosomal regions and are considered to be a prominent epigenetic signature of silenced heterochromatin (Schotta et al. 2004; Kourmouli et al. 2004; Martens et al. 2005; Jenuwein 2006). Loss of H3K9 and H4K20 trimethylation, triggered by the down-regulation of HMTs, including G9a, RIZ1, Suv39h, and Suv4-20h2, markedly compromises genome stability, diminishes the ability of cells to maintain cell cycle arrest, and severely impairs the viability of cells (Peters et al. 2001; Kondo, Shen, Cheng et al. 2008; Peng and Karpen 2009). As mentioned earlier, the decreased levels of trimethylated H3K9 and H4K20 have been shown not only in several forms of human cancer, but also at early preneoplastic stages. This led to a suggestion that low levels of H3K9 and H4K20 trimethylation may contribute to the etiology of cancer and can be used as an indicator and diagnostic marker for neoplastic transformation and tumor growth (Jenuwein 2006; Watanabe et al. 2008).

The mechanistic link between the loss of H3K9, H3K27, and H4K20 trimethylation and cancer development is associated with events that destabilize genomic integrity via chromatin decondensation, the induction of centromere and telomere abnormalities, chromosome segregation defects, and by activation of mobile repetitive DNA elements and proto-oncogenes (Martens et al. 2005; Benetti et al. 2007). The causal role of these lesions, as an integral part of neoplastic transformation in the etiology of cancer, is now commonly accepted (Coleman and Tsongalis 2006).

5.4 DIETARY MODIFICATIONS OF HISTONE LYSINE ACETYLATION AND METHYLATION PATTERNS AND CANCER PREVENTION

5.4.1 Dietary HDAC Inhibitors and Cancer Prevention

Experimental and clinical evidence has convincingly established that correcting of aberrant histone lysine acetylation by suppressing the activity of abnormally elevated HDACs is one of the main epigenetic therapeutic interventions for cancer treatment (reviewed in Glozak and Seto 2007; Lane and Chabner 2009). This is evidenced by the fact that two HDAC inhibitors, Zolinza and Istodax, have been approved by the U.S. Food and Drug Administration for clinical use. HDAC inhibitors have also been speculated to use in cancer prevention.

Several dietary components, such as sulforaphane (SFN), present in cruciferous vegetables, including broccoli, kale, and cauliflower, and diallyl disulfide (DADS), found in *Allium* vegetables, such as garlic, exhibit distinct HDAC inhibitory activity by targeting type I and type II HDAC enzymes (Myzak and Dashwood 2006a, 2006b; Dashwood et al. 2006; Dashwood and Ho 2007; Garfinkel and Ruden 2004) (Figure 5.1). Administration of SFN in the diet to APC^{min} mice, a mutant mouse strain that is highly susceptible to formation of multiple spontaneous adenomas in intestine, resulted in 50% decrease of tumor incidence (Myzak et al. 2006). The

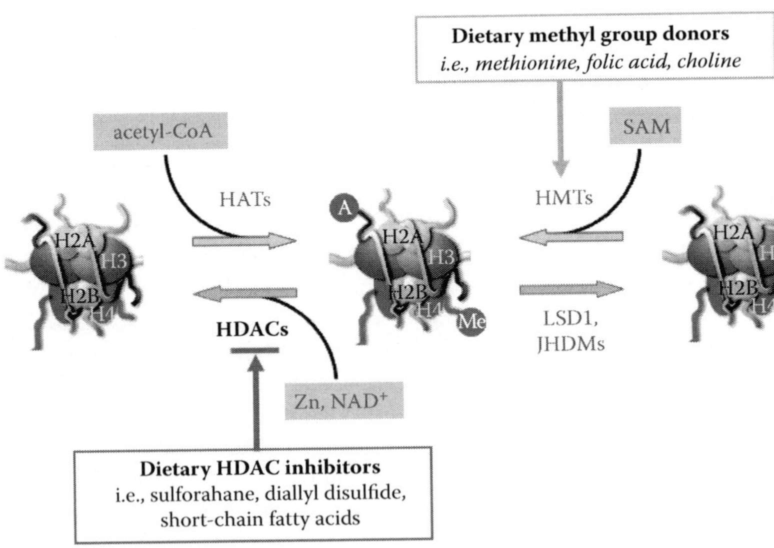

FIGURE 5.1 (Please see color insert following page 80.) Schematic model showing a dynamic state of histone lysine acetylation and methylation. The dysbalance between acetylation and deacetylation processes in cancer cells resulted in inappropriate histone lysine deacetylation during tumorigenesis. Dietary HDAC inhibitors restore the normal pattern of histone lysine acetylation and prevent and/or halt tumorigenesis. Likewise, aberrant histone lysine methylation may be corrected by adequate dietary supplementation of methyl group donors and normalization of HMT functioning.

authors correlate these results with SFN-mediated inhibition of HDAC activity and increased lysine acetylation of histones H3 and H4 (Myzak et al. 2006). Similarly, butyrate, a byproduct of dietary fiber fermentation in the colon, is one of the most effective inhibitors of HDAC activity among other short-chain fatty acids (Myzak and Dashwood 2006b) and is currently recognized as potent chemopreventive agent for colon cancer (Scharlau et al. 2009).

The newly emerged inhibitor of HDAC-1 activity, MCP30, promises to become a powerful tool in prostate cancer therapy (Xiong et al. 2009). MCP30 is a protein isolated from the bitter melon *Momordica charantia,* grown in tropical climates. Inhibition of HDAC-1 activity by MCP30 results in increased acetylation on histones 3 and 4. Remarkably, these features were paralleled by increased levels of apoptosis, observed selectively in prostate cancer cells, but not in normal prostate cells (Xiong et al. 2009). Moreover, similar results were obtained from studies using a premalignant prostate cancer cell line, suggesting that the bitter melon derivate MCP30 may be a potential cancer preventive agent.

Interestingly, the activation of sirtuins, which are class III HDACs, by dietary polyphenols, such as resveratrol, found in grapes and peanuts, have been shown to increase life span and delay cancer development (Tissenbaum and Guarente 2001; Delage and Dashwood 2008). Although the pro-longevity effects of resveratrol in mammals have been recently challenged by some studies (Bass et al. 2007; Kaeberlein 2010), the cancer therapeutic and cancer preventive abilities are well recognized and accepted (Jang et al, 1997; Carbó et al. 1999; Hecht et al. 1999; Schneider et al. 2001; Banerjee et al. 2002). Other polyphenols with anticancer activity include: butein (Y. Wang et al. 2005), which is found in the stems of *Rhus verniciflua*; quercetin (Shaik et al. 2006), which is present in berries, apples, and onions; and piceatannol (Potter et al. 2002), which is contained in blueberries.

Recent studies suggest that patterns of histone lysine acetylation can also be regulated by micronutrients in the diet. Treatment of human prostate cancer cells with selenium (Se) resulted in decreased HDAC activity, paralleled by an increase in the levels of H3K9 acetylation (Xiang et al. 2008). Moreover, treatment with selenite led to increased acetylation of glutathione-S-transferase gene (GSTP1) promoter, a phase II detoxification enzyme that protects cells from carcinogen-mediated damage. Interestingly, the aforementioned cruciferous vegetables that are rich in sulforaphane also exhibited high levels of selenium.

5.4.2 Dietary Methyl Group Donors and Cancer Prevention

The methyl groups that are needed for all cellular biological methylation reactions, including histone lysine methylation, are acquired from SAM, the primary universal donor of methyl groups in mammals, derived from methionine in the one-carbon metabolic pathway (Chiang et al. 1996). This indispensably connects the status of histone methylation to the functioning of the one-carbon metabolic pathway. There are two types of risk factors that may compromise the normal functioning of the one-carbon metabolic pathway and subsequently alter the cellular epigenetic profile. The first group consists of nonmodifiable genetic risk factors, such as genetic variations in genes encoding enzymes involved in the cellular one-carbon metabolism. The second group consists of potentially modifiable factors, such as essential nutrients, involved in the metabolism of methyl-groups, including methionine, choline, folic acid, and vitamins B_6 and B_{12}. Indeed, there is an extensive amount of data showing the regulatory effect of dietary methyl group donors on histone lysine methylation (Pogribny et al. 2006; Pogribny, Tryndyak, Woods et al., 2007; Davison et al. 2009; Mehedint et al. 2010) (Figure 5.1).

Previously, we and other researchers have demonstrated that a long-term inadequate supply of methionine, choline, folic acid, and vitamin B_{12} resulted in significant alterations in histone lysine methylation in the livers of male rats and mice and development of liver tumors (Pogribny et al. 2006; Zhou et al. 2008; Pogribny et al. 2009). In this respect, the accustomed Western diet, which primarily consists of meat and is usually low in both fruits and vegetables and therefore is imbalanced in terms of nutrients involved in one-carbon metabolism, may be one of the major causes of human cancer induced by imbalanced diet (Zhou et al. 2008). Indeed, existing evidence implicates HMTs and histone lysine methylation as potential targets for cancer preventive dietary correction. RIZ1, which exerts its tumor suppression properties by methylating H3K9, has been shown to be involved in methyl-imbalanced diet-mediated liver carcinogenesis in the rodent model (Zhou et al. 2008). It has been proposed that maintenance of RIZ1 activity may serve as a crucial cancer preventive mechanism (Huang 2008; Zhou et al. 2008).

Also, histone lysine methylation can be regulated by treatment with micronutrients, specifically selenium. Treatment of LNCaP prostate cancer cells with selenite resulted in decreased levels of methylated H3K9 (Xiang et al. 2008). In particular, implementation of Se led to reduced levels of H3K9 methylation associated with the GSTP1 promoter, which is silenced in the LNCaP prostate cancer cell line.

5.4.3 Diet and Other Histone Modifications and Cancer Prevention

Recent evidence suggests that dietary prevention of cancer is not limited to correction of aberrant histone lysine acetylation and methylation. At least two lysine residues on histone 4 (lysine 8 and lysine 12) can be modified by the covalent addition of the vitamin biotin (Camporeale et al. 2004; Hassan and Zempleni 2006). Additional biotinylation sites were more recently identified on histone 2A (H2A) at lysines 9, 13, 125, 127, and 129 (Chew et al. 2006), and histone H3 at lysines 4, 9, 18, and 23 (Kobza et al. 2005). The process of biotinylation is catalyzed by two enzymes—biotinidase and holocarboxylase synthetase; however, debiotinylases have yet to be identified (Narang et al. 2004; Camporeale et al. 2006; Hassan and Zempleni 2006; Davis and Ross 2007). Histone biotinylation is correlated with the mitotic condensation of chromatin structures, resulting in gene silencing. The biotinylation of lysine 12 on histone 4 (H4K12) is often associated with repetitive regions and has been identified as an epigenetic mechanism for the repression of long terminal repeat retrotransposons (Camporeale et al. 2007; Zempleni et al. 2009). As with methylation, biotinylation of histone residues seems to be dependent on the exogenous levels of substrate. Biotin deficiency has been assumed to be associated with abnormal chromatin structure, which implicitly links histone biotinylation to cancer. Further studies are needed to elucidate the exact mechanisms underlying histone biotinylation and its correlation with carcinogenesis.

Genistein, the isoflavone abundant in fava beans and soybeans, has been characterized as another chemopreventive agent. It is believed to have a strong potential for the significant reduction of the breast and prostate cancer rates (Lamartiniere et al. 1995; Gong et al. 2003; Molinie and Georgel 2009). One of the possible mechanisms through which genistein exhibits it chemopreventive effects is by targeting phosphorylation of H1 and H3 histones.

5.5 CONCLUSIONS AND REMARKS

Many nutrients have been reported to affect directly only one type of histone modification; however, considering a tight link between histone modifications, dietary factors may influence indirectly other types of histone modifications. The first reports suggesting such an interrelationship appeared over 30 years ago when rats treated with a biotin-deficient diet were characterized by decreased levels of phosphorylation and methylation on histone residues, in parallel with increased levels of acetylation in the liver tissue (Petrelli et al. 1978). Additionally, histone modifications and histone-modifying enzymes also interact with DNA methylation machinery, including DNA methyltransferases and methyl-binding proteins, influencing DNA methylation and vice versa (reviewed in Guil and Esteller 2009). Despite the notable progress in identifying epigenetic mechanisms of chemoprevention, and the role of histone modifications in particular, there are still a lot of gaps that need to be filled. Specifically, more research is required on the dose and timing for the dietary constituent in cancer prevention. Studies similar to those performed with drugs such as pharmacodynamics and pharmacogenomics are also necessary for dietary agents (including interactions of agent with other agents and interactions of agent with other genetic and epigenetic factors), as well as the side effects and long-term consequences of administrating a dietary compound, particularly when consumed in isolation rather than in dietary form. These studies will provide information to assist in selecting the most beneficial dietary patterns as well as dietary supplements while at the same time identify safe dietary factors for cancer prevention.

REFERENCES

Allfrey, V. G., R. Faulkner, and A. E. Mirsky. 1964. Acetylation and methylation of histones and their possible role in the regulation of RNA synthesis. *Proc Natl Acad Sci U S A* 51:786–94.

Aravin, A. A., R. Sachidanandam, D. Bouc'his, C. Schaefer, D. Pezic, K. F. F. Toth, T. Bestor, and G. J. Hannon. 2008. A piRNA pathway primed by individual transposons is linked to de novo DNA methylation in mice. *Mol Cell* 31:785–99.

Arita, K., M. Ariyoshi, H. Tochio, Y. Nakamura, and M. Shirakawa. 2008. Recognition of hemi-methylated DNA by the SRA protein UHRF1 by a base flipping mechanism. *Nature* 455:818–21.

Avvakumov, G. V., J. R. Walker, S. Xue, Y. Li, S. Duwan, C. Bronner, C. H. Arrowsmith, and S. Dhe-Paganon. 2008. Structural basis for recognition of hemimethylated DNA by the SRA domain of human UHRF1. *Nature* 455:822–25.

Banerjee, S., C. Bueso-Ramos, and B. B. Aggarwal. 2002. Suppression of 7,12-dimethylbenz(a) anthracene-induced mammary carcinogenesis in rats by resveratrol: role of nuclear factor-kappaB, cyclooxygenase 2, and matrix metalloprotease 9. *Cancer Res* 62:494–54.

Bass, T. M., D. Weinkove, K. Houthoofd, D. Gems, and L. Partrideg. 2007. Effects of resveratrol on lifespan in *Drosophila melanogaster* and *Caenorhabditis elegans*. *Mech Ageing Dev* 128:546–52.

Benetti, R., S. Gonzalo, I. Jaco, G. Schotta, P. Klatt, T. Jenuwein, and M. A. Blasco. 2007. Suv4-20h deficiency results in telomere elongation and derepression of telomere recombination. *J Cell Biol* 178:925–36.

Berger, S. L. 2007. The complex language of chromatin regulation during transcription. *Nature* 447:407–12.

Bernstein, B. E., A. Meissner, and E. S. Lander. 2007. The mammalian epigenome. *Cell* 128:669–81.

Bird, A. P., and A. P. Wolffe. 1999. Methylation-induced repression—belts, braces, and chromatin. *Cell* 99:451–54.

Brinkman, A. B., T. Roelofsen, S. W. Pennings, J. H. Martens, T. Jenuwein, and H. G. Stunnenberg. 2006. Histone modifications patterns associated with the human X chromosome. *EMBO Rep* 7:628–34.

Brown, R., and G. Strathdee. 2002. Epigenomics and epigenetic therapy of cancer. *Trends Mol Med* 8:S43–48.

Camporeale, G., E. Giordano, R. Rendina, J. Zempleni, and J. C. Eissenberg. 2006. *Drosophila* holocarboxylase synthetase is a chromosomal protein required for normal histone biotinylation, gene transcription patterns, lifespan and tolerance. *J Nutr* 136:2735–42.

Camporeale, G., A. M. Oommen, J. B. Griffin, G. Sarath, and J. Zempleni. 2007. K12-biotinylated histone H4 marks heterochromatin in human lymphoblastoma cells. *J Nutr Biochem* 18:760–68.

Camporeale, G., E. E. Shubert, G. Sarath, R. Cerny, and J. Zempleni. 2004. K8 and K12 are biotinylated in human histone H4. *Eur J Biochem* 271:2257–63.

Carbó, N., P. Costelli, F. M. Baccino, F.J. López-Soriano, and J. M. Argilés. 1999. Resveratrol, a natural product present in wine, decreases tumour growth in a rat tumour model. *Biochem Biophys Res Commun* 254:739–43.

Chew, Y. C., G. Camporeale, N. Kothapalli, G. Sarath, and J. Zempleni. 2006. Lysine residues in N- and C-terminal regions of human histone H2A are targets for biotinylation by biotinidase. *J Nutr Biochem* 17:225–33.

Chiang, P. K., R. K. Gordon, J. Tal, G. C. Zeng, B. P. Doctor, K. Pardhasaradhi, and P. P. McCann. 1996. S-adenosylmethionine and methylation. *FASEB J* 10:471–80.

Coleman, W. B., and G. J. Tsongalis. 2006. Molecular mechanisms of human carcinogenesis. *EXS* 96:321–49.

Cortez, C. C., and P. A. Jones. 2008. Chromatin, cancer and drug therapies. *Mutat Res* 647:44–51.

Dashwood, R. H., and E. Ho. 2007. Dietary histone deacetylase inhibitors: from cells to mice to man. *Semin Cancer Biol* 17:363–69.

Dashwood, R. H., M. C. Myzak, and E. Ho. 2006. Dietary HDAC inhibitors: time to rethink weak ligands in cancer chemoprevention? *Carcinogenesis* 27:344–49.

Davis, C. D., and S. A. Ross. 2007. Ditary components impact histone modifications and cancer risk. *Nutr Rev* 65:88–94.

Davison, J. M., T. J. Mellott, V. P. Kovacheva, and J. K. Blusztajn. 2009. Gestational choline supply regulates methylation of histone H3, expression of histone methyltransferases G9a (Kmt1c) and Suv39h1 (Kmt1a), and DNA methylation of their genes in rat fetal liver and brain. *J Biol Chem* 284:1982–89.

Delage, B., and R. H. Dashwood. 2008. Dietary manipulation of histone structure and function. *Annu Rev Nutr* 28:347–66.

Egger, G., G. Liang, A. Aparicio, and P. A. Jones. 2004. Epigenetics in human disease and prospects for epigenetic therapy. *Nature* 429:457–63.

Ellis, L., P. W. Atadja, and R. W. Johnstone. 2009. Epigenetics in cancer: targeting chromatin modifications. *Mol Cancer Ther* 8:1409–20.

El-Osta, A. 2003. DNMT cooperativity—the developing links between methylation, chromatin structure and cancer. *Bioessays* 25:1071–84.

Elsheikh, S. E., A. R. Green, E. A. Rakha, D. G. Powe, R. A. Ahmed, H. M. Collins, D. Soria, J. M. Garibaldi, C. E. Parish, A. A. Ammar, M. J. Grainge, G. R. Ball, M. K. Abdelghany, L. Martinez-Pomares, D. M. Heery, and I. O. Ellis. 2009. Global histone modifications in breast cancer correlate with tumor phenotypes, prognostic factors, and patient outcome. *Cancer Res* 69:3802–9.

Fabbri, M., R. Garzon, A. Cimmino, Z. Liu, N. Zenesi, E. Callegari, S. Liu, H. Alder, S. Costinean, C. Fernandez-Cymering, S. Volinia, G. Guler, C. D. Morrison, K. K. Chan, G. Marcucci, G. A. Calin, K. Huebner, and C. M. Croce. 2007. MicroRNA-29 family reverts aberrant methylation in lung cancer by targeting DNA methyltransferases 3A and 3B. *Proc Natl Acad Sci U S A* 104:15805–10.

Feinberg, A. P. 2004. The epigenetics of cancer etiology. *Semin Cancer Biol* 14:427–32.

———. 2008. Epigenetics at the epicenter of modern medicine. *JAMA* 299:1345–50.

Feinberg, A. P., R. Ohlsson, and S. Henikoff. 2006. The epigenetic progenitor origin of human cancer. *Nat Rev Genet* 7:21–33.

Fraga, M. F., E. Ballestar, A. Villar-Garea, M. Boix-Chornet, J. Espada, G. Schotta, T. Bonaldi, C. Haydon, S. Ropero, K. Petrie, N. G. Iyer, A. Pérez-Rosado, E. Calvo, J. A. Lopez, A. Cano, M. J. Calasanz, D. Colomer, M. A. Piris, N. Ahn, A., Imhof, C. Caldas, T. Jenuwein, and M. Esteller. 2005. Loss of acetylation at Lys16 and trimethylation at Lys20 of histone H4 is a common hallmark of human cancer. *Nat Genet* 37:391–400.

Garfinkel, M. D., and D. M. Ruden. 2004. Chromatin effects in nutrition, cancer and obesity. *Nutrition* 20:56–62.

Glozak, M. A., and E. Seto. 2007. Histone deacetylases and cancer. *Oncogene* 26:5420–32.

Goll, M. G., and T. H. Bestor. 2005. Eukaryotic cytosine methyltransferases. *Annu Rev Biochem* 74:481–514.

Gong, L., Y. Li, A. Nedeljkovic-Kurepa, and F. H. Sarkar. 2003. Inactivation of NF-kB by genistein is mediated via Akt signaling pathway in breast cancer cells. *Oncogene* 22:4702–9.

Gowher, H., K. Liebert, A. Hermann, G. Xu, and A. Jeltsch. 2005. Mechanism of stimulation of catalytic activity of Dnmt3A and Dnmt3B DNA-(cytosine- C5)-methyltransferases by Dnmt3L. *J Biol Chem* 280:13341–48.

Grønbaek, K., M. Treppendahl, F. Asmar, and P. Guldberg. 2008. Epigenetic changes in cancer as potential targets for prophylaxis and maintenance therapy. *Basic Clin Pharmacol Toxicol* 103:389–96.

Guil, S., and M. Esteller. 2009. DNA methylomes, histone codes and miRNAs: tying it all together. *Int J Biochem Cell Biol* 41:87–95.

Hanahan, D., and R. A. Weinberg. 2000. The hallmarks of cancer. *Cell* 100:57–70.

Hashimoto, H., J. R. Horton, X. Zhang, M. Bostick, S. E. Jacobsen, and X. Cheng. 2008. The SRA domain of UHRF1 flips 5-methylcytosine out of the DNA helix. *Nature* 455:826–29.

Hassan, Y. I., and J. Zempleni. 2006. Epigenetic regulation of chromatin structure and gene function by biotin. *J Nutr* 136:1763–65.

Hecht, S. S., P. M. Kenney, M. Wang, N. Trushin, S. Agarwal, A. V. Rao, P. Upadhyayaand. 1999. Evaluation of butylated hydroxyanisole, myo-inositol, curcumin, esculetin, resveratrol and lycopene as inhibitors of benzo[a]pyrene plus 4-(methylnitrosamino)-1-(3-pyridyl)-1-butanone-induced lung tumorigenesis in A/J mice. *Cancer Lett* 137:123–30.

Huang, S. 2008. Histone methylation and the initiation of cancer. In *Cancer epigenetics,* ed. T. Tollefsbol. New York: CRC Press, 109–50.

Hublitz, P., M. Albert, and A. H. Peters. 2009. Mechanisms of transcriptional repression by histone lysine methylation. *Int J Dev Biol* 53:335–54.

Ideraabdullah, F. Y., S. Vigneau, and M. S. Bartolomei. 2008. Genomic imprinting mechanisms in mammals. *Mutat Res* 647:77–85.

Iizuka, M., and M. M. Smith. 2003. Functional consequences of histone modifications. *Curr Opin Genet Dev* 13:154–60.

Jaffe, L. F. 2003. Epigenetic theories of cancer initiation. *Adv Cancer Res* 90:209–30.

Jang, M., L. Cai, G. O. Udeani, C. F. Thomas, C. W. Beecher, H. H. Fong, N. R. Farnsworth, A. D. Kinghorn, R. G. Mehta, R. C. Moon, and J. M. Pezzuto. 1997. Cancer chemopreventive activity of resveratrol, a natural product derived from grapes. *Science* 275:218–20.

Jenuwein, T. 2006. The epigenetic magic of histone lysine methylation. *FEBS J* 273:3121–35.

Jenuwein, T., and C. D. Allis. 2001. Translating the histone code. *Science* 293:1074–80.

Jones, P. A., and S. B. Baylin. 2007. The epigenomics of cancer. *Cell* 128:683–92.

Kaeberlein, M. 2010. Resveratrol and rapamycin: are they anti-aging drugs? *Bioessays* 32:96–99.

Karpinets, T. V., and B. D. Foy. 2005. Tumorigenesis: the adaptation of mammalian cells to sustained stress environment by epigenetic alterations and succeeding matched mutations. *Carcinogenesis* 26:1323–34.

Kim, J. K., J. Samaranayake, and S. Pradhan. 2009. Epigenetic mechanisms in mammals. *Cell Mol Life Sci* 66:596–612.

Kimura, H., and K. Shiota. 2003. Methyl-CpG-binding protein, MeCP2, is target molecule for maintenance DNA methyltransferase, Dnmt1. *J Biol Chem* 278:4806–12.

Klose, R. J., K. Yamane, Y. Bae, D. Zhang, H. Erdjument-Bromage, P. Tempst, J. Wong, and Y. Jhang. 2006. The transcriptional repressor JHDM3A demethylates trimethyl histone H3 lysine 9 and lysine 36. *Nature* 442:312–16.

Kobza, K., G. Camporeale, B. Rueckert, A. Kueh, J. B. Griffin, G. Sarath, and J. Zemplenil. 2005. K4, K9, and K18 in human histone H3 are targets for biotinylation by biotinidase. *FEBS J* 272:4249–59.

Kondo,Y., and J. P. Issa. 2004. Epigenetic changes in colorectal cancer. *Cancer Metastasis Rev* 23:29-39.

Kondo, Y., L. Shen, S. Ahmed, Y. Boumber, Y. Sekido, B. R. Haddad, and J. P. Issa. 2008. Downregulation of histone H3 lysine 9 methyltransferase G9a induces centrosome disruption and chromosome instability in cancer cells. *PLos One* 3:e2037.

Kondo, Y., L. Shen, A. S. Cheng, S. Ahmed, Y. Boumber, C. Charo, T. Yamochi, T. Urano, K. Furukawa, B. Kwabi-Addo, D. L. Gold, Y. Sekido, T. H. Huang, and J. P. Issa. 2008. Gene silencing in cancer by histone H3 lysine 27 trimethylation independent of promoter DNA methylation. *Nat Genet* 40:741–50.

Kopelovich, L., J. A. Crowell, and J. R. Fay. 2003. The epigenome as a target for cancer prevention. *J Natl Cancer Inst* 95:1747–57.

Kourmouli, N., P. Jeppesen, S. Mahadevhaiah, P. Burgoyne, R. Wu, D. M. Gilbert, S. Bongiorni, G. Prantera, L. Fanti, S. Pimpinelli, W. Shi, R. Fundele, P. B. Singh. 2004. Heterochromatin and tri-methylated lysine 20 of histone H4 in animals. *J Cell Sci* 117:2491–501.

Kouzarides, T. 2007. Chromatin modifications and their function. *Cell* 128:693–705.

Kovalchuk, O., V. P. Tryndyak, B. Montgomery, A. Boyko, K. Kutanzi, F. Zemp, A. R. Warbritton, J. R. Latendresse, I. Kovalchuk, F. A. Beland, and I. P. Pogribny. 2007. Estrogen-induced rat breast carcinogenesis is characterized by alterations in DNA methylation, histone modifications and aberrant microRNA expression. *Cell Cycle* 6:2010–18.

Lamartiniere, C. A., J. Moore, M. Holland, and S. Barnes. 1995. Neonatal genistein chemoprevents mammary cancer. *Proc Soc Exp Biol Med* 208:120–23.

Lane, A. A., and B. A. Chabner. 2009. Histone deacetylases inhibitors in cancer therapy. *J Clin Onocl* 27:5469–68.

Li, E., and A. Bird. 2007. DNA methylation in mammals. In *Epigenetics,* ed. C.D. Allis, T. Jenuwein, and D. Reinberg. Cold Spring Harbor, NY: Cold Spring Harbor Laboratory Press, 341–56.

Lennartsson, A., and K. Ekwall. 2009. Histone modifications and epigenetic codes. *Biochim Biophys Acta* 1790:863–68.

Liang, G., M. F. Chan, Y. Tomigahara, Y. C. Tsai, F. A. Gonzales, E. Li, P. W. Laird, and P. A. Jones. 2002. Cooperativity between DNA methyltransferases in the maintenance methylation of repetitive elements. *Mol Cell Biol* 22:480–91.

Mariadason, J. M. 2008. HDACs and HDAC inhibitors in colon cancer. *Epigenetics* 3:28–37.

Martens, J. H., R. J. O'Sullivan, U. Braunschweig, S. Opravil, M. Radolf, P. Steinlein, and T. Jenuwein. 2005. The profile of repeat-associated histone lysine methylation states in the mouse epigenome. *EMBO J* 24:800–12.

Mehedint, M. G., M. D. Niculescu, C. N. Craciunescu, and S. H. Zeisel. 2010. Choline deficiency alters global histone methylation and epigenetic marking at the Re1 site of the calbindin 1 gene. *FASEB J* 24:184–95.

Michan, S., and D. Sinclair. 2007. Sirtuins in mammals: insights into their biological function. *Biochem J* 404:1–13.

Miranda, T. B., and P. A. Jones. 2007. DNA methylation: the nuts and bolts of repression. *J Cell Physiol* 213:384–90.

Molinie, B., and P. Georgel. 2009. Genetic and epigenetic regulations of prostate cancer by genistein. *Drug News Perspect* 22:247–54.

Myzak, M. C., and R. H. Dashwood. 2006a. Chemoprotection by sulforaphane: keep on eye beyond keap-1. *Cancer Lett* 233:208–18.

———. 2006b. Histone deacetylases as targets for dietary cancer preventive agents: lessons learned with butyrate, diallyl disulfide and sulforaphane. *Curr Drug Targets* 7:443–52.

Myzak, M. C., W. M. Dashwood, G. A. Orner, E. Ho, and R. H. Dashwood. 2006. Sulforaphane inhibits histone deacetylases in vivo and suppresses tumorigenesis in Apc-minus mice. *FASEB J* 20:506–8.

Narang, M. A., R. Dumas, L. M.Ayer, and R. A. Gravel. 2004. Reduced histone biotinylation in multiple carboxylase deficiency patients: a nuclear role for holocarboxylase synthetase. *Hum Mol Genet* 13:15–23.

Neff, T., and S. A. Armstrong. 2009. Chromatin maps, histone modifications and leukemia. *Leukemia* 23:1243–51.

Peng, J. C., and G. H. Karpen. 2009. Heterochromatic genome stability requires regulators of histone H3K9 methylation. *PLoS Genet* 5:21000435.

Peters, A. H., D. O'Carroll, H. Scherthan, K. Mechtler, S. Sauer, C. Schöfer, K. Weipoltshammer, M. Pagani, M. Lachner, A. Kohlmaier, S. Opravil, M. Doyle, M. Sibilia, and T. Jenuwein. 2001. Loss of the Suv39h histone methyltransferases impairs mammalian heterochromatin and genome stability. *Cell* 107:323–37.

Petrelli, F., S. Coderoni, P. Moretti, and M. Paparelli. 1978. Effect of biotin phosphorylation, acetylation, methylation of rat liver histones. *Mol Biol Rep* 4:87–92.

Pogribny, I. P., L. Muskhelishvili, V. P. Tryndyak, and F. A. Beland. 2010. The role of epigenetic events in genotoxic hepatocarcinogenesis induced by 2-acetylaminofluorene. *Mutat Res* (in press).

Pogribny, I. P., S. A. Ross, V. P. Tryndyak, M. Pogribna, L. A. Porier, and T. V. Karpinets. 2006. Histone H3 lysine and H4 lysine 20 trimethylation and the expression of Suv4-20h2 and Suv-39h1histone methyltransferases in hepatocarcinogenesis induced by methyl deficiency in rats. *Carcinogenesis* 27:1180–86.

Pogribny, I. P., V. P. Tryndyak, L. Muskhelishvili, I. Rusyn, and S. A. Ross. 2007. Methyl deficiency, alterations in global histone modifications, and carcinogenesis. *J Nutr* 137:216S–22s.

Pogribny, I. P., V. P. Tryndyak, C. G. Woods, S. E. Witt, and I. Rusyn. 2007. Epigenetic effects of the continuous exposure to peroxisome proliferator Wy-13,643 in mouse liver are dependent upon peroxisome proliferator activated receptor alpha. *Mutat Res* 625:62–71.

Pogribny, I. P., V. P. Tryndyak, T. V. Bagnyukova, S. Melnyk, B. Montgomery, S. A. Ross, J. R. Latendresse, I. Rusyn, and F. A. Beland. 2009. Hepatic epigenetic phenotype predetermines individual susceptibility to hepatic steatosis in mice fed a lipogenic methyl-deficient diet. *J Hepatol* 51:176–86.

Potter, G. A., L. H. Patterson, E. Wanogho, P. J. Perry, P. C. Butler, T. Ijaz, K. C. Ruparelia, J. H. Lamb, P. B. Farmer, L. A. Stanley, and M. D. Burke. 2002. The cancer preventative agent resveratrol is converted to the anticancer agent piceatannol by the cytochrome P450 enzyme CYP1B1. *Br J Cancer* 86:774–78.

Razin, A., and H. Cedar. 1994. DNA methylation and genomic imprinting. *Cell* 77:473–76.

Razin, A., and A. D. Riggs. 1980. DNA methylation and gene function. *Science* 210:604–10.

Ropero, S., and M. Esteller. 2007. The role of histone deacetylases (HDACs) in human cancer. *Mol Oncol* 1:19–25.

Scharlau, D., A. Borowicki, N. Habermann, T. Hofmann, S. Klenow, C. Miene, U. Munjal, K. Stein, and M. Glei. 2009. Mechanisms of primary cancer prevention by butyrate and other products formed during gut flora-mediated fermentation of dietary fibre. *Mutat Res* 682:39–53.

Schlesinger, Y., R. Straussman, R., I. Keshet, S. Farkash, M. Hecht, J. Zimmerman, E. Eden, Z. Yakhini, E. Ben-Shushan, B. E. Reubinoff, Y. Bergman, I. Simon, and H. Cedar. 2007. Polycomb-mediated methylation of Lys27 of histone H3 pre-marks genes for de novo methylation in cancer. *Nat Genet* 39:232–36.

Schneider, Y., B. Duranton, F. Gossé, R. Schleiffer, N. Seiler, and F. Raul. 2001. Resveratrol inhibits intestinal tumorigenesis and modulates host-defense-related gene expression in an animal model of human familial adenomatous polyposis. *Nutr Cancer* 39:102–7.

Schotta, G., M. Lachner, K. Sarma, A. Ebert, R. Sengupta, G. Reuter, D. Reinberg, and T. Jenuwein. 2004. A silencing pathway to induce H3-K9 and H4-K20 trimethylation at constitutive heterochromatin. *Genes Dev* 18:1251–62.

Shaik, Y. B., M. L. Castellani, A. Perella, F. Conti, V. Salini, S. Tete, B. Madhappan, J. Vecchiet, M. A. De Lutiis, A. Caraffa, and G. Cerulli. 2006. Role of quercetin (a natural herbal compound) in allergy and inflammation. *J Biol Regul Homeost Agents* 20:47–52.

Sharma, S., T. K. Kelly, and P. A. Jones. 2010. Epigenetics in cancer. *Carcinogenesis* 31:27–36.

Shi, Y., F. Lan, C. Matson, P. Mulligan, J. R. Whetstine, P. A. Cole, R. A. Casero, and Y. Shi. 2004. Histone demethylation mediated by the nuclear amine oxidase homolog LSD1. *Cell* 119:941–53.

Shukla, V., T. Vaissière, and Z. Herceg. 2008. Histone acetylation and chromatin signature in stem cell identity and cancer. *Mutat Res* 637:1–15.

Singer-Sam, J., and A. D. Riggs. 1993. X-chromosome inactivation and DNA methylation. *EXS* 64:358–84.

Smith, B.C., and J. M. Denu. 2009. Chemical mechanisms of histone lysine and arginine modifications. *Biochim Biophys Acta* 1789:45–57.

Snowden, A. W., P. D. Gregory, C. C. Case, and C. O. Pabo. 2002. Gene-specific targeting of H3K9 methylation is sufficient for initiating repression in vivo. *Curr Biol* 12:2159–66.

Starlard-Davenport, A., V. P. Tryndyak, S. R. James, A. R. Karpf, J. R. Latendresse, F. A. Beland, and I. P. Pogribny. 2010. Mechanismas of epigenetic silencing of the Rassf1a gene during estrogen-induced breast carcinogenesis in ACI rats. *Carcinogenesis* (in press).

Tissenbaum, H. A., and L. Guarente. 2001. Increased dosage of a sir-2 gene extends lifespan in *Caenorhabditis elegans*. *Nature* 410:227–30.

Tryndyak, V. P., L. Muskhelishvili, O. Kovalchuk, R. Rodriguez-Juarez, B. Montgomery, M. I. Churchwell, S. A. Ross, F. A. Beland, and I. P. Pogribny. 2006. Effect of long-term tamoxifen exposure on genotoxic and epigenetic changes in rat liver: implications for tamoxifen-induced hepatocarcinogenesis. *Carcinogenesis* 27:1713–20.

Tsukada, Y., J. Fang, H. Erdjument-Bromage, M. E. Warren, C. H. Borchers, P. Tempst, and Y. Zhang. 2006. Histone demethylation by family of JnjC domain-containing proteins. *Nature* 439:811–16.

van den Broeck, A., E. Brambilla, C. Moro-Sibilot, S. Lantuejoul, C. Brambilla, B. Eymin, S. Khochbin, and S. Gazzeri. 2008. Loss of histone H4K20 trimethylation occurs in preneoplasia and influences prognosis of non-small cell lung cancer. *Clin Cancer Res* 14:7237–45.

Vogelstein, B., and K. W. Kinzler. 2004. Cancer genes and the pathways they control. *Nat Med* 10:789-99.

Wang, J., S. Hevi, J. K. Kurash, H. Lei, F. Gay, J. Bajko, H. Su, W. Sun, H. Chang, G. Xu, F. Gaudet, E. Li, and T. Chen. 2009. The lysine demethylase LSD1 (KDM1) is required for maintenance of global DNA methylation. *Nat Genet* 41:125–29.

Wang, Y., F. L. Chan, S. Chen, and L. K. Leung. 2005. The plant polyphenol butein inhibits testosterone-induced proliferation in breast cancer cells expressing aromatase. *Life Sci* 77:39–51.

Watanabe, H., K. Soejima, H. Yasuda, I. Kawada, I. Nakachi, S. Yoda, K. Naoki, and A. Ishizaka. 2008. Deregulation of histone lysine methyltransferases contributes to oncogenic transformation of human bronchoepithelial cells. *Cancer Cell Int* 8:15.

Wei, Y., W. Xia, Z. Zhang, J. Liu, H. Wang, N. V. Adsay, C. Albarracin, D. Yu, J. L. Abbruzzese, G. B. Mills, R. C. Bast Jr., G. N. Hortobagyi, and M. C. Hung. 2008. Loss of trimethylation at lysine 27 of histone H3 is a predictor of poor outcome in breast, ovarian, and pancreatic cancers. *Mol Carcinog* 47:701–6.

Xiang, N., R. Zhao, G. Song, and W. Zhong. 2008. Selenite reactivates silenced genes by modifying DNA methylation and histones in prostate cancer cells. *Carcinogenesis* 29:2175–81.

Xiong, S. D., K. Yu, X. H. Liu, L. H. Yin, A. Kirshenbaum, S. Yao, G. Narla, A. DiFeo, J. B. Wu, Y. Yuan, S. M. Ho, Y. W. Lam, and A. C. Levine. 2009. Ribosome-inactivating proteins isolated from dietary bitter melon induce apoptosis and inhibit histone deacetylases-1 selectively in premalignant and malignant prostate cancer cells. *Int J Cancer* 125:774–82.

Yang, X. J., and E. Seto. 2007. HATs and HDACs: from structure, function and regulation to novel strategies for therapy and prevention. *Oncogene* 26:5310–18.

Yoder, J. A., C. P. Walsh, and T. H. Bestor. 1997. Cytosine methylation and the ecology of intragenomic parasites. *Trends Genet* 13:335–40.

Zempleni, J., Y. C. Chew, B. Bao, V. Pestinger, and S. S. Wijeratne. 2009. Repression of transposable elements by histone biotinylation. *J Nutr* 139:2389–92.

Zhou, W., S. Alonso, D. Takai, S. C. Lu, F. Yamamoto, M. Perucho, and S. Huang. 2008. Requirement of RIZ1 for cancer prevention by methyl-balanced diet. *PLoS One* 3:e3390.

Zhu, H., T. M. Geiman, S. Xi, Q. Jhiang, A. Schmidtmann, T. Chen, E. Li, and K. Muegge. 2006. Lsh is involved in de novo methylation of DNA. *EMBO J* 25:335–45.

6 Nutrition, Epigenetics, and Vascular Function

M. Carey Satterfield, Jason R. McKnight, Xilong Li, and Guoyao Wu

CONTENTS

6.1 Introduction 125
6.2 Epigenetics and Fetal Programming 126
6.3 Impacts of Maternal Nutrition on Epigenetics and Vascular Function 128
  6.3.1 Maternal Undernutrition 128
  6.3.2 Maternal Overnutrition 130
6.4 Amino Acids as Nutrients to Regulate Epigenetics and Vascular Function 132
6.5 Conclusion and Perspectives 134
Acknowledgments 134
References 134

6.1 INTRODUCTION

Epidemiological studies have suggested a link between the fetal environment (including nutrition) and postnatal health, particularly cardiovascular function or dysfunction, in humans (Barker and Osmond 1986; Chmurzynska 2010; Hill and Duville 2000; Phillips 2007). For example, high rates of death from ischemic heart disease and the metabolic syndrome are positively correlated with low birth weight, which may result, in part, from a reduced availability of nutrients to the fetus in utero (Barker 2007; Hales and Barker 1992). The concept of fetal programming has been experimentally tested in a number of species, including nonhuman primates (Nijland et al. 2010), rats (Anderson et al. 2006), mice (Dunn and Bale 2009), cattle (Cafe et al. 2009), and sheep (Symonds et al. 2009). Of particular interest, effects of changes in nutrition or endocrine status during fetal or neonatal life can be carried forward to subsequent developmental stages (Figure 6.1). This phenomenon may be explained by epigenetics, which is defined as stable and inheritable alterations of genes through covalent modifications of DNA and core histones without changes in DNA sequences (Evertts et al. 2010).

Undernutrition due to a variety of factors (including a limited supply of food, severe nausea and vomiting, early or closely spaced pregnancies, multiple pregnancies, and placental dysfunction) is a significant problem in humans (Marsal 2002). Most of these factors are also common in livestock production (G. Wu et al. 2006).

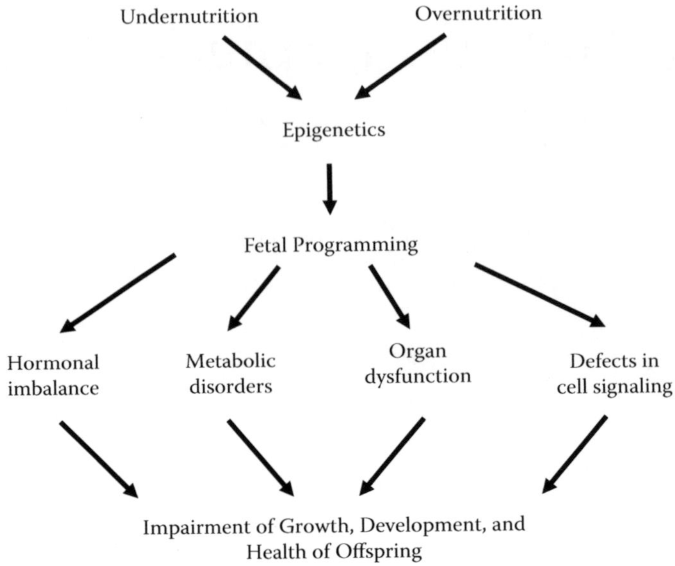

FIGURE 6.1 Impacts of maternal nutrition on fetal programming. Either undernutrition or overnutrition affects expression of the fetal genome, which may have lifelong consequences. Thus alterations in fetal nutrition may result in developmental adaptations that permanently change the structure, physiology and metabolism of the offspring, thereby predisposing individuals to metabolic, endocrine, and cardiovascular diseases in adult life.

On the other side of the nutrition spectrum, obesity is a growing problem worldwide (Hardie et al. 2006; Ogden et al. 2007). Indeed, 65% of women of reproductive age in the United States are overweight or obese (Flegal et al. 2010). Thus recent years have witnessed increasing interest in the effect of maternal obesity on fetal and postnatal health in humans (Catalano 2007; Cleal et al. 2007; Kiel et al. 2007). Likewise, increasing feed provision to dams during a short period of time (termed "flushing") around conception has been employed by livestock producers to enhance ovulation. However, this practice is known to reduce the survival and growth of embryos and fetuses in a number of species, including pigs, cattle, horses, and sheep, because of impaired secretions of progesterone (Han et al. 2004; G. Wu et al. 2006, 2010). Because of ethical concerns with obtaining human fetal tissues, use of animal models is important to understanding the mechanisms responsible for the developmental origins of health and disease in humans and to designing effective means for prevention and treatment. While multiple organs are affected by maternal nutrition during pregnancy (G. Wu et al. 2004, 2006), this review focuses on the cardiovascular system because vascular disease is the leading cause of death in the world (AHA 2010).

6.2 EPIGENETICS AND FETAL PROGRAMMING

Fetal programming can be defined as an adaptive process whereby nutrition and other environmental factors alter developmental pathways during the critical

TABLE 6.1
Effects of Birth Weight on the Hazard Ratios for Coronary Heart Disease and the Cumulative Incidence of Hypertension in Adult Men and Women

	Adult Men		Adult Women	
Birth Weight (g)	Hazard Ratios for Coronary Heart Disease[c]	Cumulative Incidence of Hypertension (%)[b]	Hazard Ratios for Coronary Heart Disease[a]	Cumulative Incidence of Hypertension (%)[d]
≤ 2500	3.63	—	1.34	—
2501–3000	1.86	19.0	1.38	21.1
3001–3500	1.99	17.0	1.24	16.3
3501–4000	2.08	14.1	1.17	13.0
> 4000	1.00	12.5	1.00	12.1
P value for trend	0.006	< 0.001	0.007[e]	< 0.001

[a] Eriksson et al. (2001). A cohort of 4630 adult men was studied.
[b] Barker et al. (2002). A cohort of 4630 adult men was studied.
[c] Forsen et al. (1999). A cohort of 3447 adult women was studied.
[d] Barker et al. (2002). A cohort of 4130 adult women was studied.
[e] Adjusted for gestation and placental weight.

period of prenatal growth, therefore inducing changes in postnatal metabolism and chronic disease susceptibility (G. Wu et al. 2004). Such changes may provide a mechanism for the adaptation of the affected offspring to the extrauterine environment that may confer an evolutionary advantage for the survival of the species. However, when environmental cues during prenatal life inappropriately program offspring, there are adverse consequences, including the increased prevalence of disease in adult life. For example, birth weight was inversely correlated with the incidences of coronary heart disease and hypertension in a cohort of 4630 adult men (Barker et al. 2002; Eriksson et al. 2001) and 3447–4130 women (Barker et al. 2002; Forsen et al. 1999) born in Helsinki, Finland, between 1924 and 1933 (Table 6.1). Moreover, among adult men and women, a small size at birth was associated with increased mortality rates from cardiovascular causes (Kajantie et al. 2005). Additionally, studies from animal models have documented that maternal undernutrition or overfeeding results in cardiovascular and renal dysfunctions as adults (Cleal et al. 2007; Han et al. 2004; Evertts et al. 2010).

The genetic code established by the DNA sequence exhibits minimal rates of change after the formation of the diploid chromatin state at fertilization (Evertts et al. 2010). Therefore a search for alternative mechanisms of gene regulation was undertaken. These studies resulted in the now widely accepted notion that changes in gene expression can be manifested by mitotically and/or meiotically heritable alterations in the DNA-protein complex without any change in the DNA sequence (C. Wu and Morris 2001). This phenomenon was originally termed "epigenetics," derived from the Greek prefix "epi," which means "over" or "above." At present, four mechanisms mediating epigenetic effects are: (1) chromatin modifications, (2) DNA methylation (occurring in 5'-positions of cytosine residues within CpG dinucleotides throughout the mammalian genome), (3) histone modifications (acetylation, methylation, and phosphorylation), and (4) RNA-based mechanisms such as noncoding RNA or inhibitory RNAs (Evertts et al. 2010). The enzymes involved in these reactions include specific DNA and protein methyltransferases, DNA demethylases, histone acetylase (lysine acetyltransferase), and GCN5-related N-acetyltransferase (a super family of acetyltransferase). DNA methylation is a reversible biological process.

DNA methylation predominantly involves the donation of a methyl group to the 5-position of cytosine linked to a guanine by a phosphodiester bond. The multiple CpG sites form clusters termed CpG islands, which are located in the regulatory regions of many genes. Due to their presence in these regulatory regions, changes in methylation status can be either stimulatory (hypomethylation) or inhibitory (hypermethylation) to the expression of that gene (Matouk and Marsden 2008). Histone modifications are posttranslational modifications that alter the properties and structure of chromatin. These posttranslational modifications, including methylation, acetylation, phosphorylation, and ubiquitination, among others, mediate the accessibility of the transcriptional machinery to the DNA binding elements, thus regulating gene expression (G. Wu, 2009). Epigenetic regulation of gene expression by RNA-based mechanisms can occur at both the posttranscriptional level and the level of chromatin. These mechanisms are mediated by small, interfering RNAs, which can act through their respective pathways to induce DNA methylation or histone modifications to silence or enhance gene expression (Thambirajah et al. 2009; Kawasaki et al. 2005).

Environmental cues appear to be the primary trigger to induce changes in the epigenome and include such factors as nutrients, stress, environmental pollutants, and toxins (Godfrey et al. 2007). The extent to which cells are able to respond to these cues lies in large part with the plasticity of cells. Thus the developing embryo/fetus is a prime target for epigenetic modifications (Evertts et al. 2010), and interventions during pregnancy may be an important strategy for ameliorating or preventing metabolic disorders (including vascular insulin resistance) in adult life (G. Wu et al. 2006). Despite a recognized role for epigenetics in fetal programming of cardiovascular disease (Turunen et al. 2009), this field is still in its infancy.

6.3 IMPACTS OF MATERNAL NUTRITION ON EPIGENETICS AND VASCULAR FUNCTION

6.3.1 Maternal Undernutrition

Severe nausea and vomiting, known as hyperemesis gravidarum, is a life-threatening disorder that occurs in 1–2% of pregnancies and generally extends beyond the 16th week of gestation (Marsal 2002). Under this condition, the mother mobilizes her own stores of nutrients (e.g., protein, lipids, and glycogen) and the fetus is malnourished. In livestock production, insufficient nutrient supply is common across all species due to poor forage quality and/or availability, heat or cold stress, increased rates of twinning in typically monotocous species, and poor management practices (G. Wu et al. 2006). In both clinical medicine and agriculture, birth weight is the predominant indicator of prior nutrient availability in utero.

Low birth weight is correlated with adult-onset diseases such as cardiovascular disease and metabolic syndrome, highlighting the importance of this simple measure on predicting and potentially managing potential health risks (Barker 2007; Kajantie et al. 2005). For example, at 4 years of age, children with low birth weight had elevated levels of glucose and insulin in response to glucose challenge (Yajnik and Deshmukh 2008). In addition, individuals exposed to the Dutch winter famine of 1944–45 in utero had higher rates of insulin resistance, vascular disease, morbidity, and mortality in adulthood (Lumey 1998). Interestingly, 60 years after birth, offspring with early prenatal experience of the famine exhibited less DNA methylation of the imprinted *IGF2* gene, in comparison with same-sex siblings without exposure to prenatal malnutrition (Heijmans et al. 2008). Furthermore, individuals with periconceptional exposure to the famine had lower methylation of the INSIGF gene but higher methylation of several genes (IL10, LEP, ABCA1, GNASAS, and MEG3) (Tobi et al. 2009). These genes are closely linked with nutrient metabolism and cardiovascular function. Interaction between undernutrition and sex affected the methylation of INSIGF, LEP, and GNASAS (Tobi et al. 2009). Moreover, a cohort study of 15,000 Swedish men and women born between 1915 and 1929 perhaps provides by far the most convincing evidence for the close association between reduced fetal growth rate and increased risk of death from ischemic heart disease (Leon et al. 1998). These observations underscore the need to monitor low-birth-weight children for the development of diabetes and to educate these children and their families about maintaining an appropriate diet to mitigate the effects of this altered metabolic profile. The same principle could be applied to nutritional management of livestock.

Experimental evidence from a variety of species and models has provided a wealth of knowledge regarding the mechanisms by which reduced nutrient availability in utero gives rise to adult disease. For example, fetal undernutrition due to placental insufficiency impaired vascular function in two generations of rats (Anderson et al. 2006), indicating an intergenerational effect. Interestingly, the effects on endothelium-dependent relaxation appear to be gender specific (Anderson et al. 2006). Also, maternal protein restriction induces hypertension and vascular dysfunction in adult female offspring of rats (Sathishkumar et al. 2009). Interestingly, these effects were maintained in the F2 generation, indicating that these alterations were likely due to epigenetic modifications inherited across generations (Harrison and Langley-Evans 2009).

Research with large animals (e.g., sheep and nonhuman primates) provides further evidence for the fetal programming of vascular dysfunction and metabolic abnormalities. For example, in sheep, maternal undernutrition induces left ventricular hypertrophy and alters gene expression in the fetal left ventricle (Han et al. 2004; Vonnahme et al. 2003). Additionally, maternal nutrient restriction results in increased myocardial lipid and altered gene expression in offspring at 1 year of age (Chan et al. 2009). Moreover, studies with sheep have demonstrated that maternal nutrient restriction impairs renal function, increases the development of glomerulosclerosis, and enhances apoptosis in kidneys, while altering the expression of proteins involved in regulating the inflammatory process (Williams et al. 2007; Sharkey et al. 2009). Interestingly, in primates a 30% reduction in maternal nutrient intake did not affect fetal weight at 0.5 and 0.9 week of gestation. However, under such conditions tissue-specific global methylation status was altered in the kidney at both 0.5 and 0.9 week of gestation (Unterberger et al. 2009) and the availability of methyl group donors was reduced (Schlabritz-Loutsevitch et al. 2009). Furthermore, emerging evidence indicates epigenetic modification of fetal baboon hepatic phosphoenolpyruvate carboxykinase (a key enzyme in gluconeogenesis) after exposure to moderately reduced nutrient availability (Nijland et al. 2010). These data indicate that even a relatively mild nutrient restriction can induce epigenetic alterations in the DNA complex and provide a potential mechanism for the incidence of endothelial dysfunction, renal impairment, and hypertension in offspring with previous experience of malnutrition during the period of fetal growth.

The consequences of intrauterine growth restriction (IUGR), defined as impaired growth and development of the fetus or its organs, have been investigated in a number of livestock species and go well beyond animal health, to factors that could potentially affect athletic performance (G. Wu et al. 2004, 2006). In both pigs and sheep, IUGR results in decreased skeletal muscle fiber number, increased deposition of adipose tissue, and increased connective tissue content (Bee 2004; Greenwood et al. 1998, 2000; Powell and Aberle 1980). Collectively, these alterations in normal development result in reduced growth performance, including both whole-body and skeletal muscle growth rates and reduced nutrient utilization. Given the combinatorial effects of impaired vascular function and muscle growth, it is possible that endurance exercise and performance of humans and animals may be highly susceptible to fetal programming. These effects likely result from genomic imprinting, which is defined as the parent-of-origin-dependent expression of a single allele of a gene in the embryo/fetus, namely parental influence on the genome of progeny (G. Wu et al. 2006).

6.3.2 Maternal Overnutrition

Almost 65% of the adult population in the United States is overweight (defined as a body mass index [BMI] > 25 kg/m^2), while 31% of the adult population is obese (defined as BMI > 30 kg/m^2) (Ogden et al. 2007). The current global obesity epidemic in humans results primarily from a chronic imbalance between energy intake and output (Jobgen et al. 2006). Overweight and obese women may unknowingly enter pregnancy and continue overeating during gestation. Unless effective interventions are adopted, these women usually gain more weight during the first pregnancy and accumulate more fat during subsequent pregnancies (Edwards et al. 1996). Maternal obesity or overnutrition before or during pregnancy may result in fetal growth restriction and increased risk of neonatal mortality and morbidity in mammals (Wu et al. 2004). Thus limited or no weight gain in obese pregnant women (<15 lb body weight gain) has favorable pregnancy outcomes (e.g., lower risk of preeclampsia, cesarean delivery, and large-for-gestational-age birth, as well as higher risk of small-for-gestational-age birth) (Kiel et al. 2007). Likewise, companion animals (e.g., cats and dogs) and livestock (e.g., pigs and sheep) can become obese if they have free access to their foods (G. Wu et al. 2006). Interestingly, the "civilization" of the horse in the Western culture has followed a similar trend to that of humans and has increased the incidence of obesity and its associated metabolic disorders (Sillence et al. 2006; Johnson et al. 2009).

Maternal obesity has been linked to metabolic perturbations and cardiovascular disorders in the offspring of a number of species (Armitage et al. 2005; Khan et al. 2005; Zhu et al. 2009). In nonhuman primates, a maternal high-fat diet has been reported to promote the development of nonalcoholic fatty liver and atherosclerosis (McCurdy et al. 2009). In these primates, a maternal high-fat diet was associated with the alteration of seven metabolites in the fetus that are related to one-carbon unit metabolism (Cox et al. 2009). Notably, alteration of the fetal metabolome results, at least in part, from an altered fetal hepatic chromatin structure that leads to aberrant hepatic gene expression (Aagaard-Tillery et al. 2008).

Feeding a high-fat diet during pregnancy or during suckling induces cardiovascular dysfunction characterized by elevated systolic blood pressure and impaired endothelium-dependent relaxation in rats (Khan et al. 2005). Importantly, response to maternal dietary treatment differed between male and female offspring in that female offspring were more susceptible to elevated systolic and diastolic blood pressure when their mothers were fed a high-fat diet during gestation and/or lactation. Despite observing sex differences for blood pressure and heart rate, endothelium-dependent relaxation in response to acetylcholine was consistently blunted in both males and females exposed to a high-fat diet either in utero or during lactation. A major underlying mechanism is likely the reduced release of nitric oxide (a major vasodilator) from endothelial cells (G. Wu and Meininger 2009). Interestingly, feeding a high-fat diet to pregnant dams reduced the mitochondrial copy number in the kidney in 1-year-old offspring, while altering the expression of a number of mitochondrial genes in the aorta (Taylor et al. 2005). These findings suggest that mitochondria may play a central role in developmental programming of vascular function. Similar observations have been observed in offspring of pregnant mice fed a high-fat diet including increased adiposity, hypertension, and insulin resistance (Samuelsson et al. 2008; Dunn and Bale 2009). In sheep, maternal obesity down-regulates expression of genes involved in myogenesis and placental angiogenesis (Zhu et al. 2009) and fetal skeletal muscle (Tong et al. 2009), while reducing the phosphorylation of AMP-activated kinase in the fetal and neonatal liver (Philp et al. 2008) as well as energy metabolism (Wallace et al. 2005).

Importantly, these changes in gene and protein expression in response to maternal overnutrition result in increased fetal and/or neonatal adiposity (Ford et al. 2009; Muhlhausler et al. 2007) and increased leptin gene expression in perirenal and subcutaneous adipose tissue depots (Muhlhausler et al. 2007). The consequences of maternal overnutrition continue to manifest in the postnatal offspring as evidenced by impaired glucose uptake on postnatal day 210 despite growing in an identical nutritional environment from birth (M. C. Satterfield and G. Wu, unpublished observation). Similarly, a maternal high-fat diet during pregnancy results in impaired glucose homeostasis in rat offspring characterized by elevated plasma insulin levels at 1 year of age (Taylor et al. 2005).

6.4 AMINO ACIDS AS NUTRIENTS TO REGULATE EPIGENETICS AND VASCULAR FUNCTION

Besides serving as building blocks of proteins, amino acids are signaling molecules, regulators of gene expression and the protein phosphorylation cascade, and key precursors for syntheses of hormones and low-molecular-weight nitrogenous substances with enormous biological importance (Figure 6.2). Amino acids also modulate cellular redox state and the secretion of hormones from endocrine organs (e.g., insulin, growth hormone, lactogen, and insulinlike factors) (G. Wu 2009). Physiological concentrations of metabolites (e.g., nitric oxide, polyamines, glutathione, taurine, thyroid hormones, and serotonin) of amino acids are required for the functions of cells and whole-body homeostasis (G. Wu 2009). It is noteworthy that maintenance and regulation of the epigenetic state, which depend on one-carbon unit metabolism, require adequate provision of methionine, serine, glycine, histidine, choline, creatine, and B vitamins (including folate, vitamin B_{12}, and vitamin B_6) (G. Wu 2009). These nutrients play an important role in regulating the availability of S-adenosylmethionine,

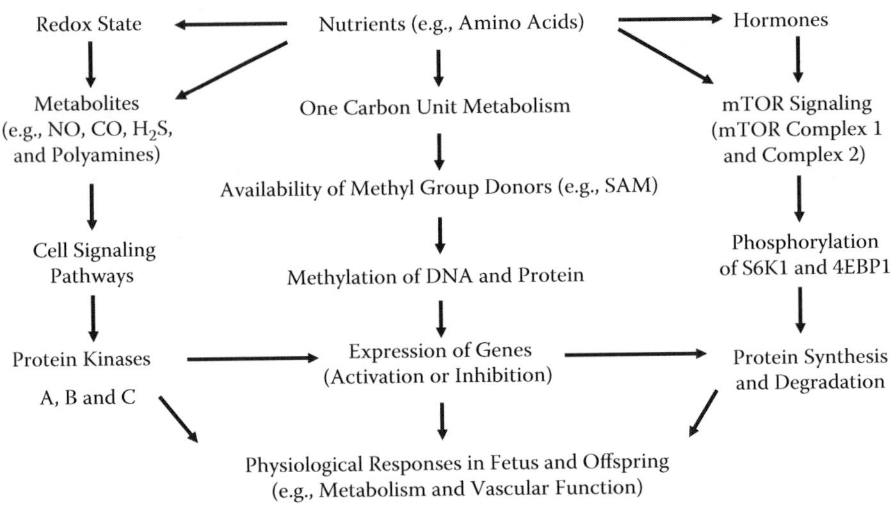

FIGURE 6.2 Roles of nutrients in epigenetics and physiological responses. Nutrients, particularly amino acids, regulate cellular redox state, the secretion of hormones (e.g., insulin and insulinlike growth factors), physiological functions, and whole-body homeostasis in humans and animals through three mechanisms: (1) the expression of genes, (2) the production of signaling gases and other metabolites, and (3) mTOR activation. S-adenosylmethionine (SAM) is the major methyl group donor in cells and its synthesis is affected by amino acids (e.g., methionine, serine, glycine, and histidine), B vitamins (including folate, vitamin B_{12}, and vitamin B_6), choline, and creatine. Methylation of DNA and protein contributes to epigenetics, which results in transcriptional activation or inhibition of select genes. Changes in intracellular protein turnover (protein synthesis and degradation) and protein kinase cascades can alter physiological responses in the fetus and offspring. CO, carbon monoxide; 4EBP1, eIF4E-binding protein-1; H_2S, hydrogen sulfide; mTOR, mammalian target of rapamycin; NO, nitric oxide; S6K1, ribosomal protein S6 kinase-1.

a major methyl donor for DNA and protein methylation by specific DNA and protein methyltransferases (G. Wu et al. 2006). Thus restriction of essential B vitamins, folate, and methionine during the periconceptional period in sheep resulted in altered DNA methylation, insulin resistance, and elevated blood pressure, observed most notably in adult male offspring (Sinclair et al. 2007).

S-adenosylmethionine is synthesized from methionine by S-adenosylmethionine synthase (also known as methionine adenosyltransferase) (G. Wu 2009). In addition to its role as a methyl donor, S-adenosylmethionine is required for the synthesis of polyamines, cysteine, taurine, and creatine. Polyamines are required for the proliferation of endothelial cells and the remodeling of the vasculature (Li et al. 2002). Importantly, the nutritional and physiological state of the animal will alter the production of these bioactive substances, thus regulating the availability of methyl donors. When cysteine or taurine is deficient in the diet, their synthesis from methionine will be increased in vivo, thus decreasing total S-adenosylmethionine availability for DNA or protein methylation. Inadequate synthesis of glycine and serine, coupled with low supplies from the diet, can also impair one-carbon unit metabolism (G. Wu et al. 2006). Therefore amino acid deficiency can alter the epigenetic code through changes in DNA methylation as well as histone modifications (Oommen et al. 2005).

Either enteral feeding or intravenous administration of amino acids (e.g., arginine and citrulline) is effective in increasing their circulating concentrations in mother and fetus (Lassala et al. 2009; G. Wu et al. 2008). Therefore these nutrients may provide an effective solution to fetal growth restriction in underfed and overfed dams and postnatal metabolic disorders. The importance of amino acids in supporting fetal growth, development, and health is an emerging area of investigation and will undoubtedly shape the future of nutritional management in medicine and animal production (G. Wu et al. 2008). The translation of basic research on amino acid nutrition into practice has yielded fruitful outcomes. For example, in pigs arginine supplementation to pregnant sows during gestation increases the placental vascularity (G. Wu et al. 2010) and embryonic/fetal survival (Mateo et al. 2007). Additionally, supplementing the gestational diet for gilts with 0.4% L-arginine plus 0.6% L-glutamine between days 30 and 114 of gestation enhanced the efficiency of nutrient utilization, reduced variation in piglet birth weight, and increased litter birth weight (G. Wu et al. 2010). In sheep, maternal arginine administration during late gestation increased the development of fetal perirenal brown adipose tissue (Satterfield et al. 2009), which is enriched with endothelial cells. The increased mass of brown adipose tissue may lead to enhancement of blood flow to tissues and the ability of neonates to combat cold exposure at birth. Moreover, intravenous administration of L-arginine-HCl (3 × 27 mg/kg body weight per day) enhanced fetal growth in ovine models of both undernutrition-induced and naturally occurring IUGR (Lassala 2008). Finally, during late (week 33) gestation, daily intravenous infusion of L-arginine (20 g/day) for 7 days to women with unknown causes of IUGR increased birth weight at term (week 39) by 6.4% (Xiao and Li 2005). The beneficial effects of arginine likely result from increases in angiogenesis, the number and size of blood vessels, and ultimately the transfer of nutrients from the mother to the embryo/fetus through utero-placental blood flows. Future studies are required to determine the cardiovascular function and health of offspring from mothers supplemented with arginine during pregnancy.

6.5 CONCLUSION AND PERSPECTIVES

Either undernutrition or overnutrition during pregnancy (particularly the periconception period), which remains a significant problem in both medicine and animal agriculture, can result in epigenetic changes of some genes in both animals and humans. These changes are affected by multiple factors (e.g., sex, gestational period, and the severity of malnutrition), may persist in offspring throughout postnatal life, and may carry on to the next generation. A unified explanation for impairment of fetal growth and development in response to both maternal undernutrition and overnutrition may be reduced utero-placental blood flow and therefore the reduced transfer of nutrients from mother to fetus (G. Wu et al. 2004). This hypothesis is gaining support from studies with rats, pigs, and sheep (Satterfield et al. 2010; G. Wu et al. 2008; Zeng et al. 2008). Nutrients, particularly amino acids, are essential for the regulation of epigenetics and vascular function. Compelling evidence indicates that the fetal and early neonatal periods of development are extremely sensitive to environmental cues, which have long-lasting consequences to postnatal growth, health, and likely athletic performance. Much research is needed with animal models (e.g., pigs, sheep, and rodents) to understand the basic mechanisms responsible for specific nutrients (e.g., arginine, citrulline, glutamine, proline, polyunsaturated fatty acids, vitamins, and minerals) in fetal programming and to design effective therapeutic means for endothelial dysfunction and metabolic abnormalities in offspring with previous experience of an adverse intrauterine environment.

ACKNOWLEDGMENTS

We thank Frances Mutscher for assistance in manuscript preparation. This work was supported, in part, by grants from National Institutes of Health (1R21 HD049449), National Research Initiative Competitive Grants from the Animal Reproduction Program (2008-35203-19120), and Animal Growth & Nutrient Utilization Program (2008-35206-18764 and 2009-35206-05211) of the USDA National Institute of Food and Agriculture, Texas AgriLife Research (H-8200), and American Heart Association (10GRNT4480020).

REFERENCES

Aagaard-Tillery, K. M., K. Grove, J. Bishop, X. Ke, Q. Fu, R. McKnight, and R. H. Lane. 2008. Developmental origins of disease and determinants of chromatin structure: maternal diet modifies the primate fetal epigenome. *J Mol Endocrinol* 41:91–102.

AHA. 2010. American Heart Association. Heart disease and stroke statistic—2010 Update. www.americanheart.org.

Anderson, C. M., F. Lopez, A. Zimmer, and J. N. Benoit. 2006. Placental insufficiency leads to developmental hypertension and mesenteric artery dysfunction in two generations of Sprague-Dawley rat offspring. *Biol Reprod* 74:538–44.

Armitage, J. A., L. Lakasing, P. D. Taylor, A. A. Balachandran, R. I. Jensen, V. Dekou, N. Ashton, J. R. Nyengaard, and L. Poston. 2005. Developmental programming of aortic and renal structure in offspring of rats fed fat-rich diets in pregnancy. *J Physiol* 565:171–84.

Barker, D. J. P. 2007. The origins of the developmental origins theory. *J Intern Med* 261:412–17.

Barker, D. J. P., T. Forsén, J. G. Eriksson, and C. Osmond. 2002. Growth and living conditions in childhood and hypertension in adult life: a longitudinal study. *J Hypertens* 20:1951–56.

Barker, D. J. P., and C. Osmond. 1986. Infant mortality, childhood nutrition, and ischaemic heart disease in England and Wales. *Lancet* 8489:1077–81.

Bee, G. 2004. Effect of early gestation feeding, birth weight, and gender of progeny on muscle fiber characteristics of pigs at slaughter. *J Anim Sci* 82:826–36.

Cafe, L. M., D. W. Hennessy, H. Hearnshaw, S. G. Morris, and P. L. Greenwood. 2009. Consequences of prenatal and preweaning growth for feedlot growth, intake and efficiency of Piedmontese- and Wagyu-sired cattle. *Anim Prod Sci* 49:461–67.

Catalano, P. M. 2007. Increasing maternal obesity and weight gain during pregnancy. *Obstet Gynecol* 110:743–44.

Chan, L. L., S. P. Sébert, M. A. Hyatt, T. Stephenson, H. Budge, M. E. Symonds, and D. S. Gardner. 2009. Effect of maternal nutrient restriction from early to midgestation on cardiac function and metabolism after adolescent-onset obesity. *Am J Physiol Regul Integr Comp Physiol* 296:R1455–63.

Chmurzynska, A. 2010. Fetal programming: link between early nutrition, DNA methylation, and complex diseases. *Nutr Rev* 68:87–98.

Cleal, J. K., K. R. Poore, J. P. Boullin O. Khan, R. Chau, O. Hambidge, C. Torrens, J. P. Newman, L. Poston, D. E. Noakes, M. A. Hanson, and L. R. Green. 2007. Mismatched pre- and postnatal nutrition leads to cardiovascular dysfunction and altered renal function in adulthood. *Proc Natl Acad Sci U S A* 104:9529–33.

Cox, J., S. Williams, K. Grove, R. H. Lane, R and K. M. Aagaard-Tillery. 2009. A maternal high-fat diet is accompanied by alterations in the fetal primate metabolome. *Am J Obstet Gynecol* 201:281 e281–89.

Dunn, G. A., and T. L. Bale. 2009. Maternal high-fat diet promotes body length increases and insulin insensitivity in second-generation mice. *Endocrinology* 150:4999–5009.

Edwards, L. E., W. L. Hellerstedt, I. R. Alton, M. Story, and J. H. Himes. 1996. Pregnancy complications and birth outcomes in obese and normal-weight women: effects of gestational weight change. *Obstet Gynecol* 87:389–94.

Eriksson, J. G., T. Forsen, J. Tuomilehto, C. Osmond, and D. J. P. Barker. 2001. Early growth and coronary heart disease in later life: longitudinal study. *Br Med J* 322:949–53.

Evertts, A. G., B. M. Zee, and B. A. Garcia. 2010. Modern approaches for investigating epigenetic signaling pathways. *J Appl Physiol.* Doi:10.1152/japplphysiol.00007.2010.

Flegal, K. M., M. D. Carroll, C. L. Ogden, and L. R. Curtin. 2010. Prevalence and trends in obesity among US adults, 1999–2008. *JAMA* 303:235–41.

Ford, S. P., L. Zhang, M. Zhu, M. M. Miller, D. T. Smith, B. W. Hess, G. E. Moss, P. W. Nathanielsz, and M. J. Nijland. 2009. Maternal obesity accelerates fetal pancreatic beta-cell but not alpha-cell development in sheep: prenatal consequences. *Am J Physiol Regul Integr Comp Physiol* 297:R835–43.

Forsen, T., J. G. Eriksson, J. Tuomilehto, C. Osmond, and D. J. P. Barker. 1999. Growth in utero and during childhood among women who develop coronary disease: longitudinal study. *Br Med J* 319:1403–7.

Godfrey, K. M., K. A. Lillycrop, G. C. Burdge, P. D. Gluckman, and M. A. Hanson. 2007. Epigenetic mechanisms and the mismatch concept of the developmental origins of health and disease. *Pediatr Res* 61:5R–10R.

Greenwood, P. L., A. S. Hunt, J. W. Hermanson, and A. W. Bell. 1998. Effects of birth weight and postnatal nutrition on neonatal sheep: I. Body growth and composition, and some aspects of energetic efficiency. *J Anim Sci* 76:2354–67.

———. 2000. Effects of birth weight and postnatal nutrition on neonatal sheep: II. Skeletal muscle growth and development. *J Anim Sci* 78:50–61.

Hales, C. N., and D. J. Barker. 1992. Type 2 (non-insulin-dependent) diabetes mellitus: the thrifty phenotype hypothesis. *Diabetologia* 35:595–601.

Han, H. C., K. J. Austin, P. W. Nathanielsz, S. P. Ford, M. J. Nijland, and T. R. Hansen. 2004. Maternal nutrient restriction alters gene expression in the ovine fetal heart. *J Physiol* 558:111–21.

Hardie, D. G., S. A. Hawley, and J. W. Scott. 2006. AMP-activated protein kinase–development of the energy sensor concept. *J Physiol* 574:7–15.

Harrison, M., and S. C. Langley-Evans. 2009. Intergenerational programming of impaired nephrogenesis and hypertension in rats following maternal protein restriction during pregnancy. *Br J Nutr* 101:1020–30.

Heijmans, B. T., E. W. Tobi, A. D. Stein, H. Putter, G. J. Blauw, E. S. Susser, P. E. Slagboom, and L. H. Lumey. 2008. Persistent epigenetic differences associated with prenatal exposure to famine in humans. *Proc Natl Acad Sci U S A* 105:17046–49.

Hill, D. J., and B. Duville. 2000. Pancreatic development and adult diabetes. *Pediatr Res* 48:269–74.

Jobgen, W. S., S. K. Fried, W. J. Fu, C. J. Meininger, and G. Wu. 2006. Regulatory role for the arginine-nitric oxide pathway in metabolism of energy substrates. *J Nutr Biochem* 17:571–588.

Johnson, P. J., C. E. Wiedmeyer, N. T. Messer, and V. K. Ganjam. 2009. Medical implications of obesity in horses—lessons for human obesity. *J Diabetes Sci Technol* 3:163–74.

Kajantie, E., C. Osmond, D. J. P. Barker, T. Forsen, D. I. W. Phillips, and J. G. Eriksson. 2005. Size at birth as a predictor of mortality in adulthood: a follow-up of 350,000 person-years. *Int J Epidemiol* 34:655–63.

Kawasaki, H., K. Taira, and K. V. Morris, K.V. 2005. siRNA induced transcriptional gene silencing in mammalian cells. *Cell Cycle* 4:442–48.

Khan, I. Y., V. Dekou, G. Douglas, R. Jensen, M. A. Hanson, L. Poston, and P. D. Taylor. 2005. A high-fat diet during rat pregnancy or suckling induces cardiovascular dysfunction in adult offspring. *Am J Physiol Regul Integr Comp Physiol* 288:R127–33.

Kiel, D. W., E. A. Dodson, R. Artal, T. K. Boehmer, and T. L. Leet. 2007. Gestational weight gain and pregnancy outcomes in obese women. *Obstet Gynecol* 110:752–58.

Lassala, A. 2008. Arginine and fetal growth in ovine models of intrauterine growth restriction. PhD dissertation, Texas A&M University, College Station, Texas.

Lassala, A., F. W. Bazer, T. A. Cudd, P. Li, X. L. Li, M. C. Satterfield, T. E. Spencer, and G. Wu. 2009. Intravenous administration of L-citrulline to pregnant ewes is more effective than L-arginine for increasing arginine availability in the fetus. *J Nutr* 139:660–65.

Leon, D. A., H. O. Lithell, D. Vagero, I. Koupilova, R. Mohsen, L. Berglund, U. B. Lithell, and P. M. McKeigue. 1998. Reduced fetal growth rate and increased risk of death from ischaemic heart disease: cohort study of 15000 Swedish men and women born 1915–29. *Br Med J* 317:241–45.

Li, H., C. J. Meininger, J. R. Hawker, K. A. Kelly, S. M. Morris, and G. Wu. 2002. Activities of arginase I and II are limiting for endothelial cell proliferation. *Am J Physiol Regul Integr Comp Physiol* 282:R64–R69.

Lumey, L. H. 1998. Reproductive outcome in women prenatally exposed to undernutrition: a review of findings from the Dutch famine birth cohort. *Proc Nutr Soc* 57:129–35.

Marsal, K. 2002. Intrauterine growth restriction. *Curr Opin Obstet Gynecol* 14:127–35.

Mateo, R. D., G. Wu, F. W. Bazer, J. C. Park, I. Shinzato, and S. W. Kim. 2007. Dietary L-arginine supplementation enhances the reproductive performance of gilts. *J Nutr* 137:652–56.

Matouk, C. C., and P. A. Marsden. 2008. Epigenetic regulation of vascular endothelial gene expression. *Circ Res* 102:873–87.

McCurdy, C. E., J. M. Bishop, S. M. Williams, B. E. Grayson, M. S. Smith, J. E. Friedman, and K. L. Grove. 2009. Maternal high-fat diet triggers lipotoxicity in the fetal livers of nonhuman primates. *J Clin Invest* 119:323–35.

Muhlhausler, B. S., J. A. Duffield, and I. C. McMillen. 2007. Increased maternal nutrition increases leptin expression in perirenal and subcutaneous adipose tissue in the postnatal lamb. *Endocrinology* 148:6157–63.

Nijland, M. J., K. Mitsuya, C. Li, S. P. Ford, T. J. McDonald, P. W. Nathanielsz, and L. A. Cox. 2010. Epigenetic modification of fetal baboon hepatic phosphoenolpyruvate carboxykinase following exposure to moderately reduced nutrient availability. *J Physiol.* Doi: 10.1113/jphysiol.2009.184168.

Ogden, C. L., S. Z. Yanovski, M. D. Carroll, and K. M. Flegal. 2007. The epidemiology of obesity. *Gastroenterology* 132:2087–2102.

Oommen, A. M., J. B. Griffin, G. Sarath, and J. Zempleni. 2005. Roles for nutrients in epigenetic events. J *Nutr Biochem* 16:74–77.

Phillips, D. I. W. 2007. Programming of the stress response: a fundamental mechanism underlying the long-term effects of the fetal environment. *J Intern Med* 261:453–60.

Philp, L. K., B. S. Muhlhausler, A. Janovska, G. A. Wittert, J. A. Duffield, and I. C. McMillen. 2008. Maternal overnutrition suppresses the phosphorylation of 5'-AMP-activated protein kinase in liver, but not skeletal muscle, in the fetal and neonatal sheep. *Am J Physiol Regul Integr Comp Physiol* 295:R1982–90.

Powell, S. E., and E. D. Aberle. 1980. Effects of birth weight on growth and carcass composition of swine. *J Anim Sci* 50:860–68.

Samuelsson, A. M., P. A. Matthews, M. Argenton, M. R. Christie, J. M. McConnell, E. H. Jansen, A. H. Piersma, S. E. Ozanne, D. F. Twinn, C. Remacle, A. Rowlerson, L. Poston, and P. D. Taylor. 2008. Diet-induced obesity in female mice leads to offspring hyperphagia, adiposity, hypertension, and insulin resistance: a novel murine model of developmental programming. *Hypertension* 51:383–92.

Sathishkumar, K., R. Elkins, U. Yallampalli, and C. Yallampalli. 2009. Protein restriction during pregnancy induces hypertension and impairs endothelium-dependent vascular function in adult female offspring. *J Vasc Res* 46:229–39.

Satterfield, M. C., F. W. Bazer, S. B. Smith, T. E. Spencer, and G. Wu. 2009. Arginine nutrition and fetal brown fat development. *Amino Acids* 37 (Suppl):S6–S7.

Satterfield, M. C., F. W. Bazer, T. E. Spencer, and G. Wu. 2010. Sildenafil citrate treatment enhances amino acid availability in the conceptus and fetal growth in an ovine model of intrauterine growth restriction. *J Nutr* 140:251–58.

Schlabritz-Loutsevitch, N., D. Farley, M. E. Tejero, A. G. Comuzzi P. B. Higgins, L. Cox, Werner, S. L. Jenkins, C. Li, J. Choj, E. J. Dick, G. B. Hubbard, P. Frost, D. D. Dudley, G. Wu, and P. W. Nathanielsz. 2009. Feto-placental adaptations to maternal obesity in the baboon. *Placenta* 30:752–60.

Sharkey, D., D. S. Gardner, M. E. Symonds, and H. Budge. 2009. Maternal nutrient restriction during early fetal kidney development attenuates the renal innate inflammatory response in obese young adult offspring. *Am J Physiol Renal Physiol* 297:F1199–1207.

Sillence, M., G. Noble, and C. McGowan. 2006. Fast food and fat fillies: the ills of Western civilisation. *Vet J* 172:396–97.

Sinclair, K. D., C. Allegrucci, R. Singh, D. S. Gardner, S. Sebastian, J. Bispham, A. Thurston, Huntley, W. D. Rees, C. A. Maloney, R. G. Lea, J. Craigon, T. G. McEvoy, and L. E. Youn. 2007. DNA methylation, insulin resistance, and blood pressure in offspring determined by maternal periconceptional B vitamin and methionine status. *Proc Natl Acad Sci U S A* 104:19351–56.

Symonds, M. E., S. P. Sebert, M. A. Hyatt, and H. Budge. 2009. Nutritional programming of the metabolic syndrome. *Nat Rev Endocrinol* 5:604–10.

Taylor, P. D., J. McConnell, I. Y. Khan, K. Holemans, K. M. Lawrence, H. Asare-Anane, S. J. Persaud, P. M. Jones, L. Petrie, M. A. Hanson, and L. Poston. 2005. Impaired glucose homeostasis and mitochondrial abnormalities in offspring of rats fed a fat-rich diet in pregnancy. *Am J Physiol Regul Integr Comp Physiol* 288:R134–39.

Thambirajah, A. A., A. Li, T. Ishibashi, and J. Ausio. 2009. New developments in post-translational modifications and functions of histone H2A variants. *Biochem Cell Biol* 87:7–17.

Tobi, E. W., L. H. Lumey, R. P. Talens, D. Kremer, H. Putter A. D. Stein, P. E. Slagboom, and B. T. Heijmans. 2009. DNA methylation differences after exposure to prenatal famine are common and timing- and sex-specific. *Hum Mol Genet* 18:4046–53.

Tong, J. F., X. Yan, M. J. Zhu, S. P. Ford, P. W. Nathanielsz, and M. Du. 2009. Maternal obesity downregulates myogenesis and beta-catenin signaling in fetal skeletal muscle. *Am J Physiol Endocrinol Metab* 296:E917–24.

Turunen, M. P., E. Aavik, and S. Yla-Herttuala. 2009. Epigenetics and atherosclerosis. *Biochim Biophys Acta* 1790:886–91.

Unterberger, A., M. Szyf, P. W. Nathanielsz, and L. A. Cox. 2009. Organ and gestational age effects of maternal nutrient restriction on global methylation in fetal baboons. *J Med Primatol* 38:219–27.

Vonnahme, K. A., B. W. Hess, B.W., T. R. Hansen, T.R., R. J. McCormick, D. C. Rule, G. E. Moss, W. J. Murdoch, M. J. Nijland, D. C. Skinner, P. W. Nathanielsz, and S. P. Ford. 2003. Maternal undernutrition from early- to mid-gestation leads to growth retardation, cardiac ventricular hypertrophy, and increased liver weight in the fetal sheep. *Biol Reprod* 69:133–40.

Wallace, J. M., J. S. Milne, and R. P. Aitken. 2005. The effect of overnourishing singleton-bearing adult ewes on nutrient partitioning to the gravid uterus. *Br J Nutr* 94:533–39.

Williams, P. J., L. O. Kurlak, A. C. Perkins, H. Budge, T. Stephenson, D. Keisler, M. E. Symonds, and D. S. Gardner. 2007. Hypertension and impaired renal function accompany juvenile obesity: the effect of prenatal diet. *Kidney Int* 72:279–89.

Wu, C., and J. R. Morris. 2001. Genes, genetics, and epigenetics: a correspondence. *Science* 293:1103-–5.

Wu, G. 2009. Amino acids: metabolism, functions, and nutrition. *Amino Acids* 37:1–17.

Wu, G., F. W. Bazer, R. C. Burghardt, G. A. Johnson, S. W. Kim, X. L. Li M. C. Satterfield, and T. E. Spencer. 2010. Impacts of amino acid nutrition on pregnancy outcome in pigs: mechanisms and implications for swine production. *J Anim Sci* 88:E195-E204.

Wu, G., F. W. Bazer, T. A. Cudd, C. J. Meininger, and T. E. Spencer. 2004. Maternal nutrition and fetal development. *J Nutr* 134:2169–72.

Wu, G., F. W. Bazer, S. Datta, G. A. Johnson, P. Li, M. C. Satterfield, and T. E. Spencer. 2008. Proline metabolism in the conceptus: implications for fetal growth and development. *Amino Acids* 35:691–702.

Wu, G., F. W. Bazer, J. M. Wallace, and T. E. Spencer. 2006. Board-invited review: intrauterine growth retardation: implications for the animal sciences. *J Anim Sci* 84:2316-–37.

Wu, G., and C. J. Meininger. 2009. Nitric oxide and vascular insulin resistance. *BioFactors* 35:21–27.

Xiao, X. M., and L. P. Li. 2005. L-arginine treatment for asymmetric fetal growth restriction. *Int J Gynecol Obstet* 88:15–18.

Yajnik, C. S., and U. S. Deshmukh. 2008. Maternal nutrition, intrauterine programming and consequential risks in the offspring. *Rev Endocr Metab Disord* 9:203–11.

Zeng, X. F., F. L. Wang, X. Fan, W. J. Yang, B. Zhou, P. F. Li, Y. L. Yin G. Wu, and J. J. Wang. 2008. Dietary arginine supplementation during early pregnancy enhances embryonic survival in rats. *J Nutr* 138:1421–25.

Zhu, M. J., M. Du, M. J. Nijland, P. W. Nathanielsz, B. W. Hess, G. E. Moss, and S. P. Ford. 2009. Down-regulation of growth signaling pathways linked to a reduced cotyledonary vascularity in placentomes of over-nourished, obese pregnant ewes. *Placenta* 30:405–10.

7 Role of Epigenetic Machinery and MicroRNAs in Diet-Induced Hepatocarcinogenesis

Kalpana Ghoshal and Tasneem Motiwala

CONTENTS

Overview 142
7.1 Hepatocellular Carcinoma 143
7.2 Folate Deficiency 144
7.2.1 Rodent Models of Folate Deficiency and Hepatocellular Carcinoma 145
7.2.2 Epigenetic Alterations in Rodent Models of Diet-Induced HCC 147
7.3 Introduction to MicroRNA 154
7.3.1 MicroRNA Genes and Biogenesis of MicroRNAs 155
7.3.2 Function of MicroRNAs 156
7.3.3 MicroRNAs Expressed in the Liver 157
7.4 Aberrations in MicroRNA Expression Occur in Hepatocellular Carcinomas 157
7.4.1 MicroRNAs Dysregulated in the Rat Model of Folate Methyl-Deficient (FMD) Diet-Induced HCC 157
7.4.2 MicroRNAs Dysregulated in the Mouse Model of NASH-Induced Hepatocarcinogenesis Developed by Feeding CDAA Diet 162
7.4.3 MicroRNA Alteration in Human HCC 164
7.4.4 Hepatic MicroRNAs Dysregulated in NASH Patients 165
7.5 MicroRNAs with Important Function in Liver and HCC 166
7.5.1 miR-122 166
7.5.2 miR-181 167
7.5.3 miR-155 168
7.6 Epigenetic Regulation of MicroRNA Expression in Hepatocellular Cancer 168
7.7 Concluding Remarks 169
Acknowledgment 170
References 170

OVERVIEW

It is now well established that cancer is both a genetic and epigenetic disease. Although existence of DNA methylation and histone modifications in biological systems was known for decades, only studies in the last 15 years have recognized that these postreplication modifications play key roles in epigenetic regulation of gene expression (Figure 7.1). Emerging studies now support the notion that noncoding RNAs will be major players in the regulation of expression of coding regions by affecting chromatin structure, transcription, mRNA stability, and translation, thereby adding to the ensemble of epigenetic "equipment." Indeed, a surprising revelation of the postgenomic era is that only a miniscule amount of the mammalian genome (<5%) codes for proteins, while the major part of the once thought to be "junk DNA" codes for noncoding RNAs ranging in size from a few nucleotides to

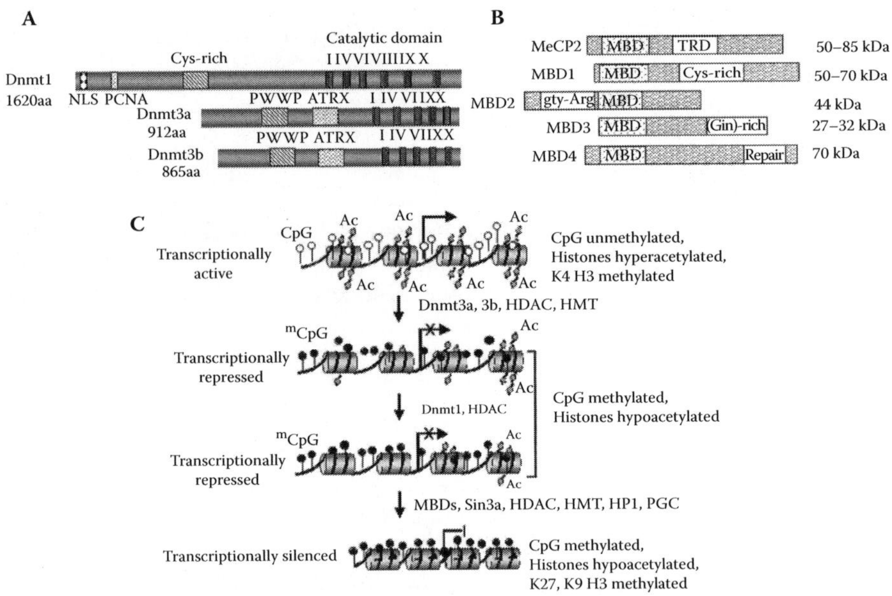

FIGURE 7.1 Schematic representation of Dnmts and MBDs involved in epigenetic silencing of genes. A. Three functional Dnmts are highly conserved in mammals. Simplified mechanism of methylation mediated silencing. B. Schematic diagram of MBD family members. MBD and TRD stand for methyl CpG binding domain and transcriptional repressor domain, respectively. C. Schematic representation of methylation-mediated silencing of genes. Nucleosomes wrapped around transcriptionally active promoters are relaxed, with CpG base pairs unmethylated and core nucleosomal histones acetylated. Dnmt3a or Dnmt3b (de novo methyl transferase) initiate methylation of CpG base pairs, which are maintained postreplication by Dnmt1. Methyl CpG binding proteins then bind to the methylated CpGs and recruit different corepressors, resulting in nucleosomal condensation and epigenetic silencing. Open and filled lollipops denote methylated and unmethylated CpG, respectively. MBD and TRD stand for methyl CpG binding domain and transcriptional repressor domain, respectively. (Modified from Ghoshal, K., Li, X., Datta, J., et al., *J Nutr,* 136, 1522–27, 2006. With permission.)

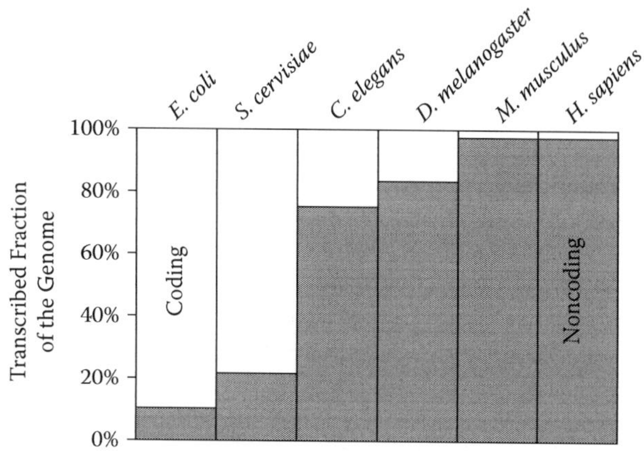

FIGURE 7.2 The protein-coding fraction of the genome is minimal in mammals, whereas the noncoding region of the genome expanded enormously during evolution. Based on the genome and transcriptome sequence data from different organisms including bacteria, lower eukaryote, worm, fly, and mammals.

several hundred kb (Figure 7.2). Among these, microRNAs (miRs) are small (21–25 nt) noncoding RNAs that negatively regulate expression of protein-coding genes primarily at the posttranscriptional level in animals. These tiny RNAs are essential for animal development, and aberrations in their expression lead to different diseased states such as cancer, viral infection, inflammation, diabetes, and cardiovascular neuronal disorders. Recent studies have shown that in addition to genetic factors and lifestyle, diet plays a major causal role in metabolic syndromes that increase the risk of several diseases including cardiovascular disorders, diabetes, and cancer in humans. MicroRNAs have tremendous therapeutic potential since they are involved in every aspect of biology including metabolic disorders that often lead to liver cancer. In this chapter, we discuss how dietary manipulations modulate epigenetic machinery and the expression of cancer-causing protein-coding and noncoding (microRNA) genes during multistage hepatocarcinogenesis in animal models. We also discuss how epigenetic mechanisms modulate microRNA expression in hepatocellular cancer. Finally, we conclude with the potential of epigenetic and microRNA therapy against this disease with increasing mortality.

7.1 HEPATOCELLULAR CARCINOMA

Hepatocellular carcinoma (HCC) is the fifth most prevalent cancer in the world and the third leading cause of cancer-related death with the annual death rate exceeding 500,000 (1–3). Primary hepatocellular carcinoma, the most common primary malignant tumor, accounts for >90% of all primary liver cancer. Metastatic liver tumors are often disseminated from primary colon, prostate, or breast carcinomas. Liver cancer may be a single localized mass or contain multiple cancerous lesions resulting from intrahepaic metastasis. The disease is progressive, and death usually occurs within 10 months of initial diagnosis. The high mortality rate is because of lack of biomarkers for early detection of this cancer and ineffectiveness of available therapies at later stages. The 5-year survival rate for this cancer is only 5%, and the death rate is expected to rise in the next 20 years because of increase in epidemic proportion in metabolic syndromes that constitute obesity, cardiovascular diseases, hyperlipidemia, diabetes, and insulin resistance. This is projected to be a major epidemic, especially in people on a Western diet that is enriched in fats and carbohydrates (4, 5). Major risk factors for HCC include alcohol consumption, hepatitis C infection, diabetes (6), and certain toxins. Indeed, alcohol is the most common cause of HCC in the United States, accounting for 32–45% of HCC cases. Further, alcohol use by individuals exposed to chronic hepatitis C doubles the risk for HCC as compared to the risk due to hepatitis C alone (2, 7). On the other hand, the majority of the population in sub-Sahara Africa and Asia is highly exposed to aflatoxin (8, 9). This toxin, which is ingested with food contaminated by *Aspergillus flavus,* significantly enhances the carcinogenic effects of viral hepatitis. There is also mounting evidence that the liver is highly susceptible to tobacco carcinogenicity. Another common liver disease is nonalcoholic steatohepatitis (NASH). Although this "silent" disease resembles alcoholic liver disease, it occurs in people with little or no alcohol consumption (10–12). It affects 2–5% of Americans and is becoming more common, possibly due to a dramatic increase in obesity, a risk factor for NASH. NASH can lead to cirrhosis, which increases the risk of HCC. It is therefore predicted to be a major etiological factor for HCC in the not so distant future. However, unlike HBV/HCV-induced HCCs, there is very little knowledge about the pathogenesis of NASH-associated HCC, which has been increasing rapidly in the past 10 years. At the molecular level, the development of hepatocellular carcinoma is a complex, multistep process involving both genetic and epigenetic aberrations (13, 14). Animal models are invaluable resources that enable development of biomarkers for early detection, assist in identification of molecular targets for therapy, and provide a preclinical platform for testing emerging drugs for preclinical trials. Although several rodent models are widely used for these purposes, we will limit our discussion here to observations made in rodent models of diet-induced HCC.

7.2 FOLATE DEFICIENCY

Although nutritional deficiency and imbalance are prevalent in third world countries, their occurrence is not uncommon in the Western world due to various reasons. In the United States, folate insufficiency occurs particularly during pregnancy and lactation, and among alcoholics, smokers, AIDS patients, adolescent females, and low-income elderly (15). Epidemiological and clinical studies have clearly demonstrated that folate deficiency in humans could lead to susceptibility to certain types of cancers, e.g., colorectal adenoma, colorectal carcinoma, cervical dysplasia, and liver and esophageal cancers (16–21). Premalignant dysplasia of cervical, bronchial, and colonic epithelial cells could be reversed by folate supplementation, implying that folate deficiency may have a casual role in the process. Insufficient folate intake has been shown to increase the frequency of somatic mutations caused by chemotherapy (22, 23), the level of uracil and micronuclei in erythroblasts (24–27), and heritable fragile sites at translocation hot spots in tumors (16, 28) (Figure 7.2). The importance of sufficient folate intake in reducing colon cancer risk is further substantiated in individuals with biallelic C to T mutation in the methylene tetrahydrofolate reductase gene (21, 27, 29–33). Studies have shown that moderate folate deficiency potentiates tumor induction by chemical carcinogens (19, 23, 34). Emerging studies indicate that folate status also modulates mitochondrial DNA (mtDNA) stability (35–38). A recent study with human HCC patients has shown a correlation of low blood folate with increased risks of liver damage and HCC (26), suggesting a possible role of folate deficiency in the human liver carcinogenesis (36). The genetic instability that results in a high rate of mtDNA deletions in lymphocytes of HCC patients correlated inversely with serum folate level in humans (36). Low folate status or large deletions in mtDNA may result in mitochondrial dysfunction, causing excessive ROS generation and leading to apoptotic cell death (39), which may ultimately lead to cancer development. Rats fed folic acid supplementation have reduced mtDNA deletions in the liver upon exposure to chemotherapeutic agents or at old age (40, 41).

Choline, a quaternary amine, is an essential nutrient that is required for important cellular processes like neurotransmission (acetylcholine synthesis), cell membrane signaling (phospholipids synthesis), transport of lipids (lipoprotein synthesis), and as a source of methyl group for biological methylation reactions (42, 43). Although choline is made available through diet and de novo biosynthesis, increasing demand during pregnancy and lactation could result in depletion of tissue stores (44). It is therefore considered a required dietary nutrient since 1998 by the U.S. Institute of Medicine's Food and Nutrition Board. Further, almost 50% of the U.S. population is thought to have genetic polymorphisms that make them susceptible to choline deficiency due to increase in their dietary methyl requirements (32, 45). Choline deficiency is thought to cause nonalcoholic fatty liver disease, atherosclerosis, and neurological disorders (46). A study of healthy adult subjects has demonstrated that 77% of men and 80% of postmenopausal women developed fatty liver and muscle damage upon dietary choline deprivation (42). Interestingly, another study has shown that individuals with inadequate choline intake develop fatty liver despite adequate folate and methionine intake. This liver damage was resolved upon dietary supplementation with choline (42). These studies reinforce the importance of methyl sources in the diet.

7.2.1 Rodent Models of Folate Deficiency and Hepatocellular Carcinoma

Rodent models have been extremely valuable for elucidating the mechanisms by which folic acid deficiency causes tumor induction. It is noteworthy that while humans depend almost exclusively on dietary folate for synthesis of methionine and for the conversion of folate into metabolically active forms, rats can efficiently synthesize methionine de novo from choline due to higher activity of betaine-homocysteine methyltransferase (47). Rats are therefore less sensitive to folate deficiency alone and have to be fed a diet deficient in folate as well as choline and methionine to mimic metabolic alterations caused by folate/choline/methionine deficiency in humans (30, 48–52). This "lipotrope deficient" or "folate-methionine-choline deficient" (FMD, folate and methyl deficient) diet consists of 6% casein and 6% gelatin not supplemented with methionine, choline, and folate). This diet model (48–50, 53) is extensively studied as a rat model of HCC because nutritional deprivation of methyl groups rather than exogenous carcinogens reproducibly causes preneoplastic nodule formation after 32 weeks and HCC formation after 54 weeks (48–50, 54–59). More importantly, the underlying pathological changes during FMD diet-induced hepatocarcinogenesis mimic those of human HCCs associated with HBV and HCV infections, alcohol consumption, and metabolic syndrome (60). Therefore the lipotrope-deficient model is an ideal system to study the mechanisms by which nutritional imbalance and deficiency, quite common among malnourished humans, can lead to human cancers (Figure 7.3). The liver-specific carcinogenesis is consistent with the accumulation of lipid within a few days of this dietary regimen, followed by development of fibrosis and cirrhosis of the liver at a later period (61, 62). One advantage of this animal model is that tumor progression occurs slowly and involves different stages that include steatosis, apoptosis, fibrosis, cirrhosis, and formation of adenomas and carcinomas.

Similar to rats, a semisynthetic, choline-deficient, L-amino acid-defined (CDAA) diet, lacking choline and low in methionine, induces HCCs in mice (63–67). This diet that consists of Lombardi's choline-deficient (0g/Kg), low-methionine (1.7g/Kg), and amino acid-defined diet (CDAA diet, #518753 from Dyets Inc., Philadelphia). The control is a choline-sufficient (14.48g/Kg), amino acid–defined diet (CSAA diet, #518754) fortified with methionine (4g/Kg). Both diets were supplemented with AIN-76A vitamin mix (#300050), providing 0.2% folic acid (w/w) to the diet (68). Feeding

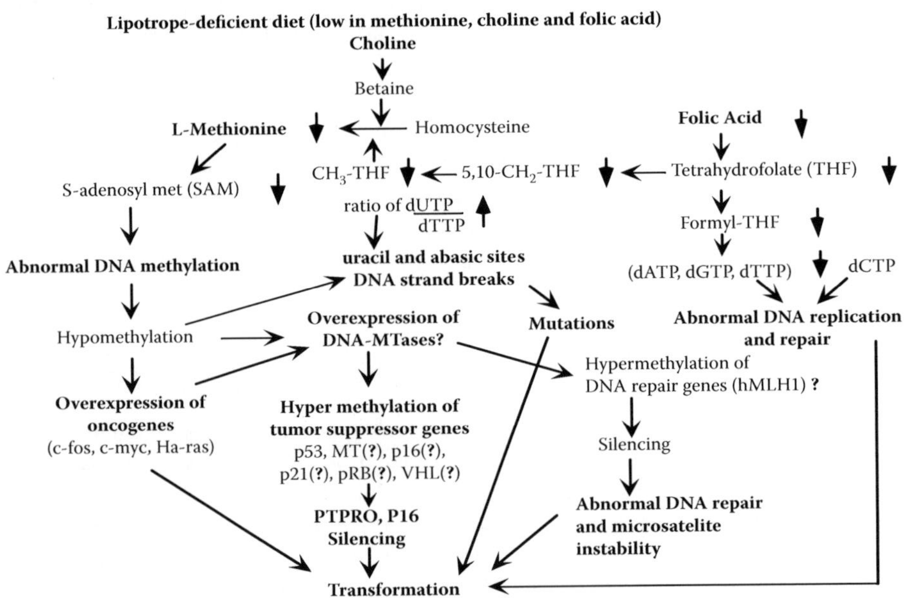

FIGURE 7.3 Schematic diagram depicting biochemical, genetic, and epigenetic changes that occur in the livers of rodents upon feeding rats folate and methyl-deficient diet.

the CDAA diet reproducibly triggers severe and persistent fatty liver coupled with extensive cell death, a combination that is frequently considered to be appropriate for the induction of "micronodular" (fatty) cirrhosis in humans. In male mice, fatty change and fibrosis occur at 22 weeks, and preneoplastic foci are evident in 100% of the animals at a multiplicity of 6.6 +/– 4.0 per mouse at 65 weeks. Hepatocellular adenomas and carcinomas develop at incidences of 66% and 21%, at multiplicities of 1.42 +/– 1.32 and 0.29 +/– 0.62, respectively, at 84 weeks of feeding the CDAA diet (69). Dietary supplementation with methionine completely prevents the development of both preneoplastic nodules and carcinomas (70, 71). As in humans, the female mice are resistance to development of these lesions because the CDAA diet increases the levels of 8-hydroxydeoxy guanine (8-OHdG), a marker of DNA damage only in male mice.

7.2.2 Epigenetic Alterations in Rodent Models of Diet-Induced HCC

Both genetic and epigenetic changes occur during multistage hepatocarcinogenesis in rats fed the FMD diet. Several hypotheses have been put forward to explain carcinogenesis induced by this diet. These include DNA hypomethylation leading to enhanced expression of oncogenes, DNA damage mediated by free radicals, altered membrane phospholipid metabolism, receptor and protein kinase C-mediated signal transduction, abnormal deoxynucleotide metabolism that promotes DNA base transition mutations, and oncogene and tumor suppressor gene aberrations (30, 72–75). The most promising hypothesis that has received substantial experimental support involves global DNA hypomethylation and regional hypermethylation of certain genes, particularly those of tumor suppressors, in a variety of preneoplastic and transformed cells (52, 76). It appears that carcinogenesis caused by folate deficiency is largely due to depletion of S-adenosyl methionine (SAM) (49, 50, 53, 55, 57, 77), the cofactor for DNA methyltransferase (Figures 7.3 and 7.4). This model is based on some published results and some hypothesized consequences of SAM deficiency.

It is now well established that global hypomethylation leading to genomic instability and silencing of tumor suppressor genes (TSGs) by hypermethylation of CpG island located in their promoters plays a key role in carcinogenesis (78–80). DNA hypomethylation can alter the conformation and stability of the chromatin structure that results in aberrant expression of growth-promoting genes and exposure of the affected regions to DNA-damaging agents as well as DNA methyltransferase. Hypermethylation usually silences the affected genes, and their functions are stably lost in a clonally propagated fashion (81–86). The latter gene modification has received considerable attention in recent years, as hypermethylation of many tumor suppressor genes is responsible for their silencing. In fact, at least half of the genes suffer loss of function through epigenetic modification rather than through genetic defects. Interestingly, many epigenetically silenced genes are not mutated, and these genes can be reactivated by DNA hypomethylating agents. Indeed, more than 100 clinical trials with these agents (5-azacytidine or 5-deoxy-azycytidine) have been reported in the National Cancer Institute database. It now appears that promoter hypermethylation, like mutation, can be at least one of the "hits" in Knudson's two-hit theory of carcinogenesis, one "hit" triggering initiation and the other "hit" causing

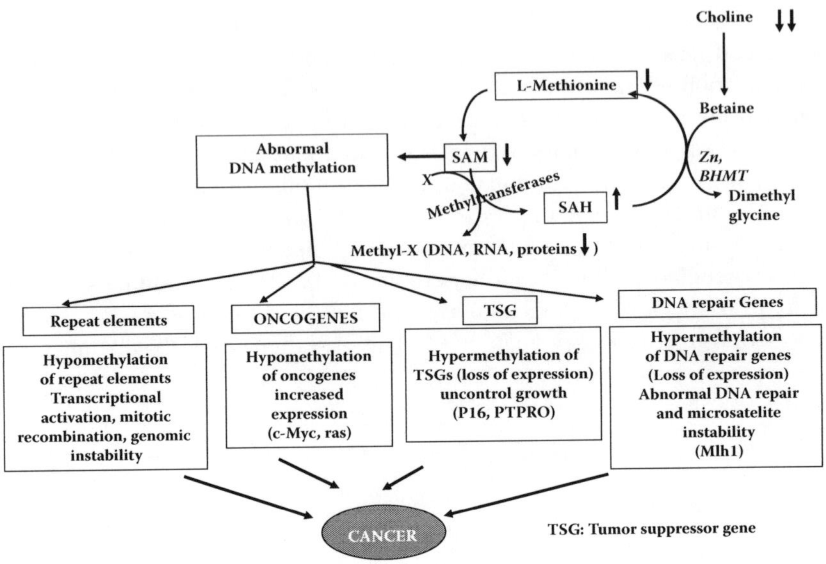

FIGURE 7.4 Schematic representation of the potential role of DNA methylation in activating oncogenes and suppressing tumor suppressors during folate-methyl deficient diet-induced hepatocarcinogenesis. SAH, S-adenosyl homocysteine; SAM, S-adenosyl methionine; BHMT, betaine-homocysteine-S-methyltransferase.

progression of tumorigenesis (79, 85, 87). Some examples of genes methylated and in most cases silenced in specific types of cancer include RB, P15, BRAC1, E-CAD, GSTP1, DAPK1, ID4, C/EBPa, and P16 (84).

With the methyl-deficient diet where tumor progression can be followed stepwise, it was possible to discern how "methyl" deficiency alters methylation status of different genes, which probably contribute to carcinogenesis. As a first step to elucidate the role of DNMTs in FMD diet-induced hepatocarcinogenesis, we monitored temporal changes in the mRNA and protein levels of three major DNA methyltransferases (Dnmt1, Dnmt3a, and Dnmt3b). DNA methyltransferases heritably maintain DNA methylation that is normally required for the silencing of spurious retroviral promoters, transposable elements in the genome, and regulating expression of imprinted genes and of genes on the inactive X chromosome (88). DNA methylation is initiated by Dnmt3a and Dnmt3b, which is maintained in the newly synthesized DNA strand by Dnmt1 (89). At the C-terminal these enzymes harbor methyltransferase domain, which is homologous to bacterial CpG methylase. The N-terminal regions that harbor several domains involved in interaction with DNA and protein are absent in bacterial methyltransferases (Figure 7.1A). These enzymes also act as transcriptional repressors by recruiting corepressors like histone deacetylase, histone methyltransferase, and the polycomb group of proteins (90). The Dnmt1 and Dnmt3a mRNA and protein levels were significantly elevated ($P < 0.001$) as early as 9 weeks following feeding the diet (Figure 7.5). Maximal changes occurred at 9 weeks (two- to fourfold), which persisted even after 36 weeks. It is likely that the elevated levels of these enzymes could play a critical role in de novo methylation

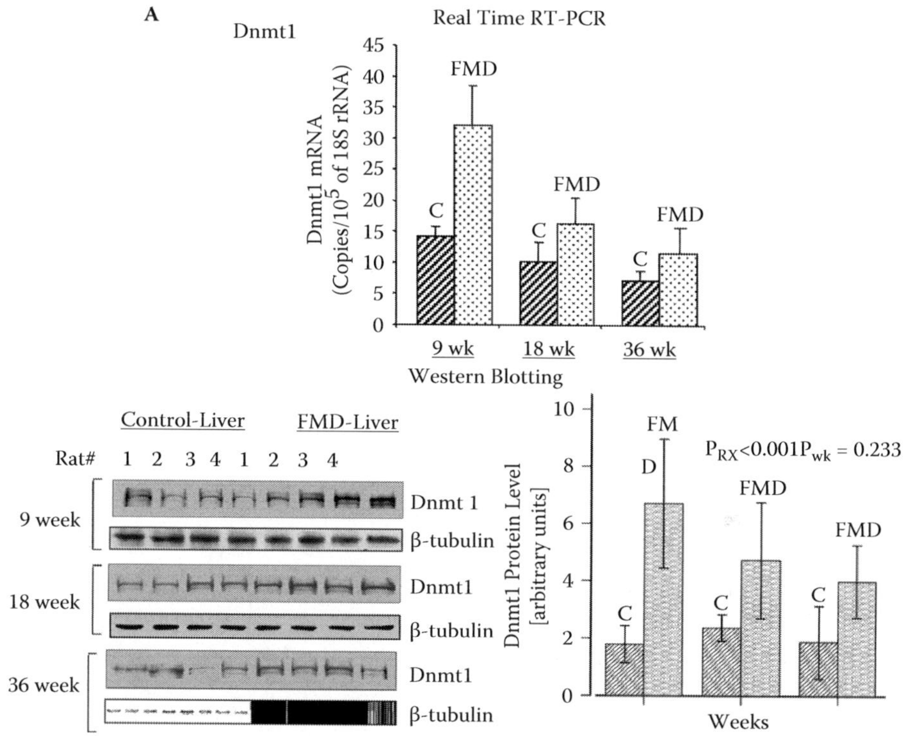

FIGURE 7.5A Temporal changes in expression of hepatic Dnmt1, Dnmt3a, and Dnmt3b RNA and protein levels in rats fed the FMD diet or a methyl-adequate diet for different time periods. Total liver RNA was subjected to real-rime RT-PCR with rat gene-specific primers, and data was normalized to 18S rRNA. The nuclear extracts were subjected to immunoblot analysis with respective antibodies. The blot was reprobed with β-tubulin antibody. HRP-conjugated antirabbit was used as a secondary antibody, and the signal was developed using ECL reagent; the scanned x-ray film was quantified by Kodak Imaging software. (Modified from Ghoshal, K., Li, X., Datta, J., et al, *J Nutr,* 136, 1522–27, 2006. With permission.)

(Continued)

of tumor suppressor genes (Figure 7.5). Surprisingly, Dnmt3b expression was not altered in response to FMD diet (91). Methyl CpG binding domain proteins (MBDs) are mediators (92, 93) of the DNA methylation signal. Five MBDs that harbor characteristic methyl CpG binding domain have been identified in mammals (Figure 7.1B). These proteins bind symmetrically methylated 5-methyl-CpG through their conserved methyl CpG binding domain. MBDs suppress methylated promoters by recruiting a variety of histone-modifying enzymes and chromatin remodelers. Methylated DNA acquires a chromatin configuration that is refractory to transcription factors, resulting in gene silencing (Figure 7.1C) (94–96). We therefore also studied alteration in the expression profile of MBDs in response to folate deficiency. The mRNA levels of MBD1, MBD2, MBD3, and MeCP2 were significantly ($P < 0.001$) higher in the livers of rats fed FMD diet than in controls (Figure 7.6). The increase in the expression of these proteins was observed as early

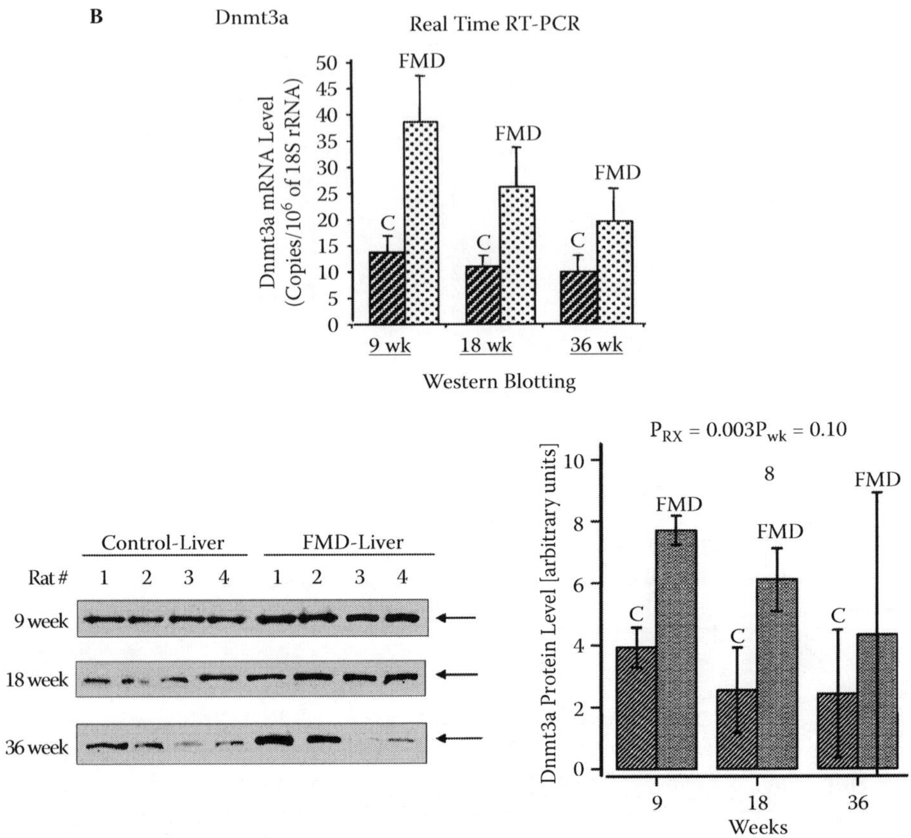

FIGURE 7.5B (Continued) Temporal changes in expression of hepatic Dnmtl, Dnmt3a, and Dnmt3b RNA and protein levels in rats fed the FMD diet or a methyl-adequate diet for different time periods. Total liver RNA was subjected to real-rime RT-PCR with rat gene-specific primers, and data was normalized to 18S rRNA. The nuclear extracts were subjected to immunoblot analysis with respective antibodies. The blot was reprobed with β-tubulin antibody. HRP-conjugated antirabbit was used as a secondary antibody, and the signal was developed using ECL reagent; the scanned x-ray film was quantified by Kodak Imaging software. (Modified from Ghoshal, K., Li, X., Datta, J., et al, *J Nutr,* 136, 1522–27, 2006. With permission.) (Continued)

as 9 weeks at both the RNA and protein levels. Thus FMD diet causes coordinate induction of DNMTs and MBDs (53). The lack of correlation between RNA and protein levels of MeCP2 implicates involvement of posttranscriptional regulation probably through microRNAs.

Next, to identify differentially methylated genes during hepatocarcinogenesis in rats fed the FMD diet we used a genomewide screening approach called Restriction Landmark Genomic Scanning. This two-dimensional gel electrophoresis technique allows detection of DNA rearrangements and altered DNA methylation patterns (97, 98). Methylation analysis using this technique (RLGS-M) involves digestion of

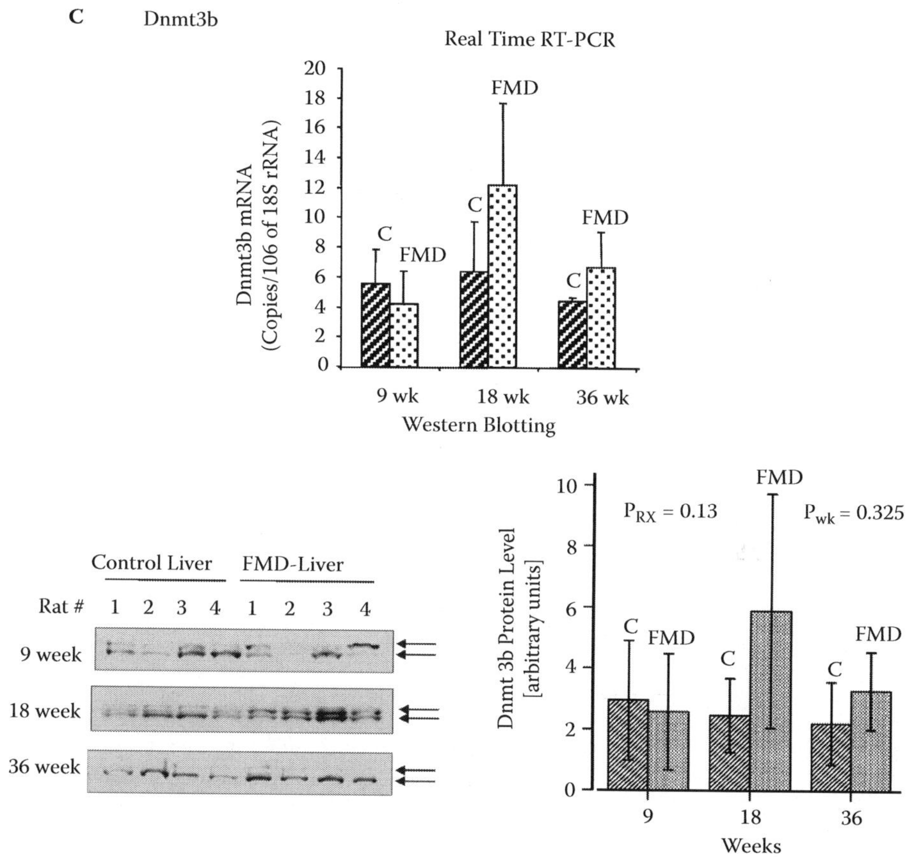

FIGURE 7.5C (Continued) Temporal changes in expression of hepatic Dnmt1, Dnmt3a, and Dnmt3b RNA and protein levels in rats fed the FMD diet or a methyl-adequate diet for different time periods. Total liver RNA was subjected to real-rime RT-PCR with rat gene-specific primers, and data was normalized to 18S rRNA. The nuclear extracts were subjected to immunoblot analysis with respective antibodies. The blot was reprobed with β-tubulin antibody. HRP-conjugated antirabbit was used as a secondary antibody, and the signal was developed using ECL reagent; the scanned x-ray film was quantified by Kodak Imaging software. (Modified from Ghoshal, K., Li, X., Datta, J., et al, *J Nutr,* 136, 1522–27, 2006. With permission.)

genomic DNA with a rare-cutting methylation-sensitive enzyme and ^{32}P labeling of resulting ends. The digested, end-labeled DNA is then digested with another enzyme (e.g., *Eco*R V), separated by cylindrical agarose gel electrophoresis, digested in-gel with *Hin*f I and separated in a second direction on an acrylamide gel. The gel is dried and subjected to autoradiography. If a genomic *Not* I site is methylated, the enzyme cannot cleave, the site will not be end-labeled, and the spot will be missing (Figure 7.7A). On the other hand, if the *Not* I site is unmethylated, the site will be cut and the restriction ends will be end-labeled. The resulting RLGS profile displays

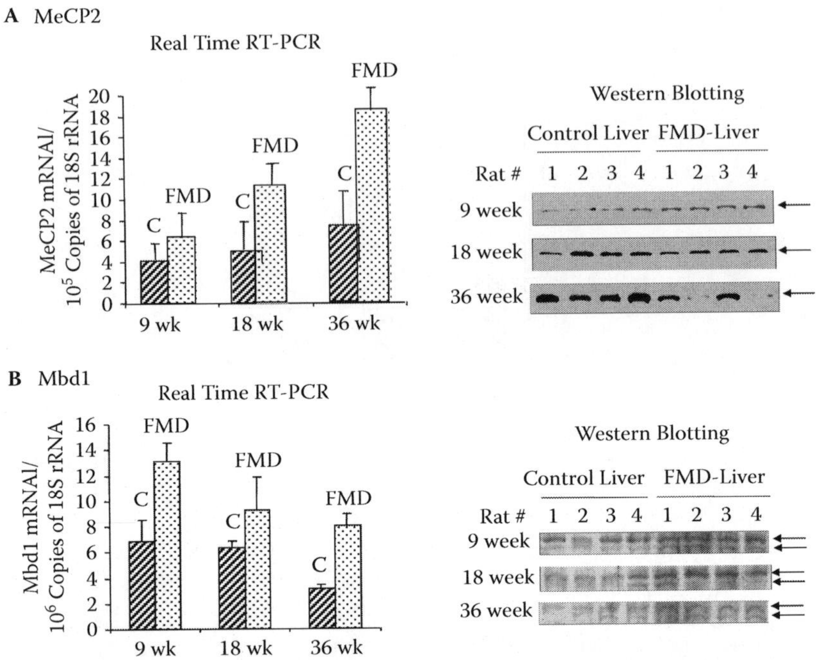

FIGURE 7.6A,B Temporal changes in expression of hepatic MeCP2 and Mbd1-4 RNA and protein levels in rats fed the FMD diet or a methyl-adequate diet for different time periods. RNA and protein levels of 5 methyl CpG binding proteins with signature MBD (shown in Figure 7.1) were measured by real-time RT-PCR using SYBR Green method and Western blot analysis with specific antibodies, respectively. (Modified from Ghoshal, K., Li, X., Datta, J., et al., *J Nutr,* 136, 1522–27, 2006. With permission.) (Continued)

a highly reproducible spot pattern in which each spot represents an end-labeled *Not* I site. Up to 2000 end-labeled rare-cutting restriction sites are displayed in a single RLGS profile. Most of the rare-cutting restriction enzyme sites with GC-rich recognition sequences are located in the promoter region of genes, resulting in a selective display of gene sequences rather than random genomic sequences. Several novel tumor suppressor genes (TSGs) such as TCF21, ID4, TWIST2, DAPK1, and FOXD3 have been identified using RLGS-M analysis of different human malignancies (80, 99–104).

RLGS-M analysis of preneoplastic nodules (PNNs) and HCCs developed in Fisher 344 rats fed FMD diet for 36 and 54 weeks, respectively, revealed 6 methylated fragments in the PNNs with an additional 33 methylated fragments in the tumors (Table 7.1) (77). In contrast, the appearance of several new spots indicative of hypomethylation at the *Not* I were also observed. One of the spots lost in tumors (Figure 7.7A) was identified as protein tyrosine phosphatase receptor-type O (PTPRO) after cloning the spot from a rat *Not* I-*Eco*R V library and sequencing (77). Southern blot analysis of genomic DNA confirmed that this gene was methylated in PNNs and tumors (Figure 7.7B). Subsequent functional analysis has shown

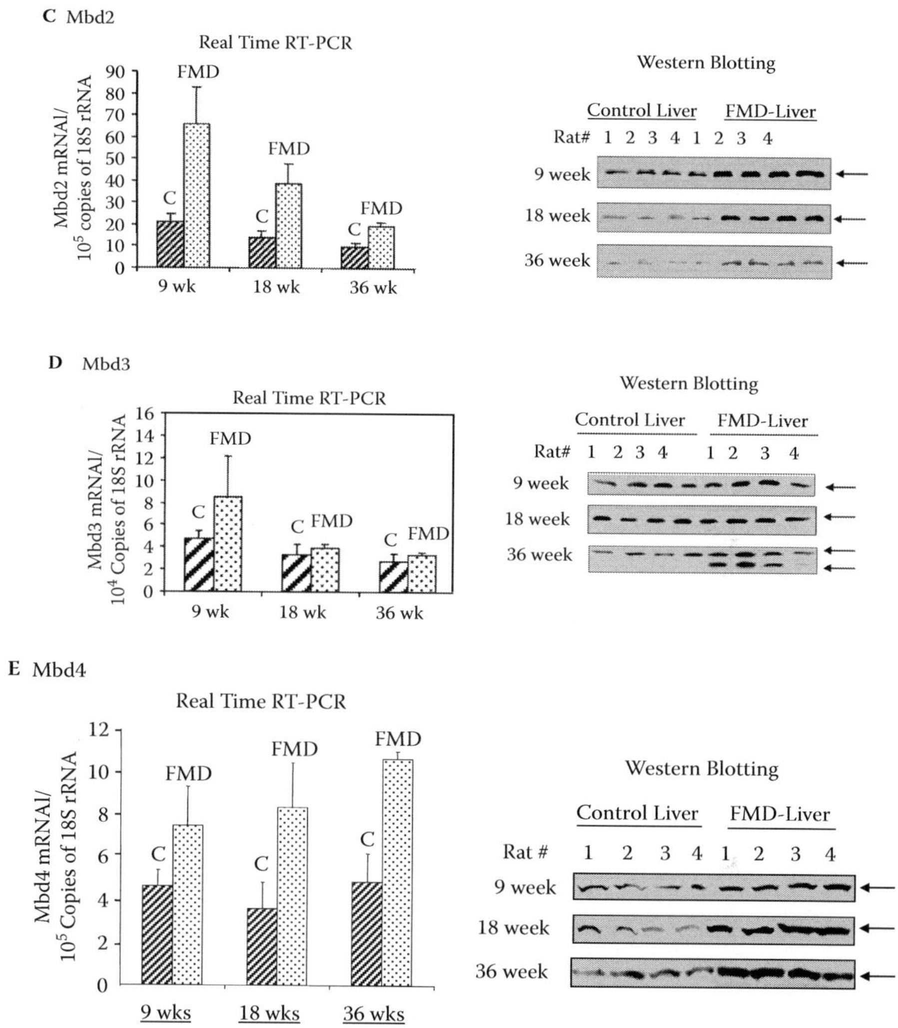

FIGURE 7.6C,D,E (Continued) Temporal changes in expression of hepatic MeCP2 and Mbd1-4 RNA and protein levels in rats fed the FMD diet or a methyl-adequate diet for different time periods. RNA and protein levels of 5 methyl CpG binding proteins with signature MBD (shown in Figure 7.1) were measured by real-time RT-PCR using SYBR Green method and Western blot analysis with specific antibodies, respectively. (Modified from Ghoshal, K., Li, X., Datta, J., et al., *J Nutr,* 136, 1522–27, 2006. With permission.)

that this protein exhibits tumor suppressor characteristics (105). Similarly, one of the amplified spots was identified as c-Myc (Figure 7.7). Gain of this spot (c-Myc) was, however, not due to hypomethylation but due to gene amplification in the liver tumors formed feeding folate-depleted diet. Thus animal models are valuable tools to identify novel cancer causing genes. The identity and functions of other spots lost or gained in RLGS gel remains to be established.

TABLE 7.1
RLGS Analysis of Liver DNA Identified Several Spots (*Not* I-*Eco*R V Fragment) Lost or Gained in the Preneoplastic Nodules and HCCs Developed in Rats after Feeding FMD Diet for 36 and 54 Weeks, Respectively

	Control	Preneoplastic	Tumor
Total spots counted	1020	1020	1020
Spot lost (total)		3	32
Spot lost (partial)		3	7
Spot gained (total)		7	14
Spot gained (partial)		1	10

Source: Wang, B., S. Majumder, G. Nuovo, H., et al. 2009. *Hepatology* 50:1152–61.
Note: DNA from age-matched rat livers fed chow diet was used as controls.

7.3 INTRODUCTION TO MicroRNA

MicroRNAs (miRs) are small noncoding RNAs that negatively regulate expression of protein-coding genes. After sequencing of the genomes of different organisms, it has become clear that, unlike prokaryotes and lower eukaryotes, less than 5% of the animal genome codes for protein (Figure 7.2) (106–108). Recent high-throughput sequencing of transcriptome from animals and plants revealed that most of the genome is transcribed into nonprotein coding RNAs (ncRNAs) that include microRNAs (miRNA), piwi-interacting RNAs, PASR (polymerase-associated small RNA), small nuclear RNAs (snRNAs), small nucleolar RNAs (snoRNAs), small interfering RNAs (siRNAs), long noncoding RNAs, and antisense RNAs (109, 110). MicroRNAs are short (20–25 nucleotide) single-stranded RNAs, identified in plants, animals, and some viruses but not in yeasts. Since their first discovery in 1993 (111, 112), more than a thousand miRs have been identified by high-throughput sequencing and bioinformatic analysis, among which several hundreds are experimentally validated by Northern blot and/or real-time RT-PCR analysis (113–115). Subsequent cloning and deep sequencing of small RNAs, once thought to be "degraded RNAs" (reviewed in 116), identified hundreds of conserved miRNAs in animals. The majority of the microRNAs are ubiquitously expressed, while some are tissue specific. For example, miR-1 and miR-133 are predominantly expressed in the heart (117, 118), whereas miR-122 is liver specific (119). The version 14 of miRbase database contains 10,883 entries representing hairpin precursor miRNAs, expressing 10,581 mature miRNA products, in 115 species (http://www.mirbase.org/). Many of these microRNAs are highly conserved. Furthermore, some of these noncoding RNAs are developmentally regulated and are critical for organismal development. It is likely that the ncRNAs evolved to regulate the complexities of higher organism. In animals it is believed that ~5% of genes encode miRNAs, which in turn regulate ~30% of protein-coding genes (120). Although individual

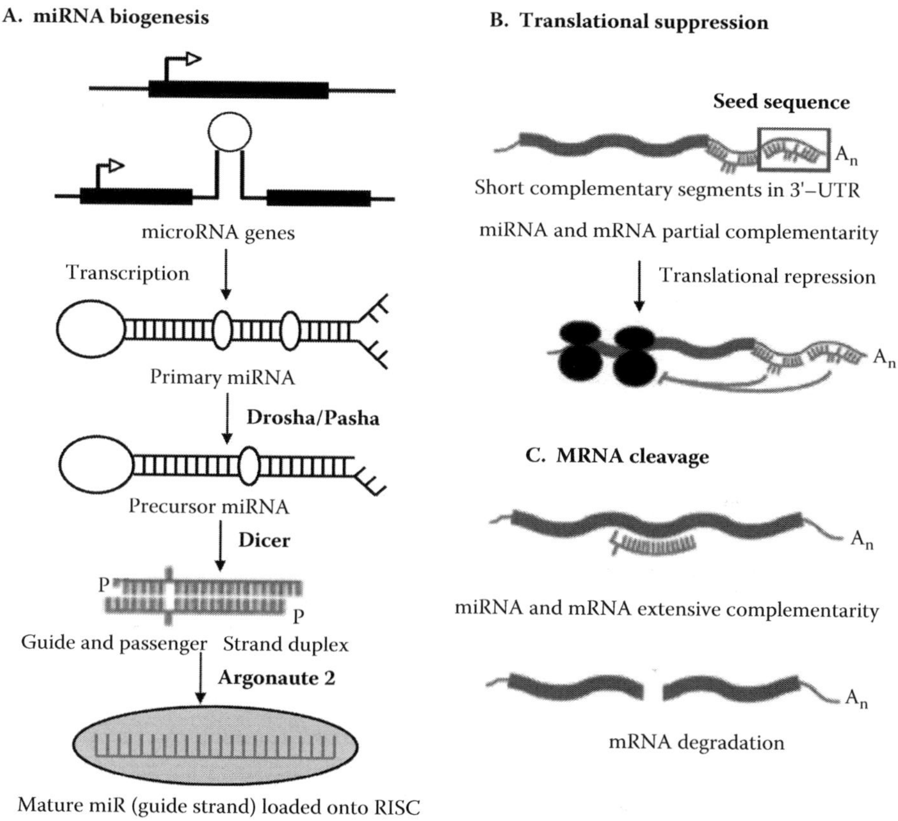

FIGURE 7.7 Schematic presentation of microRNA biogenesis and function. Primary microRNAs are transcribed and processed in the nucleus to form precursor miRs, which are then transported to the cytoplasm for further processing to imperfect duplex. The mature miR is then recognized by miRISC to exert their functions. The seed sequence (nucleotides at the 5′ end of the miR that form Watson-Crick base pairing with 3′-UTR) is critical for base pairing with target mRNA.

microRNAs affect the expression of target mRNAs only modestly (121, 122), the cumulative effect of different miRs on target mRNAs may be significant because microRNAs act in combinatorial fashion to regulate protein expression.

7.3.1 MicroRNA Genes and Biogenesis of MicroRNAs

As of March 1, 2010, the miRNA database (www.mirbase.org) contains the following numbers of miRNAs: 174 *Caenorhabditis elegans,* 190 *Arabidopsis thaliana,* 157 *Drosophila melanogaster,* 360 *Danio rerio* (zebrafish), 479 *Gallus gallus* (chicken), 721 human, 325 *Rattus norvegicus,* 579 *Mus musculus,* and 7 *Xenopus levis.* Among these, certain microRNAs such as miR-122 are conserved across species, whereas a large number of microRNAs are species specific. The existence of highly conserved miRNAs in animals indicates that miRNAs have performed critical functions throughout evolution. The genes coding miRs can be intergenic, intragenic, or intronic, and approximately 50% are in clusters (123, 124). The majority of miRs are transcribed in the nucleus by RNA polymerase II into primary miRs (pri-miRs) with 5′-Cap and 3′-poly(A) tails (Figure 7.7A), whereas those associated with repeat elements are generally transcribed by pol III. Intronic miRs are processed from lariant intermediates. The largest known human miRNA gene cluster, C19MC (chromosome 19 miRNA cluster) harboring 46 miRNA genes, is transcribed by RNA polymerase III (123). Pri-miRs with characteristic stem-loop structure of variable sizes are processed by RNase III like enzyme Drosha and its partner, double-stranded RNA binding protein Pasha (DGCR8) microprocessor complex in the nucleus into ~70-nucleotide precursor miRs (pre-miRs) that have a 2-nucleotide 3′ overhang (Figure 7.7A). Half-lives of pri-miRs are very short, and processing of intronic miRs appears to occur before splicing (124). Pre-miRs transported into the cytoplasm by exportin 5/Ran GTPase are further processed by RNase III like enzyme Dicer to cleave the stem-loop structure generating imperfectly base-paired double-stranded miR (sense) and miR* (antisense). This is followed by dissociation of the two strands and degradation of the antisense (miR*) strand, by an unidentified mechanism, provided it does not function as a microRNA itself. The mature miRs are then loaded into Argounaute proteins and incorporated into miR-induced gene-silencing complex (miRISC), which then interact with specific target mRNAs to cause either degradation and/or translation suppression. Mammalian miRNAs interact with target mRNAs by forming imperfect base pairing to sequences located predominantly in the 3′-untranslated region (UTR) and in some cases with the 5′-UTR or coding regions (125). The interaction of the seed sequence (nucleotides 2–8 on the 5′ end) of the miRNA with its cognate site on its target mRNA is critical for its function. Extensive complementarity between the miRs and mRNA results in degradation of the mRNA by Argonaute 2, whereas imperfect base pairing occurring in a majority of cases leads to translational suppression that sometimes results in mRNA destabilization (Figure 7.8B, C). At least 30% of protein-coding genes are regulated by the concerted action of different miRs.

7.3.2 Function of MicroRNAs

The inability of Dicer null mice to survive demonstrates that miRs are essential for mammalian development (126). The development of hepatitis and spontaneous liver tumors in adult mice deleted of Dicer in the adult liver indicates that miRs are required for normal liver function (127, 128). In animals, miRs are involved in various biological processes such as organ development, cell proliferation, cell fate determination, cell cycle regulation, differentiation, apoptosis, immune response, metabolism, signal transduction, hematopoietic lineage differentiation, viral infection, and energy metabolism, including fat metabolism and glucose homeostasis (reviewed in 109, 129–134). For example, miR-375 is involved in glucose homeostasis by inhibiting insulin secretion in pancreatic β-cells in response to feeding glucose. Similarly miR-122, a miRNA abundantly expressed in the liver, regulates plasma cholesterol level in mice by an as yet unidentified mechanism (135, 136).

7.3.3 MicroRNAs Expressed in the Liver

The majority of the microRNAs are ubiquitously expressed, whereas some are tissue specific. miR-122 was identified as the most abundant liver-specific RNA by Lagos-Quintana et al. (137). In addition to miR-122, miR-192, 194, and 148 are also liver-specific miRs. Ubiquitous miRNAs, such as miR-192, miR-194, miR-221, miR-223, miR-26a, miR-16, the miR-17-92 family, miR-27b, miR-30d, miR-126, miR-miR-143, and the let-7 family members, are also highly expressed in the adult liver tissue. Unlike miR-122, miR-92a and miR-483 are specifically expressed in the fetal liver (138). Thus expression of certain miRNAs is developmentally regulated. Loss of liver functions and development of spontaneous HCCs in mice lacking Dicer 1 in adult hepatocytes indicate that microRNAs play essential functions in the adult liver (128).

7.4 ABERRATIONS IN MICRORNA EXPRESSION OCCUR IN HEPATOCELLULAR CARCINOMAS

A link between miRNA and cancer was the seminal observation from Croce's group. They identified that two miRNAs (miR-15 and miR-16) localized in the region of chromosome 13 (13q14), frequently deleted in chronic lymphocytic leukemia (CLL), function as tumor suppressors (139–141). Extensive profiling studies of different human cancers have revealed that expression of miRNAs is altered in almost all cancers including hepatocellular carcinoma, and that miR signature is a more reproducible and reliable marker for neoplastic cells than the mRNA profile. Here we discuss microRNAs that are dysregulated in two different animal models of liver cancer.

7.4.1 MicroRNAs Dysregulated in the Rat Model of Folate Methyl-Deficient (FMD) Diet-Induced HCC

To understand the role of microRNA in FMD diet-induced HCC model, RNA isolated from HCCs developed in rats on FMD diet for 54 weeks and liver RNA from control animals were subjected to microarray analysis using custom-made microRNA microarray (142). Among 245 miRs analyzed, expression of 23 miRs was elevated whereas the expression of only 3 miRs was suppressed (Table 7.2). Among these up-regulated miRs, 9 (miR-17-92, -23a,b, -24, -101b, -130, -172a, -219, and -328) were elevated twofold or higher in all three tumors. Among the down-regulated miRs only miR-122, a liver-specific miR, was reduced by 50% in all three tumors, whereas miR-123 and miR-215 levels decreased in two out of three tumors. Northern blot analysis was performed to validate microarray data and to measure miR expression also at 36 weeks when PNNs are formed (53). The results revealed reduced expression of miR-122 in all tumors but not in PNNs (Figure 7.8A, B) suggesting

TABLE 7.2
MicroRNAs Dysregulated in the HCCs Developed in Fisher Rats Fed FMD Diet for 54 Weeks

miRNA	T1/N	T2/N	T2/N
miR-101b	3.7	2.7	3.1
miR-130	5.9	2.5	2
miR-130a	5.7	2.6	2.2
miR-172a	3.6	2.4	3.5
miR-219	2.9	2	2.4
miR-23a	4.1	2.8	2.2
miR-23b	4.6	3.8	2.6
miR-24	3.4	2.9	3.1
miR-328	3.8	2.9	2.9
Let-7a	3.9	1.8	3.7
miR-103	3.1	2.2	1
miR-106	3.6	2.7	1.3
miR-106a	3.8	2.8	1.4
miR-106b	2.9	2.2	1.3
miR-130a	5.4	2.2	1.9
miR-17	4.4	3.7	1.5
miR-20	4.7	3.9	1.7
miR-21	2.4	1.8	2.4
miR-320	2.3	1.6	2.4
miR-93	3	2.1	1.2
miR-99b	4.8	3.8	1.2
miR-122	0.4	0.3	0.2
miR-123	0.9	0.5	0.4
miR-215	0.5	0.9	0.4

Source: Reproduced from Kutay, H., Bai, S., Datta, et al., *J Cell Biochem,* 99, 671–78, 2006.

Note: Age-matched control livers from rat fed chow diet were used as controls. Total RNA from three HCC samples and three age-matched livers from rats on normal diet were used for microarray analysis. RNA labeling and hybridization on miR microarray chips was done as described (194). Five μg of RNA from each sample was biotin-labeled during reverse transcription using random hexamers. Hybridization was carried out on miR microarray chip (KCI version 1.0), which contains 368 probes, including 245 human and mouse miR genes (both precursors and mature), in duplicate. Each sample was hybridized to duplicate array. Hybridization signals were detected by binding of a streptavidin-Alexa 647 conjugate and detected using a Perkin-Elmer ScanArray XL5K. Scanned images were quantified by the Quantarray software (Perkin-Elmer, Wellesley, Massachusetts). The miR signal in each tumor was normalized to the average signal in three normal livers. T and N denote HCCs developed in rats fed FMD diet and livers from rats fed control diet, respectively.

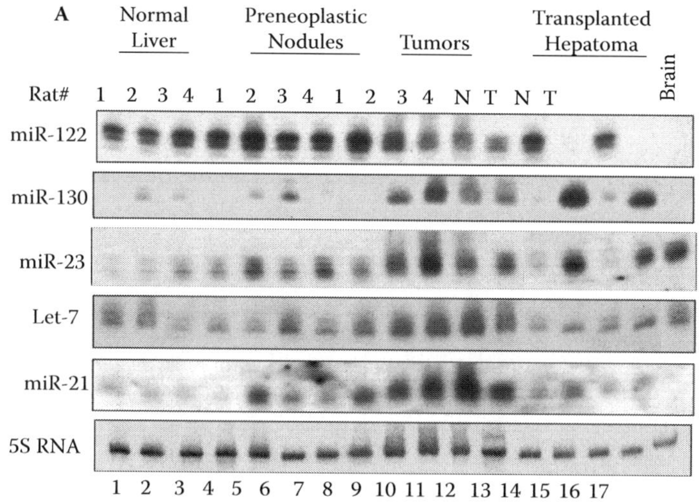

FIGURE 7.8A A. Northern blot analysis confirmed down-regulation of miR-122 and up-regulation of miR-23, miR-21, miR-130, and let-7 in HCC compared with the controls. Total RNA was isolated from the livers of four rats fed normal diet and four rats fed FMD diet for 36 and 54 weeks, respectively. An aliquot (30 μg) of the total RNA was separated by denaturing PAGE, transferred to a nylon membrane, and subjected to Northern blot analysis with ^{32}P-labeled deoxyoligonucleotide antisense to specific miRs. The blot was reprobed with oligo antisense to 5S rRNA, and the ratio of miR signals to that of 5S rRNA were determined. B. Quantitative analysis of Northern blot data showed significant down-regulation of miR-122 and up-regulation of miR-21, miR-23, miR-130, and let-7 in rat HCCs. ^{32}P-signal was measured using Imagequant software and quantified using volume analysis program. PNNs denote preneoplastic nodules. The results represent the average signal of each miR normalized to that of 5S rRNA SD. The p-value of ≤ 0.05 was considered significant. Asterisks indicate significant changes. (Reproduced from Kutay, H., Bai, S., Datta, et al., *J Cell Biochem*, 99, 671–78, 2006.) (Continued)

that the down-regulation of miR-122 starts during neoplastic transformation. miR-122 expression was not detectable in Morris hepatoma, a transplanted liver tumor initially generated by treatment with the carcinogen methylmethane sulfonate (143), implicating that down-regulation of miR-122 is not restricted to the diet model. It is noteworthy that miR-122 level reversed to the control level when rats were provided folate- and methyl-adequate diet for 18 weeks after 36 weeks of feeding FMD diet (Figure 7.9A, B). Animals switched to folate/methyl-adequate diet after 36 weeks of feeding deficient diet did not develop hepatomas at 54 weeks. Thus dietary intervention at preneoplastic stage can prevent or delay down-regulation of miR-122 and development of liver tumors. These observations support the notion that miR-122 down-regulation is associated with neoplastic changes in hepatocytes. Among the up-regulated miRs, miR-21 and miR-23 are likely to be involved in the initiation of preneoplastic transformation as evident from their increase at both 36 weeks and 54 weeks, whereas miR-130 and let-7 were up-regulated only in tumors (Figure 7.8A, B). miR-23 and miR-130 were also induced in the transplanted rat hepatoma.

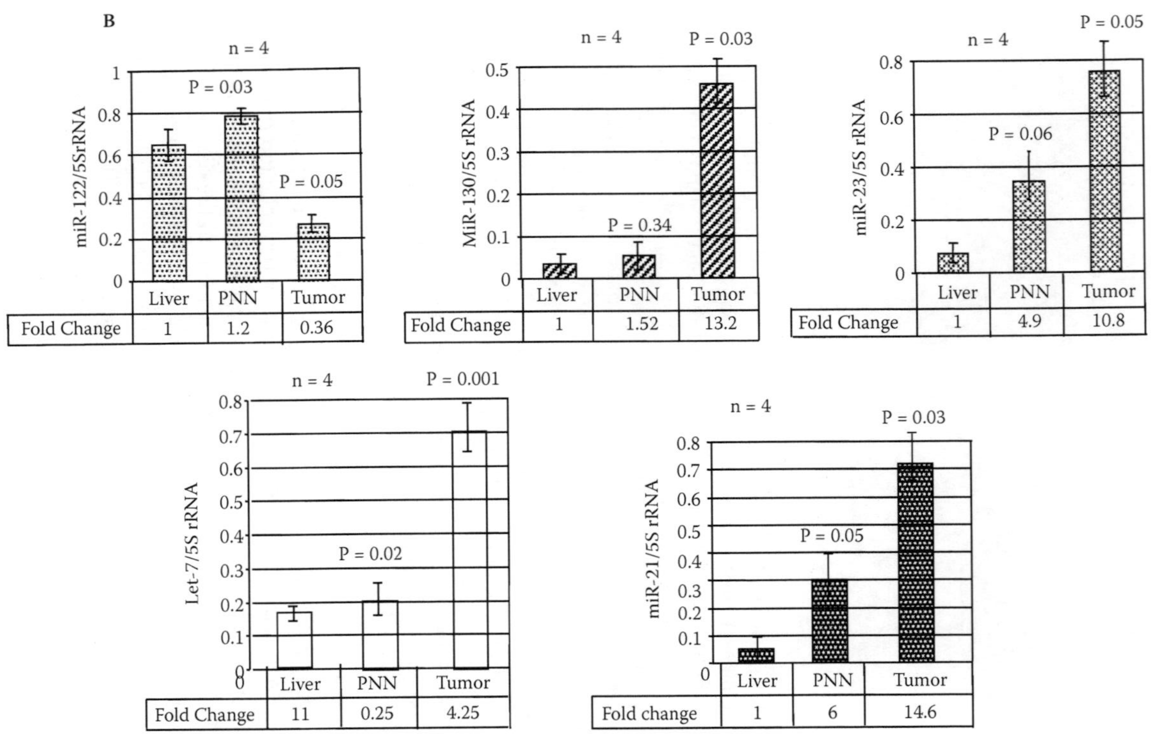

FIGURE 7.8B (Continued) A. Northern blot analysis confirmed down-regulation of miR-122 and up-regulation of miR-23, miR-21, miR-130, and let-7 in HCC compared with the controls. Total RNA was isolated from the livers of four rats fed normal diet and four rats fed FMD diet for 36 and 54 weeks, respectively. An aliquot (30 μg) of the total RNA was separated by denaturing PAGE, transferred to a nylon membrane, and subjected to Northern blot analysis with ^{32}P-labeled deoxyoligonucleotide antisense to specific miRs. The blot was reprobed with oligo antisense to 5S rRNA, and the ratio of miR signals to that of 5S rRNA were determined. B. Quantitative analysis of Northern blot data showed significant down-regulation of miR-122 and up-regulation of miR-21, miR-23, miR-130, and let-7 in rat HCCs. ^{32}P-signal was measured using Imagequant software and quantified using volume analysis program. PNNs denote preneoplastic nodules. The results represent the average signal of each miR normalized to that of 5S rRNA SD. The *p*-value of ≤0.05 was considered significant. Asterisks indicate significant changes. (Reproduced from Kutay, H., Bai, S., Datta, et al., *J Cell Biochem*, 99, 671–78, 2006.)

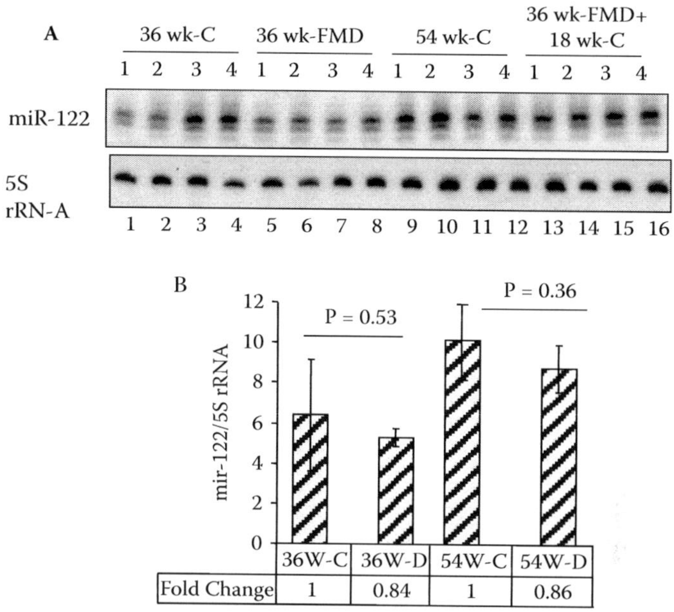

FIGURE 7.9 Down-regulation of miR-122 does not occur in the livers of rats switched to folate/methyl adequate diet after 36 weeks on the FMD diet. The control animals were on adequate diet for 54 weeks, whereas deficient animals were on FMD diet for the same time period. The third group was provided FMD diet for 36 weeks followed by adequate diet for another 18 weeks. A. Total RNA isolated from the livers was analyzed by Northern blotting. B. The quantitative representation of the data in A. The results represent the average signal of each miR normalized to that of 5S rRNA ± SD. The p-value of ≤0.05 was considered significant. The ratio of miR-122 to 5S rRNA in the controls (36 wk-C and 54 wk-C) was assigned a value of 1. (Reproduced from Kutay, H., Bai, S., Datta, J., et al., *J Cell Biochem*, 99, 671–78, 2006.)

Suppression of miR-122 in rat primary liver tumors raised the obvious question: What happens to its expression in human primary hepatocellular cancer? To address this question we performed Northern blot analysis of RNA from tumors and matching liver tissues from 20 samples (Figure 7.10A, B). The results showed significant decrease in miR-122 levels in almost all HCC compared to the surrounding liver tissues. This was the first report of down-regulation of the most abundant liver specific microRNA in hepatocellular cancer. Notably, as observed in rat primary HCCs (Figure 7.8), miR-21 was up-regulated in human HCC samples compared to matching livers (142). These observations validate animal models for the discovery of novel biomarkers for human disease.

Recently, Pogribny et al. analyzed miR profile in Fisher 344 rats fed FMD diet starting from 9 weeks when widespread metabolic changes occur in the liver (144). They measured selected miR levels by qRT-PCR after 9, 18, 36, and 60 weeks on FMD or control diet. They observed down-regulation of hepatic miR-34a, miR-127, and miR-200b by 80–90% after 9 and 18 weeks on the diet, which correlated with extensive steatosis, fibrosis, hepatocyte regeneration, and proliferation of oval (liver

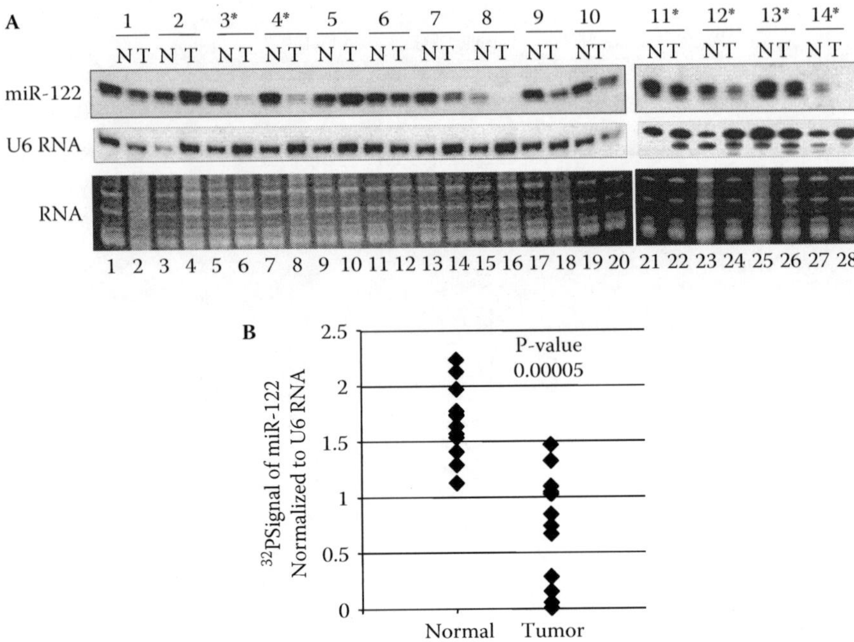

FIGURE 7.10 miR-122 is suppressed in human primary HCCs. A. An aliquot (5 μg) of total RNA from tumor (T) and matching normal (N) tissues was subjected to Northern blot analysis. Asterisks denote human primary HCCs in which miR-122 is down-regulated. The signal of miR-122 normalized to that of 5S rRNA is presented below each sample. Asterisks indicate HCCs with decrease in miR-122 compared with the respective matching controls. B. Quantitative analysis in 20 samples showed that the decrease in miR-122 in HCCs was statistically significant. (Reproduced from Kutay, H., Bai, S., Datta, J. et al., *J Cell Biochem,* 99, 671–78, 2006.)

progenitor) cells in animals fed deficient diet. FMD diet also induced loss of miR-16 expression, albeit at a less pronounced level. They also observed concomitant up-regulation of some oncogenic or antiapoptotic predicted targets of these miRs, e.g., E2F3, NOTCH1, BCL6, ZFHX1B, and BCL2 proteins of these miRNAs. Thus early onset of dysregulation of these miRNAs is likely to play a causal role in liver pathogenesis by altering expression of various targets involved in different biological processes.

7.4.2 MicroRNAs Dysregulated in the Mouse Model of NASH-Induced Hepatocarcinogenesis Developed by Feeding CDAA Diet

In order to address the role of microRNAs in NASH-induced hepatocarcinogenesis, we fed C57BL6 mice (which are naturally resistant to hepatocarcinogenesis) the choline-deficient L-amino acid-defined (CDAA) diet that has been shown to mimic human NASH in both mice and rats by causing steatohepatitis, liver fibrosis, and liver cancer (71). Microarray analysis of liver RNAs of mice on tumorigenic diet for different time periods demonstrated temporal changes in miR expression (Table 7.3) (66, 67). We

TABLE 7.3
Hepatic MicroRNAs Dysregulated in Mice Fed CDAA Diet for 6, 18, 32, and 65 Weeks

Gene Symbol	Fold Change				Parametric *p*-Value	FDR False Discovery Rate
	6 Week	18 Week	32 Week	65 Week		
miR-200c	1.00	0.84	1.00	2.67	0.000	0.004
miR-200b	1.00	1.00	1.00	2.53	0.001	0.010
miR-181d	0.77	0.45	4.21	2.40	0.000	0.005
miR-155	0.87	1.43	2.81	1.90	0.000	0.000
miR-487a	3.04	0.92	0.46	1.78	0.006	0.040
miR-181b	0.70	0.92	3.20	1.73	< 1e-07	< 1e-07
miR-223	0.47	1.84	2.82	1.62	< 1e-07	< 1e-07
miR-342-3p	0.51	1.54	1.73	1.40	0.000	0.000
miR-150	0.99	0.92	2.79	1.39	0.000	0.005
miR-99b	0.76	1.00	1.68	1.33	0.000	0.001
miR-214	0.80	0.98	1.73	1.32	0.002	0.017
miR-221	0.80	1.19	1.51	1.32	0.009	0.057
miR-195	0.90	1.02	1.67	1.29	0.000	0.000
miR-142-5p	0.54	1.03	2.19	1.28	0.005	0.035
miR-222	0.84	1.17	1.37	1.28	0.002	0.017
miR-34a	0.57	1.22	1.62	1.26	0.004	0.032
miR-16	0.88	1.12	1.72	1.16	0.000	0.000
miR-107	1.01	0.69	0.67	0.92	0.001	0.006
miR-30a	1.02	0.92	0.81	0.88	0.005	0.038
let-7a	1.10	0.66	0.62	0.85	0.005	0.038
miR-103	1.00	0.71	1.08	0.81	0.009	0.054
miR-30b	0.99	0.73	0.96	0.74	0.004	0.032
miR-30e	0.96	0.83	0.70	0.70	0.010	0.059
miR-323-5p	1.34	0.75	0.52	0.60	0.001	0.008
miR-27a	1.06	1.01	0.97	0.40	0.007	0.048
miR-802	1.00	0.74	0.71	0.28	0.000	0.000
miR-32	0.98	0.87	0.79	0.21	0.000	0.001
miR-17	0.83	0.99	1.45	0.90	0.000	0.003
miR-346	1.50	1.19	1.34	0.41	0.000	0.001
miR-20b	0.77	0.92	1.20	0.76	0.002	0.016

Source: Wang, B., Majumder, S., Nuovo, G., et al., *Hepatology,* 50, 1152–61, 2009.

Note: Analysis of variance with randomized block design identified 30 hepatic miRNAs that are altered ($p < 0.01$) upon feeding mice CDAA diet compared to CSAA diet. RNA from four mice on CSAA diet and five mice on CDAA diet at each time point was used for microarray analysis. Average values of the replicate spots of each miRNA were background-subtracted and subjected to further analysis. Data normalization was performed by using quantiles. Minimal miRNAs expression was set to 75. Genes showing minimal variation across the set of arrays were excluded from the analysis. Accordingly, 175 miRNAs whose expression differed by at least 1.5-fold from the median in at least 5% of the arrays were subjected to further statistical analysis. This filtering was applied to limit the number of false positive findings.

identified deregulation of 30 miRNAs ($p < 0.01$) after 6, 18, 32, and 65 weeks of feeding CDAA diet among which 21 were up-regulated ≥1.5-fold or down-regulated ≥50% in at least one time point. Based on temporal pattern of expression, the up-regulated miRs in mice fed CDAA diet were grouped into four different classes:

1. miR-155, miR-221, miR-222, miR-34a, miR-223, miR-342, and miR-16 were consistently up-regulated and remained high from 18 weeks to 65 weeks.
2. miR-181, miR-150, miR-99b, miR-214, miR-142, and miR-195 were up-regulated from 32 weeks to 65 weeks.
3. miR-17 and miR-346 up-regulated transiently after 32 weeks.
4. miR-200 up-regulated only after 65 weeks.

It is notable that several oncogenic microRNAs such as miR-155 and miR-221/222, which are known to be elevated in a variety of cancers including HCC (153), were up-regulated as early as 18 weeks compared to the age-matched mice fed control (choline sufficient and amino acid defined, CSAA) diet (Table 7.3). In situ hybridization with LNA probe demonstrated that miR-155 was up-regulated in the hepatocytes of mice fed CDAA diet and correlated with increased NAS score (145), a cumulative marker for steatosis, inflammation, and fibrosis, in mice fed CDAA diet. Thus miR-155, a mediator of inflammatory signal (146), can be a potential marker for NASH.

Analysis of microarray data for each time point between the two diet groups showed more microRNAs to be dysregulated with increasing time of feeding CDAA diet. For example, after 6 weeks, most miRs were down-regulated (16 out of 19) whereas only 3 miRs were up-regulated (miR-487a, miR-383, and miR-96), which is probably due to dramatic alteration in metabolic activity in the liver upon switching from low-fat protein-rich chow diet to CDAA diet. Interestingly, more miRs were up-regulated with prolonged consumption of CDAA diet, e.g., 7 out of 18 after 18 weeks, 17 out of 20 after 32 weeks, and 16 out of 32 after 65 weeks. Many more miRs were deregulated after 65 weeks when preneoplastic changes occur, implicating their role in hepatocyte transformation. Up-regulation of miR-181b/d variants occurred from early stages of feeding CDAA diet (Table 7.3). The level of Let-7a, a well-known tumor suppressor (147, 148), was reduced after 18 weeks of feeding the deficient diet. As observed in rats fed FMD diet, we observed down-regulation of miR-122 by Northern blot analysis in mouse liver only at 65 weeks when hepatocytes undergo preneoplastic changes (68). Similarly, up-regulation of miR-21 in CDAA livers was demonstrated by Northern blot analysis. Taken together, the CDAA diet model is an ideal system to identify microRNAs that may play a causal role in NASH-induced hepatocarcinogenesis.

7.4.3 MicroRNA Alteration in Human HCC

Recently, hundreds of dysregulated miRNAs have been identified in HCCs by different investigators (reviewed in [149]). Some of these miRNAs may be potential biomarkers for HCCs and play key roles in the pathogenesis of HCC (150). Among these miRNAs, some are up-regulated and function as oncogene. For example, miR-21 is up-regulated in HCCs and has been validated to target PTEN (151). PTEN is a negative regulator of Akt signaling pathway, which phosphorylates several protein targets and promotes cell survival. Other studies have also shown that miR-21 can target PDCD4, TIMP1, and maspin, thereby promoting invasion and metastasis of HCC cells (152). miR-221/222, two oncogenic miRs transcribed from the same primary transcript, are reported to be the most up-regulated miRs in primary human HCCs (153–156). These miRs directly target cyclin-dependent kinase inhibitors CDKN1B/p27 and CDKN1C/p57, DNA damage inducible protein DDIT4, a regulator of mTOR pathway (153–156), and well-known tumor suppressors PTEN and TIMP3 (63–67). Thus miR-221/222 promotes cell cycle progression and imparts resistance to TRAIL-induced apoptosis. Several other miRNAs up-regulated in HCCs were also reported recently, such as miR-224 targeting apoptosis inhibitor-5 (API-5) (157), miR-18a targeting estrogen receptor-alpha (158), and miR-143 targeting fibronectin (159). On the other hand, some miRs are identified to be decreased in HCCs and function as tumor suppressor. miR-122, the most abundant miR in liver, was found to be down-regulated in most of HCCs, and cyclin G1 (160), Bcl-w (161), ADAM17 (162), SRF, ADAM10, and IGF1R (163) are its validated targets. miR-223, another miR that is commonly repressed in HCCs, targets STMN1 (164). Met, a tyrosine kinase receptor of HGF that plays an important role in the tumorigenesis and metastasis of HCC, has been confirmed to be a direct target of miR-1 (165) and miR-199a* (166, 167), both of which are down-regulated in HCCs. Recently, miR-101 was reported to be a tumor suppressor in HCC by targeting oncogene FOS and antiapoptotic Mcl-1 (168). miR-195 was found to inhibit tumorigenesis and promote G1/S transition of human hepatocellular carcinoma cells by targeting cyclin D1, CDK6, and E2F3 (169).

7.4.4 Hepatic MicroRNAs Dysregulated in NASH Patients

Nonalcoholic fatty liver disease (NAFLD) is emerging as a major health problem affecting 70 million adults (30% of the adult population) in the United States. It is projected that 20% of these individuals will develop NASH, which may result in cirrhosis and HCC. NASH can also be associated with obesity, diabetes, and insulin resistance, all of which can contribute to an increased risk of HCC (170–172). Although the etiology underlying the development and progression of NASH is not well understood, factors such as diet, lifestyle, and genetic predisposition appear to play a causal role. It is characterized by abnormal lipid metabolism. Depending upon its severity, NAFLD can be classified into different types: (a) steatosis alone, (b) steatosis plus inflammation, (c) steatosis plus hepatocyte injury or ballooning degeneration, and (d) steatosis plus sinusoidal fibrosis, Mallory bodies (4, 173). Severe forms of NAFLD (types c and d) lead to NASH and often lead to cirrhosis and hepatocellular carcinoma. Insulin resistance is currently thought to be a key factor in the development of both NAFLD and NASH. Since dietary effects on lipid and glucose metabolic pathways play a critical role in NAFLD, a key strategy in affected patients is the control of diet to decrease body weight, which in turn will improve insulin resistance and dyslipidemia, reduce cardiovascular risks, and treat fatty liver. A recent study has analyzed the expression of 474 human microRNAs in subjects with the metabolic syndrome and NASH (characterized by steatosis, cytoloplasmic ballooning, and inflammation in liver histology) with those in control subjects with normal liver histology (174). Their results showed that a total

TABLE 7.4
miRs Significantly (P<0.05) Dysregulated in NASH Patients

Down-Regulated miRs	26b, 28, 30d, 92b, **122,** 126, 139, 145, 188, 191*, 198, 203, 223, 361, 375, 563, 574, 601, 617, 641, 671, 765, 768-5p
Up-Regulated miRs	**16, 21, 23a, 23b, 24,** 27b, **34a, 99b,** 100, 125b, 127, 128a, 128b, 146b, **181b,** 199a*, 199a, **200a, 221, 222, 214,** 224, 455

Note: These miRs were identified by microarray analysis of 15 control (normal liver histology) and 15 NASH patients. Those in bold are also similarly affected during diet-induced hepatocarcinogenesis in rodents (described in Tables 7.2 and 7.3).

of 46 microRNAs were dysregulated, among which 50% were down-regulated in NASH patients (Table 7.4). Interestingly, among the up-regulated microRNAs, miR-16, 21, 23a, 23b, 24, 34a, 99b, 221, 222, and 214 are also elevated in rodent livers upon feeding FMD or CDAA diets that cause NASH (compare Tables 7.2 and 7.3 with Table 7.4). Similar to mice fed CDAA diet, which causes NASH, miR-122 level decreased ~63% in the livers of NASH patients ($p < 1 \times 10^5$). Among the up-regulated miRs were well-known oncogenes like miR-221/222, miR-21, and miR-181b. Down-regulated tumor suppressor miRs included miR-34a and miR-199/199*. Notably, the only common microRNA down-regulated in rodent and human NASH livers is miR-122, suggesting that its loss may play a critical role in the initiation/progression of NASH.

7.5 MICRORNAS WITH IMPORTANT FUNCTION IN LIVER AND HCC

7.5.1 miR-122

miR-122, a developmentally regulated liver-specific miR, functions as a tumor suppressor. miR-122 is the most abundant liver-specific microRNA, constituting 70% of total hepatic miRs (119, 137). It is transcribed by pol II into a noncoding primary RNA of 4.6kb (mouse) and 4.9kb (human) that is first processed into a 66-nt long pre-miRNA with a characteristic hairpin structure and ultimately processed by Dicer to generate the mature miR-122 (119). Expression of this evolutionarily conserved microRNA starts during gestation and attains maximal level in the adult liver. Transcription of miR-122 gene follows circadian rhythm and is regulated by the orphan nuclear receptor REV-ERBalpha (175). miR-122 facilitates replication (176) and translation (177) of HCV RNA in HCC cells in culture. Depletion of miR-122 with antisense-miR in mice results in decrease in serum cholesterol and triglyceride levels (135, 178). Microarray analysis of hepatic RNA from mice depleted of miR-122 by antagomirs showed that the majority of the genes induced upon miR-122 depletion are normally repressed in the liver, implicating that it is involved in the maintenance of differentiated state of hepatocytes (179). In contrast, genes involved in lipid synthesis were predominantly up-regulated by an unknown mechanism (136). miR-122 level is very high in mouse and human hepatocytes, but it is either silent or expressed at a very low level in most HCC and transformed cell lines (180). As described earlier, we first identified the down-regulation of miR-122 in the rodent model of diet-induced multistage hepatocarcinogenesis (142). Analysis of human primary HCCs has also shown suppression of miR-122 compared to the matching liver tissues (142, 160, 162, 181) (Figure 7.8). It is also repressed in patients with NAFLD (Table 7.4) (182), which often leads to liver cancer. Subsequent functional analyses to understand the consequence of its loss in HCC demonstrated that miR-122 exhibits tumor suppressor characteristics (162, 163). Its ectopic expression in nonexpressing cells inhibited growth, clonogenic survival, migration, invasion, and tumor growth in nude mice (163). Furthermore, ectopic expression of its oncogenic targets SRF, ADAM-10, and Igf1R could partially reverse its antitumorigenic properties. Additionally, it has been reported to inhibit intrahepatic metastasis in human HCC patients and also in an orthotopic model of HCC by suppressing its target ADAM-17 (162). It would be of interest to test therapeutic potential of miR-122 against NASH-related HCCs at least in animal models.

7.5.2 miR-181

Recently there has been considerable interest in understanding the roles of miR-181 family of miRNAs in cancer. These studies have suggested that miR-181 family members function as both oncogenes and tumor suppressors depending upon the cellular context. These miRNAs are elevated in colon tumors (183) and pancreatic cancer (184) but are reduced in gliomas (185). They play an inhibitory role in hematopoietic differentiation, and their increased levels are associated with acute myeloid leukemia (186). In CLL, the relationship between miR-181 expression and disease phenotype is complex. For example, in CLL patients harboring chromosome 17p deletion, low expression of miR-181 family is strongly associated with disease progression, whereas in those harboring only trisomy 12, high expression of the miR-181a is associated with more aggressive disease (187). miR-181 family members are also up-regulated in HCCs and are associated with EpCAM+ (a marker for hepatic stem cells) HCC cells (188). Thus the role of miR-181 in tumorigenesis depends not only on the tissue type but also on genetic aberrations in the tumors, which is probably because of differential expression of the targets in different tumors. For example, miR-181b functions as a tumor suppressor in glioma cells (189) whereas it enhances tumorigenesis of HCC cells (188). We first identified up-regulation of miR-181 family members in the livers of mice fed CDAA diet for different time periods (Table 7.3) (67). Extension of this observation to human primary HCCs confirmed induction of these miRs in human tumors. Expression of miR-181 inversely correlated with its tumor suppressor target Timp3. Our study also showed that up-regulation of TGFβ and its downstream mediators Smad2, 3, and 4 contributed to induction of miR-181b/d in mice fed CDAA diet. Reduced miR-181 expression upon siRNA-mediated depletion of Smad4 confirmed the involvement of TGFβ pathway in the up-regulation of miR-181b. Furthermore, by ectopic expression and depletion studies we have confirmed that miR-181b functions as an oncogene by enhancing matrix metallopeptidase activity, promoting growth, clonogenic survival, migration, and invasion of hepatocellular carcinoma (HCC) cells. Additionally, the ability of anti-miR-181b in inhibiting ex vivo growth of HCC cells in nude mice and in sensitizing HCC cells to Doxorubicin, a potent anticancer drug, suggest its therapeutic potential.

7.5.3 miR-155

miR-155 is primarily involved in normal physiological processes like hematopoietic lineage differentiation and immune response (190, 191). It also contributes to various pathological processes such as inflammation, cancer, and cardiovascular diseases (reviewed in 192, 193). miR-155 is expressed at low levels in most cell types unless these cells are activated by inflammatory cytokines, which rapidly increase miR-155 expression (146). The available experimental evidence indicates that miR-155 is overexpressed in a variety of solid tumors (194) and hematopoietic malignancies (195). Elevated expression of miR-155 along with microRNAs 203, 210, and 222 in pancreatic tumors is associated with poorer survival (196). The increased level of miR-155 is also a candidate biomarker of diffuse large B cell lymphomas and many solid tumors (141, 194, 197, 198). Development of lymphoma in transgenic mice overexpressing miR-155 underscores its role in tumorigenesis (199). Our studies on the mouse model of CDAA diet-induced HCC indicated that up-regulation of hepatic miR-155 correlated with NASH as early as 18 weeks of feeding the deficient diet (66). Since chronic inflammation can lead to cancer, consistent up-regulation of miR-155 with reciprocal decrease in its target C/EBPβ from early stages could play a causal role in hepatocarcinogenesis. We also observed significant up-regulation of miR-155 in human primary HCCs and its growth stimulatory property in HCC cells (200). Therefore anti-miR-155 therapy that can potentially inhibit liver cancer by blocking proliferation of hepatocytes as well as inflammatory cells like lymphocytes and Kupffer cells in the tumor microenvironment could be extremely beneficial for therapy against HCC.

7.6 EPIGENETIC REGULATION OF MicroRNA EXPRESSION IN HEPATOCELLULAR CANCER

Several mechanisms such as chromosomal deletion/translocation/amplification, mutations in the miR genes or factors involved in their processing, DNA methylation, histone modifications, and polymorphisms (SNPs) may play a causal role in dysregulation of miR expression in diseased states including cancer (187). Since ~70% of the protein-coding genes are associated with CpG islands (CGI) in their promoters and many are silenced due to promoter methylation in cancer cells, it is likely that differential DNA methylation of certain microRNA genes embedded in CGIs is likely to regulate their tissue-specific expression or alter their expression in cancer cells. Bioinformatic analysis of chromosomal regions harboring miR genes revealed that almost half of these genes identified to date are surrounded by CGIs, suggesting their potential regulation by DNA methylation (for review, see [187]). Indeed, miR-127, that targets BCL6 (201), miR-124a that targets CDK6 (202), and let-7a3 that targets RAS are differentially methylated (203) in different cancers.

To identify tumor suppressor microRNAs that are epigenetically silenced in liver cancer, the microRNA expression profile of hepatocarcinoma (HCC) cell lines treated with 5-azacytidine (DNA hypomethylating agent) and/or trichostatin A (histone deacetylase inhibitor) was studied (165). The results showed that these epigenetic drugs differentially regulated the expression of a few miRs, particularly *miR-1-1,* in several HCC cell lines tested. This gene, located in the intron of an open reading frame, is embedded in several CGIs. Our studies showed that the CGI spanning exon 1 and intron 1 was tumor-specifically methylated. Up-regulation of miR-1-1 upon treatment with DNA hypomethylating agent 5-AzaC was accompanied by down-regulation of its oncogenic targets like FoxP1, MET, and HDAC4 (165). Functional analysis demonstrated that miR-1 indeed functions as a tumor suppressor. The induction of miR-1 is likely to be one of the mechanisms by which DNA hypomethylating agents inhibit growth of liver cancer. It is possible that some key factors (protein or microRNA) are also repressed in other liver diseases by epigenetic mechanism. It would therefore be of interest to test the therapeutic potential of epigenetic drugs not only against cancer but also against other liver diseases at least in animal models.

7.7 CONCLUDING REMARKS

Therapeutic intervention against hepatocellular carcinoma is very limited and depends on several factors, with the underlying liver disease being the key determinant. Although surgical resection and liver transplantation are by far the best options, with a 5-year survival of >75% and a tumor recurrence rate of <15% and 5 years (204), their applicability is limited to about 5% of patients based on liver function. Other therapies include chemoembolization and radiotherapy for local tumors, whereas systemic chemotherapy is the only option for advanced HCC. Currently used chemotherapeutic drugs against HCC are doxorubicin, cisplatin, and fluorouracil. These drugs, however, have very low response rates with no benefit to overall survival (205). Unfortunately, chemotherapy is not very well tolerated and is not very effective, especially in patients with hepatic dysfunction. Further, HCC is generally a highly chemoresistant tumor. More recently, antiangiogenic agent (e.g., bevacizumab) and multikinase inhibitor (e.g., sorafenib) have shown promise for HCC patients. There is, however, an immense requirement for the development of novel, more effective therapeutic strategies for HCC.

In this era of personalized medicine, the advances made in the identification of molecular diagnostic, prognostic, and therapeutic markers will play an important role in dictating the future of drug development, especially against chemo-resistant tumors like HCC. Two such areas with potential to advance to the clinical setting are epigenetics and microRNAs. While genetic aberrations like chromosomal translocations or deletions are nonreversible, epigenetic alterations that lie DNA methylation and histone posttranslational modifications are "temporary" reversible changes. With the elucidation of the role of epigenetics in cancer, lots of interest in the last decade has focused on developing epigenetic drugs for therapeutic intervention. DNA hypomethylating agents Vidaza (5-azacytidine) and Dacogen (5-aza-2′-de-oxycytidine, or decitabine) are currently approved by the Food and Drug Administration (FDA) for treatment of myelodysplastic syndrome (for review, see 206). These drugs are also undergoing clinical trials against a variety of leukemia and solid tumors. In a rat model of hepatocellular cancer, we first showed that 5-AzaC completely regressed tumor growth (207). However, it remains to be seen whether these hypomethylating agents have therapeutic potential in human patients suffering from different liver diseases including NAFLD, NASH, cirrhosis, and HCC.

Another approach to target specific genetic alterations has been gene therapy. Despite considerable efforts at developing gene therapy, several drawbacks like short-lived nature, immune response, risk with viral vectors, and the multigene nature of most disorders have restricted its effectiveness. There is, therefore, no current FDA-approved gene therapy against cancer. Because expression of microRNAs is dysregulated in almost all types of cancer, and certain miRs play a causal role in tumorigenesis by targeting classic oncogenes or tumor suppressors (reviewed in 208, 209), they have become important targets for therapy. They gain advantage over gene therapy due to their small size, stability, and ability to target multiple genes/proteins. Further, microRNA levels can be manipulated with relative ease, essentially following RNA interference technology. MicroRNA therapy against human disease seems quite feasible based on the success of anti-miR-122 in lowering blood cholesterol in primates (210). For their use in therapy, the effect of chemical modification to siRNAs on their activity in vivo and stability in biological fluids have been extensively studied (for review, see 211). LNA-, 2′-O-methyl ribose substitution, phosphorothioate linkages at the 3′-end and/or conjugation of 3′-end to cholesterol are likely to enhance stability and activity of siRNA. A major problem with the delivery of high molecular weight and highly charged oligonucleotides is the lack of a natural mechanism for targeted delivery to epithelial cells such as hepatocytes or HCC. The best way to accomplish this is likely by combining chemical modification of the miRNA (sense or antisense) backbones and encapsulation of miRNAs into targeted liposomal nanoparticles. It has been shown that introduction of galactosylation selectively targets siRNA nanoparticles to HCC cells, which express the asialoglycoprotein receptor (ASGR) (212). The advantage of miR-based therapy is that multiple miRs can be simultaneously targeted using combinations. Thus the liver represents an ideal organ target for miR therapy. Since microRNAs regulate numerous targets regulating different cellular processes, development of resistance to these agents is unlikely. It is hoped that the extensive basic knowledge on microRNAs can be extended to patients' bedside in the near future.

ACKNOWLEDGMENT

This study was supported by the Grant CA086978 from NIH.

REFERENCES

1. El-Serag, H. B., and K. L. Rudolph. 2007. Hepatocellular carcinoma: epidemiology and molecular carcinogenesis. *Gastroenterology* 132:2557–76.

2. Levrero, M. 2006. Viral hepatitis and liver cancer: the case of hepatitis C. *Oncogene* 25:3834–47.
3. Minguez, B., V. Tovar, D. Chiang, A. Villanueva, and J. M. Llovet. 2009. Pathogenesis of hepatocellular carcinoma and molecular therapies. *Curr Opin Gastroenterol* 25:186–94.
4. Zivkovic, A. M., J. B. German, and A. J. Sanyal. 2007. Comparative review of diets for the metabolic syndrome: implications for nonalcoholic fatty liver disease. *Am J Clin Nutr* 86:285–300.
5. Siegel, A. B., and A. X. Zhu. 2009. Metabolic syndrome and hepatocellular carcinoma: two growing epidemics with a potential link. *Cancer* 115:5651–61.
6. El-Serag, H. B., H. Hampel, and F. Javadi. 2006. The association between diabetes and hepatocellular carcinoma: a systematic review of epidemiologic evidence. *Clin Gastroenterol Hepatol* 4:369–80.
7. Jamal, M. M., Z. Saadi, and T. R. Morgan. 2005. Alcohol and hepatitis C. *Dig Dis* 23:285–96.
8. Bruix, J., A. J. Hessheimer, A. Forner, L. Boix, R. Vilana, and J. M. Llovet. 2006. New aspects of diagnosis and therapy of hepatocellular carcinoma. *Oncogene* 25:3848–56.
9. Yu, M. C., and J. M. Yuan. 2004. Environmental factors and risk for hepatocellular carcinoma. *Gastroenterology* 127:S72–78.
10. Bugianesi, E. 2007. Non-alcoholic steatohepatitis and cancer. *Clin Liver Dis* 11:191–207, x–xi.
11. Schreuder, T. C., B. J. Verwer, C. M. van Nieuwkerk, and C. J. Mulder. 2008. Nonalcoholic fatty liver disease: An overview of current insights in pathogenesis, diagnosis and treatment. *World J Gastroenterol* 14:2474–86.
12. Erickson, S. K. 2008. Nonalcoholic fatty liver disease (NAFLD). *J Lipid Res* (e-pub).
13. Breuhahn, K., R. Longerich, and P. Schirmacher. 2006. Dysregulation of growth factor signaling in human hepatocellular carcinoma. *Oncogene* 25:3787–3800.
14. Thorgeirsson, S. S., J. S. Lee, and J. W. Grisham. 2006. Functional genomics of hepatocellular carcinoma. *Hepatology* 43:S145–50.
15. Glynn, S. A., and D. Albanes. 1994. Folate and cancer: a review of the literature. *Nutr Cancer* 22:101–19.
16. Duthie, S. J., S. Narayanan, L. Sharp, J. Little, G. Basten, and H. Powers. 2004. Folate, DNA stability and colo-rectal neoplasia. *Proc Nutr Soc* 63:571–78.
17. Schernhammer, E., B. Wolpin, N. Rifai, B. Cochrane, J. A. Manson, J. Ma, E. Giovannucci, C. Thomson, M. J. Stampfer, and C. Fuchs. 2007. Plasma folate, vitamin B6, vitamin B12, and homocysteine and pancreatic cancer risk in four large cohorts. *Cancer Res* 67:5553–60.
18. Giovannucci, E., J. Chen, S. A. Smith-Warner, E. B. Rimm, C. S. Fuchs, C. Palomeque, W. C. Willett, and D. J. Hunter. 2003. Methylenetetrahydrofolate reductase, alcohol dehydrogenase, diet, and risk of colorectal adenomas. *Cancer Epidemiol Biomarkers Prev* 12:970–79.
19. Giovannucci, E. 2002. Epidemiologic studies of folate and colorectal neoplasia: a review. *J Nutr* 132:2350S–55S.
20. Platz, E. A., W. C. Willett, G. A. Colditz, E. B. Rimm, D. Spiegelman, and E. Giovannucci. 2000. Proportion of colon cancer risk that might be preventable in a cohort of middle-aged US men. *Cancer Causes Control* 11:579–88.
21. Ma, J., M. J. Stampfer, E. Giovannucci, C. Artigas, D. J. Hunter, C. Fuchs, W. C. Willett, J. Selhub, C. H. Hennekens, and R. Rosen. 1997. Methylenetetrahydrofolate reductase polymorphism, dietary interactions, and risk of colorectal cancer. *Cancer Res* 57:1098–1102.

22. Trentin, G. A., J. Moody, and J. A. Heddle. 1998. Effect of maternal folate levels on somatic mutation frequency in the developing colon. *Mutat Res* 405:81–87.
23. Branda, R. F., and D. B. Blickensderfer. 1993. Folate deficiency increases genetic damage caused by alkylating agents and gamma-irradiation in Chinese hamster ovary cells. *Cancer Res* 53:5401–8.
24. Teo, T., and M. Fenech. 2008. The interactive effect of alcohol and folic acid on genome stability in human WIL2-NS cells measured using the cytokinesis-block micronucleus cytome assay. *Mutat Res* 657:32–38.
25. Lindberg, H. K., X. Wang, H. Jarventaus, G. C. Falck, H. Norppa, and M. Fenech. 2007. Origin of nuclear buds and micronuclei in normal and folate-deprived human lymphocytes. *Mutat Res* 617:33–45.
26. Kazimirova, A., M. Barancokova, K. Krajcovicova-Kudlackova, M. Volkovova, Staruchova, M. Valachovicova, V. Paukova, P. Blazicek, L. Wsolova, and M. Dusinska. 2006. The relationship between micronuclei in human lymphocytes and selected micronutrients in vegetarians and non-vegetarians. *Mutat Res* 611:64–70.
27. Leopardi, P., F. Marcon, S. Caiola, A. Cafolla, E. Siniscalchi, A. Zijno, and R. Crebelli. 2006. Effects of folic acid deficiency and MTHFR C677T polymorphism on spontaneous and radiation-induced micronuclei in human lymphocytes. *Mutagenesis* 21:327–33.
28. Fenech, M. 2001. The role of folic acid and vitamin B12 in genomic stability of human cells. *Mutat Res* 475:57–67.
29. Fang, J. Y., and S. D. Xiao. 2003. Folic acid, polymorphism of methyl-group metabolism genes, and DNA methylation in relation to GI carcinogenesis. *J Gastroenterol* 38:821–29.
30. Friso, S., and S. W. Choi. 2002. Gene-nutrient interactions and DNA methylation. *J Nutr* 132:2382S–87S.
31. Friso, S., D. Girelli, E. Trabetti, O. Olivieri, P. Guarini, P. F. Pignatti, R. Corrocher, and S. W. Choi. 2005. The MTHFR 1298A>C polymorphism and genomic DNA methylation in human lymphocytes. *Cancer Epidemiol Biomarkers Prev* 14:938–43.
32. Kohlmeier, M., K. A. da Costa, L. M. Fischer, and S. H. Zeisel. 2005. Genetic variation of folate-mediated one-carbon transfer pathway predicts susceptibility to choline deficiency in humans. *Proc Natl Acad Sci U S A* 102:16025–30.
33. Mason, J. B., and S. W. Choi. 2005. Effects of alcohol on folate metabolism: implications for carcinogenesis. *Alcohol* 35:235–41.
34. Kim, Y. I., S. Shirwadkar, S. W. Choi, M. Puchyr, Y. Wang, and J. B. Mason. 2000. Effects of dietary folate on DNA strand breaks within mutation-prone exons of the p53 gene in rat colon. *Gastroenterology* 119:151–61.
35. Chou, Y. F., and R. F. Huang. 2009. Mitochondrial DNA deletions of blood lymphocytes as genetic markers of low folate-related mitochondrial genotoxicity in peripheral tissues. *Eur J Nutr* 48:429–36.
36. Wu, M. Y., C. S. Kuo, C. Y. Lin, D. L. Lu, and R. F. Syu Huang. 2009. Lymphocytic mitochondrial DNA deletions, biochemical folate status and hepatocellular carcinoma susceptibility in a case-control study. *Br J Nutr* 102:715–21.
37. Chang, C. M., C. C. Yu, H. T. Lu, Y. F. Chou, and R. F. Huang. 2007. Folate deprivation promotes mitochondrial oxidative decay: DNA large deletions, cytochrome c oxidase dysfunction, membrane depolarization and superoxide overproduction in rat liver. *Br J Nutr* 97:855–63.
38. Ross, C. M. 2005. Folate, mitochondria, ROS, and the aging brain. *Am J Med* 118:1174; author reply:5.

39. Amarzguioui, M., J. J. Rossi, and D. Kim. 2005. Approaches for chemically synthesized siRNA and vector-mediated RNAi. *FEBS Lett* 579:5974–81.
40. Crott, J. W., S. W. Choi, R. F. Branda, and J. B. Mason. 2005. Accumulation of mitochondrial DNA deletions is age, tissue and folate-dependent in rats. *Mutat Res* 570:63–70.
41. Branda, R. F., Z. Chen, E. M. Brooks, S. J. Naud, T. D. Trainer, and J. J. McCormack. 2002. Diet modulates the toxicity of cancer chemotherapy in rats. *J Lab Clin Med* 140:358–68.
42. Zeisel, S. H., and K. A. da Costa. 2009. Choline: an essential nutrient for public health. *Nutr Rev* 67:615–23.
43. Buchman, A. L. 2009. The addition of choline to parenteral nutrition. *Gastroenterology* 137:S119–28.
44. Steegers-Theunissen, R. P, W. O. Renier, G. F. Borm, C. M. Thomas, H. M. Merkus, D. A. Op de Coul, P. A. DeJong, H. P. van Geijn, M. Wouters, and T. K. Eskes. 1994. Factors influencing the risk of abnormal pregnancy outcome in epileptic women: a multi-centre prospective study. *Epilepsy Res* 18:261–69.
45. Song, J., K. A. da Costa, L. M. Fischer, M. Kohlmeier, L. Kwock, S. Wang, and S. H. Zeisel. 2005. Polymorphism of the PEMT gene and susceptibility to nonalcoholic fatty liver disease (NAFLD). *FASEB J* 19:1266–71.
46. Li, Z., and D. E. Vance. 2008. Phosphatidylcholine and choline homeostasis. *J Lipid Res* 49:1187–94.
47. McKeever, M. P., D. G. Weir, A. Molloy, and J. M. Scott. 1991. Betaine-homocysteine methyltransferase: organ distribution in man, pig and rat and subcellular distribution in the rat. *Clin Sci (Lond)* 81: 551–56.
48. James, S. J., I. P. Pogribny, M. Pogribna, B. J. Miller, S. Jernigan, and S. Melnyk. 2003. Mechanisms of DNA damage, DNA hypomethylation, and tumor progression in the folate/methyl-deficient rat model of hepatocarcinogenesis. *J Nutr* 133:3740S–47S.
49. Pogribny, I. P., and S. J. James. 2002. De novo methylation of the p16INK4A gene in early preneoplastic liver and tumors induced by folate/methyl deficiency in rats. *Cancer Lett* 187:69–75.
50. Pogribny, I. P., S. J. James, S. Jernigan, and M. Pogribna. 2004. Genomic hypomethylation is specific for preneoplastic liver in folate/methyl deficient rats and does not occur in non-target tissues. *Mutat Res* 548:53–59.
51. Kim, Y. I. 2004. Folate, colorectal carcinogenesis, and DNA methylation: lessons from animal studies. *Environ Mol Mutagen* 44:10–25.
52. Purohit, V., M. R. Abdelmalek, S. Barve, N. J. Benevenga, C. H. Halsted, N. Kaplowitz, K. K. Kharbanda, Q. Y. Liu, S. C. Lu, C. J. McClain, C. Swanson, and S. Zakhari. 2007. Role of S-adenosylmethionine, folate, and betaine in the treatment of alcoholic liver disease: summary of a symposium. *Am J Clin Nutr* 86:14–24.
53. Ghoshal, K., X. Li, J. Datta, S. Bai, I. Pogribny, M. Pogribny, Y. Huang, D. Young, and S. T. Jacob 2006. A folate- and methyl-deficient diet alters the expression of DNA methyltransferases and methyl CpG binding proteins involved in epigenetic gene silencing in livers of F344 rats. *J Nutr* 136:1522–27.
54. James, S. J., S. Melnyk, M. Pogribna, I. P. Pogribny, and M. A. Caudill. 2002. Elevation in S-adenosylhomocysteine and DNA hypomethylation: potential epigenetic mechanism for homocysteine-related pathology. *J Nutr* 132:2361S–66S.
55. Pogribny, I. P., B. J. Miller, and S. J. James. (1997) Alterations in hepatic p53 gene methylation patterns during tumor progression with folate/methyl deficiency in the rat. *Cancer Lett* 115:31–38.

56. Pogribny, I. P., L. Muskhelishvili, B. J. Miller, and S. J. James. 1997. Presence and consequence of uracil in preneoplastic DNA from folate/methyl-deficient rats. *Carcinogenesis* 18:2071–76.
57. Pogribny, I. P., S. A. Ross, C. Wise, I. Rusyn, and S. A. Ross. 2005. Irreversible global DNA hypomethylation as a key step in hepatocarcinogenesis induced by dietary methyl deficiency. *Mutat Res* 3:3.
58. Steinmetz, K. L., I. P. Pogribny, S. J. James, and H. C. Pitot. 1998. Hypomethylation of the rat glutathione S-transferase pi (GSTP) promoter region isolated from methyl-deficient livers and GSTP-positive liver neoplasms. *Carcinogenesis* 19:1487–94.
59. Motiwala, T., K. Ghoshal, A. Das et al. 2003. Suppression of the protein tyrosine phosphatase receptor type O gene (PTPRO) by methylation in hepatocellular carcinomas. *Oncogene* 22:6319–31.
60. Kim, C. H., and Z. M. Younossi. 2008. Nonalcoholic fatty liver disease: a manifestation of the metabolic syndrome. *Cleve Clin J Med* 75:721–28.
61. Christman, J. K. 1995. Dietary effects on DNA methylation: do they account for the hepatocarcinogenic properties of lipotrope deficient diets? *Adv Exp Med Biol* 369:141–54.
62. Ghoshal, A. K. 1995. New insight into the biochemical pathology of liver in choline deficiency. *Crit Rev Biochem Mol Biol* 30:263–73.
63. Denda, A., A. Kitayama, H. Kishida, M. Nao, T. Masahiro, T. Toshifumi, N. Dai, and K. Yoichi. 2002. Development of hepatocellular adenomas and carcinomas associated with fibrosis in C57BL/6J male mice given a choline-deficient, L-amino acid-defined diet. *Jpn J Cancer Res* 93:125–32.
64. Nakae, D., F. Uematsu, H. Kishida, K. Osamu, K. Shin-Ichi, Y. Modiri, T. Masakazu, M. Akihiko, D. Ayumi, K. Yoichi, K. Yashige, and R. A. Floyd. 2004. Inhibition of the development of hepatocellular carcinomas by phenyl N-tert-butyl nitrone in rats fed with a choline-deficient, L-amino acid-defined diet. *Cancer Lett* 206:1–13.
65. Bai, S., M. W. M. W. Nasser, B. Wang, S.-H. Hsu, J. Datta, H. Kutay, A. Yadav, G. Nuovo, P. Kumar, and K. Ghoshal. 2009. MicroRNA-122 inhibits tumorigenic properties of hepatocellular carcinoma cells and sensitizes these cells to Sorafenib. *J Biol Chem* 284:32015–27.
66. Wang, B., S. Majumder, G. Nuovo, H. Kutay, S. Volinia, T. Patel, T. D. Schmittgen, C. Croce, K. Ghoshal, and S. T. Jacob. 2009. Role of microRNA-155 at early stages of hepatocarcinogenesis induced by choline-deficient and amino acid-defined diet in C57BL/6 mice. *Hepatology* 50:1152–61.
67. Wang, B., S. H. Hsu, S. Majumder, H. Kutay, W. Huang, S. T. Jacob, and K. Ghoshal. 2009. TGFbeta-mediated upregulation of hepatic miR-181b promotes hepatocarcinogenesis by targeting TIMP3. *Oncogene* 29(12):1787–97.
68. Wang, B., S. Majumder, G. Nuovo, H. Kutay, S. Volinia, T. Patel, T. D. Schmittgen, C. Croce, K. Ghosahl, and S. T. Jacob. 2009. Role of microRNA-155 at early stages of hepatocarcinogenesis induced by choline-deficient and amino acid-defined diet in C57BL/6 mice. *Hepatology* 50 (4):1152–61.
69. Nakae, D., H. Yoshiji, Y. Mizumoto, K. Horiguchi, T. Kazutoshi, and Y. Konishi. 1992. High incidence of hepatocellular carcinomas induced by a choline deficient L-amino acid defined diet in rats. *Cancer Res* 52:5042–45.
70. Tsujiuchi, T., Kobayashi, Nakae, Y. Mizumoto, N. Andoh, H. Kitada, K. Ohashi, T. Fukuda, A. Kodo, M. Tsutsumi, A. Denda, and Y. Konishi. 1995. Prevention by methionine of enhancement of hepatocarcinogenesis by coadministration of a choline-deficient L-amino acid-defined diet and ethionine in rats. *Jpn J Cancer Res* 86:1136–42.
71. Nakae, D. 1999. Endogenous liver carcinogenesis in the rat. *Pathol Int* 49:1028–42.

72. Brunaud, L., J. M. Alberto, A. Ayav et al. 2003. Effects of vitamin B12 and folate deficiencies on DNA methylation and carcinogenesis in rat liver. *Clin Chem Lab Med* 41:1012–19.
73. Fenech, M. 1998. Chromosomal damage rate, aging, and diet. *Ann N Y Acad Sci* 854:23–36.
74. Halsted, C. H., J. A. Villanueva, and A. M. Devlin. 2002. Folate deficiency, methionine metabolism, and alcoholic liver disease. *Alcohol* 27:169–72.
75. Li, G. M., S. R. Presnell, and L. Gu. 2003. Folate deficiency, mismatch repair-dependent apoptosis, and human disease. *J Nutr Biochem* 14:568–75.
76. Kim, Y. I. 2004. Folate and DNA methylation: a mechanistic link between folate deficiency and colorectal cancer? *Cancer Epidemiol Biomarkers Prev* 13:511–19.
77. Motiwala, T., K. Ghoshal, A. Das et al. 2003. Suppression of the protein tyrosine phosphatase receptor type O gene (PTPRO) by methylation in hepatocellular carcinomas. *Oncogene* 22:6319–31.
78. Feinberg, A. P., and G. Tycko. 2004. The history of cancer epigenetics. *Nat Rev Cancer* 4:143–53.
79. Esteller, M. 2005. Aberrant DNA methylation as a cancer-inducing mechanism. *Annu Rev Pharmacol Toxicol* 45:629–56.
80. Raval, A., S. M. Tanner, H. C. Byrd, E. G. Angerman, J. D. Perko, S. S. Chen, B. Hackanson, M. R. Grever, D. M. Lucas, J. J. Matkovic, T. S. Lin, T. J. Kipps, F. Murray, D. Weisenburger, W. Sanger, J. Lynch, P. Watson, M. Jansen, Y. Yoshinaga, R. Rosenquist, P. J. de Jong, P. Coggill, S. Beck, H. Lynch, A. de la Chapelle, and C. Plass. 2007. Downregulation of death-associated protein kinase 1 (DAPK1) in chronic lymphocytic leukemia. *Cell* 129:879–90.
81. Baylin, S. B. 2005. DNA methylation and gene silencing in cancer. *Nat Clin Pract Oncol* 2 Suppl 1:S4–11.
82. Baylin, S. B., and J. E. Ohm. 2006. Epigenetic gene silencing in cancer—a mechanism for early oncogenic pathway addiction? *Nat Rev Cancer* 6:107–16.
83. Galm, O., J. G. Herman, and S. B. Baylin. 2006. The fundamental role of epigenetics in hematopoietic malignancies. *Blood Rev* 20:1–13.
84. Herman, J.. G., and S. B. Baylin. 2003. Gene silencing in cancer in association with promoter hypermethylation. *N Engl J Med* 349:2042–54.
85. Jones, P. A., and S. B. Baylin. 2002. The fundamental role of epigenetic events in cancer. *Nat Rev Genet* 3:415–28.
86. Jacob, S. T., and T. Motiwala. 2005. Epigenetic regulation of protein tyrosine phosphatases: potential molecular targets for cancer therapy. *Cancer Gene Ther* 12:665–72.
87. Finnell, R. H., O. Spiegelstein, B. Wlodarczyk et al. 2002. DNA methylation in Folbp1 knockout mice supplemented with folic acid during gestation. *J Nutr* 132:2457S–61S.
88. Goll, M. G., and T. H. Bestor. 2005. Eukaryotic cytosine methyltransferases. *Annu Rev Biochem* 74:481–514.
89. Okano, M., D. W. Bell, D. A. Haber, and E. Li. 1999. DNA methyltransferases Dnmt3a and Dnmt3b are essential for de novo methylation and mammalian development. *Cell* 99:247–57.
90. Vire, E., Brenner, Deplus, L. Blanchon, M. Fraga, C. Didelot, L. Morey, A. van Eynde, D. Bernard, J.-M. Vanderwinden, M. Bollen, M. Esteller, L. Di Croce, Y. de Launoit, and F. Fuks. 2006. The polycomb group protein EZH2 directly controls DNA methylation. *Nature* 439:871–74.
91. Bai, S., K. Ghoshal, and S. T. Jacob. 2006. Identification of T-cadherin as a novel target of DNA methyltransferase 3B and its role in the suppression of nerve growth factor-mediated neurite outgrowth in PC12 cells. *J Biol Chem* 281:13604–11.

92. Dhasarathy, A., and P. A. Wade. 2008. The MBD protein family-reading an epigenetic mark? *Mutat Res* 647:39–43.
93. Wade, P. A. 2001. Methyl CpG-binding proteins and transcriptional repression. *Bioessays* 23:1131–37.
94. Wang, S., Y. Yan-Neale, M. Zeremski, and D. Cohen. 2004. Transcription regulation by histone deacetylases. *Novartis Found Symp* 259:238–45; discussion 45–48, 85–88.
95. Ballestar, E., and M. Esteller. 2008. Epigenetic gene regulation in cancer. *Adv Genet* 61:247–67.
96. Esteller, M. 2007. Epigenetic gene silencing in cancer: the DNA hypermethylome. *Hum Mol Genet* 16 (spec no 1):R50–59.
97. Rush, L. J., and C. Plass. 2002. Restriction landmark genomic scanning for DNA methylation in cancer: past, present, and future applications. *Anal Biochem* 307:191–201.
98. Smiraglia, D. J., and C. Plass. 2002. The study of aberrant methylation in cancer via restriction landmark genomic scanning. *Oncogene* 21:5414–26.
99. Bennett, K. L., B. Hackanson, L. T. Smith, C. D. Morrison, J. C. Lang, D. E. Schuller, F. Weber, C. Eng, and C. Plass. 2007. Tumor suppressor activity of CCAAT/enhancer binding protein alpha is epigenetically down-regulated in head and neck squamous cell carcinoma. *Cancer Res* 67:4657–64.
100. Liu, T. H., A. Raval, S. S. Chen, J. J. Matkovic, J. C. Byrd, and C. Plass. 2006. CpG island methylation and expression of the secreted frizzled-related protein gene family in chronic lymphocytic leukemia. *Cancer Res* 66:653–58.
101. Smith, L.T., M. Lin, R. M. Brena, J. C. Lang, D. E. Schuller, G. A. Otterson, C. D. Morrison, D. J. Smiraglia, and C. Plass. 2006. Epigenetic regulation of the tumor suppressor gene TCF21 on 6q23–q24 in lung and head and neck cancer. *Proc Natl Acad Sci U S A* 103:982–87.
102. Yu, L., C. Liu C, J. Vandeusen, B. Becknell, Z. Dai, Y.-Z. Wu, A. Raval, T.-H. Liu, W. Ding, C. Mao, S. Liu, L. T. Smith, S. Lee, L. Rassenti, G. Marcucci, J. Byrd, M. A. Caligiuri, and C. Plass. 2005. Global assessment of promoter methylation in a mouse model of cancer identifies ID4 as a putative tumor-suppressor gene in human leukemia. *Nat Genet* 37:265–74 (e-pub 2005 Feb 20).
103. Chen, S. S., A. Raval, A. J. Johnson, E. Hertlein, T.-H. Liu, V. X. Jin, M. H. Sherman, S.-J. Liu, D. W. Dawson, K. E. Williams, M. Lanasa, S. Liyanarachchi, T. S. Lin, G. Marcucci, Y. Pekarsky, R. Davulur, C. M. Croce, D. C. Guttridge, M. A. Teitell, J. C. Byrd, and C. Plass. 2009. Epigenetic changes during disease progression in a murine model of human chronic lymphocytic leukemia. *Proc Natl Acad Sci U S A* 106:13433–38.
104. Raval, A., D. M. Lucas, J. J. Matkovic, K. L. Bennette, S. Liyanarachchi, D. C. Young, L. Rassenti, T. J. Kipps, M. R. Grever, J. C. Byrd, and C. Plass. 2005. TWIST2 demonstrates differential methylation in immunoglobulin variable heavy chain mutated and unmutated chronic lymphocytic leukemia. *J Clin Oncol* 23:3877–85.
105. Motiwala, T., H. Kutay, K. Ghoshal, S. Bai, H. Seimiya, T. Tsuruo, S. Suster, C. Morrison, and S. T. Jacob. 2004. Protein tyrosine phosphatase receptor-type O (PTPRO) exhibits characteristics of a candidate tumor suppressor in human lung cancer. *Proc Natl Acad Sci U S A* 101:13844–49 (e-pub 2004 Sep 8).
106. Washietl, S., J. S. Pedersen, J. O. Korbel, C. Stocsits, A. R. Gruber, J. Hackermüller, J. Hertel, M. Lindemeyer, K. Reiche, A. Tanzer, C. Ucla, C. Wyss, S. E. Antonarakis, F. Denoeud, J. Lagarde, J. Drenkow, P. Kapranov, T. R. Gingeras, R. Guigó, M. Snyder, M. B. Gerstein, A. Reymond, I. L. Hofacker, and P. F. Stadler. 2007. Structured RNAs in the ENCODE selected regions of the human genome. *Genome Res* 17:852–64.
107. Willingham, A. T., and T. R. Gingeras. 2006. TUF love for "junk" DNA. *Cell* 125:1215–20.

108. Fejes-Toth, K., V. Sotirova, R. Sachidanandam, G. Assaf, G. J. Hannon, P. Kapranov, S. Fosiac, A. T. Willingham, R. Duttagupta, E. Dumais, and T. R. Gingeras. 2009. Post-transcriptional processing generates a diversity of 5′-modified long and short RNAs. *Nature* 457:1028–32.
109. Filipowicz, W., S. N. Bhattacharyya, and N. Sonenberg. 2008. Mechanisms of post-transcriptional regulation by microRNAs: are the answers in sight? *Nat Rev Genet* 9:102–14.
110. Kim, V. N., J. Han, and M. S. Siomi. 2009. Biogenesis of small RNAs in animals. *Nat Rev Mol Cell Biol* 10:126–39.
111. Lee, R. C., R. L. Feinbaum, and V. Ambros. 1993. The *C. elegans* heterochronic gene lin-4 encodes small RNAs with antisense complementarity to lin-14. *Cell* 75:843–54.
112. Wightman, B., I. Ha, and G. Ruvkun. 1993. Posttranscriptional regulation of the heterochronic gene lin-14 by lin-4 mediates temporal pattern formation in *C. elegans*. *Cell* 75:855–62.
113. Lagos-Quintana, M., R. Rauhut, W. Lendeckel, and T. Tuschl. 2001. Identification of novel genes coding for small expressed RNAs. *Science* 294:853–58.
114. Lau, N. C., L. P. Lim, E. G. Weinstein, and D. P. Bartel. 2001. An abundant class of tiny RNAs with probable regulatory roles in *Caenorhabditis elegans*. *Science* 294:858–62.
115. Lee, R. C., and V. Ambros. 2001. An extensive class of small RNAs in *Caenorhabditis elegans*. *Science* 294:862–64.
116. Schmittgen, T. D. 2008. Regulation of microRNA Processing in development, differentiation and cancer. *J Cell Mol Med* 12 (5B):1811–19.
117. Yang, B., H. Lin, J. Xiao, Y. Lu, X. Luo, L. Baoxin, Y. Zhang, C. Xu, Y. Bai, H. Wang, G. Chen, and Z. Wang. 2007. The muscle-specific microRNA miR-1 regulates cardiac arrhythmogenic potential by targeting GJA1 and KCNJ2. *Nat Med* 13:486–91.
118. Chen, J. F., E. M. Mandel, J. M. Thomson, Q. Wu, T. E. Callis, S. M. Hammond, F. L. Conlon, and D.-Z. Wang. 2006. The role of microRNA-1 and microRNA-133 in skeletal muscle proliferation and differentiation. *Nat Genet* 38:228–33.
119. Chang, J., E. Nicolas, D. Marks, C. Sander, A. Lerro, M. A. Buendia, W. S. Mason, T. Moloshok, R. Bort, K. S. Zaret, and J. M. Taylor. 2004. miR-122, a mammalian liver-specific microRNA, is processed from hcr mRNA and may downregulate the high affinity cationic amino acid transporter CAT-1. *RNA Biol* 1:106–13.
120. Bartel, D. P. 2009. MicroRNAs: target recognition and regulatory functions. *Cell* 136:215–33.
121. Selbach, M., B. Schwanhausser, N. Thierfelder, Z. Fang, R. Khanin, and N. Rajewsky. 2008. Widespread changes in protein synthesis induced by microRNAs. *Nature* 455:58–63.
122. Baek, D., J. Villen, C. Shin, F. D. Camargo, S. P. Gygi, and D. P. Bartel. 2008. The impact of microRNAs on protein output. *Nature* 455:64–71.
123. Kim, V. N. 2005. Small RNAs: classification, biogenesis, and function. *Mol Cells* 19:1–15.
124. Kim, V. N. 2005. MicroRNA biogenesis: coordinated cropping and dicing. *Nat Rev Mol Cell Biol* 6:376–85.
125. Ragan, C., N. Cloonan, S. M. Grimmond, M. Zuker, and M. A. Ragan. 2009. Transcriptome-wide prediction of miRNA targets in human and mouse using FASTH. *PLoS One* 4:e5745.
126. Kanellopoulou, C., S. A. Muljo, A. L. Kung, S. Ganesan, R. Drapkin, T. Jenuwein, D. M. Livingston, and K. Rajewsky. 2005. Dicer-deficient mouse embryonic stem cells are defective in differentiation and centromeric silencing. *Genes Dev* 19:489–501.
127. Hand, N. J., Z. R. Master, J. Le Lay, and J. R. Friedman. 2009. Hepatic function is preserved in the absence of mature microRNAs. *Hepatology* 49:618–26.

128. Sekine, S., R. Ogawa, R. Ito, N. Hiraoka, M. T. McManus, Y. Kanai, and M. Hebrok. 2009. Disruption of Dicer1 induces dysregulated fetal gene expression and promotes hepatocarcinogenesis. *Gastroenterology* 136:2304–15, e1–4.
129. Ambros, V. 2003. MicroRNA pathways in flies and worms: growth, death, fat, stress, and timing. *Cell* 113:673–76.
130. Bartel, D. P. 2004. MicroRNAs: genomics, biogenesis, mechanism, and function. *Cell* 116:281–97.
131. Wienholds, E., and R. H. Plasterk. 2005. MicroRNA function in animal development. *FEBS Lett* 16:16.
132. Pillai, R. S. 2005. MicroRNA function: multiple mechanisms for a tiny RNA? *RNA* 11:1753–61.
133. Miska, E. A. 2005. How microRNAs control cell division, differentiation and death. *Curr Opin Genet Dev* 15:563–68.
134. Bushati, N., and S. M. Cohen 2007. microRNA functions. *Annu Rev Cell Dev Biol* 23:175–205.
135. Esau, C., S. Davis, S. F. Murray, X. X. Yu, S. K. Pandey, M. Pear, L. Watts, S. L. Booten, M. Graham, R. McKay, A. Subramaniam, S. Propp, B. A. Lollo, S. Freier, C. F. Bennett, S. Bhanot, and B. P. Monia. 2006. miR-122 regulation of lipid metabolism revealed by in vivo antisense targeting. *Cell Metab* 3:87–98.
136. Krutzfeldt, J., N. Rajewsky, R. Braich, K. G. Rajeev, T. Tuschl, M. Manoharan, and M. Stoffel. 2005. Silencing of microRNAs in vivo with "antagomirs." *Nature* 438:685–89.
137. Lagos-Quintana, M., R. Rauhut, A. Yalcin, J. Meyer, W. Lendeckel, and T. Tuschl. 2002. Identification of tissue-specific microRNAs from mouse. *Curr Biol* 12:735–39.
138. Fu, H., Y. Tie, C. Xu, Z. Zhang, Z. Zhu, Y. Shi, H. Jiang, Z. Sun, and X. Zheng. 2005. Identification of human fetal liver miRNAs by a novel method. *FEBS Lett* 579:3849–54.
139. Calin, G. A., C. D. Dumitru, M. Shimizu, R. Bichi, S. Zupo, E. Noch, H. Aldler, S. Rattan, M. Keating, K. Rai, L. Rassenti, T. Kipps, M. Negrini, F. Bullrich, and C. M. Croce. 2002. Frequent deletions and down-regulation of micro- RNA genes miR15 and miR16 at 13q14 in chronic lymphocytic leukemia. *Proc Natl Acad Sci U S A* 99:15524–29.
140. Cimmino, A., G. A. Calin, M. Fabbri, M. V. Iorio, M. Ferracin, M. Shimizu, S. E. Wojcik, R. I. Ageilan, S. Zupo, M. Dono, L. Rassenti, H. Alder, S. Volinia, C. G. Liu, T. J. Kipps, M. Negrini, and C. M. Croce. 2005. miR-15 and miR-16 induce apoptosis by targeting BCL2. *Proc Natl Acad Sci U S A* 102: 13944–49.
141. Calin, G. A., M. Ferracin, A. Cimmino, G. DiLeva, M. Shimizu, S. E. Wojcik, M. V. Iorio, R. Visone, N. I. Sever, M. Fabbri, R. Iuliano, T,. Palumbo, F. Picchiorri, C. Roldo, R. Garzon, C. Sevignani, L. Rassenti, H. Alder, S. Volinia, C. Liu, T. J. Kipps, M. Negrini, and C. M. Croce. 2005. A MicroRNA signature associated with prognosis and progression in chronic lymphocytic leukemia. *N Engl J Med* 353:1793–1801.
142. Kutay, H., S. Bai, J. Datta, T. Motiwala, I. Pogribny, W. Frankel, S. T. Jacob, and K. Ghoshal. 2006. Downregulation of miR-122 in the rodent and human hepatocellular carcinomas. *J Cell Biochem* 99:671–78.
143. Majumder, S., K. Ghoshal, J. Datta, S. Bai, X. Dong, N. Quan, C. Plass, and S. T. Jacob. 2002. Role of de novo DNA methyltransferases and methyl CpG-binding proteins in gene silencing in a rat hepatoma. *J Biol Chem* 277:16048–58.
144. Tryndyak, V. P., S. A. Ross, F. A. Beland, and I. P. Pogribny. 2009. Down-regulation of the microRNAs miR-34a, miR-127, and miR-200b in rat liver during hepatocarcinogenesis induced by a methyl-deficient diet. *Mol Carcinog* 48:479–87.

145. Kleiner, D. E., E. M. Brunt, M. van Natta, C. Behling, M. J. Contos, O. W. Cummings, L. D. Ferrell, Y. C. Liu, M. S. Torbensen, A. Unalp-Arida, M. Yeh, A. J. McCullough, A. J. Sanyal, and Nonalcoholic Steatohepatitis Clinical Research Network. 2005. Design and validation of a histological scoring system for nonalcoholic fatty liver disease. *Hepatology* 41:1313–21.
146. O'Connell, R. M., K. D. Taganov, M. P. Boldin, G. Cheng, and D. Baltimore. 2007. MicroRNA-155 is induced during the macrophage inflammatory response. *Proc Natl Acad Sci U S A* 104:1604–9.
147. Johnson, C. D., A. Esquela-Kerscher, G. Stefani, M. Byrom, K. Kelnar, D. Ovcharenko, M. Wilson, X. Wang, J. Shelton, J. Shingara, L. Chin, D. Brown, and F. J. Slack. 2007. The let-7 microRNA represses cell proliferation pathways in human cells. *Cancer Res* 67:7713–22.
148. Takamizawa, J., H. Konishi, K. Yanagisawa, S. Tomida, H. Osada, H. Endoh, T. Harano, Y. Yatabe, M. Nagino, Y. Nimura, T. Mitsudomi, and T. Takahashi. 2004. Reduced expression of the let-7 microRNAs in human lung cancers in association with shortened postoperative survival. *Cancer Res* 64:3753–56.
149. Varnholt, H., U. Drebber, F. Schulze, I. Wedemeyer, P. Schirmacher, H. P. Dienes, and M. Odenthal. 2008. MicroRNA gene expression profile of hepatitis C virus-associated hepatocellular carcinoma. *Hepatology* 47:1223–32.
150. Ladeiro, Y., G. Couchy, C. Balabaud, P. Bioulac-Sage, L. Pelletier, S. Rebouissou, and J. Zucman-Rossi. 2008. MicroRNA profiling in hepatocellular tumors is associated with clinical features and oncogene/tumor suppressor gene mutations. *Hepatology* 47:1955–63.
151. Meng, F., R. Henson, H. Wehbe-Janek, K. Ghoshal, S. T. Jacob, and T. Patel. 2007. MicroRNA-21 regulates expression of the PTEN tumor suppressor gene in human hepatocellular cancer. *Gastroenterology* 133:647–58.
152. Selaru, F. M., A. V. Olaru, T. Kan, S. David, Y. Cheng, Y. Mori, J. Yang, B. Paun, Z. Jin, R. Agarwal, J. P. Hamilton, J. Abraham, C. Georgiades, H. Alvarez, P. Vivekanandan, W. Yu, A. Maitra, M. Torbenson, P. J. Thuluvath, G. J. Gores, N. F. LaRusso, R. Hruban, and S. J. Meltzer. 2009. MicroRNA-21 is overexpressed in human cholangiocarcinoma and regulates programmed cell death 4 and tissue inhibitor of metalloproteinase 3. *Hepatology* 49:1595–601.
153. Pineau, P., S. Volinia, K. McJunkin, A. Marchio, C. Battiston, B. Terri, V. Mazzaferro, S. W. Lowe, C. M. Croce, and A. Dejean. 2010. miR-221 overexpression contributes to liver tumorigenesis. *Proc Natl Acad Sci U S A* 107:264–69.
154. Garofalo, M., G. Di Leva, G. Romano, G. Nuovo, S. S. Suh, A. Ngankeu, C. Taccioli, F. Pichiorri, H. Alder, P. Secchioro, P. Gasparini, A. Gonelli, S. Costinean, M. Acunzo, G. Condorelli, and C. M. Croce. 2009. miR-221&222 regulate TRAIL resistance and enhance tumorigenicity through PTEN and TIMP3 downregulation. *Cancer Cell* 16:498–509.
155. Gramantieri, L., F. Fornari, M. Ferracin, A. Veronese, S. Sabbioni, A. G. Calin, G. L. Grazi, C. M. Croce, L. Bolondi, and M. Negrini. 2009. MicroRNA-221 targets Bmf in hepatocellular carcinoma and correlates with tumor multifocality. *Clin Cancer Res* 15:5073–81.
156. Fornari, F., L. Gramantieri, M. Ferracin, A. Veronese, S. Sabbioni, G. A. Calin, G. L. Grazi, C. Giovannini, C. M. Croce, L. Bolondi, and M. Negrini. 2008. MiR-221 controls CDKN1C/p57 and CDKN1B/p27 expression in human hepatocellular carcinoma. *Oncogene* 27:5651–61.
157. Wang, Y., A. T. Lee, J. Z. Ma, J. Wang, J. Ren, Y. Yang, E. Tantoso, K.-B. Li, L. L. P. J. Ooi, P. Tan, and C. G. L. Lee. 2008. Profiling microRNA expression in hepatocellular carcinoma reveals microRNA-224 up-regulation and apoptosis inhibitor-5 as a microRNA-224-specific target. *J Biol Chem* 283:13205–15.

158. Liu, W. H., S. H. Yeh, C. C. Lu, S. L. Yu, H. Y. Chen, C. Y. Lin, D. S. Chen, and P. J. Chen. 2009. MicroRNA-18a prevents estrogen receptor-alpha expression, promoting proliferation of hepatocellular carcinoma cells. *Gastroenterology* 136:683–93.
159. Zhang, X., S. Liu, T. Hu, S. Liu, Y. He, and S. Sun. 2009. Up-regulated microRNA-143 transcribed by nuclear factor kappa B enhances hepatocarcinoma metastasis by repressing fibronectin expression. *Hepatology* 50:490–99.
160. Gramantieri, L., M. Ferracin, F. Fornari, A. Veronese, S. Sabbioni, C.-G. Liu, G. A. Calin, C. Giovannini, E. Ferrazzi, G. L. Grazi, C. M. Croce, L. Bolondi, and M. Negrini. 2007. Cyclin G1 is a target of miR-122a, a microRNA frequently down-regulated in human hepatocellular carcinoma. *Cancer Res* 67:6092–99.
161. Lin, C. J., H. Y. Gong, H. C. Tseng, W. L. Wang, and J. L. Wu. 2008. miR-122 targets an anti-apoptotic gene, Bcl-w, in human hepatocellular carcinoma cell lines. *Biochem Biophys Res Commun* 375:315–20.
162. Tsai, W. C., P. W. Hsu, T. C. Lai, S.-H. Hsu, J. Datta, H. Kutay, A. Yadav, G. Nuovo, P. Kumar, and K. Ghoshal. 2008. MicroRNA-122, a tumor suppressor microRNA that regulates intrahepatic metastasis of hepatocellular carcinoma. *Hepatology* 49:1571–82.
163. Bai, S., M. W. Nasser, B. Wang, S.-H. Hsu, J. Datta, H. Kutay, A. Yadav, G. Nuovo, P. Kumar, and K. Ghoshal. 2009. MicroRNA-122 inhibits tumorigenic properties of hepatocellular carcinoma cells and sensitizes these cells to sorafenib. *J Biol Chem* 284:32015–27.
164. Wong, Q. W., R. W. Lung, P. T. Law, P. B. Lai, K. Y. Chan, K. F. To, and N. Wong. 2008. MicroRNA-223 is commonly repressed in hepatocellular carcinoma and potentiates expression of Stathmin1. *Gastroenterology* 135:257–69.
165. Datta, J., H. Kutay, M. W. Nasser, G. J. Nuovo, B. Wang. S. Majumder, S. G. Liu, S. Volinia, C. M. Croce, T. D. Schmittgen, K. Ghosahl, and S. T. Jacob. 2008. Methylation mediated silencing of MicroRNA-1 gene and its role in hepatocellular carcinogenesis. *Cancer Res* 68:5049–58.
166. Migliore, C., A. Petrelli A, E. Ghiso, S. Corso, L. Capparuccia, A. Eramo, P. M. Comoglio, and S. Giordano. 2008. MicroRNAs impair MET-mediated invasive growth. *Cancer Res* 68:10128–36.
167. Kim, S., U. J. Lee, M. N. Kim, E. J. Lee, K. Y. Kim, M. Y. Lee, S. Choung, Y. J. Kim, and Y. C. Choi. 2008. MicroRNA miR-199a* regulates the MET proto-oncogene and the downstream extracellular signal-regulated kinase 2 (ERK2). *J Biol Chem* 283:18158–66.
168. Su, H., J. R. Yang, T. Xu, J. Huang, L. Xu, Y. Yuan, and S. M. Zhuang. 2009. MicroRNA-101, down-regulated in hepatocellular carcinoma, promotes apoptosis and suppresses tumorigenicity. *Cancer Res* 69:1135–42.
169. Xu, T., Y. Zhu, Y. Xiong, Y. Y. Ge, J. P. Yun, and S. M. Zhuang. 2009. MicroRNA-195 suppresses tumorigenicity and regulates G1/S transition of human hepatocellular carcinoma cells. *Hepatology* 50:113–21.
170. Caldwell, S., and S. H. Park. 2009. The epidemiology of hepatocellular cancer: from the perspectives of public health problem to tumor biology. *J Gastroenterol* 44 (Suppl 19):96–101.
171. Cuadrado, A., A. Orive, C. Garcia-Suarez et al. 2005. Non-alcoholic steatohepatitis (NASH) and hepatocellular carcinoma. *Obes Surg* 15:442–46.
172. Cheung, O., and A. J. Sanyal. 2009. Recent advances in nonalcoholic fatty liver disease. *Curr Opin Gastroenterol* 25:230–37.
173. Sanyal, A. J. 2005. Mechanisms of disease: pathogenesis of nonalcoholic fatty liver disease. *Nat Clin Pract Gastroenterol Hepatol* 2:46–53.
174. Cheung, O., P. Puri, C. Eicken, M. J. Contos, F. Mirshahi, J. W. Maher, J. M. Kellum, H. Min, V. A. Luketic, and A. J. Sanyal. 2008. Nonalcoholic steatohepatitis is associated with altered hepatic MicroRNA expression. *Hepatology* 48:1810–20.

175. Gatfield, D., G. Le Martelot, C. E. Vejnar, O. Schaad, F. Fleury-Olala, A. L. Ruskeepää,M. Oresic, C. C. Esau, E. M. Zdobnov, and U. Schibler. 2009. Integration of microRNA miR-122 in hepatic circadian gene expression. *Genes Dev* 23:1313–26.
176. Jopling, C. L. 2008. Regulation of hepatitis C virus by microRNA-122. *Biochem Soc Trans* 36:1220–23.
177. Henke, J. I., D. Goergen, J. Zheng, Y. Song, C. G. Schüttler, C. Fehr, C. Jünemann, and M. Niepmann. 2008. microRNA-122 stimulates translation of hepatitis C virus RNA. *Embo J* 27:3300–3310.
178. Krutzfeldt, J., M. N. Poy, and M. Stoffel. 2006. Strategies to determine the biological function of microRNAs. *Nat Genet* 38 (Suppl):S14–19.
179. Krutzfeldt, J., and M. Stoffel. 2006. MicroRNAs: a new class of regulatory genes affecting metabolism. *Cell Metab* 4:9–12.
180. Chang, J., J. T. Guo, D. Jiang, H. Guo, J. M. Taylor, and T. M. Block. 2008. Liver-specific microRNA miR-122 enhances the replication of hepatitis C virus in nonhepatic cells. *J Virol* 82:8215–23.
181. Coulouarn, C., V. M. Factor, J. B. Andersen, M. E. Durkin, and S. S. Thorgeirsson. 2009. Loss of miR-122 expression in liver cancer correlates with suppression of the hepatic phenotype and gain of metastatic properties. *Oncogene* 28:3526–36.
182. Xiao, J., B. Yang, H. Lin, Y. Lu, X. Luo, and Z. Wang. 2007. Novel approaches for gene-specific interference via manipulating actions of microRNAs: examination on the pacemaker channel genes HCN2 and HCN4. *J Cell Physiol* 212:285–92.
183. Nakajima, G., K. Hayashi, Y. Xi, K. Kudo, C. Kittas, V. G. Gorgoulis, and G. T. Tsangaris. 2006. Non-coding microRNAs hsa-let-7g and hsa-miR-181b are associated with chemoresponse to S-1 in colon cancer. *Cancer Genomics Proteomics* 3:317–24.
184. Lee, E. J., Y. Gusev, J. Jiang, G. J. Nuovo, M. R. Lerner, W. L. Frankel, D. L. Morgan, R. G. Postier, D. J. Brackett, and T. D. Schmittgen. 2007. Expression profiling identifies microRNA signature in pancreatic cancer. *Int J Cancer* 120:1046–54.
185. Shi, L., Z. Cheng, J. Zhang, R. Li, P. Zhao, F. Zu, and Y. You. 2008. hsa-mir-181a and hsa-mir-181b function as tumor suppressors in human glioma cells. *Brain Res* 1236:185–93.
186. Marcucci, G., K. Maharry, M. D. Radmacher, K. Mrózek, T. Vukosavljevic, P. Paschka, S. P. Whitman, C. Langer, C. D. Baldus, C.-G. Liu, A. S. Ruppert, B. L. Powell, A. J. Carroll, M. A. Caligiuri, J. E. Kolitz, R. A. Larson, and C. D. Bloomfield. 2008. Prognostic significance of, and gene and microRNA expression signatures associated with, CEBPA mutations in cytogenetically normal acute myeloid leukemia with high-risk molecular features: a Cancer and Leukemia Group B Study. *J Clin Oncol* 26:5078–87.
187. Visone, R., L. Z. Rassenti , A. Veronese, C. Taccioli, S. Costinean, B. D. Aguda, S. Volinia, M. Ferracin, J. Palatini, V. Balatti, H. Alder, M. Negrini, T. J. Kipps, and C. M. Croce. 2009. Karyotype specific microRNA signature in chronic lymphocytic leukemia. *Blood* 114 (18):3672–79.
188. Ji, J., T. Yamashita, A. Budhu, M. Forgues, H.-L. Jia, C. Li, C. Deng, E. Wauthier, L. M. Reid, Q.-H. Ye, L.-X. Qin, W. Yang, Z.-Y. Tang, C. M. Croce, and X. W. Wang. 2009. Identification of microRNA-181 by genome-wide screening as a critical player in EpCAM-positive hepatic cancer stem cells. *Hepatology* 50:472–80.
189. Shi, L., Z. Cheng, J. Zhang, R. Li, P. Zhao, Z. Fu, and Y. You. 2008. hsa-mir-181a and hsa-mir-181b function as tumor suppressors in human glioma cells. *Brain Res* 1236:185–93.
190. Thai, T H., D. P. Calado, S. Casola, K. M. Ansel, C. Xiao, Y. Yue, A. Murphy, D. Frendeway, D. Valenzuela, J. L. Kutok, M. Schmidt, Supprian, N. Rajewsky, G. Yancopoulos, A. Rao, and K. Rajewsky. 2007. Regulation of the germinal center response by microRNA-155. *Science* 316:604–8.

191. Vigorito, E., K. L. Perks, C. Abreu-Goodger, S. Bunting, Z. Xiang, S. Kohlhaas, P. P. Das, E. A. Miska, A. Rodriguez, A. Bradley, K. G. Smith, C. Rada, A. J. Enright, K.-M. Toellner, I. C. Maclennan, and M. Turner. 2007. microRNA-155 regulates the generation of immunoglobulin class-switched plasma cells. *Immunity* 27:847–59.
192. Faraoni, I., F. R. Antonetti, J. Cardone, and E. Bonmassar. 2009. miR-155 gene: a typical multifunctional microRNA. *Biochim Biophys Acta* 1792:497–505.
193. Tili, E., C. M. Croce, and J. J. Michaille. 2009. miR-155: on the crosstalk between inflammation and cancer. *Int Rev Immunol* 28:264–84.
194. Volinia, S., G. A. Calin, C. G. Liu, S. Ambs, A. Cimmino, F. Petrocca, R. Visone, M. Iorio, C. Roldo, M. Ferracin, R. L. Prueitt, N. Yanaihara, G. Lanza, A. Scarpa, A. Vecchione, M. Negrini, C. C. Harris, and C. M. Croce. 2006. A microRNA expression signature of human solid tumors defines cancer gene targets. *Proc Natl Acad Sci U S A* 103: 2257–61.
195. Garzon, R., and C. M. Croce. 2008. MicroRNAs in normal and malignant hematopoiesis. *Curr Opin Hematol* 15:352–58.
196. Greither, T., L. F. Grochola, A. Udelnow, C. Lautenschlager, P. Wurl, and H. Taubert. 2009. Elevated expression of microRNAs 155, 203, 210 and 222 in pancreatic tumors is associated with poorer survival. *Int J Cancer* 126 (1):76–80.
197. Eis, P. S., W. Tam, L. Sun, A. Chadburn, Z. Li, M. F. Gomez, E. Lund, and J. E. Dahlberg. 2005. Accumulation of miR-155 and BIC RNA in human B cell lymphomas. *Proc Natl Acad Sci U S A* 102:3627–32.
198. Habbe, N., J. B. Koorstra, J. T. Mendell, J. Offerhaus, J. K. Ryu, G. Feldmann, M. E. Mullendore, M. G. Goggins, S.-M. Hong, and A. Maitra. 2009. MicroRNA miR-155 is a biomarker of early pancreatic neoplasia. *Cancer Biol Ther* 8:340–46.
199. Costinean, S., N. Zanesi, Y. Pekarsky et al. 2006. Pre-B cell proliferation and lymphoblastic leukemia/high-grade lymphoma in E(mu)-miR155 transgenic mice. *Proc Natl Acad Sci U S A* 103:7024–29.
200. Wang, B., S. Majumder, G. Nuovo, H. Kutay, S. Volinia, T. Patel, T. D. Schmittgen, C. Croce, K. Ghoshal, and S. T. Jacob. 2009. Role of miR-155 at early stages of hepatocarcinogenesis induced by choline-deficient and amino acid defined diet in C57BL/6 mice. *Hepatology* (in press).
201. Saito, Y., and P. A. Jones. 2006. Epigenetic activation of tumor suppressor microRNAs in human cancer cells. *Cell Cycle* 5:2220–22.
202. Brueckner, B., C. Stresemann, R. Kuner, C. Mund, T. Musch, M. Meister, H. Sültmann, and F. Lyko. 2007. The human let-7a-3 locus contains an epigenetically regulated microRNA gene with oncogenic function. *Cancer Res* 67:1419–23.
203. Lujambio, A., S. Ropero, E. Ballestar, M. F. Fraga, C. Cerrato, F. Setién, F. Casado, A. Suarez-Gauthier, M. Sanchez-Cespedes, A. Gitt, I. Spiteri, P. Das, C. Caldas, E. Miska, and M. Esteller. 2007. Genetic unmasking of an epigenetically silenced microRNA in human cancer cells. *Cancer Res* 67:1424–29.
204. Rougier, P., E. Mitry, J. C. Barbare, and J. Taieb. 2007. Hepatocellular carcinoma (HCC): an update. *Semin Oncol* 34:S12–20.
205. Thomas, M. B., R. Chadha, K. Glover, X. Wang, J. Morris, T. Brown, A. Rashid, J. Dancey, and J. L. Abbruzzese. 2007. Phase 2 study of erlotinib in patients with unresectable hepatocellular carcinoma. *Cancer* 110:1059–67.
206. Ghoshal, K., and S. Bai. 2007. DNA methyltransferases as targets for cancer therapy. *Drugs Today (Barc)* 43:395–422.
207. Ghoshal, K., S. Majumder, Z. Li, X. Dong, and S. T. Jacob. (2000) Suppression of metallothionein gene expression in a rat hepatoma because of promoter-specific DNA methylation. *J Biol Chem* 275:539–47.
208. Tili, E., J. J. Michaille, V. Gandhi, W. Plunkett, D. Sampath, and G. A. Calin. 2007. miRNAs and their potential for use against cancer and other diseases. 2007. *Future Oncol* 3:521–37.

209. Visone, R., and C. M. Croce. 2009. MiRNAs and cancer. *Am J Pathol* 174:1131–38.
210. Lanford, R. E., E. S. Hildebrandt-Eriksen, A. Petri, R. Persson, M. Lindow, M. E. Munk, S. Kaupinnen, and H. Orum. 2010. Therapeutic silencing of microRNA-122 in primates with chronic hepatitis C virus infection. *Science* 327:198–201.
211. Watts, G. 2006. Work on RNA interference brings Nobel triumph. *BMJ* 333:717.
212. Sato, A., A. Takagi, A. Shimamoto, S. Kawakami, and M. Hashida. 2007. Small interfering RNA delivery to the liver by intravenous administration of galactosylated cationic liposomes in mice. *Biomaterials* 28:1434–42.

8 Epigenetic Mechanisms in Lung Inflammation and Chronic Airway Diseases and Intervention by Dietary Polyphenols

Irfan Rahman

CONTENTS

8.1 Cigarette Smoking and Pathophysiology of Chronic Airway Diseases Including COPD 186
8.2 Epigenetics, Environmental Factors, and Predisposition to Chronic Inflammatory Lung Disorders Including COPD 187
8.2.1 Asthma: Link with Environmental Factors and Effects on Epigenetic Changes 189
8.2.1.1 Epigenetic Mechanisms in Asthma 190
8.3 Dietary Polyphenols and Modulation/Intervention of Epigenetic Changes Associated with Chronic Lung Diseases 191
8.3.1 Deacetylases (SIRT1 and HDACs), Epigenetic Changes, and Polyphenols 191
8.3.1.1 HDAC2 and Epigenetic Changes 192
8.3.1.2 Modulation of HDACs by Dietary Polyphenols 192
8.3.2 Sirtuins and Epigenetic Changes 195
8.3.2.1 Modulation of SIRT1 by Dietary Polyphenols 196
8.3.3 Nrf2, Epigenetic Changes, and Dietary Polyphenols 198
8.4 Beneficial Effects of Dietary Polyphenols in Improving Lung Function 199
8.5 Adverse Effects of Polyphenols 201
8.6 Bioavailability and Effective Doses of Dietary Polyphenols 202
8.7 Conclusions and Future Perspectives 203
Acknowledgments 204
References 204

Oxidative stress as a result of cigarette smoking is an important etiological factor in the pathogenesis of chronic obstructive pulmonary disease (COPD). Various reports are available on the imbalance of oxidant-antioxidant in the lungs as a result of cigarette smoke and airborne particulate matter. Abnormal inflammation is the hallmark of chronic airway diseases (COPD and asthma), which is caused by oxidative stress. Keeping in mind that only about 15–20% of smokers develop COPD, a new paradigm in COPD research has now emerged that suggests that epigenetic alterations might be a clue for understanding the pathophysiology of COPD and other chronic lung disorders such as asthma. At present, no effective treatment exists to halt the progressive decline in lung function of smokers who get the disease. Since COPD is considered to be a result of oxidative stress imposed by smoking, treatment with dietary antioxidants has been in vogue for some time. Other than thiols, and vitamins such as C and E, the most prominent antioxidant candidate of choice has been dietary polyphenols. Besides thiols, which are mucolytic, vitamins C and E are less efficacious as antioxidants in the lung. Recently, a unique role of dietary polyphenols is described via modulation of redox signaling in inflammation. In this chapter, a brief overview of the pathophysiology of COPD and asthma is given in light of oxidative stress and inflammatory response, followed by information on epigenetic changes due to environmental factors linked to oxidative stress and inflammation, and mechanisms of epigenetic alterations in chronic lung diseases. Furthermore, how polyphenols may be beneficial and have influence in modulating/intervening the epigenetic changes and associated pathophysiological alterations in chronic airway disorders is explored.

8.1 CIGARETTE SMOKING AND PATHOPHYSIOLOGY OF CHRONIC AIRWAY DISEASES INCLUDING COPD

COPD/chronic airway obstruction disease is a condition characterized by largely irreversible airflow limitation (Pauwels et al. 2001). The airflow limitation is usually progressive and is associated with an abnormal inflammatory response of the lungs to noxious particles or gases. The diagnosis is usually based on the history of exposure to toxic stimuli (mainly tobacco smoke) and abnormal lung function tests. Cigarette smoking has been strongly implicated as the major etiological cause of the disease. Only about 15–20% of smokers develop the disease, and they exhibit a rapid decline in forced expiratory volume in one second (FEV_1). Factors such as air pollutants, indoor cooking/biomass fuel burning, infections, and occupational dusts may cause and/or exacerbate COPD.

Neutrophils, macrophages, and T cells (CD8+) play an important role in the pathophysiology of COPD. In view of their ability to release a wide array of inflammatory mediators and tissue-degrading enzymes, neutrophils and macrophages can orchestrate tissue destruction and chronic inflammation (K. F. Chung 2001; Stockley 2002). Infiltration of neutrophils and macrophages and their activation in the pulmonary microvasculature could lead to damage of lung microenvironment, due to the release of reactive oxygen species (ROS) and proteases, thus leading to destruction of the alveolar wall, which occurs in emphysema. COPD is characterized by increased oxidative stress in the lungs, resulting in oxidant/antioxidant imbalance. Lipid peroxidation products, such as 4-hydroxy-2-nonenal (4-HNE) and F_2-isoprostanes, are increased in response to cigarette smoking in bronchiolar epithelial, alveolar-type cells and macrophages, and in lungs of patients with COPD (Aoshiba et al. 2003; Rahman and Adcock 2006). Chronic smokers exhibit decreased total antioxidant capacity, which is mainly associated with depletion of major plasma antioxidants such as ascorbic acid, vitamin E, β-carotene, and selenium in the serum (Mezzetti et al. 1995). Glutathione (GSH), an important thiol antioxidant, has been shown to be decreased in lungs of patients with COPD. Asthma is characterized by increased CD4+ cells, mast cells, and eosinophils along with other inflammatory/immune cells associated with increased oxidative stress and inflammatory response in the lung (reviewed by Jeffery 2001).

Nuclear factor kappa B (NF-κB) is an inducible pleiotropic transcription factor that plays a key role in the expression of multiple genes, leading to transcription of proinflammatory mediators such as cytokines and chemokines. It has been shown that cells obtained from bronchoalveolar lavage fluid (BALF) of smokers exhibit 10-fold higher activation of NF-κB in response to lipopolysaccharide (LPS) as compared to that of nonsmokers (Mochida-Nishimura et al. 2001). NF-κB is activated in lungs of patients with COPD and plays a pivotal role in inducing the chronic proinflammatory response seen in lungs of patients with COPD and asthma (Di Stefano et al. 2002; Caramori et al. 2003; Rajendrasozhan, Yang, Edirisinghe et al. 2008; Szulakowski et al. 2006; Ito et al. 2002). Another important facet of NF-κB is its ability to bring about epigenetic modifications by altering the histones upon binding to the DNA, leading to proinflammatory gene transcription (discussed later in this chapter).

COPD is associated with airway/airspace inflammation and is characterized by the presence of inflammatory biomarkers, such as IL-8 and TNF-α, which are elevated in patients with COPD (Keatings et al. 1996). Increased leukocyte adhesion to vascular endothelium has been attributed to cigarette smoking due to up-regulation of NF-κB-dependent cell adhesion molecules such as intercellular adhesion molecule (ICAM-1), endothelial leukocyte adhesion molecule 1 (ELAM-1), and vascular cell adhesion molecule (VCAM-1). For further reading the reviewers are directed to several exhaustive reviews describing oxidative stress, inflammation, and other aspects of pathogenesis of asthma and COPD that are relevant to this chapter (Rahman and Adcock 2006; Rajendrasozhan, Yang, Edirisinghe et al. 2008; Rajendrasozhan et al. 2009; Yoshida and Tuder 2007; K. F. Chung and Adcock 2008; Jeffery 2001).

8.2 EPIGENETICS, ENVIRONMENTAL FACTORS, AND PREDISPOSITION TO CHRONIC INFLAMMATORY LUNG DISORDERS INCLUDING COPD

Recent research has mostly focused on chronic airways disease therapy based on palliative cure. However, much remains to be understood as to the molecular mechanisms involved and the genetic pattern/epigenetic modifications that make people susceptible to such chronic diseases due to environmental stimuli. There is a recent surge in the belief that many diseases start early in life, possibly *in utero* due to epigenetic modifications in the epigenome (Castro et al. 2009).

The lung is one of the organs that develops and matures very rapidly in the fetus in order to support breathing at birth. Abnormal development and maturation may therefore lead to chronic lung diseases later in life. There are now increasing reports that pre- and postnatal environmental exposures can induce long-term effects on lung structure and function related to cellular and molecular events that may vary from individual to individual. Studies have revealed that epigenetic phenomenon associated with nutritional/dietary and environmental factors have a role in accelerated decline in lung function, and hence in pathogenesis of chronic lung diseases. Several epidemiologic studies have identified various environmental factors that alter lung development and have been well characterized in animal models. For example, the *in utero* and postnatal effect of environmental tobacco smoke as well as the effect of nutritional deficiencies of vitamins A and D or a high methyl donor diet (Litonjua and Weiss 2007) during lung development are predictive of chronic diseases later in life (Wang and Pinkerton 2008; Doherty et al. 2009; Lim and Kobzik 2009). However, very little information is available on other environmental factors that might influence fetal and postnatal lung development. It is now suggested that the epigenome might be affected in the early critical periods of lung development by environmental factors, such as diet, drugs, toxins, pollutants, and interestingly, the socioeconomic environment and lifestyle of an individual (Jaenisch and Bird 2003). It might, therefore, be prudent to investigate whether such environmental factors can cause subtle epigenetic effects that might influence developmental genes during gestation and predispose an individual to a particular postnatal or adult lung disease(s) (Gicquel, et al. 2008). However, no such information is currently available linking these factors to epigenetics of chronic lung diseases except the epidemiological aspects of environment to lung diseases.

Although regulation of gene expression by epigenetic mechanisms has been reported to influence early embryonic development (Bernstein et al. 2007), little is known as to how epigenetic changes regulate cell differentiation during pre- and postnatal lung development (Cortese et al. 2008; Millien et al. 2008). There are emerging reports that epigenetic alterations might be associated with chronic lung diseases. One of the best characterized epigenetic mechanisms of lung gene regulation is the methylation pattern of the regions of genomic DNA rich in CG content, a phenomenon now associated with regulation of tumor suppressor genes (Hsu et al. 2007), fibrosis (Sanders et al. 2008), and asthma-related genes (Adcock et al. 2007; Bousquet et al. 2004; Y. Yang et al. 2008). Furthermore, hypermethylation of CpG islands in the promoter region of the tumor suppressor genes is now considered to be a key factor in many types of cancer associated with tobacco smoking (Meissner et al. 2008; Zochbauer-Muller et al. 2001). Such epigenetic changes can now be detected for a candidate tumor suppressor gene, genomewide or in localized regions of the genome, which would enable early prognosis and treatment of various types of cancer (Meissner et al. 2008).

Methylation in the promoter region of several genes in patients with stage I non-small cell lung cancer has also been recently associated with early recurrence of the disease (Brock et al. 2008). Determination of specific DNA methylation of genes may thus prove to be useful clinical markers for early detection and/or chemoprotective intervention in lung cancer. Although reports of DNA methylation pattern are now emerging in lung cancer (Brock et al. 2008; Meissner et al. 2008; Zochbauer-Muller et al. 2001), sporadic reports exist for histone/DNA methylation in cigarette smoke-induced chronic lung inflammatory diseases. Available reports have demonstrated frequent methylation of p16 (INK4a) promoter, a tumor suppressor gene in sputum of patients with COPD, which correlated significantly with heavy cigarette smoking (Georgiou et al. 2007). Similarly, hypermethylation of p16-regulating lung cancer was reported to be a result of exposure to environmental tobacco smoke (ETS) (Digel and Lubbert 2005). To date, very little data are available regarding histone/DNA methylation in cigarette smoke-mediated chronic lung inflammatory diseases. Detailed investigations regarding methylation pattern of histone/DNA and other modifications may yield new biomarkers and therapeutic targets for the treatment of inflammatory airway diseases.

Epigenetic alterations can also be due to covalent modifications of the histones by various post-translational modifications, such as histone acetylation, deacetylation, phosphorylation, ubiquitination, sumoylation, and methylation mechanisms (Rajendrasozhan et al. 2009). These chromatin/epigenetic mechanisms may play an important role in disease development, which is evident by the report that asthmatic and COPD lungs exhibit abnormally high regions of acetylated histones and reduced deacetylates (Islam and Mendelson 2008; Rahman and Adcock 2006; Ito et al. 2002, 2005; Szulakowski et al. 2006; Coward et al. 2009). Certainly, the role of these epigenetic modifications in modulation of gene expression in lung development and predisposition to environmental mediated chronic lung diseases are areas of future research. Following are some current thoughts in light of the recent research on epigenetics of asthma, but no other such studies are available linking lung at birth with environmental effects on epigenetic modifications in COPD.

8.2.1 Asthma: Link with Environmental Factors and Effects on Epigenetic Changes

Asthma is a complex genetic disease triggered by a variety of environmental factors. Individual asthmatics are unique in their clinical presentation, which is suggestive of a specific genetic-environmental interaction in individual cases. The onset of asthma also differs, as some develop the condition at an early age, while others do so in adulthood (Hoppin et al. 2004). Furthermore, occupational exposures are responsible for the asthmatic attack in some individuals, whereas urban air, pollens, pet dandruffs, and allergens can induce the disease in others (Venn et al. 2001). Therefore, it appears that early specific epigenetic alterations in the fetal stages due to specific environmental exposures might be responsible for clinical heterogeneity among asthmatics.

The history of asthma varies from one individual to another and is often a result of early or later environmental exposure that is responsible for specific epigenetic alterations (Miller and Ho 2008). There is now a spate of evidence that has revealed that there are specific time periods, such as prenatal development, early childhood, and adolescence, when an individual is more susceptible to the effects of environmental exposures and other asthma-inducing factors, as highlighted earlier (Kurukulaaratchy et al. 2004; Mandhane et al. 2005). These are the time periods when the cells are also more prone to epigenetic alterations that lead to altered antigen processing, and once these cells (memory T-cells) are epigenetically imprinted, then these modifications persist. A large number of studies have revealed that impaired respiratory function, wheezing, asthma, and other respiratory problems in infants, young children, and adolescents may be associated with exposures to environmental tobacco smoke (ETS) (Alati et al. 2006). For example, altered airway structures in the prenatal lungs have been observed upon exposure to ETS (Elliot et al. 2003). Therefore, early prenatal environmental exposures might play a role in the subsequent development of asthma or even other chronic lung diseases in adult life. Additional factors other than ETS that may lead to epigenetic changes in the prenatal stage include low maternal intake of vitamin E and zinc (Devereux et al. 2006), and use of antibiotics and steroids during pregnancy (Jedrychowski et al. 2006). These are the factors that also render an individual susceptible to childhood asthma, which can be prevented by maternal intake of probiotic foods and other foods having higher nutritional value (Fitzsimon et al. 2007). Hence, maternal history of asthma is implicated in the postnatal consequences that, if present during pregnancy, may influence the fetal epigenetic response to environmental triggers and hence the postnatal asthma risk.

8.2.1.1 Epigenetic Mechanisms in Asthma

It is normally believed that interactions between genetic polymorphisms and environmental factors are the sole determinants of inter-individual risk for asthma (Miller and Ho 2008). However, recent advances in asthma research have suggested that epigenetic mechanisms, such as genomic imprinting, histone modifications, DNA methylation of regulatory sequences, and regulation by microRNA also contribute to susceptibility, predisposition, and complexity of the disease (Tang and Ho 2007). Maternal and paternal alleles have been reported to express unequally as a consequence of genomic imprinting. Establishing a specific genomic imprint might involve differential DNA methylation of the target genes (Jiang et al. 2004). This is supported by the observation that there is a vertical transmission of genomic imprints of infections and associated inflammation to the offspring when a mother is infected during pregnancy (Finch and Crimmins 2004). This has been shown to make the offspring increasingly susceptible to environmental pathogens, and in turn leads to morbidity and mortality (Finch and Crimmins 2004). However, more in-depth studies are required to establish a direct epigenetic mechanism(s) linking to pathogenesis of asthma.

DNA methylation and consequent alterations in chromatin structure lead to loss of plasticity of Th helper cells and also their ability to differentiate into Th1 versus the pro-allergic Th2 pattern of cytokine gene expression (Wilson et al. 2009). In contrast, increased IL-4 production and differentiation into Th2 cells takes place when the promoter and conserved intronic regulatory element regions of the first intron of IL-4 gene are demethylated, and concomitantly counter regulatory interferon (IFN)-γ promoter is hypermethylated (Jones and Chen 2006). It is now known that in Th2 cells, many genes have multiple CpG islands, which are highly conserved across the species (Finch and Crimmins 2004). Methylation of CpG253 of the activator protein-1 (AP-1)-binding site leads to the alteration in the binding of the transcription factor (Young et al. 1994). Furthermore, there was a polarization of the

Th2 cells when the CpG253 regions in the IFN-γ promoter was methylated, cAMP response element binding protein (CREB) was inhibited, and activating transcription factor 2 (ATF2)/c-Jun binding to the CpG-containing AP1 site was activated (Young et al. 1994). It is interesting to note that IL-4, AP-1, CREB, and ATF2/c-Jun are all active components of the inflammatory machinery in various chronic respiratory diseases (Rahman and Adcock 2006; Barnes 2008). Atopic development of polarized Th2 phenotype therefore is now suggested to be a consequence of altered methylation patterns of genes regulating differentiation of Th2 cells. This suggestion is supported by a study wherein mice exposed to diesel and *Aspergillus fumigatus* led to hypermethylation of the IFN-γ promoter and hypomethylation of the interleukin (IL)-4 promoter, which in turn were associated with significant alterations in IgE levels (Liu et al. 2008). Similar observations were found when mice were fed with high-methyl donor diet at pregnancy, and the progeny had more airway hyperresponsiveness due to hypermethylation of various promoters (Hollingsworth et al. 2008). Epidemiological studies have also linked maternal smoking to asthma incidence in children, which in turn also provided evidence for higher susceptibility in grandchildren of women who smoked during pregnancy as well (Li et al. 2005). In addition, a direct link between prenatal dietary exposure to a high-methyl donor diet and increased severity of asthma symptoms upon challenge with ovalbumin has been demonstrated in mice (Hollingsworth et al. 2008). Therefore, the methylation of Th2 genes and IgE production may be directly regulated *in vivo* by inhaled environmental factors and dietary components. However, the association of the phenomena just discussed with clinical asthma needs further validation and experimental support.

Induction of reversible histone modifications is another mechanism that can bring about chromatin remodeling and proinflammatory gene transcription. Acetylation of histones is now reported in asthma, which involves generation of reactive oxygen intermediates and release of inflammatory mediators (Ito et al. 2002; Jeffery 2001). It has been reported that untreated asthmatic subjects exhibited elevated levels of histone acetyltransferase (HAT) activity and decreased histone deacetylase (HDAC) activity in bronchial samples (Ito et al. 2002).

The aforementioned sections clearly highlighted the role of epigenetics in chronic inflammatory diseases in relation to environmental factors and dietary components. This chapter now provides details on how polyphenols may be beneficial in modulating/intervening the epigenetic changes (histone acetylation/deacetylation and protein deacetylation by SIRT1) and associated steroid resistance in COPD and asthma.

8.3 DIETARY POLYPHENOLS AND MODULATION/INTERVENTION OF EPIGENETIC CHANGES ASSOCIATED WITH CHRONIC LUNG DISEASES

8.3.1 Deacetylases (SIRT1 and HDACs), Epigenetic Changes, and Polyphenols

Polyphenols are secondary metabolites of plants and represent a vast group of compounds having aromatic ring(s), characterized by presence of one or more hydroxyl groups with varying structural complexities. The most widely distributed group of plant phenolics are flavonoids. The flavonoids subclasses comprise flavonols, flavones, flavanols, isoflavones, anthocyanins, and others. Dietary polyphenols, such as resveratrol, curcumin, quercetins, and catechins, are shown to modulate NF-κB activation and chromatin remodeling through modulation of deacetylase activity attenuating inflammatory gene expression in lung epithelium and macrophages (Aggarwal and Shishodia 2004; Biswas et al. 2005; Kode et al. 2008; Rahman et al. 2006; Sharafkhaneh et al. 2007). NF-κB (due to intrinsic HAT activity) can lead to acetylation of histones, thus causing epigenetic effects (Rahman et al. 2002). Dietary polyphenols potentially have beneficial effects in lung cancer and COPD, particularly in reversing the epigenetic modifications (Lawless et al. 2009). The following sections deal with the biological properties with special reference to epigenetics and inflammation of some well-known and well-studied polyphenols, such as resveratrol (present in red wine), curcumin (active ingredient in turmeric), quercetins (present in onion), and the catechins (present in green tea).

8.3.1.1 HDAC2 and Epigenetic Changes

Acetylation of core histone proteins on the lysine residues in the N-terminal tails by HAT led to uncoiling of the DNA, allowing increased accessibility to transcription factors. While HAT acetylates the lysine residues, HDACs remove the acetyl group leading to DNA recoiling. HATs and HDACs also target nonhistone proteins, particularly transcription factors (RelA/p65 at lysine 310) to modify their activity and hence proinflammatory gene expression. HDACs can also deacetylate nonhistone proteins such as NF-κB, thus regulating NF-κB-dependent proinflammatory gene transcription (Sengupta and Seto 2004; S. R. Yang et al. 2006). Corticosteroids have been one the major modes of therapy against various respiratory diseases, such as asthma and COPD. Chronic resistance to glucocorticoids is observed in patients with moderate to severe COPD and in asthmatics who smoke. HDAC2 is an important mediator of glucocorticoid activity and is found to be reduced in lungs of patients with COPD and cells exposed to oxidative stress or cigarette smoke (Adenuga et al. 2009; Culpitt, Rogers, Shah et al. 2003; Ito et al. 2001; Meja et al. 2008; S. R. Yang et al. 2006). Glucocorticoids suppress expression of inflammatory genes by recruiting HDAC2 to the transcription complex via the glucocorticoid receptor (Barnes 2009), leading to deacetylation of histones and hence decreasing inflammatory gene transcription. It has been demonstrated that cigarette smoke solution treatment led to release of IL-8 and GM-CSF, a phenomenon that was not inhibited by dexamethasone in alveolar macrophages obtained from patients with COPD compared to that of smokers (Culpitt, Rogers, Shah et al. 2003). Failure of corticosteroids to ameliorate such disease conditions has been attributed to their failure to recruit HDAC2 to the presence of an oxidatively modified HDAC2 in lungs of asthmatics and COPD patients. Hence, as discussed previously, modulation of HDAC2 by dietary polyphenolic compounds may be useful in overcoming the steroid resistance in patients with severe asthma and COPD.

8.3.1.2 Modulation of HDACs by Dietary Polyphenols

Restoration of glucocorticoid function by dietary polyphenol curcumin is mediated through up-regulation of HDAC2 activity and restoration of HDAC1 and HDAC3

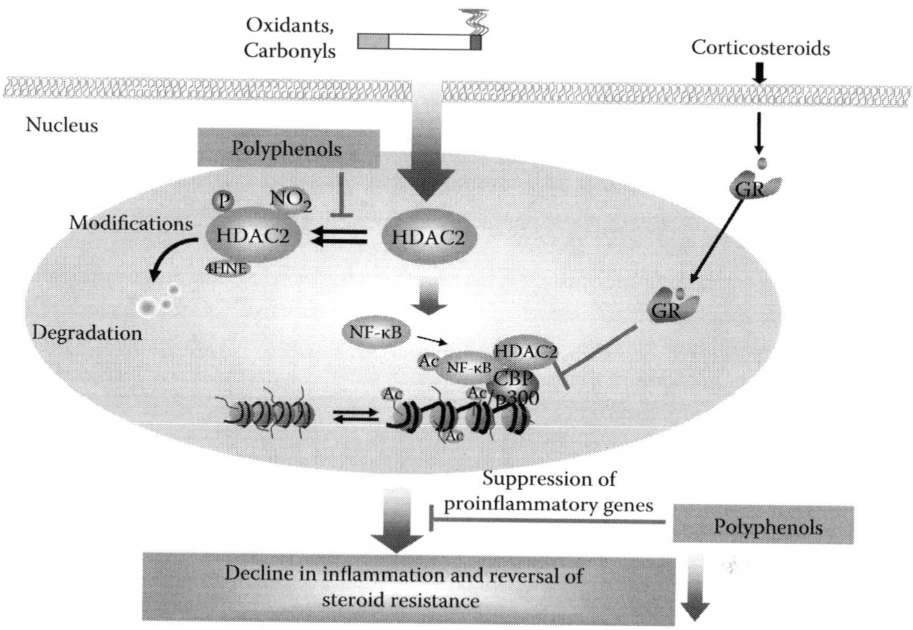

FIGURE 8.1 HDAC2 and regulation of lung inflammation by polyphenols. Dietary polyphenols modulate cigarette smoke-mediated human lung inflammation by regulation of histone modifications. DNA is coiled around the histone proteins and forms chromatin structure. Cigarette smoke inhibits histone deacetylases, such as HDAC2 (by posttranslational modifications), and/or triggers cellular signaling process leading to histone modifications. HDAC2 is modified by phosphorylation, carbonylation, aldehyde adducts formation, and nitration, leading to proteasomal degradation of HDAC2. Increased acetylation on histones (on proinflammatory genes) leads to increased accessibility of transcription factors and thereby culminates in increased proinflammatory gene transcription. These epigenetic changes/chromatin modifications can cause abnormal induction of proinflammatory genes. Dietary polyphenols inhibit degradation of HDAC2 and restore glucocorticoid efficacy, leading to inhibition of chronic inflammatory response in the lung. HDAC: histone deacetylases; P: phosphate; Ac: acetylation; 4HNE: 4-hydroxynonenal; NO_2: nitric oxide; GR; glucocorticoid receptor; CBP; CREB (cAMP response element binding protein)-binding protein.

levels (Figure 8.1) (Meja et al. 2008; S. R. Yang et al. 2006). Curcumin restored both HDAC2 activity and corticosteroid resistance in a concentration-dependent manner with an EC_{50} of 15 nM and 200–300 nM respectively in the monocytes and macrophages. Interestingly, it has recently been suggested that the anti-inflammatory actions of curcumin at 50 μM are propagated through inhibition of HAT activity, preventing NF-κB-mediated chromatin acetylation. Histones H3 and H4 are acetylated on specific lysine residues in rodent lungs in response to cigarette smoke, and in lungs of smokers/patients with COPD (Szulakowski et al. 2006; Ito et al. 2005; S. R. Yang et al. 2009; S. R. Yang, Valvo et al. 2008). Therefore polyphenol-dependent restoration of ROS-induced HAT/HDAC imbalance can have a significant impact on epigenome, therefore down-regulating inflammation, a concept that is corroborated by other reports on the ability of curcumin to inhibit HAT activity at very high concentrations and stalling NF-κB-mediated chromatin acetylation (Kang et al. 2005). Alternative mechanisms of polyphenol-mediated inhibition of inflammatory response could be through the quenching ROS or reversing posttranslational protein modifications caused by oxidants and reactive aldehydes.

Cigarette smoke-induced reduction in HDAC2 was associated with increased levels of total and acetylated RelA/p65 (Yao et al. 2008). Furthermore, recent studies revealed that RelA/p65 interacts with HDAC2, and RelA/p65 is retained in the nucleus, leading to activation of pro-inflammatory gene transcription when HDAC2 is reduced (S. R. Yang et al. 2006; Yao et al. 2008). In light of this, it is important to note that there is a significant decrement in the expression/activity of HDAC2 in lung parenchyma, bronchial biopsies, and alveolar macrophages of COPD patients. Such a decrease in the expression/activity of HDAC2 agrees well with the disease severity and the intensity of lung inflammation and hypoxic environment of the tissue (Adenuga et al. 2009; Ito et al. 2005). In contrast to increased HAT activity in bronchial biopsies and alveolar macrophages of asthmatics (Barnes et al. 2005), there was no observed alteration in HAT activity in the lungs of patients with COPD (Ito et al. 2005). HDAC2 is required for the anti-inflammatory effects of glucocorticoids in COPD patients (Ito et al. 2005). Reduction in the levels/activity of HDAC2 leads to corticosteroid resistance in such patients (Ito et al. 2005). Polyphenolic compounds, such as theophylline, have been shown to significantly increase HDAC2 activity, thereby enhancing dexamethasone-induced suppression of IL-8 release in alveolar macrophage of COPD subjects (Cosio et al. 2004). However, theophylline has a narrow window of efficacy and in higher doses it is toxic; it may also have several other biological effects, e.g., prevention of NAD^+ depletion via poly (ADP-ribose) polymerase (PARP)1 inhibition and associated sparing SIRT1 activity in macrophages and lung cells of patients with COPD (Moonen et al. 2005; Hageman et al. 2003; Weseler et al. 2009). Another polyphenolic compound, epigallocatechin gallate (EGCG), given in drinking water at a dose of 50 mg/ml to mice, induced lung levels of HDAC2 (Adenuga et al. 2009). Furthermore, the ability of HDAC2 to deacetylate glucocorticoid receptor (GR) and enable GR and RelA/p65 association, which leads to the attenuation of proinflammatory gene transcription, is described (Ito et al. 2006). Therefore, therapeutic restoration of HDAC2-dependent deacetylation of RelA/p65 and GR appears to be a good strategy for enhancing glucocorticoid sensitivity in COPD. As glucocorticoids are the main thrust of anti-inflammatory treatment, any therapeutic agent that can be used as an add-on to improve steroid responsiveness in COPD would be of significant clinical benefit. Clearly, clinical trials by using a combination approach of a steroid with polyphenols are warranted in patients with COPD.

In contrast to COPD, lung cancer inhibition of HDACs is now believed to be a new concept in cancer chemoprevention. Of the many HDAC inhibitors known, butyrate, diallyl disulfide (DADS), present in garlic, and sulforaphane (SFN), an active ingredient in broccoli/cruciferous vegetables, are reported to exhibit anticancer properties (Myzak and Dashwood 2006). However, in contrast to the traditional HDAC inhibitors such as trichostatin A or SAHA, which are effective at lower concentrations (nanomolar range), the new range of HDAC inhibitors are required in greater concentrations (micromolar range) (R. H. Dashwood et al. 2006). Therefore it is important to determine whether or not the concentrations of the new class of inhibitors are achievable under normal physiological conditions despite their low bioavailability.

DADS, found in garlic, is another global HDAC inhibitor (Druesne et al. 2004). *In vivo*, it is metabolically converted to S-allylmercaptocysteine, and its structure is similar to butyrate except that it has a "spacer" ending with a carboxylate group (Guyonnet et al. 2004). SFN-cysteine (SFN-Cys) contains a similar spacer and is a metabolite of SFN found in cruciferous vegetables, such as broccoli and broccoli sprouts. In the concentration range of 3–15 mM, SFN-Cys significantly inhibits HDAC activity (Myzak et al. 2004). In contrast, the parent compound SFN alone had no effect on HDAC activity. However, little is known about the distribution and concentrations/bioavailability of SFN and its active form(s) in different tissues. Although there are many dietary compounds having HDAC inhibitory properties, more investigations are required in order to understand their bioavailability and the achievable concentrations of these compounds within the body.

EGCG predominates among the various tea polyphenols and is considered to down-regulate the expression of genes related to angiogenesis and metastasis (Fassina et al. 2004). Similar to curcumin, green tea polyphenols also modulate a myriad of inflammatory signaling pathways and perhaps similarly the epigenetic modifications (Biesalski 2007; Di Paola et al. 2005), and therefore a single pathway cannot be assigned to the anti-inflammatory/anticancer properties of these dietary natural compounds as major therapeutic agents (C. S. Yang et al. 2008). Various studies on the effect of EGCG on cancer cells have revealed that EGCG can induce apoptosis, cell-growth arrest, and deregulation of the cyclin kinase inhibitor $p21^{WAF}$ possibly due to inhibition of HDACs (Ahmad et al. 2000; Qin et al. 2008; Raza and John 2008).

8.3.2 Sirtuins and Epigenetic Changes

Sirtuins (SIRT) and the mammalian equivalent SIRT1 belong to class III HDACs. SIRT1 was the first to be shown in determining the life span in yeast, flies, and the nematode. Unlike class I and II deacetylases, sirtuins are NAD^+-dependent and are not inhibited by trichostatin A or SAHA (Imai et al. 2000). Since sirtuins require NAD^+ coming from metabolic reactions, it is hypothesized that sirtuins might act as a molecular link between cellular metabolic status (expressed by the NAD^+/NADH levels) and cellular transcription (Bordone and Guarente 2005). The best characterized and studied among the sirtuins is SIRT1. It is a nuclear deacetylase that primarily but not exclusively deacetylates proteins involved in transcriptional regulation. SIRT1 can therefore influence a wide range of physiological aspects such as apoptosis/cell survival, autophagy, chromatin remodeling, gene transcription, senescence, endocrine signaling, and differentiation (Anastasiou and Krek 2006).

The epigenetic effects of SIRT1 can be appreciated in view of the ability of SIRT1 to deacetylate various transcription factors such as p53, FOXO (drosophila forkhead transcription factor), and RelA/p65 subunit of NF-κB. Some of the physiological phenomena regulated by these transcription factors in response to environmental and genotoxic challenge include stress resistance, apoptosis, and inflammation (T. Yang and Sauve 2006). While acetylation of FOXO3 leads to its inactivation, deacetylation by SIRT1 leads to its activation. Therefore, it can be surmised that SIRT1-mediated deacetylation of FOXO3 can induce cell cycle arrest, a phenomenon altered in cancer cells (Motta et al. 2004). Furthermore, increased transcription of GADD45 (DNA repair system) and MnSOD (reactive oxygen detoxification) is a direct physiological consequence of SIRT1 deacetylation of FOXO3, a modulator of stress-positive pathways (Kobayashi et al. 2005).

The proapoptotic role of p53 is governed by the acetylation status of its C-terminal regulatory domain (Vaziri et al. 2001). While increased acetylation of p53 leads to apoptosis, removal of the acetyl moieties by SIRT1 decreases the proapoptotic properties of p53. Hence SIRT1 may play an important role in cellular defense against stress-induced apoptosis (Vaziri et al. 2001). It has been reported that accelerated accumulation of acetylated p53 due to oxidative stress may lead to augmented cellular senescence (Furukawa et al. 2007). This phenomenon has been associated with a concomitant drop in SIRT1 activity and depleted levels of NAD^+ during oxidative/genotoxic stress (Furukawa et al. 2007), wherein NAD^+ is consumed by ROS due to activation of PARP-1. Further support to this phenomenon has been lent by findings of decreased SIRT1 levels in the nucleus in response to cigarette smoke exposure, both *in vivo* in mouse lung and *in vitro* in epithelial cells and macrophages (S. R. Yang et al. 2007). However, whether or not SIRT1-dependent deacetylation of FOXO3 plays a role in cigarette smoke-mediated apoptosis and senescence still remains obscure. It is also known that SIRT1 when localized in the cytoplasm induces apoptosis, while in the nucleus it acts as an antiapoptotic factor (Jin et al. 2007). Therefore, nucleus-cytoplasmic translocation of SIRT1 might be an important mechanism regulating cell death and differentiation (Vaziri et al. 2001).

A series of reports have now emphasized the role of SIRT1 in epigenetic regulation of gene expression in cancer cells. Hyperacetylation of H4-K16 and decreased trimethylation of H3-K9 and H4-K20 have been observed after down-regulation of SIRT1 by siRNA in mammalian cells (Vaquero et al. 2004). SIRT1 preferentially deacetylates H4-K16 *in vitro* (Vaquero et al. 2004). In addition, loss of H4-K16 acetylation and H4-K20 trimethylation has been a hallmark in various tumors and tumor-derived cell lines, suggesting that these modifications may be characteristic epigenetic markers of cancer (Fraga et al. 2005). Promoter regions of tumor suppressor genes, whose DNA are hypermethylated and are silenced in many types of cancers, are characteristic sites of localization of SIRT1 (Pruitt et al. 2006). Such silenced genes were up-regulated in breast and colon cancer cells by down-regulation of SIRT1 levels/activity via increased H4-K16 as well as H3-K9 acetylation in such promoters (Pruitt et al. 2006). Thus SIRT1-mediated epigenetic changes may play an important role in the modulation of various types of cancers. However, such modulation during lung inflammation/carcinogenesis is not known. Nevertheless, SIRT1-mediated epigenetic changes may play an important role in modulation of various types of cancers and in pathogenesis of chronic lung disorders and modulation of SIRT1 by polyphenols may serve as a chemopreventive agent (Figure 8.2).

8.3.2.1 Modulation of SIRT1 by Dietary Polyphenols

A wide variety of natural compounds have now been identified that can inhibit and/or activate sirtuins. Resveratrol (3,5,40-trihydroxystilbene—a flavonoid) is a

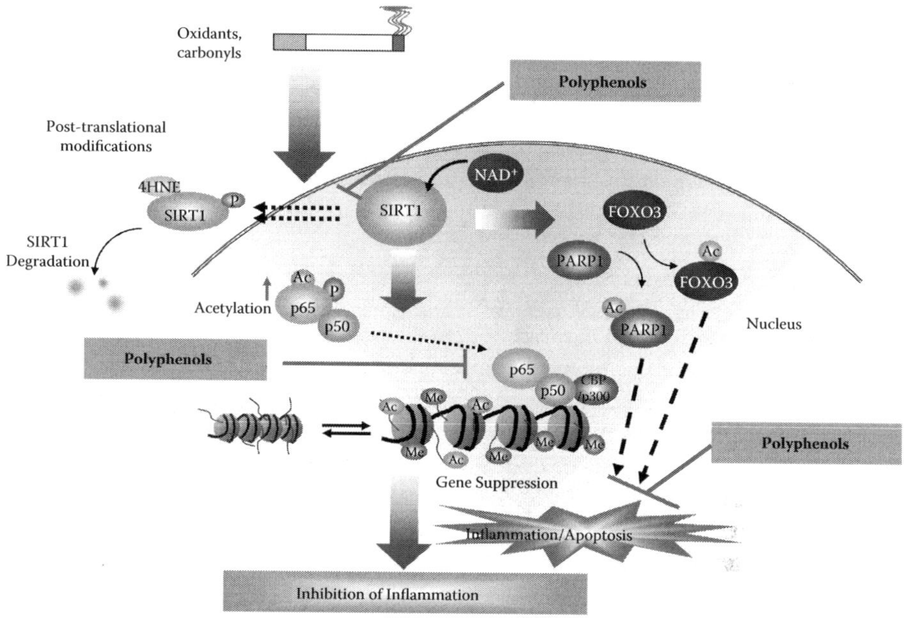

FIGURE 8.2 (Please see color insert following page 80.) SIRT1 and regulation of lung inflammation by polyphenols. Cigarette smoke oxidants increased abnormal human lung inflammation by degradation of SIRT1, a class III HDAC. SIRT1 is modified by phosphorylation, carbonylation, and aldehyde adducts formation, leading to export of modified SIRT1 in the cytosol and subsequent degradation. SIRT1 suppresses transcription factor NF-κB by deacetylation of RelA/p65 at Lys 310 residue. Acetylation of RelA/p65 by cigarette smoke induces the expression of proinflammatory mediators. Decreased SIRT1 leads to increased histone acetylation, which causes increased chromatin modifications and release of proinflammatory mediators. CBP/p300 functions as intrinsic histone acetyltransferases (HATs). Polyphenol, such as resveratrol, can restore the activity of SIRT1, leading to repression of proinflammatory response. SIRT1 regulates FOXO transcription factor acetylation/deacetylation, which is involved in DNA repair, cell cycle, oxidative stress, and apoptosis. Acetylation of FOXO is associated with inflammatory gene induction, stress response genes, and cellular apoptosis. Under the resting condition, inflammatory-related genes are controlled by SIRT1 deacetylase. However, when NF-κB and FOXO3 are acetylated by CBP/p300, transcription of pro-inflammatory and apoptotic/stress genes are up-regulated. Polyphenols inhibit the acetylation of FOXO3 via activation of SIRT1 by increasing NAD^+ or inhibiting PARP1. SIRT1: sirtuin1; P: phosphate; Ac: acetylation; 4HNE: 4-hydroxy-2-nonenal; CBP; CREB (cAMP response element binding protein)-binding protein; PARP: poly (ADP-ribose) polymerase; FOXO: drosophila forkhead transcription factor.

phytoalexin present in red wine (grapes) that activates SIRT1 and stalls p53 activation by significantly inhibiting p53 acetylation or by increased deacetylation of p53 (Howitz et al. 2003). Thus resveratrol can protect from p53-mediated cellular apoptosis. In addition, resveratrol can also impart protection against Bax induced by favoring SIRT1-induced formation of Ku70-Bax complex (Cohen et al. 2004). Resveratrol has also been reported to increase DNA repair capacity and stress resistance via FOXO1/3-dependent expression of GADD43 and $p27^{kip1}$ (Daitoku et al. 2004). Such an effect has also been reported for other sirtuin-activating compounds. Thus resveratrol can impart cellular protection via modulating multiple targets (Middleton et al. 2000; Csiszar et al. 2008).

Alternatively, cancer cells might be targeted using sirtuin inhibitors. These inhibitors induce cell damage by sensitizing the cells to p53-dependent apoptosis (Howitz et al. 2003). Pharmacological inhibition of SIRT1 decreases cellular resistance to stress and hence promotes cellular apoptosis due to reduced constraint on FOXO3/4 otherwise inhibited by SIRT1 (Brunet et al. 2004). SIRT1 is known to sensitize tumor cells to tumor necrosis factor (TNF)-α-induced cell death via inhibiting transactivation of NF-κB (Yeung et al. 2004). Thus it appears that SIRT1 inhibitors might yield cytoprotective effects by desensitizing the cells to TNF-α and therefore preventing cell death. Recently, it has been shown that SIRT1 levels are decreased in lungs of patients with COPD (Nakamaru et al. 2009; Rajendrasozhan, Yang, Kinnula et al. 2008). SIRT1 activators inhibit the release of NF-κB-mediated inflammatory mediators and possibly serve to overcome steroid resistance in response to oxidative stress (Nakamaru et al. 2009; Rajendrasozhan, Yang, Kinnula et al. 2008; S. R. Yang et al. 2007). Therefore the dietary modulators of sirtuins might act as novel immunomodulatory drugs via modulation of NF-κB. Activation of SIRT1 by SRT1720 and resveratrol inhibited the release of proinflammatory mediators in response to cigarette smoke exposure (Rajendrasozhan, Yang, Kinnula et al. 2008; Csiszar et al. 2008), suggesting that modulation of SIRT1 with activators or endogenous regulators (Milne and Denu 2008; Milne et al. 2007) would be an approach for the intervention of COPD. As reports emerge, it is becoming increasingly attractive to consider whether a combination of sirtuin inhibitors and DNA-damaging antitumor drugs might offer a novel strategy for effective chemotherapeutic cancer therapy.

8.3.3 Nrf2, Epigenetic Changes, and Dietary Polyphenols

Nrf2 is expressed in a wide range of cells/tissues, many of which are sites of expression for phase 2 detoxification genes. Nrf2, a member of the "cap 'n' collar" family of transcription factors, binds to the nuclear factor-erythroid derived 2 (NF-E2) binding sites (GCTGAGTCA), which consists of a subset of antioxidant response elements (ARE) having the sequence GCNNNGTCA. ARE regulates the expression of several phase 2 detoxification genes that are inducible by xenobiotics and antioxidants and thereby provide cytoprotection against chemically induced oxidative/electrophilic stress (Dinkova-Kostova et al. 2002). A recent study has shown that Nrf2 is acetylated by its interaction with CBP/p300, implicating the importance of deacetylases in regulation of Nrf2 (Sun et al. 2009). NF-κB/RelA antagonizes Nrf2 pathway by depriving CREB (cAMP response element binding protein)-binding protein (CBP) from Nrf2. The imbalance of Nrf-RelA, which occurs in patients with COPD (Malhotra et al. 2008; Rajendrasozhan, Yang, Kinnula et al. 2008; Csiszar et al. 2008), may lead to initiation of inflammatory response (G. Liu et al. 2008; Katoh et al. 2001).

Sulforaphane (present in broccoli and cruciferous vegetables), curcumin, resveratrol, caffeic acid phenethyl ester (CAPE), 4′-bromoflavone, and other polyphenols are potential chemopreventive agents and are first known to be selective activators of

Nrf2-Keap1-ARE (Dinkova-Kostova et al. 2002; J. S. Lee and Surh 2005; Prestera and Talalay 1995). Mechanistically, the interaction of Nrf2 with Keap1 enables Nrf2 to translocate into the nucleus, binds to the ARE, and initiates the transcription of detoxifying and cytoprotective genes. Such cellular responses are also triggered by other electrophilic compounds including polyphenols and plant-derived constituents. Curcumin, resveratrol, and CAPE have been identified to induce heme oxygenase-1 (HO-1) via Nrf2 and provide protection against various forms of stresses (Balogun et al. 2003). Similarly, curcumin induces HO-1 gene expression by promoting dissociation of the Nrf2-Keap1 complex and increased binding of Nrf2 to the HO-1 AREs.

Since sulforaphane, curcumin, and resveratrol are also known to activate Nrf2 expression (Balogun et al. 2003; J. S. Lee and Surh 2005; H. Yang et al. 2005; C. Y. Chen et al. 2005), it appears that the antioxidant function of curcumin/resveratrol may be mediated via Nrf2-ARE-GCLC axis, which may, therefore, increase the levels of GSH (a major cellular thiol antioxidant) by modulating the activity status of Nrf2. EGCG also induced transcriptional activation of phase II detoxifying enzymes through ARE/EpRE (Yu et al. 1997). It is evident from the preceding discussion that polyphenols (alone or in combination), in addition to their antioxidant function, can also modulate a wide range of signaling processes in different types of cells. Hence it is interesting to consider the therapeutic potential of dietary polyphenols in reversing epigenetic changes seen in patients with COPD and lung cancer.

8.4 BENEFICIAL EFFECTS OF DIETARY POLYPHENOLS IN IMPROVING LUNG FUNCTION

Several epidemiological studies have established a beneficial link between dietary polyphenol intake and reduced risk of disease, which were attributed to both the antioxidant and anti-inflammatory properties of polyphenols (Arts and Hollman 2005). The beneficial anti-inflammatory effect of polyphenols was demonstrated by two separate studies: (1) a Finnish study that involved over 10,000 participants, wherein a significant inverse correlation was observed between polyphenol intake and the incidence of asthma (Knekt et al. 2002; M. Lee et al. 2009), and (2) a study encompassing over 13,000 adults, wherein similar beneficial associations were also observed for COPD. This study reported that increased polyphenols, such as catechin (e.g., green tea polyphenols, epigallocatechin gallate), flavonol (e.g., quercetin and kaempferol), and flavone (such as apigenin and luteolin) intake correlated with improved symptoms, as assessed by cough, phlegm production, and breathlessness as well as improved lung function as measured by FEV_1 (Tabak et al. 2001). Two other studies appeared to corroborate these findings. One study showed a beneficial protective effect against COPD symptoms for increased fruit intake, high in polyphenol and vitamin E content (Walda et al. 2002). In the second study, a standardized polyphenol extract administered orally was shown to be effective in reducing oxidant stress and increasing arterial oxygen saturation (PaO_2), as well as improvements in FEV_1 between initial enrollment and the end of the study (Santus et al. 2005). More importantly, while single-component intake, such as catechin, was independently associated with FEV_1 and all three COPD symptoms, flavonol and flavone intake was independently associated with chronic cough only. The importance of this study was further substantiated by Walda and colleagues (2002), who showed the protective effect of fruit containing polyphenols and vitamin E intake against COPD symptoms in a 20-year COPD mortality study from three European countries consisting of Finnish, Italian, and Dutch cohorts. These important studies certainly encourage carrying out further multinational clinical studies to demonstrate the beneficial effects of a high intake of polyphenols against epigenetic modifications related to COPD symptoms/preventing the progression of COPD. Recent studies have shown that dietary polyphenol alone or in combinations can have beneficial chemopreventive effects in lung carcinogenesis in smokers (reviewed by Hecht et al. 2009).

While the aforementioned studies would appear to demonstrate an epidemiological link between polyphenol intake and the clinical benefit in asthma and COPD, other studies have demonstrated a direct impact of specific polyphenolic compounds on inflammation *in vitro* and *in vivo*. For example, the flavonoid resveratrol, a constituent of red wine, inhibits inflammatory cytokine release from macrophages isolated from COPD patients (Culpitt et al. 2003). Similarly, in another study resveratrol was shown to act as an anti-inflammatory in rat lungs challenged with LPS (Birrell et al. 2005). Furthermore, in both monocytic U937 cells and alveolar epithelial A549 cells, resveratrol inhibits NF-κB and AP-1 activation (Manna et al. 2000; Donnelly et al. 2004). Therefore, it appears that resveratrol can modulate a variety of proinflammatory pathways via inhibiting NF-κB activation (Leiro et al. 2005; Biesalski 2007). It has been recently shown that resveratrol, a red wine polyphenol, induces GSH synthesis via activation of Nrf2 in human lung epithelial cells (Kode et al. 2008). However, the clinical utility of resveratrol in terms of inhibiting inflammation and improving lung function by epigenetic modifications in smokers and in patients with COPD is not known. Curcumin, the active constituent of *Curcuma longa,* commonly known as turmeric, inhibits NF-κB activation and suppresses IL-8 release, COX-2 expression, and neutrophil recruitment in the lungs (Biswas et al. 2005). Curcumin inhibits cigarette smoke-induced NF-κB activation by inhibiting IκBα kinase in human lung epithelial cells (Biswas et al. 2005; Shishodia et al. 2003). Curcumin also has been shown to reverse the oxidative posttranslational modifications on HDAC2 and have an impact on epigenetic modifications on proinflammatory genes (Meja et al 2008). Therefore, inhibition of NF-κB and restoration of HDAC2 levels by curcumin may be considered to be a potential therapeutic strategy against chronic inflammatory diseases (Nanji et al. 2003; Rajendrasozhan, Yang, Edrisinghe et al. 2008). A recent study has shown that curcumin inhibits COPD-like airway inflammation and lung cancer progression in mice via suppression of k-ras and inflammation (Moghaddam et al. 2009). Green tea polyphenol EGCG has been shown to inhibit cigarette smoke extract-induced proinflammatory cytokine release in lung epithelial cells (Syed et al. 2007). EGCG has also been shown to modulate NF-κB/AP-1 activity in PMA-stimulated mouse epidermal JB6 cells via inactivation of AP-1 (Dong et al. 1997) and/or NF-κB (Nomura, Ma, Chen et al. 2000). Importantly, EGCG has been shown to reverse promoter methylation of Wnt inhibitory factor-1, which is a fundamental mechanism of epigenetic silencing in human cancers (Gao et al. 2009).

Despite many successful reports of the beneficial effects of polyphenols *in vitro*, it is important to note that many polyphenols are either very less bioavailable *in vivo* and/or are biotransformed in the gastrointestinal tract into compounds that may lose the bioactivity or may also be toxic. Future studies employing polyphenols should be designed keeping in view of the preceding observations and limitations.

8.5 ADVERSE EFFECTS OF POLYPHENOLS

There have been no serious adverse effects noticed with high dietary intake of polyphenols reported. This is possibly due to the relatively low bioavailability and rapid metabolism and elimination of most dietary (fruits and vegetables) flavonoids and polyphenols. Moreover, no long-term studies on the effect of polyphenols and flavonoids are available in humans; therefore it would be premature to consider that all the polyphenols and flavonoids (either taken alone or in combinations) do not have any adverse effects. Despite the several beneficial effects attributed to polyphenols/flavonoids, sporadic reports of ill effects of polyphenols and flavonoids are available. For example, people consuming up to 1000 mg/day of quercetin (present in onion) for 1 month reported nausea, headache, or tingling of the extremities (Dong et al. 1997). Some cancer patients administered intravenous quercetin in phase I clinical trial reported nausea, vomiting, sweating, flushing, and dyspnea (Nomura, Ma, Huang et al. 2000). In another trial on cancer patients, caffeinated green tea extracts (6 g/day in 3–6 divided doses) caused mild to moderate gastrointestinal discomfort, including nausea, vomiting, abdominal pain, and diarrhea (J. Y. Chung et al. 1999; G. Y. Yang et al. 2000). However, the side effects were associated with the caffeine in the green tea extract (Nomura, Ma, Huang et al. 2000), which were greatly alleviated in a trial using decaffeinated green tea extracts (800 mg/day of EGCG) (Nomura et al. 2001). Similarly, it has been reported that a higher intake of EGCG present in green tea is linked to the development of asthma in susceptible populations (Shirai et al. 2003), whereas another study has shown no beneficial effects of dietary intake of flavonoids in asthmatics (Garcia et al. 2005). This simply may be due to low bioavailability, interference of signaling pathways, and interaction with other mediators, oxidants, and aldehydes.

Polyphenols and flavonoids have also been reported to interfere with the drug metabolism ability in humans. For example, cytochrome P450 (CYP 3A4), an important intestinal drug detoxifying system, was found to be irreversibly inhibited by grapefruit juice (Pianetti et al. 2002). Furanocoumarins, particularly dihydroxybergamottin, were found to be the active components of grapefruit juice effecting such inhibition. However, certain flavonoids (naringenin and quercetin) have also been found to inhibit CYP3A4 *in vitro*. Inhibition of CYP3A4 can in turn increase the bioavailability and toxicity of a number of drugs, such as HIV protease inhibitors, immunosuppressants, HMG-CoA reductase inhibitors, calcium channel antagonists, antiarrhythmic agents, antihistamines, anticonvulsants, anxiolytics, and serotonin specific reuptake inhibitors for a long period of time (Yu et al. 1997). Therefore, subjects using any of the aforementioned therapies are advised against consuming grapefruit juice during the course of treatment (Pianetti et al. 2002). No other reports are available for dietary intake of higher doses of other polyphenols and flavonoids.

Toxic effects of drugs are reduced by their efflux and decreased absorption mediated by P-glycoproteins in the intestinal tract. Polyphenols such as quercetin, naringenin, and the green tea flavanol (EGCG) have been reported to inhibit the efflux activity of P-glycoprotein in cultured cells (C. Chen et al. 2000). Several drugs such as digoxin, antihypertensive, antiarrhythmic, chemotherapeutic, antifungal agents, HIV protease inhibitors, immunosuppressive agents, H_2 receptor antagonists, and some antibiotics are known to be substrates of P-glycoproteins (W. M. Dashwood et al. 2002). Hence, high intake of such flavonoids/polyphenols may adversely increase the bioavailability and the risk of toxicity of such drugs. Certain flavanols present in purple grapes and dark chocolate are reported to inhibit platelet aggregation *in vitro* (Caporali et al. 2004). Therefore high intakes of flavonoids could increase the risk of bleeding when taken with anticoagulant, antiplatelet, and nonsteroidal anti-inflammatory drugs (NSAIDs), but no data are available for combination of polyphenols/flavonoids with endogenous steroids. Flavonoids can also bind nonheme iron and inhibit intestinal absorption of iron from food (Lin and Lin 1997). Flavonoids have also been reported to inhibit the transport of vitamin C into cells, which may cause oxidation of vitamin C (Wheeler et al. 2004). More in-depth investigations are needed, however, to determine the significance of the preceding findings in humans, particularly in smokers or in susceptible population, and whether the beneficial effects outweigh the harmful effects of dietary polyphenols/flavonoids.

8.6 BIOAVAILABILITY AND EFFECTIVE DOSES OF DIETARY POLYPHENOLS

Pharmacologically, curcumin is reported to be safe, and human clinical trials indicated no toxicity even when administered at doses up to 10–15 g/day (Cheng et al. 2001; Aggarwal and Sung 2009). Curcumin has low oral toxicity in humans, but has low oral bioavailability (500–1000 nM after 8 g/day oral dose). However, consumption of curcumin (2 g) along with piperin (20 mg) (an active ingredient in peppers), a known inhibitor of intestinal and hepatic β-glucuronidation of curcumin, may improve the oral bioavailability to 2000-fold, but with a half-life of 0.25–1 hr (Aggarwal and Sung 2009; Shoba et al. 1998). Curcumin, after metabolism in the liver, is mainly excreted through bile. Curcumin may be distributed in only a limited number of organs/tissues and that, too, in small amounts (due to low bioavailability and absorption). Therefore it is important to investigate which tissues/organs are amenable to curcumin and the average half-life of curcumin in these tissues. Dosage studies with resveratrol in humans have shown that a person of 70 kg can safely consume a minimum of 14 mg resveratrol per day. It is also calculated that daily consumption of pure resveratrol and its analog piceatannol with a dose of 25–50 mg daily leads to nM concentration (which is beneficial) of circulatory levels of resveratrol. It is estimated that 20 glasses of red wine provide 25 mg of resveratrol, which of course differs from region to region where the grapes are grown. Since polyphenols are poorly absorbed and undergo extensive biotransformation, clinical studies have demonstrated that it is safe to consume EGCG or polyphenol E (a defined, decaffeinated green tea polyphenol mixture) in amounts equivalent to the EGCG content in 8–16 cups of green tea once a day, or in divided doses twice a day for 4 weeks. About three cups of green tea per day provide 240–320 mg of EGCG, which is effective in producing anti-inflammatory response *in vivo* (Feng 2006; Scholz and Williamson 2007).

8.7 CONCLUSIONS AND FUTURE PERSPECTIVES

Epigenetic changes are increasingly believed to modulate the initiation and progression of many diseases including cancer and respiratory disorders. Although it is important to identify the gene(s) undergoing epigenetic alterations, it is more important to understand the epigenome, which reflects the overall epigenetic state of a cell. Obviously, it is imperative to map the pattern of the entire epigenetic alteration of a cell in order to identify disease mechanism and therapeutic targets. Another area of importance will be to understand whether or not a common target is shared by HDACs and sirtuins, so that a common therapeutic agent may be designed. Recent reports highlight the pharmacological significance of HDAC and SIRT1-modulating drugs, and also suggest that identification of substrates, specifically targeted by HDAC and SIRT1, would have considerable therapeutic implications in chronic inflammatory diseases.

Polyphenols and flavonoids seem to be important metabolic modulators by virtue of their ability to modulate and influence several cellular processes such as signaling, proliferation, apoptosis, redox balance, and differentiation. Dietary polyphenols such as curcumin, resveratrol, and catechins have been reported to modulate epigenetic alterations in various experimental models. In addition, dietary polyphenols, such as curcumin and resveratrol, have been reported to overcome glucocorticoid resistance in COPD subjects. The anti-inflammatory property of curcumin, resveratrol, and catechins may be due to their ability to induce HDAC2 activity thereby restoring the efficacy of glucocorticoids or overcome its resistance. Thus these polyphenolic compounds have a therapeutic value as adjuvant therapy with steroids against chronic inflammatory epigenetically regulated diseases.

Although polyphenols are abundant in most dietary sources such as fruits, vegetables, tea, and wine, more detailed studies are still required to determine their true absorption and bioavailability. It is important to note that most of the beneficial effects have been obtained from *in vitro* studies. In addition, most human exposure studies using polyphenols have been on a short-term basis; therefore more studies are required on an extended basis in order to determine the long-term effects of these diverse compounds. Given the fact that polyphenols undergo a considerable degree of chemical modifications during digestion and absorption, and that the modified forms may have altered biological properties and potencies, it is extremely important to practice caution before claiming any definite pharmacological and epigenetic applications for these compounds. Moreover, despite their beneficial health effects, polyphenols have also been shown to have adverse effects. In view of their anti-inflammatory and antioxidant abilities and their capacity to modulate important inflammatory and anti-inflammatory signaling pathways, epigenetic modifications, glucocorticoid efficacy, polyphenols and flavonoids hold great promise as potential therapeutic strategies for controlling epigenetic modifications seen in chronic lung inflammation and airway diseases. Thus, regulation of inflammatory response by dietary polyphenols and restoration of glucocorticoid efficacy at the molecular level are possible ways forward to design therapeutic strategies in the treatment of chronic inflammatory diseases. However, future studies on bioavailability, absorption, and tissue distribution, and understanding of the *in vivo* molecular effects of curcumin, resveratrol, quercetins, and cathechins are needed in order to consider these dietary polyphenols as natural therapy "nutraceuticals" for chronic inflammatory disorders.

ACKNOWLEDGMENTS

Supported by the NIH-NHLBI Grant RO1-HL-085613 and NIEHS Environmental Health Sciences Center Grant ES-01247. I thank Dr. Saibal Biswas for useful discussions and proofreading, and Dr. S. W. Chung for organizing the manuscript.

REFERENCES

Adcock, I. M., L. Tsaprouni, P. Bhavsar, and K. Ito. 2007. Epigenetic regulation of airway inflammation. *Curr Opin Immunol* 19:694–700.

Adenuga, D., H. Yao, T. H. March, J. Seagrave, and I. Rahman. 2009. Histone deacetylase 2 is phosphorylated, ubiquitinated, and degraded by cigarette smoke. *Am J Respir Cell Mol Biol* 40:464–73.

Aggarwal, B. B., and S. Shishodia. 2004. Suppression of the nuclear factor-kappaB activation pathway by spice-derived phytochemicals: reasoning for seasoning. *Ann N Y Acad Sci* 1030:434–41.

Aggarwal, B. B., and B. Sung. 2009. Pharmacological basis for the role of curcumin in chronic diseases: an age-old spice with modern targets. *Trends Pharmacol Sci* 30:85–94.

Ahmad, N., S. Gupta, and H. Mukhtar. 2000. Green tea polyphenol epigallocatechin-3-gallate differentially modulates nuclear factor kappaB in cancer cells versus normal cells. *Arch Biochem Biophys* 376:338–46.

Alati, R., A. Al Mamun, M. O'Callaghan, J. M. Najman, and G. M. Williams. 2006. In utero and postnatal maternal smoking and asthma in adolescence. *Epidemiology* 17:138–44.

Anastasiou, D., and W. Krek. 2006. SIRT1: linking adaptive cellular responses to aging-associated changes in organismal physiology. *Physiology (Bethesda)* 21:404–10.

Aoshiba, K., M. Koinuma, N. Yokohori, and A. Nagai. 2003. Immunohistochemical evaluation of oxidative stress in murine lungs after cigarette smoke exposure. *Inhal Toxicol* 15:1029–38.

Arts, I. C., and P. C. Hollman. 2005. Polyphenols and disease risk in epidemiologic studies. *Am J Clin Nutr* 81:317S–325S.

Balogun, E., M. Hoque, P. Gong, E. Killeen, C. J. Green, R. Foresti, J. Alam, and R. Motterlini. 2003. Curcumin activates the haem oxygenase-1 gene via regulation of Nrf2 and the antioxidant-responsive element. *Biochem J* 371:887–95.

Barnes, P. J. 2008. The cytokine network in asthma and chronic obstructive pulmonary disease. *J Clin Invest* 118:3546–56.

Barnes, P. J. 2009. Role of HDAC2 in the pathophysiology of COPD. *Annu Rev Physiol* 71:451–64.

Barnes, P. J., I. M. Adcock, and K. Ito. 2005. Histone acetylation and deacetylation: importance in inflammatory lung diseases. *Eur Respir J* 25:552–63.

Bernstein, B. E., A. Meissner, and E. S. Lander. 2007. The mammalian epigenome. *Cell* 128:669–81.

Biesalski, H. K. 2007. Polyphenols and inflammation: basic interactions. *Curr Opin Clin Nutr Metab Care* 10:724–28.

Birrell, M. A., K. McCluskie, S. Wong, L. E. Donnelly, P. J. Barnes, and M. G. Belvisi. 2005. Resveratrol, an extract of red wine, inhibits lipopolysaccharide induced airway neutrophilia and inflammatory mediators through an NF-kappaB-independent mechanism. *FASEB J* 19:840–41.

Biswas, S. K., D. McClure, L. A. Jimenez, I. L. Megson, and I. Rahman. 2005. Curcumin induces glutathione biosynthesis and inhibits NF-kappaB activation and interleukin-8 release in alveolar epithelial cells: mechanism of free radical scavenging activity. *Antioxid Redox Signal* 7:32–41.

Bordone, L., and L. Guarente. 2005. Calorie restriction, SIRT1 and metabolism: understanding longevity. *Nat Rev Mol Cell Biol* 6:298–305.

Bousquet, J., W. Jacot, H. Yssel, A. M. Vignola, and M. Humbert. 2004. Epigenetic inheritance of fetal genes in allergic asthma. *Allergy* 59:138–47.

Brock, M. V., C. M. Hooker, E. Ota-Machida, Y. Han, M. Guo, S. Ames, S. Glockner, S. Piantadosi, E. Gabrielson, G. Pridham, K. Pelosky, S. A. Belinsky, S. C. Yang, S. B. Baylin, and J. G. Herman. 2008. DNA methylation markers and early recurrence in stage I lung cancer. *N Engl J Med* 358:1118–28.

Brunet, A., L. B. Sweeney, J. F. Sturgill, K. F. Chua, P. L. Greer, Y. Lin, H. Tran, S. E. Ross, R. Mostoslavsky, H. Y. Cohen, L. S. Hu, H. L. Cheng, M. P. Jedrychowski, S. P. Gygi, D. A. Sinclair, F. W. Alt, and M. E. Greenberg. 2004. Stress-dependent regulation of FOXO transcription factors by the SIRT1 deacetylase. *Science* 303:2011–15.

Caporali, A., P. Davalli, S. Astancolle, D. D'Arca, M. Brausi, S. Bettuzzi, and A. Corti. 2004. The chemopreventive action of catechins in the TRAMP mouse model of prostate carcinogenesis is accompanied by clusterin over-expression. *Carcinogenesis* 25:2217–24.

Caramori, G., M. Romagnoli, P. Casolari, C. Bellettato, G. Casoni, P. Boschetto, K. F. Chung, P. J. Barnes, I. M. Adcock, A. Ciaccia, L. M. Fabbri, and A. Papi. 2003. Nuclear localisation of p65 in sputum macrophages but not in sputum neutrophils during COPD exacerbations. *Thorax* 58:348–51.

Castro, M., M. I. Ramirez, J. E. Gern, G. Cutting, G. Redding, J. S. Hagood, J. Whitsett, S. Abman, J. U. Raj, R. Barst, G. J. Kato, D. Gozal, G. G. Haddad, N. R. Prabhakar, E. Gauda, F. D. Martinez, R. Tepper, R. E. Wood, F. Accurso, W. G. Teague, J. Venegas, F. S. Cole, and R. J. Wright. 2009. Strategic plan for pediatric respiratory diseases research: an NHLBI working group report. *Proc Am Thorac Soc* 6:1–10.

Chen, C., R. Yu, E. D. Owuor, and A. N. Kong. 2000. Activation of antioxidant-response element (ARE), mitogen-activated protein kinases (MAPKs) and caspases by major green tea polyphenol components during cell survival and death. *Arch Pharm Res* 23:605–12.

Chen, C. Y., J. H. Jang, M. H. Li, and Y. J. Surh. 2005. Resveratrol upregulates heme oxygenase-1 expression via activation of NF-E2-related factor 2 in PC12 cells. *Biochem Biophys Res Commun* 331:993–1000.

Cheng, A. L., C. H. Hsu, J. K. Lin, M. M. Hsu, Y. F. Ho, T. S. Shen, J. Y. Ko, J. T. Lin, B. R. Lin, W. Ming-Shiang, H. S. Yu, S. H. Jee, G. S. Chen, T. M. Chen, C. A. Chen, M. K. Lai, Y. S. Pu, M. H. Pan, Y. J. Wang, C. C. Tsai, and C. Y. Hsieh. 2001. Phase I clinical trial of curcumin, a chemopreventive agent, in patients with high-risk or pre-malignant lesions. *Anticancer Res* 21:2895–2900.

Chung, J. Y., C. Huang, X. Meng, Z. Dong, and C. S. Yang. 1999. Inhibition of activator protein 1 activity and cell growth by purified green tea and black tea polyphenols in H-ras-transformed cells: structure-activity relationship and mechanisms involved. *Cancer Res* 59:4610–17.

Chung, K. F. 2001. Cytokines in chronic obstructive pulmonary disease. *Eur Respir J Suppl* 34:50s–59s.

Chung, K. F., and I. M. Adcock. 2008. Multifaceted mechanisms in COPD: inflammation, immunity, and tissue repair and destruction. *Eur Respir J* 31:1334–56.

Cohen, H. Y., C. Miller, K. J. Bitterman, N. R. Wall, B. Hekking, B. Kessler, K. T. Howitz, M. Gorospe, R. de Cabo, and D. A. Sinclair. 2004. Calorie restriction promotes mammalian cell survival by inducing the SIRT1 deacetylase. *Science* 305:390–92.

Cortese, R., O. Hartmann, K. Berlin, and F. Eckhardt. 2008. Correlative gene expression and DNA methylation profiling in lung development nominate new biomarkers in lung cancer. *Int J Biochem Cell Biol* 40:1494–508.

Cosio, B. G., L. Tsaprouni, K. Ito, E. Jazrawi, I. M. Adcock, and P. J. Barnes. 2004. Theophylline restores histone deacetylase activity and steroid responses in COPD macrophages. *J Exp Med* 200:689–95.

Coward, W. R., K. Watts, C. A. Feghali-Bostwick, A. Knox, and L. Pang. 2009. Defective histone acetylation is responsible for the diminished expression of cyclooxygenase 2 in idiopathic pulmonary fibrosis. *Mol Cell Biol* 29:4325–39.

Csiszar, A., N. Labinskyy, A. Podlutsky, P. M. Kaminski, M. S. Wolin, C. Zhang, P. Mukhopadhyay, P. Pacher, F. Hu, R. de Cabo, P. Ballabh, and Z. Ungvari. 2008. Vasoprotective effects of resveratrol and SIRT1: attenuation of cigarette smoke-induced oxidative stress and proinflammatory phenotypic alterations. *Am J Physiol Heart Circ Physiol* 294:H2721–35.

Culpitt, S. V., D. F. Rogers, P. S. Fenwick, P. Shah, C. De Matos, R. E. Russell, P. J. Barnes, and L. E. Donnelly. 2003. Inhibition by red wine extract, resveratrol, of cytokine release by alveolar macrophages in COPD. *Thorax* 58:942–46.

Culpitt, S. V., D. F. Rogers, P. Shah, C. De Matos, R. E. Russell, L. E. Donnelly, and P. J. Barnes. 2003. Impaired inhibition by dexamethasone of cytokine release by alveolar macrophages from patients with chronic obstructive pulmonary disease. *Am J Respir Crit Care Med* 167:24–31.

Daitoku, H., M. Hatta, H. Matsuzaki, S. Aratani, T. Ohshima, M. Miyagishi, T. Nakajima, and A. Fukamizu. 2004. Silent information regulator 2 potentiates Foxo1-mediated transcription through its deacetylase activity. *Proc Natl Acad Sci U S A* 101:10042–47.

Dashwood, R. H., M. C. Myzak, and E. Ho. 2006. Dietary HDAC inhibitors: time to rethink weak ligands in cancer chemoprevention? *Carcinogenesis* 27:344–49.

Dashwood, W. M., G. A. Orner, and R. H. Dashwood. 2002. Inhibition of beta-catenin/Tcf activity by white tea, green tea, and epigallocatechin-3-gallate (EGCG): minor contribution of H(2)O(2) at physiologically relevant EGCG concentrations. *Biochem Biophys Res Commun* 296:584–88.

Devereux, G., S. W. Turner, L. C. Craig, G. McNeill, S. Martindale, P. J. Harbour, P. J. Helms, and A. Seaton. 2006. Low maternal vitamin E intake during pregnancy is associated with asthma in 5-year-old children. *Am J Respir Crit Care Med* 174:499–507.

Digel, W., and M. Lubbert. 2005. DNA methylation disturbances as novel therapeutic target in lung cancer: preclinical and clinical results. *Crit Rev Oncol Hematol* 55:1–11.

Dinkova-Kostova, A. T., W. D. Holtzclaw, R. N. Cole, K. Itoh, N. Wakabayashi, Y. Katoh, M. Yamamoto, and P. Talalay. 2002. Direct evidence that sulfhydryl groups of Keap1 are the sensors regulating induction of phase 2 enzymes that protect against carcinogens and oxidants. *Proc Natl Acad Sci U S A* 99:11908–13.

Di Paola, R., E. Mazzon, C. Muia, T. Genovese, M. Menegazzi, R. Zaffini, H. Suzuki, and S. Cuzzocrea. 2005. Green tea polyphenol extract attenuates lung injury in experimental model of carrageenan-induced pleurisy in mice. *Respir Res* 6:66.

Di Stefano, A., G. Caramori, T. Oates, A. Capelli, M. Lusuardi, I. Gnemmi, F. Ioli, K. F. Chung, C. F. Donner, P. J. Barnes, and I. M. Adcock. 2002. Increased expression of nuclear factor-kappaB in bronchial biopsies from smokers and patients with COPD. *Eur Respir J* 20:556–63.

Doherty, S. P., J. Grabowski, C. Hoffman, S. P. Ng, and J. T. Zelikoff. 2009. Early life insult from cigarette smoke may be predictive of chronic diseases later in life. *Biomarkers* 14 (Suppl 1):97–101.

Dong, Z., W. Ma, C. Huang, and C. S. Yang. 1997. Inhibition of tumor promoter-induced activator protein 1 activation and cell transformation by tea polyphenols, (-)-epigallocatechin gallate, and theaflavins. *Cancer Res* 57:4414–19.

Donnelly, L. E., R. Newton, G. E. Kennedy, P. S. Fenwick, R. H. Leung, K. Ito, R. E. Russell, and P. J. Barnes. 2004. Anti-inflammatory effects of resveratrol in lung epithelial cells: molecular mechanisms. *Am J Physiol Lung Cell Mol Physiol* 287:L774–83.

Druesne, N., A. Pagniez, C. Mayeur, M. Thomas, C. Cherbuy, P. H. Duee, P. Martel, and C. Chaumontet. 2004. Diallyl disulfide (DADS) increases histone acetylation and p21(waf1/cip1) expression in human colon tumor cell lines. *Carcinogenesis* 25:1227–36.

Elliot, J. G., N. G. Carroll, A. L. James, and P. J. Robinson. 2003. Airway alveolar attachment points and exposure to cigarette smoke in utero. *Am J Respir Crit Care Med* 167:45–49.

Fassina, G., R. Vene, M. Morini, S. Minghelli, R. Benelli, D. M. Noonan, and A. Albini. 2004. Mechanisms of inhibition of tumor angiogenesis and vascular tumor growth by epigallocatechin-3-gallate. *Clin Cancer Res* 10:4865–73.

Feng, W. Y. 2006. Metabolism of green tea catechins: an overview. *Curr Drug Metab* 7:755–809.

Finch, C. E., and E. M. Crimmins. 2004. Inflammatory exposure and historical changes in human life-spans. *Science* 305:1736–39.

Fitzsimon, N., U. Fallon, D. O'Mahony, B. G. Loftus, G. Bury, A. W. Murphy, and C. C. Kelleher. 2007. Mothers' dietary patterns during pregnancy and risk of asthma symptoms in children at 3 years. *Ir Med J* 100:suppl 27–32.

Fraga, M. F., E. Ballestar, A. Villar-Garea, M. Boix-Chornet, J. Espada, G. Schotta, T. Bonaldi, C. Haydon, S. Ropero, K. Petrie, N. G. Iyer, A. Perez-Rosado, E. Calvo, J. A. Lopez, A. Cano, M. J. Calasanz, D. Colomer, M. A. Piris, N. Ahn, A. Imhof, C. Caldas, T. Jenuwein, and M. Esteller. 2005. Loss of acetylation at Lys16 and trimethylation at Lys20 of histone H4 is a common hallmark of human cancer. *Nat Genet* 37:391–400.

Furukawa, A., S. Tada-Oikawa, S. Kawanishi, and S. Oikawa. 2007. H2O2 accelerates cellular senescence by accumulation of acetylated p53 via decrease in the function of SIRT1 by NAD+ depletion. *Cell Physiol Biochem* 20:45–54.

Gao, Z., Z. Xu, M. S. Hung, Y. C. Lin, T. Wang, M. Gong, X. Zhi, D. M. Jablon, and L. You. 2009. Promoter demethylation of WIF-1 by epigallocatechin-3-gallate in lung cancer cells. *Anticancer Res* 29:202–30.

Garcia, V., I. C. Arts, J. A. Sterne, R. L. Thompson, and S. O. Shaheen. 2005. Dietary intake of flavonoids and asthma in adults. *Eur Respir J* 26:449–52.

Georgiou, E., R. Valeri, G. Tzimagiorgis, J. Anzel, D. Krikelis, C. Tsilikas, G. Sarikos, C. Destouni, A. Dimitriadou, and S. Kouidou. 2007. Aberrant p16 promoter methylation among Greek lung cancer patients and smokers: correlation with smoking. *Eur J Cancer Prev* 16:396–402.

Gicquel, C., A. El-Osta, and Y. Le Bouc. 2008. Epigenetic regulation and fetal programming. *Best Pract Res Clin Endocrinol Metab* 22:1–16.

Guyonnet, D., R. Berges, M. H. Siess, M. F. Pinnert, M. C. Chagnon, M. Suschetet, and A. M. Le Bon. 2004. Post-initiation modulating effects of allyl sulfides in rat hepatocarcinogenesis. *Food Chem Toxicol* 42:1479–85.

Hageman, G. J., I. Larik, H. J. Pennings, G. R. Haenen, E. F. Wouters, and A. Bast. 2003. Systemic poly(ADP-ribose) polymerase-1 activation, chronic inflammation, and oxidative stress in COPD patients. *Free Radic Biol Med* 35:140–48.

Hecht, S. S., F. Kassie, and D. K. Hatsukami. 2009. Chemoprevention of lung carcinogenesis in addicted smokers and ex-smokers. *Nat Rev Cancer* 9:476–88.

Hollingsworth, J. W., S. Maruoka, K. Boon, S. Garantziotis, Z. Li, J. Tomfohr, N. Bailey, E. N. Potts, G. Whitehead, D. M. Brass, and D. A. Schwartz. 2008. In utero supplementation with methyl donors enhances allergic airway disease in mice. *J Clin Invest* 118:3462–69.

Hoppin, J. A., D. M. Umbach, S. J. London, M. C. Alavanja, and D. P. Sandler. 2004. Diesel exhaust, solvents, and other occupational exposures as risk factors for wheeze among farmers. *Am J Respir Crit Care Med* 169:1308–13.

Howitz, K. T., K. J. Bitterman, H. Y. Cohen, D. W. Lamming, S. Lavu, J. G. Wood, R. E. Zipkin, P. Chung, A. Kisielewski, L. L. Zhang, B. Scherer, and D. A. Sinclair. 2003. Small molecule activators of sirtuins extend *Saccharomyces cerevisiae* lifespan. *Nature* 425:191–96.

Hsu, H. S., T. P. Chen, C. H. Hung, C. K. Wen, R. K. Lin, H. C. Lee, and Y. C. Wang. 2007. Characterization of a multiple epigenetic marker panel for lung cancer detection and risk assessment in plasma. *Cancer* 110:2019–26.

Imai, S., C. M. Armstrong, M. Kaeberlein, and L. Guarente. 2000. Transcriptional silencing and longevity protein Sir2 is an NAD-dependent histone deacetylase. *Nature* 403:795–800.

Islam, K. N., and C. R. Mendelson. 2008. Glucocorticoid/glucocorticoid receptor inhibition of surfactant protein-A (SP-A) gene expression in lung type II cells is mediated by repressive changes in histone modification at the SP-A promoter. *Mol Endocrinol* 22:585–96.

Ito, K., G. Caramori, S. Lim, T. Oates, K. F. Chung, P. J. Barnes, and I. M. Adcock. 2002. Expression and activity of histone deacetylases in human asthmatic airways. *Am J Respir Crit Care Med* 166:392–96.

Ito, K., M. Ito, W. M. Elliott, B. Cosio, G. Caramori, O. M. Kon, A. Barczyk, S. Hayashi, I. M. Adcock, J. C. Hogg, and P. J. Barnes. 2005. Decreased histone deacetylase activity in chronic obstructive pulmonary disease. *N Engl J Med* 352:1967–76.

Ito, K., S. Lim, G. Caramori, K. F. Chung, P. J. Barnes, and I. M. Adcock. 2001. Cigarette smoking reduces histone deacetylase 2 expression, enhances cytokine expression, and inhibits glucocorticoid actions in alveolar macrophages. *FASEB J* 15:1110–12.

Ito, K., S. Yamamura, S. Essilfie-Quaye, B. Cosio, M. Ito, P. J. Barnes, and I. M. Adcock. 2006. Histone deacetylase 2-mediated deacetylation of the glucocorticoid receptor enables NF-kappaB suppression. *J Exp Med* 203:7–13.

Jaenisch, R., and A. Bird. 2003. Epigenetic regulation of gene expression: how the genome integrates intrinsic and environmental signals. *Nat Genet* 33 (Suppl):245–54.

Jedrychowski, W., A. Galas, R. Whyatt, and F. Perera. 2006. The prenatal use of antibiotics and the development of allergic disease in one year old infants. A preliminary study. *Int J Occup Med Environ Health* 19:70–76.

Jeffery, P. K. 2001. Remodeling in asthma and chronic obstructive lung disease. *Am J Respir Crit Care Med* 164:S28–38.

Jiang, Y. H., J. Bressler, and A. L. Beaudet. 2004. Epigenetics and human disease. *Annu Rev Genomics Hum Genet* 5:479–510.

Jin, Q., T. Yan, X. Ge, C. Sun, X. Shi, and Q. Zhai. 2007. Cytoplasm-localized SIRT1 enhances apoptosis. *J Cell Physiol* 213:88–97.

Jones, B., and J. Chen. 2006. Inhibition of IFN-gamma transcription by site-specific methylation during T helper cell development. *EMBO J* 25:2443–52.

Kang, J., J. Chen, Y. Shi, J. Jia, and Y. Zhang. 2005. Curcumin-induced histone hypoacetylation: the role of reactive oxygen species. *Biochem Pharmacol* 69:1205–13.

Katoh, Y., K. Itoh, E. Yoshida, M. Miyagishi, A. Fukamizu, and M. Yamamoto. 2001. Two domains of Nrf2 cooperatively bind CBP, a CREB binding protein, and synergistically activate transcription. *Genes Cells* 6:857–68.

Keatings, V. M., P. D. Collins, D. M. Scott, and P. J. Barnes. 1996. Differences in interleukin-8 and tumor necrosis factor-alpha in induced sputum from patients with chronic obstructive pulmonary disease or asthma. *Am J Respir Crit Care Med* 153:530–34.

Knekt, P., J. Kumpulainen, R. Jarvinen, H. Rissanen, M. Heliovaara, A. Reunanen, T. Hakulinen, and A. Aromaa. 2002. Flavonoid intake and risk of chronic diseases. *Am J Clin Nutr* 76:560–68.

Kobayashi, Y., Y. Furukawa-Hibi, C. Chen, Y. Horio, K. Isobe, K. Ikeda, and N. Motoyama. 2005. SIRT1 is critical regulator of FOXO-mediated transcription in response to oxidative stress. *Int J Mol Med* 16:237–43.

Kode, A., S. Rajendrasozhan, S. Caito, S. R. Yang, I. L. Megson, and I. Rahman. 2008. Resveratrol induces glutathione synthesis by activation of Nrf2 and protects against cigarette smoke-mediated oxidative stress in human lung epithelial cells. *Am J Physiol Lung Cell Mol Physiol* 294:L478–88.

Kurukulaaratchy, R. J., S. Matthews, and S. H. Arshad. 2004. Does environment mediate earlier onset of the persistent childhood asthma phenotype? *Pediatrics* 113:345–50.

Lawless, M. W., K. J. O'Byrne, and S. G. Gray. 2009. Oxidative stress induced lung cancer and COPD: opportunities for epigenetic therapy. *J Cell Mol Med* 9A:2800–21.

Lee, J. S., and Y. J. Surh. 2005. Nrf2 as a novel molecular target for chemoprevention. *Cancer Lett* 224:171–84.

Lee, M., S. Kim, O. K. Kwon, S. R. Oh, H. K. Lee, and K. Ahn. 2009. Anti-inflammatory and anti-asthmatic effects of resveratrol, a polyphenolic stilbene, in a mouse model of allergic asthma. *Int Immunopharmacol* 9:418–24.

Leiro, J., J. A. Arranz, N. Fraiz, M. L. Sanmartin, E. Quezada, and F. Orallo. 2005. Effect of cis-resveratrol on genes involved in nuclear factor kappa B signaling. *Int Immunopharmacol* 5:393–406.

Li, Y. F., B. Langholz, M. T. Salam, and F. D. Gilliland. 2005. Maternal and grandmaternal smoking patterns are associated with early childhood asthma. *Chest* 127:1232–41.

Lim, R. H., and L. Kobzik. 2009. Maternal transmission of asthma risk. *Am J Reprod Immunol* 61:1–10.

Lin, Y. L., and J. K. Lin. 1997. (-)-Epigallocatechin-3-gallate blocks the induction of nitric oxide synthase by down-regulating lipopolysaccharide-induced activity of transcription factor nuclear factor-kappaB. *Mol Pharmacol* 52:465–72.

Litonjua, A. A., and S. T. Weiss. 2007. Is vitamin D deficiency to blame for the asthma epidemic? *J Allergy Clin Immunol* 120:1031–35.

Liu, G. H., J. Qu, and X. Shen. 2008. NF-kappaB/p65 antagonizes Nrf2-ARE pathway by depriving CBP from Nrf2 and facilitating recruitment of HDAC3 to MafK. *Biochim Biophys Acta* 1783:713–27.

Liu, J., M. Ballaney, U. Al-alem, C. Quan, X. Jin, F. Perera, L. C. Chen, and R. L. Miller. 2008. Combined inhaled diesel exhaust particles and allergen exposure alter methylation of T helper genes and IgE production in vivo. *Toxicol Sci* 102:76–81.

Malhotra, D., R. Thimmulappa, A. Navas-Acien, A. Sandford, M. Elliott, A. Singh, L. Chen, X. Zhuang, J. Hogg, P. Pare, R. M. Tuder, and S. Biswal. 2008. Decline in NRF2-regulated antioxidants in chronic obstructive pulmonary disease lungs due to loss of its positive regulator, DJ-1. *Am J Respir Crit Care Med* 178:592–604.

Mandhane, P. J., J. M. Greene, J. O. Cowan, D. R. Taylor, and M. R. Sears. 2005. Sex differences in factors associated with childhood- and adolescent-onset wheeze. *Am J Respir Crit Care Med* 172:45–54.

Manna, S. K., A. Mukhopadhyay, and B. B. Aggarwal. 2000. Resveratrol suppresses TNF-induced activation of nuclear transcription factors NF-kappa B, activator protein-1, and apoptosis: potential role of reactive oxygen intermediates and lipid peroxidation. *J Immunol* 164:6509–19.

Meissner, A., T. S. Mikkelsen, H. Gu, M. Wernig, J. Hanna, A. Sivachenko, X. Zhang, B. E. Bernstein, C. Nusbaum, D. B. Jaffe, A. Gnirke, R. Jaenisch, and E. S. Lander. 2008. Genome-scale DNA methylation maps of pluripotent and differentiated cells. *Nature* 454:766–70.

Meja, K. K., S. Rajendrasozhan, D. Adenuga, S. K. Biswas, I. K. Sundar, G. Spooner, J. A. Marwick, P. Chakravarty, D. Fletcher, P. Whittaker, I. L. Megson, P. A. Kirkham, and I. Rahman. 2008. Curcumin restores corticosteroid function in monocytes exposed to oxidants by maintaining HDAC2. *Am J Respir Cell Mol Biol* 39:312–23.

Mezzetti, A., D. Lapenna, S. D. Pierdomenico, A. M. Calafiore, F. Costantini, G. Riario-Sforza, T. Imbastaro, M. Neri, and F. Cuccurullo. 1995. Vitamins E, C and lipid peroxidation in plasma and arterial tissue of smokers and non-smokers. *Atherosclerosis* 112:91–99.

Middleton, E., Jr., C. Kandaswami, and T. C. Theoharides. 2000. The effects of plant flavonoids on mammalian cells: implications for inflammation, heart disease, and cancer. *Pharmacol Rev* 52:673–751.

Miller, R. L., and S. M. Ho. 2008. Environmental epigenetics and asthma: current concepts and call for studies. *Am J Respir Crit Care Med* 177:567–73.

Millien, G., J. Beane, M. Lenburg, P. N. Tsao, J. Lu, A. Spira, and M. I. Ramirez. 2008. Characterization of the mid-foregut transcriptome identifies genes regulated during lung bud induction. *Gene Expr Patterns* 8:124–39.

Milne, J. C., and J. M. Denu. 2008. The Sirtuin family: therapeutic targets to treat diseases of aging. *Curr Opin Chem Biol* 12:11–17.

Milne, J. C., P. D. Lambert, S. Schenk, D. P. Carney, J. J. Smith, D. J. Gagne, L. Jin, O. Boss, R. B. Perni, C. B. Vu, J. E. Bemis, R. Xie, J. S. Disch, P. Y. Ng, J. J. Nunes, A. V. Lynch, H. Yang, H. Galonek, K. Israelian, W. Choy, A. Iffland, S. Lavu, O. Medvedik, D. A. Sinclair, J. M. Olefsky, M. R. Jirousek, P. J. Elliott, and C. H. Westphal. 2007. Small molecule activators of SIRT1 as therapeutics for the treatment of type 2 diabetes. *Nature* 450:712–16.

Mochida-Nishimura, K., K. Surewicz, J. V. Cross, R. Hejal, D. Templeton, E. A. Rich, and Z. Toossi. 2001. Differential activation of MAP kinase signaling pathways and nuclear factor-kappaB in bronchoalveolar cells of smokers and nonsmokers. *Mol Med* 7:177–85.

Moghaddam, S. J., P. Barta, S. G. Mirabolfathinejad, Z. Ammar-Aouchiche, N. Torres Garza, T. T. Vo, R. A. Newman, B. B. Aggarwal, C. M. Evans, M. J. Tuvim, R. Lotan, and B. F. Dickey. 2009. Curcumin inhibits COPD-like airway inflammation and lung cancer progression in mice. *Carcinogenesis* 11:1949–56.

Moonen, H. J., L. Geraets, A. Vaarhorst, A. Bast, E. F. Wouters, and G. J. Hageman. 2005. Theophylline prevents NAD+ depletion via PARP-1 inhibition in human pulmonary epithelial cells. *Biochem Biophys Res Commun* 338:1805–10.

Motta, M. C., N. Divecha, M. Lemieux, C. Kamel, D. Chen, W. Gu, Y. Bultsma, M. McBurney, and L. Guarente. 2004. Mammalian SIRT1 represses forkhead transcription factors. *Cell* 116:551–63.

Myzak, M. C., and R. H. Dashwood. 2006. Histone deacetylases as targets for dietary cancer preventive agents: lessons learned with butyrate, diallyl disulfide, and sulforaphane. *Curr Drug Targets* 7:443–452.

Myzak, M. C., P. A. Karplus, F. L. Chung, and R. H. Dashwood. 2004. A novel mechanism of chemoprotection by sulforaphane: inhibition of histone deacetylase. *Cancer Res* 64:5767–74.

Nakamaru, Y., C. Vuppusetty, H. Wada, J. C. Milne, M. Ito, C. Rossios, M. Elliot, J. Hogg, S. Kharitonov, H. Goto, J. E. Bemis, P. Elliott, P. J. Barnes, and K. Ito. 2009. A protein deacetylase SIRT1 is a negative regulator of metalloproteinase-9. *FASEB J* 9:2810–19.

Nanji, A. A., K. Jokelainen, G. L. Tipoe, A. Rahemtulla, P. Thomas, and A. J. Dannenberg. 2003. Curcumin prevents alcohol-induced liver disease in rats by inhibiting the expression of NF-kappa B-dependent genes. *Am J Physiol Gastrointest Liver Physiol* 284:G321–27.

Nomura, M., A. Kaji, W. Ma, K. Miyamoto, and Z. Dong. 2001. Suppression of cell transformation and induction of apoptosis by caffeic acid phenethyl ester. *Mol Carcinog* 31:83–89.

Nomura, M., W. Ma, N. Chen, A. M. Bode, and Z. Dong. 2000. Inhibition of 12-O-tetradecanoylphorbol-13-acetate-induced NF-kappaB activation by tea polyphenols, (-)-epigallocatechin gallate and theaflavins. *Carcinogenesis* 21:1885–90.

Nomura, M., W. Y. Ma, C. Huang, C. S. Yang, G. T. Bowden, K. Miyamoto, and Z. Dong. 2000. Inhibition of ultraviolet B-induced AP-1 activation by theaflavins from black tea. *Mol Carcinog* 28:148–55.

Pauwels, R. A., A. S. Buist, P. M. Calverley, C. R. Jenkins, and S. S. Hurd. 2001. Global strategy for the diagnosis, management, and prevention of chronic obstructive pulmonary disease. NHLBI/WHO Global Initiative for Chronic Obstructive Lung Disease (GOLD) Workshop summary. *Am J Respir Crit Care Med* 163:1256–76.

Pianetti, S., S. Guo, K. T. Kavanagh, and G. E. Sonenshein. 2002. Green tea polyphenol epigallocatechin-3 gallate inhibits Her-2/neu signaling, proliferation, and transformed phenotype of breast cancer cells. *Cancer Res* 62:652–55.

Prestera, T., and P. Talalay. 1995. Electrophile and antioxidant regulation of enzymes that detoxify carcinogens. *Proc Natl Acad Sci U S A* 92:8965–69.

Pruitt, K., R. L. Zinn, J. E. Ohm, K. M. McGarvey, S. H. Kang, D. N. Watkins, J. G. Herman, and S. B. Baylin. 2006. Inhibition of SIRT1 reactivates silenced cancer genes without loss of promoter DNA hypermethylation. *PLoS Genet* 2 (3):e40.

Qin, J., H. G. Chen, Q. Yan, M. Deng, J. Liu, S. Doerge, W. Ma, Z. Dong, and D. W. Li. 2008. Protein phosphatase-2A is a target of epigallocatechin-3-gallate and modulates p53-Bak apoptotic pathway. *Cancer Res* 68:4150–62.

Rahman, I., and I. M. Adcock. 2006. Oxidative stress and redox regulation of lung inflammation in COPD. *Eur Respir J* 28:219–42.

Rahman, I., S. K. Biswas, and P. A. Kirkham. 2006. Regulation of inflammation and redox signaling by dietary polyphenols. *Biochem Pharmacol* 72:1439–52.

Rahman, I., P. S. Gilmour, L. A. Jimenez, and W. MacNee. 2002. Oxidative stress and TNF-alpha induce histone acetylation and NF-kappaB/AP-1 activation in alveolar epithelial cells: potential mechanism in gene transcription in lung inflammation. *Mol Cell Biochem* 234:239–48.

Rajendrasozhan, S., S. R. Yang, I. Edirisinghe, H. Yao, D. Adenuga, and I. Rahman. 2008. Deacetylases and NF-kappaB in redox regulation of cigarette smoke-induced lung inflammation: epigenetics in pathogenesis of COPD. *Antioxid Redox Signal* 10:799–811.

Rajendrasozhan, S., S. R. Yang, V. L. Kinnula, and I. Rahman. 2008. SIRT1, an antiinflammatory and antiaging protein, is decreased in lungs of patients with chronic obstructive pulmonary disease. *Am J Respir Crit Care Med* 177:861–70.

Rajendrasozhan, S., H. Yao, and I. Rahman. 2009. Current perspectives on role of chromatin modifications and deacetylases in lung inflammation in COPD. *COPD* 6:291–97.

Raza, H., and A. John. 2008. In vitro effects of tea polyphenols on redox metabolism, oxidative stress, and apoptosis in PC12 cells. *Ann N Y Acad Sci* 1138:358–65.

Sanders, Y. Y., A. Pardo, M. Selman, G. J. Nuovo, T. O. Tollefsbol, G. P. Siegal, and J. S. Hagood. 2008. Thy-1 promoter hypermethylation: a novel epigenetic pathogenic mechanism in pulmonary fibrosis. *Am J Respir Cell Mol Biol* 39:610–18.

Santus, P., A. Sola, P. Carlucci, F. Fumagalli, A. Di Gennaro, M. Mondoni, C. Carnini, S. Centanni, and A. Sala. 2005. Lipid peroxidation and 5-lipoxygenase activity in chronic obstructive pulmonary disease. *Am J Respir Crit Care Med* 171:838–43.

Scholz, S., and G. Williamson. 2007. Interactions affecting the bioavailability of dietary polyphenols in vivo. *Int J Vitam Nutr Res* 77:224–35.

Sengupta, N., and E. Seto. 2004. Regulation of histone deacetylase activities. *J Cell Biochem* 93:57–67.

Sharafkhaneh, A., S. Velamuri, V. Badmaev, C. Lan, and N. Hanania. 2007. The potential role of natural agents in treatment of airway inflammation. *Ther Adv Respir Dis* 1:105–20.

Shirai, T., K. Reshad, A. Yoshitomi, K. Chida, H. Nakamura, and M. Taniguchi. 2003. Green tea-induced asthma: relationship between immunological reactivity, specific and non-specific bronchial responsiveness. *Clin Exp Allergy* 33:1252–55.

Shishodia, S., P. Potdar, C. G. Gairola, and B. B. Aggarwal. 2003. Curcumin (diferuloylmethane) down-regulates cigarette smoke-induced NF-kappaB activation through inhibition of IkappaBalpha kinase in human lung epithelial cells: correlation with suppression of COX-2, MMP-9 and cyclin D1. *Carcinogenesis* 24:1269–79.

Shoba, G., D. Joy, T. Joseph, M. Majeed, R. Rajendran, and P. S. Srinivas. 1998. Influence of piperine on the pharmacokinetics of curcumin in animals and human volunteers. *Planta Med* 64:353–56.

Stockley, R. A. 2002. Neutrophils and the pathogenesis of COPD. *Chest* 121:151S–155S.

Sun, Z., Y. E. Chin, and D. D. Zhang. 2009. Acetylation of Nrf2 by p300/CBP augments promoter-specific DNA binding of Nrf2 during the antioxidant response. *Mol Cell Biol* 29:2658–72.

Syed, D. N., F. Afaq, M. H. Kweon, N. Hadi, N. Bhatia, V. S. Spiegelman, and H. Mukhtar. 2007. Green tea polyphenol EGCG suppresses cigarette smoke condensate-induced NF-kappaB activation in normal human bronchial epithelial cells. *Oncogene* 26:673–82.

Szulakowski, P., A. J. Crowther, L. A. Jimenez, K. Donaldson, R. Mayer, T. B. Leonard, W. MacNee, and E. M. Drost. 2006. The effect of smoking on the transcriptional regulation of lung inflammation in patients with chronic obstructive pulmonary disease. *Am J Respir Crit Care Med* 174:41–50.

Tabak, C., I. C. Arts, H. A. Smit, D. Heederik, and D. Kromhout. 2001. Chronic obstructive pulmonary disease and intake of catechins, flavonols, and flavones: the MORGEN Study. *Am J Respir Crit Care Med* 164:61–64.

Tang, W. Y., and S. M. Ho. 2007. Epigenetic reprogramming and imprinting in origins of disease. *Rev Endocr Metab Disord* 8:173–82.

Vaquero, A., M. Scher, D. Lee, H. Erdjument-Bromage, P. Tempst, and D. Reinberg. 2004. Human SirT1 interacts with histone H1 and promotes formation of facultative heterochromatin. *Mol Cell* 16:93–105.

Vaziri, H., S. K. Dessain, E. Ng Eaton, S. I. Imai, R. A. Frye, T. K. Pandita, L. Guarente, and R. A. Weinberg. 2001. hSIR2(SIRT1) functions as an NAD-dependent p53 deacetylase. *Cell* 107:149–59.

Venn, A. J., S. A. Lewis, M. Cooper, R. Hubbard, and J. Britton. 2001. Living near a main road and the risk of wheezing illness in children. *Am J Respir Crit Care Med* 164:2177–80.

Walda, I. C., C. Tabak, H. A. Smit, L. Rasanen, F. Fidanza, A. Menotti, A. Nissinen, E. J. Feskens, and D. Kromhout. 2002. Diet and 20-year chronic obstructive pulmonary disease mortality in middle-aged men from three European countries. *Eur J Clin Nutr* 56:638–43.

Wang, L., and K. E. Pinkerton. 2008. Detrimental effects of tobacco smoke exposure during development on postnatal lung function and asthma. *Birth Defects Res C Embryo Today* 84:54–60.

Weseler, A. R., L. Geraets, H. J. Moonen, R. J. Manders, L. J. van Loon, H. J. Pennings, E. F. Wouters, A. Bast, and G. J. Hageman. 2009. Poly (ADP-ribose) polymerase-1-inhibiting flavonoids attenuate cytokine release in blood from male patients with chronic obstructive pulmonary disease or type 2 diabetes. *J Nutr* 139:952–57.

Wheeler, D. S., J. D. Catravas, K. Odoms, A. Denenberg, V. Malhotra, and H. R. Wong. 2004. Epigallocatechin-3-gallate, a green tea-derived polyphenol, inhibits IL-1 beta-dependent proinflammatory signal transduction in cultured respiratory epithelial cells. *J Nutr* 134:1039–44.

Wilson, C. B., E. Rowell, and M. Sekimata. 2009. Epigenetic control of T-helper-cell differentiation. *Nat Rev Immunol* 9:91–105.

Yang, C. S., M. Fang, J. D. Lambert, P. Yan, and T. H. Huang. 2008. Reversal of hypermethylation and reactivation of genes by dietary polyphenolic compounds. *Nutr Rev* 66 (Suppl 1):S18–20.

Yang, G. Y., J. Liao, C. Li, J. Chung, E. J. Yurkow, C. T. Ho, and C. S. Yang. 2000. Effect of black and green tea polyphenols on c-jun phosphorylation and H_2O_2 production in transformed and non-transformed human bronchial cell lines: possible mechanisms of cell growth inhibition and apoptosis induction. *Carcinogenesis* 21:2035–39.

Yang, H., N. Magilnick, C. Lee, D. Kalmaz, X. Ou, J. Y. Chan, and S. C. Lu. 2005. Nrf1 and Nrf2 regulate rat glutamate-cysteine ligase catalytic subunit transcription indirectly via NF-kappaB and AP-1. *Mol Cell Biol* 25:5933–46.

Yang, S. R., A. S. Chida, M. R. Bauter, N. Shafiq, K. Seweryniak, S. B. Maggirwar, I. Kilty, and I. Rahman. 2006. Cigarette smoke induces proinflammatory cytokine release by activation of NF-kappaB and posttranslational modifications of histone deacetylase in macrophages. *Am J Physiol Lung Cell Mol Physiol* 291:L46–57.

Yang, S. R., S. Valvo, H. Yao, A. Kode, S. Rajendrasozhan, I. Edirisinghe, S. Caito, D. Adenuga, R. Henry, G. Fromm, S. Maggirwar, J. D. Li, M. Bulger, and I. Rahman. 2008. IKK alpha causes chromatin modification on pro-inflammatory genes by cigarette smoke in mouse lung. *Am J Respir Cell Mol Biol* 38:689–98.

Yang, S. R., J. Wright, M. Bauter, K. Seweryniak, A. Kode, and I. Rahman. 2007. Sirtuin regulates cigarette smoke-induced proinflammatory mediator release via RelA/p65 NF-kappaB in macrophages in vitro and in rat lungs in vivo: implications for chronic inflammation and aging. *Am J Physiol Lung Cell Mol Physiol* 292:L567–76.

Yang, S. R., H. Yao, S. Rajendrasozhan, S. Chung, I. Edirisinghe, S. Valvo, G. Fromm, M. J. McCabe Jr., P. J. Sime, R. P. Phipps, J. D. Li, M. Bulger, and I. Rahman. 2009. RelB is differentially regulated by IkappaB Kinase-alpha in B cells and mouse lung by cigarette smoke. *Am J Respir Cell Mol Biol* 40:147–58.

Yang, T., and A. A. Sauve. 2006. NAD metabolism and sirtuins: metabolic regulation of protein deacetylation in stress and toxicity. *AAPS J* 8:E632–43.

Yang, Y., H. M. Haitchi, J. Cakebread, D. Sammut, A. Harvey, R. M. Powell, J. W. Holloway, P. Howarth, S. T. Holgate, and D. E. Davies. 2008. Epigenetic mechanisms silence a disintegrin and metalloprotease 33 expression in bronchial epithelial cells. *J Allergy Clin Immunol* 121:1393–99, 1399, e1–14.

Yao, H., I. Edirisinghe, S. Rajendrasozhan, S. R. Yang, S. Caito, D. Adenuga, and I. Rahman. 2008. Cigarette smoke-mediated inflammatory and oxidative responses are strain-dependent in mice. *Am J Physiol Lung Cell Mol Physiol* 294:L1174–86.

Yeung, F., J. E. Hoberg, C. S. Ramsey, M. D. Keller, D. R. Jones, R. A. Frye, and M. W. Mayo. 2004. Modulation of NF-kappaB-dependent transcription and cell survival by the SIRT1 deacetylase. *EMBO J* 23:2369–80.

Yoshida, T., and R. M. Tuder. 2007. Pathobiology of cigarette smoke-induced chronic obstructive pulmonary disease. *Physiol Rev* 87:1047–82.

Young, H. A., P. Ghosh, J. Ye, J. Lederer, A. Lichtman, J. R. Gerard, L. Penix, C. B. Wilson, A. J. Melvin, M. E. McGurn, and et al. 1994. Differentiation of the T helper phenotypes by analysis of the methylation state of the IFN-gamma gene. *J Immunol* 153:3603–10.

Yu, R., J. J. Jiao, J. L. Duh, K. Gudehithlu, T. H. Tan, and A. N. Kong. 1997. Activation of mitogen-activated protein kinases by green tea polyphenols: potential signaling pathways in the regulation of antioxidant-responsive element-mediated phase II enzyme gene expression. *Carcinogenesis* 18:451–56.

Zochbauer-Muller, S., K. M. Fong, A. K. Virmani, J. Geradts, A. F. Gazdar, and J. D. Minna. 2001. Aberrant promoter methylation of multiple genes in non-small cell lung cancers. *Cancer Res* 61:249–55.

9 Glycemic Memory and Epigenetic Changes

Andrew L. Siebel, Ana Z. Fernandez, and Assam El-Osta

CONTENTS

9.1 Glycemic Variability and Diabetic Complications 215
9.2 Epigenetic Mechanisms—Back to Basics 216
9.3 Nutritional Intervention and Epigenetic Mechanisms 220
9.4 Epigenetic Regulation of Diabetic Complications 221
9.5 The Role of Histone-Modifying Enzymes in Glycemic Memory 221
9.6 The Current Landscape and Perspectives—Where to Next? 223
References 224

9.1 GLYCEMIC VARIABILITY AND DIABETIC COMPLICATIONS

Cardiovascular complications remain the major cause of morbidity and mortality in the diabetic population (Cooper and Johnston 2000). Patients with type 1 or type 2 diabetes have a two- to fourfold higher risk of cardiovascular disease compared to healthy individuals (Hu et al. 2001; Fox et al. 2004), and those with impaired glucose tolerance alone have a cardiovascular disease risk comparable to type 2 diabetics (Qiao et al. 2002). It is increasingly appreciated that exposure to high glucose is the major factor leading to these complications. Furthermore, there appears to be a "metabolic memory" (Nathan et al. 2005) or "legacy effect" (Chalmers and Cooper 2008) whereby diabetic complications, particularly vascular events, continue to develop and progress even in individuals who have returned to normal glycemic control after a period of transient hyperglycemia. In the latest follow-up from the Diabetes Control Complications Trial (DCCT) and the Epidemiology of Diabetes Interventions and Complications (EDIC) trial, it is evident that the deleterious effects in the vasculature of both conventional and intensified glycemic control continue to operate more than 5 years after the patients have returned to normoglycemia (Nathan et al. 2005). Further to this, the results of the action in diabetes and vascular disease—preterax and diamicron MR controlled evaluation (ADVANCE) and Action to Control Cardiovascular Risk in Diabetes (ACCORD) trials—raise the debate about whether tight glucose control is beneficial at all in diabetes (Patel et al. 2008; Gerstein et al. 2008). Whereas the initial interpretation of the results from the UK Prospective Diabetes Study (UKPDS) showed no significant effect of strict glycemic control on myocardial infarction, meta-analysis of these data show significant reductions in all diabetes-related endpoints (Nathan 1998). A recent follow-up in type 2 diabetics from this study argues for the utility of long-term hyperglycemic control in preventing cardiovascular disease (Holman et al. 2008).

Although the underlying molecular explanation for this "metabolic memory" has not yet been clearly defined, this phenomenon is not new, as it was first observed more than 20 years ago in the retina of diabetic dogs exposed to high glucose for 2.5 years, then allowed normal glucose control for a further 2.5 years (Engerman and Kern 1987). Those animals that returned to normoglycemia for 2.5 years had a similar incidence of retinopathy to that of their control counterparts who had poor glucose control throughout the 5-year study, suggesting that these cells retained a "memory" of the early hyperglycemic episodes (Engerman and Kern 1987). Soon after this study, another group showed that there was a persistent up-regulation of extracellular matrix (ECM)-related gene expression in isolated endothelial cells and kidneys of diabetic rats 1 week after glucose normalization following 2 weeks of hyperglycemia (Roy et al. 1990). Recent in vitro studies by our laboratory suggest an important role for epigenetic modification as a result of both ambient and prior hyperglycemia, in primary human aortic endothelial cells (El-Osta et al. 2008).

Further elucidation of the regulatory mechanisms associated with hyperglycemic memory and the role of histone methyltransferases, particularly SET domain-containing protein 7/9 (Set7/9) and suppressor of variegation 3–9 homologue 1 (Suv39h1)/SuVa39, was demonstrated in human microvascular endothelial cells (Brasacchio et al. 2009). These results indicate that glucose is conferring gene-activating events that are associated with diabetic complications in specific cell types (see Figure 9.1). The ability of epigenetic changes to confer sustained effects on the vasculature, as a result of acute hyperglycemia, emphasizes the potential deleterious long-term effects of "metabolic memory." This could have direct implications for clinicians, emphasizing the importance of achieving tight metabolic control in an attempt to avoid episodic hyperglycemia. In addition, it may provide a potential mechanism for the adverse outcomes that have been suggested in various studies after institution of strict glycemic control. This includes the adverse effects on retinopathy in the initial phases of the early insulin pump studies in the Kroc Collaborative Study Group or the recent unexpected findings of increased mortality in the ACCORD study (Gerstein et al. 2007). Finally, the delineation of the key epigenetic events in vivo that are involved in glucose-induced gene modulation in blood vessels will assist in identifying appropriate targets for new treatments to reduce, reverse, or prevent diabetic complications.

9.2 EPIGENETIC MECHANISMS—BACK TO BASICS

Epigenetics describes the study of gene regulation events independent of changes to nucleotide sequence; this may include heritable changes in gene activity and expression but also long-term alterations in the transcriptional potential of a cell that are not heritable (Bird 2007). These epigenetic changes are potentially reversible and modulated by the environment, diet, or pharmacological intervention (Figure 9.2). This in turn may affect genomic stability and gene expression (Kouzarides 2007;

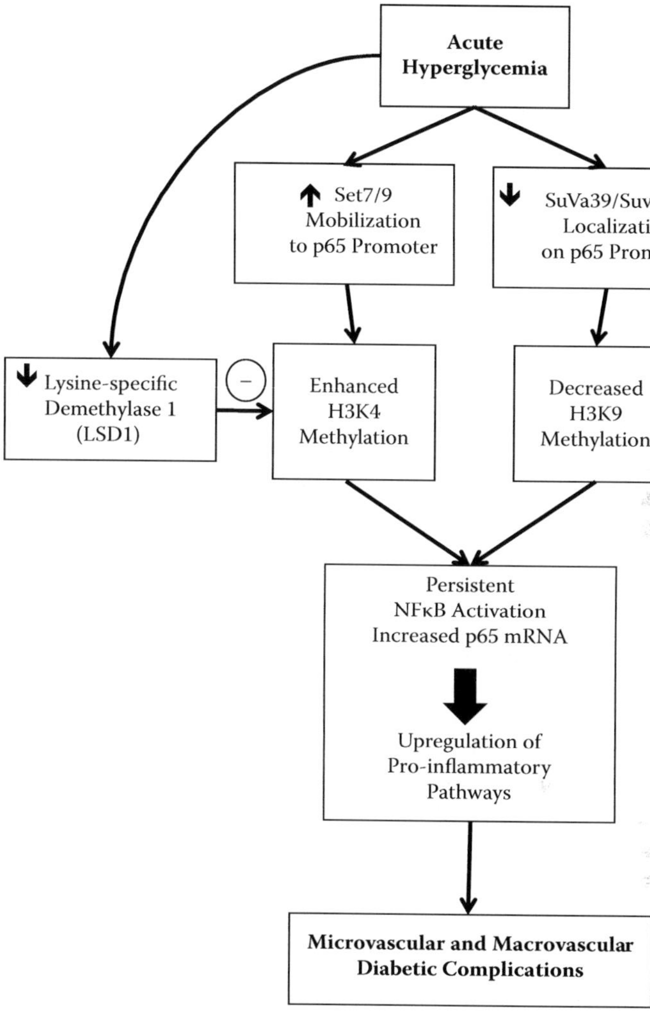

FIGURE 9.1 Proposed mechanism—epigenetic effects of hyperglycemia on vascular cells. Hyperglycemia is associated with epigenetic changes leading to transcriptional activation of the proinflammatory transcription factor NFκB. Increased mobilization of Set7/9 to the p65 promoter results in enhanced H3K4me (active mark) leading to transcriptional activation, whereas decreased localization of SuVa39/Suv39h1 to the gene promoter results in decreased H3K9me (repressive mark) and transcriptional inactivation. Relative levels of the histone demethylase LSD1 are also decreased in response to hyperglycemia, therefore blocking its demethylase activity on H3K4. These epigenetic modifications and gene-activating events can persist in the absence of sustained hyperglycemia, establishing a hyperglycemic memory in vascular cells.

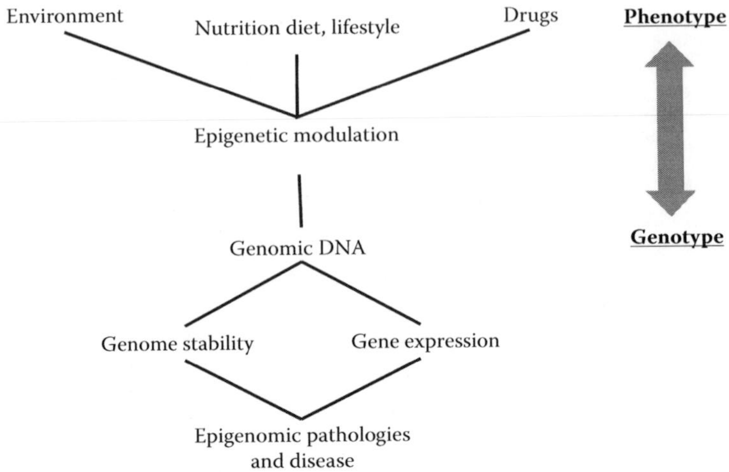

FIGURE 9.2 Schematic of how various stimuli, epigenetic mechanisms, and gene-activating events can lead to epigenomic pathologies. Epigenetic changes can be modulated by the environment, diet, or pharmacological intervention. As outlined, epigenetics essentially provides a link between genotype and phenotype, which can help explain how cells carrying identical DNA differentiate into different cell types with distinct functions. Nutrient-driven epigenetic changes are also involved in the development of disease.

Ozanne and Constancia 2007; Talbert and Henikoff 2006). Essentially the field of epigenetics provides a link between genotype and phenotype, which can help explain how cells carrying identical DNA differentiate into different cell types with distinct functions (Jaenisch and Young 2008). The principal mechanisms of epigenetic changes in mammals are DNA methylation (not covered in this chapter) and modifications of histone tails, which result mostly in altered chromatin structure (Matouk and Marsden 2008).

Genomic DNA is packaged in eukaryotic cells with histone proteins to form a protein/DNA complex known as chromatin. The fundamental unit of chromatin is the nucleosome and is composed of an octamer of the four core histones (H2A, H2B, H3, and H4) around which approximately 146 base pairs (bp) of DNA are wrapped (Klug et al. 1980) (see Figure 9.3). The core histones are predominantly globular except for their amino-terminal tails, which are accessible to histone modifying enzymes (Luger et al. 1997). Histones can be modified at many sites with over 60 amino acid residues currently identified and detected by specific antibodies or by mass spectrometry (Kouzarides 2007). The timing of appearance of a particular posttranslational modification will depend on the signaling conditions within the nucleus of the cell. Histone modifications of different classes (e.g., acetylation, methylation, ubiquitination, and phosphorylation) can define epigenetic regulation of a variety of biological functions (Jenuwein and Allis 2001; B. Li et al. 2007; Mack 2008). Histone modifications help to partition the genome into distinct domains such as active euchromatin, where DNA remains in an open conformation accessible for transcription and silent heterochromatin, where chromatin is condensed

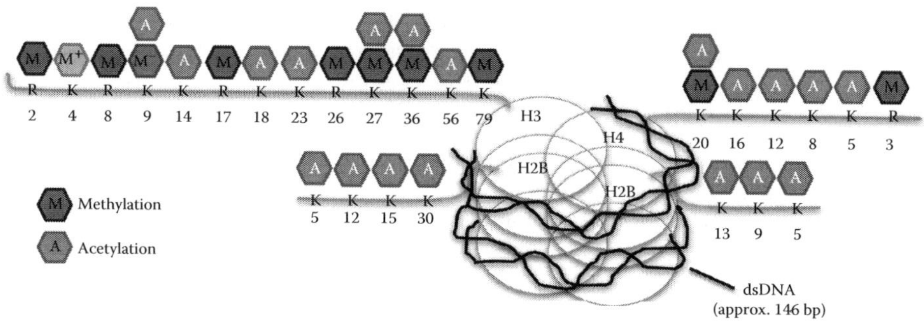

FIGURE 9.3 (Please see color insert following page 80.) Nucleosome structure with known histone modifications on specific amino acid residues. A single nucleosome is composed of approximately 146 bp of double-stranded DNA (dsDNA) wrapped around an octamer core of histone proteins from which histone N-terminal tails protrude. Specific amino acids (R: arginine; K: lysine) are subjected to different posttranslational modifications. Here only acetylation (A: blue) and methylation (M: red) marks are included, the most relevant to diabetes and hyperglycemic response being: H3K4 methylation (active mark: green +) and H3K9 methylation (repressive mark: purple –).

and inaccessible to transcriptional machinery (Jenuwein and Allis 2001; Birney et al. 2007). Gene transcription and activation are dynamic processes involving the conversion of compact heterochromatin into transcription factor-accessible euchromatin (Berger 2007). Euchromatin represents a large proportion of the genome allowing DNA flexibility to turn genes on or off and allow DNA repair or replication.

The term "histone code" has been used to describe the role of modifications to enable DNA-related functions (Jenuwein and Allis 2001). Heterochromatin plays an important role in protecting chromosome ends, controlling chromosome stability, and preventing mutations and translocations (Talbert and Henikoff 2006; Huang et al. 2004). In mammals, demarcation between the different environments is set up by boundary elements, which recruit enzymes to modify the chromatin. The regulation of gene expression within euchromatin requires the delivery of chromatin-modifying enzymes by DNA-bound transcription factors (B. Li et al. 2007). Following an environmental stimulus, such as glucose, and in the presence of essential transcriptional machinery, transcription factors bind to the promoter of specific genes and initiate a cascade of modification events, which result in the activation or silencing of the gene (B. Li et al. 2007). Generally, modifications are divided into those that correlate with gene activation and those that correlate with gene repression.

There are two established mechanisms for the function of modifications. The first is the disruption of contacts between nucleosomes in order to "unravel" chromatin and the second is the recruitment of nonhistone proteins (Kouzarides 2007). Modifications may affect higher-order chromatin structure by affecting the contact between different histones in adjacent nucleosomes or the interaction of histones with DNA. Acetylation has the greatest potential to unfold chromatin, since its major action is to neutralize the basic charge of the lysine residue (Jenuwein and Allis 2001). Histone acetyltransferases (HATs), such as p300 and CREB-binding protein, have been shown to modify a variety of lysine residues on H3 and H4. The histone deacetylases (HDACs) remove acetyl groups and inhibit transcription factor binding and transcription. Binding proteins are recruited to modifications via specific domains. Unlike histone acetylation, histone methylation can be associated with both gene activation and inactivation, and the effects of methylation are essentially mediated by recruitment of additional positive or negative transacting factors. The recent isolation of several proteins that recognize Histone-3-Lysine-4 methylation (H3K4me) has highlighted the fact that their purpose is to tether enzymatic activities onto chromatin. In mammals, the silent heterochromatic state is associated with low levels of acetylation and high levels of specific methylated sites at H3K9, H3K27, and H4K20 (Ruthenburg et al. 2007). Further enzymatic activities are required for transcription to take place, which is typically characterized by high levels of acetylation and trimethylation of lysines at H3K4, H3K36, and H3K79 (Koch et al. 2007). Additional complexity arises from the fact that methylation occurs not only at lysines but also at arginines, and these may be mono- (me), di- (me2), or trimethylated (me3) for lysines and mono- or dimethylated for arginines, being that these dynamic modifications are not uniformly distributed (reviewed in Mack 2008).

9.3 NUTRITIONAL INTERVENTION AND EPIGENETIC MECHANISMS

Many studies have examined the intimate links between obesity, energy metabolism, nutrient balance, and epigenetic modifications (Tateishi et al. 2009; Milagro et al. 2009; Khan et al. 2005; Armitage et al. 2003; Taylor et al. 2005), the majority of which have used rodent models to examine these links and interrogate the mechanisms in more detail. Obesity is associated with loss of function of the histone demethylase, Jhdm2a, resulting in decreased expression of the metabolically active gene peroxisome proliferator-activated receptor-alpha (PPAR-alpha) in skeletal muscle and impaired cold-induced uncoupling protein 1 expression in brown adipose tissue, suggesting a relationship between epigenetic mechanisms and obesity (Tateishi et al. 2009). The nicotinamide adenine dinucleotide (NAD+)-dependent sirtuins (class III HDACs) target both histone and nonhistone proteins in another example of epigenetic control of metabolic pathways, including adipogenesis, glucose utilization, and insulin secretion (Schwer and Verdin 2008). It has also been recently recognized that glucose availability can affect histone acetylation in an ATP-citrate lyase-dependent manner, further linking energy metabolism to epigenetic regulation.

An elegant example of how environmental exposure to nutrients may change gene expression and alter phenotype through epigenetic modifications is the agouti mouse. The agouti gene encodes a paracrine-signaling molecule that promotes melanocytes to produce a yellow pigment rather than black, altering their coat color and making these yellow mice prone to develop obesity and diabetes (Morgan et al. 1999; Duhl et al. 1994; Wolff et al. 1998). The coat color and subsequent disease susceptibility is essentially controlled by the degree of methylation on the agouti gene. Supplementation of pregnant mice diets with folic acid, a known methyl donor, increases DNA methylation of the agouti gene in offspring, resulting in suppressed gene expression and a brown coat color (Jirtle and Skinner 2007). Interestingly, this phenomenon can also be inherited in the next generation via germline transmission (Cropley et al. 2006). Nutrient-driven epigenetic changes are also involved in the development of disease as outlined in Figure 9.2. For example, rats exposed to a high-fat diet during pregnancy are associated with impaired glucose tolerance, as well as mitochondrial and cardiovascular dysfunction in adult offspring, possibly due to epigenetic modifications (Khan et al. 2005; Armitage et al. 2003; Taylor et al. 2005).

9.4 EPIGENETIC REGULATION OF DIABETIC COMPLICATIONS

Oxidative stress, dyslipidemia, and hyperglycemia are thought to be associated with the development of diabetic complications. The major event in the progression of diabetic complications is vascular inflammation, triggered by a cascade of mediators to enhance inflammatory signaling. Nuclear factor-κB (NF-κB) is the predominant transcription factor activated under diabetic conditions that regulates expression of genes in the inflammatory pathway, leading to recruitment of monocytes and macrophages to the vessel and macrovasculature atherosclerosis (Miao et al. 2004). In fact, poor glycemic control increases NF-κB activity in monocytes and in turn up-regulates gene expression of inflammatory cytokines (Shanmugam et al. 2003; Hofmann et al. 1998). This involves an interaction between NF-κB and HATs, resulting in hyperacetylation of target genes including tumor necrosis factor-alpha (TNF-α) and cyclooxygenase-2 promoters (Miao et al. 2004). Hyperglycemia-induced oxidative stress and the formation of advanced glycation end products (AGEs) leads to the release of cytokines, cell adhesion molecules, and ECM-modifying genes that facilitate lymphocyte activation and invasion (Brownlee 2001; Libby and Plutzky 2002; Dragomir and Simionescu 2006; Hansson 2005).

9.5 THE ROLE OF HISTONE-MODIFYING ENZYMES IN GLYCEMIC MEMORY

Intensive research has gone into identification and characterization of the enzymes that control and direct histone modifications (Dillon et al. 2005; Martin and Zhang 2005), of which the methyltransferases are most specific (see Table 9.1). Lysine methyltransferases have enormous specificity compared to acetyltransferases, usually modifying one single lysine on a single histone (Bannister and Kouzarides 2005; Kouzarides 2007). The existence of lysine demethylases remained contentious for many years following the discovery of histone methyltransferases (HMTase). The first of these discovered was lysine-specific demethylase 1 (LSD1), which acts to demethylate H3K4 and repress transcription (Shi et al. 2004). This effectively dispelled the myth that histone methylation is a permanent mark (Shi and Whetstine 2007).

To explore the effects of glycemic variability, we have specifically developed an in vitro model to determine gene-activating events that are associated with epigenetic modifications. Primary human aortic endothelial cells were incubated in high glucose (HG, 30 mM) for 16 hours then returned to physiological levels (LG, 5 mM) for 6 days. Analyses revealed a persistent increase in expression of the NFκB subunit

TABLE 9.1
Histone-Modifying Enzymes Related to Diabetes and Glycemic Variability

Enzyme	Target	Function/Family
Set7/9	H3K4	Histone methyltransferase
LSD1	H3K4	Lysine demethylase
SuVa39/Suv39h1	H3K9	Histone methyltransferase
Jhdm2a	H3K9	Histone demethylase
p300	Multiple	Histone acetyltransferase
CBP	Multiple	Histone acetyltransferase
PRMT5	H4R3	Arginine methyltransferase

Note: Set7/9: SET domain-containing protein 7/9; LSD1: Lysine-specific demethylase 1; Suv39h1: Suppressor of variegation 3–9 homologue 1; Jhdm2a: JmjC domain-containing histone demethylation protein 2a; p300: Histone acetyltransferase p300; CBP: CREB-binding protein; PRMT5: Protein arginine N-methyltransferase 5.

p65 gene, despite a return to normoglycemia (El-Osta et al. 2008). Parallel experiments were performed in human microvascular endothelial cells, confirming a similar upregulation of p65 as well as proteins linked to ECM accumulation, such as fibronectin (El-Osta et al. 2008). Thus transient hyperglycemia is capable of inducing persistent gene-activating events associated with epigenetic change. Further studies revealed that this was as a result of histone modifications, specifically H3K4me, associated with activation of the p65 gene (El-Osta et al. 2008). Furthermore, increased p65 gene expression was associated with NFκB activation as determined by gel shift analyses and upregulation of the NFκB-dependent chemokine, monocyte chemoattractant protein-1 (Mcp-1), implicated in diabetes-associated vascular injury (El-Osta et al. 2008; Dragomir and Simionescu 2006).

In our laboratory we are interested in understanding the function of histone methyltransferases (HMTase), and we have recently demonstrated a central role for the HMTase enzyme, Set7/9, in promoting H3K4 methylation in endothelial cells, both from the macrovasculature and in primary aortic endothelial cells (El-Osta et al. 2008; Brasacchio et al. 2009). Set7/9 has also been shown to influence the recruitment of NFκB p65 to gene promoters and thereby its regulation of proinflammatory genes in macrophages from diabetic mice (Li et al. 2008). In order to assess whether Set7/9 is mobilized to maintain the active transcriptional state, immunopurified chromatin from microvascular endothelial cells exposed to acute hyperglycemia for 16 hours was enriched for Set7/9 on the p65 promoter (Brasacchio et al. 2009). Furthermore, gene silencing of Set7/9 with small interfering RNA (siRNA) in monocytes significantly inhibited TNF-induced inflammatory genes and H3K4 methylation on these promoters, as well as monocyte adhesion to endothelial or smooth muscle cells (Y. Li et al. 2008).

Further evidence of the involvement of Set7/9 in glucose-stimulated insulin secretion is that the Set7/9 gene promoter contains an islet-specific enhancer located 5–6 kb downstream of the transcriptional start site that exhibits pancreatic and duodenal homeobox 1 (Pdx1)-responsive activation in β-cells within the pancreas (Deering et al. 2009). Interestingly, siRNA knockdown in insulinoma and mouse islets suppressed genes in glucose-stimulated insulin secretion, including insulin 1/2 (Ins1/2), glucose transporter 2 (Glut2), and v-maf musculoaponeurotic fibrosarcoma oncogene homolog A (MafA). These changes in expression were correlated with reduced H3K4me2 and RNA Polymerase II recruitment on these gene sequences. In fact, Set7/9 knockdown in primary mouse islets led to defects in glucose-stimulated Ca^{2+} mobilization and insulin secretion (Deering et al. 2009).

Hyperglycemia was also associated with reduced localization of the HMTase SuVa39/Suv39h1 to the p65 promoter, and this was indirectly but tightly correlated with reduced H3K9 methylation on gene sequences (Brasacchio et al. 2009), suggesting that the sustained increase in p65 gene expression is linked to specific epigenetic modifications that are typically associated with increased gene transcription. Furthermore, vascular smooth muscle cells isolated from diabetic mice have reduced levels of H3K9me3 and elevated levels of H3K4me2 at the promoters of inflammatory genes interleukin-6 (IL-6) and Mcp-1 in parallel with decreased levels of H3K9me3 methyltransferase Suv39h1 and the histone demethylase LSD1 (Reddy et al. 2008; Villeneuve et al. 2008). Taken together, these studies suggest that hyperglycemia may induce epigenetic modifications on proinflammatory genes, which subsequently regulate gene expression and lead to the development of vascular inflammation. Figure 9.1 outlines a proposed mechanism, by which acute hyperglycemia can lead to diabetic complications in specific vascular cells.

9.6 THE CURRENT LANDSCAPE AND PERSPECTIVES—WHERE TO NEXT?

Investigators in our laboratory are interested in examining the dynamic state of epigenetic changes in response to environmental stimuli. Specifically, hyperglycemia and the persistence of epigenetic phenomena are a primary focus of some of the research currently investigated. In fact, many researchers are attempting to unravel the molecular determinants associated with recognizing the chromatin template and that regulate the histone code. The observation that H3K4me3 patterns persist to subsequent generations or transmits to daughter chromatin gives it true epigenetic status (Kouzarides 2007). However, further elucidation of the exact changes in histone methylation, the enzymes responsible for these posttranslational modifications, and the direct in vivo characterization of these changes in response to glucose in specific cell types is paramount. Future research may focus on the translation of established in vitro findings to an in vivo setting, such as the investigation of glucose on the chromatin template in specific vascular cell types, the transcriptional decisions that histone-modifying enzymes mediate, and the location of such spatial modifications. The recent advent of massive parallel sequencing now allows investigators to examine in greater depth genomewide associations and unravel the locality of histone modifications. Coupled with the characterization of changes in the transcriptome, this will allow us to identify gene targets for pharmaceutical or dietary intervention. It may also be possible that drugs that mediate epigenetic changes, such as HDAC inhibitors, will be used in the treatment of diabetic complications (Bieliauskas and Pflum 2008; Szyf 2009; Haberland et al. 2009). In support of this concept are the recent experimental findings indicating that HDAC inhibitors during myocardial infarction can reduce infarct area as well as programmed cell death (Granger et al. 2008). Several drugs that currently target epigenetic changes in malignant cells might also be tested for their effects on atherosclerotic plaque formation. The role of epigenetics in the pathogenesis of cardiovascular diseases represents an essentially unexplored territory that importantly may suggest new therapeutic possibilities.

REFERENCES

1984. Blood glucose control and the evolution of diabetic retinopathy and albuminuria. A preliminary multicenter trial. The Kroc Collaborative Study Group. *N Engl J Med* 311:365–72.

1998. Intensive blood-glucose control with sulphonylureas or insulin compared with conventional treatment and risk of complications in patients with type 2 diabetes (UKPDS 33). UK Prospective Diabetes Study (UKPDS) Group. *Lancet* 352:837–53.

Armitage, J. A., A. D. Pearce, A. J. Sinclair, A. J. Vingrys, R. S. Weisinger, and H. S. Weisinger. 2003. Increased blood pressure later in life may be associated with perinatal n-3 fatty acid deficiency. *Lipids* 38:459–64.

Bannister, A. J., and T. Kouzarides. 2005. Reversing histone methylation. *Nature* 436:1103–6.

Berger, S. L. 2007. The complex language of chromatin regulation during transcription. *Nature* 447:407–12.

Bieliauskas, A. V., and M. K. Pflum. 2008. Isoform-selective histone deacetylase inhibitors. *Chem Soc Rev* 37:1402–13.

Bird, A. 2007. Perceptions of epigenetics. *Nature* 447:396–98.

Birney, E., J. A. Stamatoyannopoulos, A. Dutta, R. Guigo, T. R. Gingeras, E. H. Margulies, Z. Weng, M. Snyder, E. T. Dermitzakis, R. E. Thurman et al. 2007. Identification and analysis of functional elements in 1% of the human genome by the ENCODE pilot project. *Nature* 447:799–816.

Brasacchio, D., J. Okabe, C. Tikellis, A. Balcerczyk, P. George, E. K. Baker, A. C. Calkin, M. Brownlee, M. E. Cooper, and A. El-Osta. 2009. Hyperglycemia induces a dynamic cooperativity of histone methylase and demethylase enzymes associated with gene-activating epigenetic marks that coexist on the lysine tail. *Diabetes* 58:1229–36.

Brownlee, M. 2001. Biochemistry and molecular cell biology of diabetic complications. *Nature* 414:813–20.

Chalmers, J., and M. E. Cooper. 2008. UKPDS and the legacy effect. *N Engl J Med* 359:1618–20.

Cooper, M. E., and C. I. Johnston. 2000. Optimizing treatment of hypertension in patients with diabetes. *JAMA 283*:3177–79.

Cropley, J. E., C. M. Suter, K. B. Beckman, and D. I. Martin. 2006. Germ-line epigenetic modification of the murine A vy allele by nutritional supplementation. *Proc Natl Acad Sci U S A* 103:17308–12.

Deering, T. G., T. Ogihara, A. P. Trace, B. Maier, and R. G. Mirmira. 2009. Methyltransferase Set7/9 maintains transcription and euchromatin structure at islet-enriched genes. *Diabetes* 58:185–93.

Dillon, S. C., X. Zhang, R. C. Trievel, and X. Cheng. 2005. The SET-domain protein superfamily: protein lysine methyltransferases. *Genome Biol* 6:227.

Dragomir, E., and M. Simionescu. 2006. Monocyte chemoattractant protein-1—a major contributor to the inflammatory process associated with diabetes. *Arch Physiol Biochem* 112:239–44.

Duhl, D. M., H. Vrieling, K. A. Miller, G. L. Wolff, and G. S. Barsh. (1994). Neomorphic agouti mutations in obese yellow mice. *Nat Genet* 8:59–65.

El-Osta, A., D. Brasacchio, D. Yao, A. Pocai, P. L. Jones, R. G. Roeder, M. E. Cooper, and M. Brownlee. 2008. Transient high glucose causes persistent epigenetic changes and altered gene expression during subsequent normoglycemia. *J Exp Med* 205:2409–17.

Engerman, R. L., and T. S. Kern. 1987. Progression of incipient diabetic retinopathy during good glycemic control. *Diabetes* 36:808–12.

Fox, C. S., S. Coady, P. D. Sorlie, D. Levy, J. B. Meigs, R. B. D'Agostino Sr., P. W. Wilson, and P. J. Savage. 2004. Trends in cardiovascular complications of diabetes. *JAMA* 292:2495–99.

Gerstein, H. C., M. E. Miller, R. P. Byington, D. C. Goff Jr., J. T. Bigger, J. B. Buse, W. C. Cushman, S. Genuth, F. Ismail-Beigi, R. H. Grimm Jr. et al. 2008. Effects of intensive glucose lowering in type 2 diabetes. *N Engl J Med* 358:2545–59.

Gerstein, H. C., M. C. Riddle, D. M. Kendall, R. M. Cohen, R. Golan M. N. Feinglos, J. K. Kirk, B. P. Hamilton, F. Ismail-Beigi, and P. Feeney. 2007. Glycemia treatment strategies in the Action to Control Cardiovascular Risk in Diabetes (ACCORD) trial. *Am J Cardiol* 99:34i–43i.

Granger, A., I. Abdullah, F. Huebner, A. Stout, T. Wang, T. Huebner, J. A. Epstein, and P. J. Gruber. 2008. Histone deacetylase inhibition reduces myocardial ischemia-reperfusion injury in mice. *FASEB J* 22:3549–60.

Haberland, M., R. L. Montgomery, and E. N. Olson. 2009. The many roles of histone deacetylases in development and physiology: implications for disease and therapy. *Nat Rev Genet* 10:32–42.

Hansson, G. K. 2005. Inflammation, atherosclerosis, and coronary artery disease. *N Engl J Med* 352:1685–95.

Hofmann, M. A., S. Schiekofer, M. Kanitz, M. S. Klevesath, M. Joswig, V. Lee, M. Morcos, H. Tritschler, R. Ziegler, P. Wahl et al. 1998. Insufficient glycemic control increases nuclear factor-kappa B binding activity in peripheral blood mononuclear cells isolated from patients with type 1 diabetes. *Diabetes Care* 21:1310–16.

Holman, R. R., S. K. Paul, M. A. Bethel, H. A. Neil, and D. R. Matthews. 2008. Long-term follow-up after tight control of blood pressure in type 2 diabetes. *N Engl J Med* 359:1565–76.

Hu, F. B., M. J. Stampfer, C. G. Solomon, S. Liu, W. C. Willett, F. E. Speizer, D. M. Nathan, and J. E. Manson. 2001. The impact of diabetes mellitus on mortality from all causes and coronary heart disease in women: 20 years of follow-up. *Arch Intern Med* 161:1717–23.

Huang, J., T. Fan, Q. Yan, H. Zhu, S. Fox, H. J. Issaq, L. Best, L. Gangi, D. Munroe, and K. Muegge. 2004. Lsh, an epigenetic guardian of repetitive elements. *Nucleic Acids Res* 32:5019–28.

Jaenisch, R., and R. Young. 2008. Stem cells, the molecular circuitry of pluripotency and nuclear reprogramming. *Cell* 132:567–82.

Jenuwein, T., and C. D. Allis. 2001. Translating the histone code. *Science* 293:1074–80.

Jirtle, R. L., and M. K. Skinner. 2007. Environmental epigenomics and disease susceptibility. *Nat Rev Genet* 8:253–62.

Khan, I. Y., V. Dekou, G. Douglas, R. Jensen, M. A. Hanson, L. Poston, and P. D. Taylor. 2005. A high-fat diet during rat pregnancy or suckling induces cardiovascular dysfunction in adult offspring. *Am J Physiol Regul Integr Comp Physiol* 288:R127–33.

Klug, A., D. Rhodes, J. Smith, J. T. Finch, and J. O. Thomas. 1980. A low resolution structure for the histone core of the nucleosome. *Nature* 287:509–16.

Koch, C. M., R. M. Andrews, P. Flicek, S. C. Dillon, U. Karaoz, G. K. Clelland, S. Wilcox, D. M. Beare, J. C. Fowler, P. Couttet et al. 2007. The landscape of histone modifications across 1% of the human genome in five human cell lines. *Genome Res* 17:691–707.

Kouzarides, T. 2007. Chromatin modifications and their function. *Cell* 128:693–705.

Li, B., M. Carey, and J. L. Workman. 2007. The role of chromatin during transcription. *Cell* 128:707–19.

Li, Y., M. A. Reddy, F. Miao, N. Shanmugam, J. K. Yee, D. Hawkins, B. Ren, and R. Natarajan. 2008. Role of the histone H3 lysine 4 methyltransferase, SET7/9, in the regulation of NF-kappaB-dependent inflammatory genes. Relevance to diabetes and inflammation. *J Biol Chem* 283: 26771–81.

Libby, P., and J. Plutzky. 2002. Diabetic macrovascular disease: the glucose paradox? *Circulation* 106:2760–63.

Luger, K., A. W. Mader, R. K. Richmond, D. F. Sargent, and T. J. Richmond. 1997. Crystal structure of the nucleosome core particle at 2.8 A resolution. *Nature* 389:251–60.

Mack, C. P. 2008. An epigenetic clue to diabetic vascular disease. *Circ Res* 103:568–70.

Martin, C., and Y. Zhang, 2005. The diverse functions of histone lysine methylation. *Nat Rev Mol Cell Biol* 6:838–49.

Matouk, C. C., and P. A. Marsden. 2008. Epigenetic regulation of vascular endothelial gene expression. *Circ Res* 102:873–87.

Miao, F., I. G. Gonzalo, L. Lanting, and R. Natarajan. 2004. In vivo chromatin remodeling events leading to inflammatory gene transcription under diabetic conditions. *J Biol Chem* 279:18091–97.

Milagro, F. I., J. Campion, D. F. Garcia-Diaz, E. Goyenechea, L. Paternain, and J. A. Martinez. 2009. High fat diet-induced obesity modifies the methylation pattern of leptin promoter in rats. *J Physiol Biochem* 65:1–9.

Morgan, H. D., H. G. Sutherland, D. I. Martin, and E. Whitelaw. 1999. Epigenetic inheritance at the agouti locus in the mouse. *Nat Genet* 23:314–18.

Nathan, D. M. 1998. Some answers, more controversy, from UKPDS. United Kingdom Prospective Diabetes Study. *Lancet* 352:832–33.

Nathan, D. M., P. A. Cleary, J. Y. Backlund, S. M. Genuth, J. M. Lachin, T. J. Orchard, P. Raskin, and B. Zinman. 2005. Intensive diabetes treatment and cardiovascular disease in patients with type 1 diabetes. *N Engl J Med* 353:2643–53.

Ozanne, S. E., and M. Constancia. 2007. Mechanisms of disease: the developmental origins of disease and the role of the epigenotype. *Nat Clin Pract Endocrinol Metab* 3:539–46.

Patel, A., S. MacMahon, J. Chalmers, B. Neal, L. Billot, M. Woodward, M. Marre, M. Cooper, P. Glasziou, D. Grobbee et al. 2008. Intensive blood glucose control and vascular outcomes in patients with type 2 diabetes. *N Engl J Med* 358:2560–72.

Qiao, Q., K. Pyorala, M. Pyorala, A. Nissinen J. Lindstrom, R. Tilvis, and J. Tuomilehto. 2002. Two-hour glucose is a better risk predictor for incident coronary heart disease and cardiovascular mortality than fasting glucose. *Eur Heart J* 23:1267–75.

Reddy, M. A., L. M. Villeneuve, M. Wang, L. Lanting, and R. Natarajan. 2008. Role of the lysine-specific demethylase 1 in the proinflammatory phenotype of vascular smooth muscle cells of diabetic mice. *Circ Res* 103:615–23.

Roy, S., R. Sala, E. Cagliero, and M. Lorenzi. 1990. Overexpression of fibronectin induced by diabetes or high glucose: phenomenon with a memory. *Proc Natl Acad Sci U S A* 87:404–8.

Ruthenburg, A. J., H. Li, D. J. Patel, and C. D. Allis. 2007. Multivalent engagement of chromatin modifications by linked binding modules. *Nat Rev Mol Cell Biol* 8:983–94.

Schwer, B., and E. Verdin, E. 2008. Conserved metabolic regulatory functions of sirtuins. *Cell Metab* 7:104–12.

Shanmugam, N., M. A. Reddy, M. Guha, and R. Natarajan. 2003. High glucose-induced expression of proinflammatory cytokine and chemokine genes in monocytic cells. *Diabetes* 52:1256–64.

Shi, Y., F. Lan, C. Matson, P. Mulligan, J. R. Whetstine, P. A. Cole, and R. A. Casero. 2004. Histone demethylation mediated by the nuclear amine oxidase homolog LSD1. *Cell* 119:941–53.

Shi, Y., and J. R. Whetstine. 2007. Dynamic regulation of histone lysine methylation by demethylases. *Mol Cell* 25:1–14.

Szyf, M. 2009. Epigenetics, DNA methylation, and chromatin modifying drugs. *Annu Rev Pharmacol Toxicol* 49:243–63.

Talbert, P. B., and S. Henikoff. 2006. Spreading of silent chromatin: inaction at a distance. *Nat Rev Genet* 7:793–803.

Tateishi, K., Y. Okada, E. M. Kallin, and Y. Zhang. 2009. Role of Jhdm2a in regulating metabolic gene expression and obesity resistance. *Nature* 458:757–61.

Taylor, P. D., J. McConnell, I. Y. Khan, K. Holemans, K. M. Lawrence, H. Asare-Anane, S. J. Persaud, P. M. Jones, L. Petrie, M. A. Hanson et al. 2005. Impaired glucose homeostasis and mitochondrial abnormalities in offspring of rats fed a fat-rich diet in pregnancy. *Am J Physiol Regul Integr Comp Physiol* 288:R134–39.

Villeneuve, L. M., M. A. Reddy, L. L. Lanting, M. Wang, L. Meng, and R. Natarajan. 2008. Epigenetic histone H3 lysine 9 methylation in metabolic memory and inflammatory phenotype of vascular smooth muscle cells in diabetes. *Proc Natl Acad Sci U S A* 105:9047–52.

Wolff, G. L., R. L. Kodell, S. R. Moore, and C. A. Cooney. 1998. Maternal epigenetics and methyl supplements affect agouti gene expression in Avy/a mice. *FASEB J* 12:949–57.

Writing Team for the Diabetes Control and Complications Trial/Epidemiology of Diabetes Interventions and Complications Research Group. 2003. Sustained effect of intensive treatment of type 1 diabetes mellitus on development and progression of diabetic nephropathy: the Epidemiology of Diabetes Interventions and Complications (EDIC) study. *JAMA* 290:2159–67.

10 Maternal Nutrition, Intrauterine Development, and Disease Risks in the Offspring through Epigenetic Regulation of Gene Expression

Yuan-Xiang Pan, Rita Strakovsky, and Shasha Zheng

CONTENTS

10.1 Introduction 230
10.2 Placental Development and Gene Expression 231
10.3 Maternal Nutrition and Placental Development 233
10.4 Maternal Nutrition and Obesity 236
10.5 Maternal Nutrition and Type 2 Diabetes 239
  10.5.1 Recommendations 245
10.6 Maternal Nutrition and Cancers 245
  10.6.1 Breast Cancer 246
  10.6.2 Other Cancers 249
  10.6.3 Recommendations 250
10.7 Summary and Future Directions 250
Acknowledgments 250
References 250

10.1 INTRODUCTION

The importance of maternal nutrition during gestation with regard to pregnancy outcome has long been acknowledged. The tendency of mothers to briefly overhaul their lifestyle and diet is a classic one, with many women changing their habits and abandoning many of their vices, such as alcohol and caffeine drinking or cigarette smoking. This importance has been further emphasized in the recent new understanding of fetal programming on adult outcomes demonstrated by numerous laboratory and epidemiological studies. We are only beginning to understand how maternal nutrition and intrauterine environment may impact not only immediate pregnancy outcomes, but also the life and health course of the offspring. We need a better understanding of diet and dietary supplements during pregnancy and of whether diets are particularly low in some nutrients. Also, we need to understand how common epigenetic variations influence nutrient requirements during these periods. Associations between maternal nutrition and infant growth and development suggest that improving the diets of women of child-bearing age might be an important component of public health strategies aimed at improving the health, nutrition, and well-being of women themselves, as well as reducing the burden of chronic disease in their offspring.

The concept that epigenetic alterations occur during gestation and may have the ability to program adult diseases has led to a multitude of studies focusing on maternal behavior and health during pregnancy. Epigenetics is the study of heritable alterations in gene expression patterns not caused by changes in genomic DNA sequence. Genetic imprinting and X chromosome inactivation are two well-studied examples of epigenetic mechanisms related to gene expression regulation. Genomic imprinting controls gene expression depending on the parent of origin, and X chromosome inactivation controls gene expression by silencing one of the two copies of the X chromosome within each cell in females. Maintenance of different gene expression patterns among the diverse cell types also relies on epigenetic modifications using the molecular mechanisms such as DNA methylation, microRNA expression, and covalent modifications of the histone proteins that package DNA into chromatin structure in the nucleus.

Epigenetic alterations of chromatin via covalent modifications allow for heritable gene regulation without altering the DNA sequence. Chromatin comprises DNA, linker histones H1 and H5, core histones H2A, H2B, H3, and H4, and nonhistone proteins. Histones are strongly alkaline proteins packaging the DNA into structural units called nucleosomes. The nucleosome core is formed of two H2A-H2B dimers and an H3-H4 tetramer. The H3 and H4 histones have long tails protruding from the nucleosome that can be covalently modified at several places. Covalent modifications of the H3 and H4 histone tails, such as acetylation, biotinylation, methylation, phosphorylation, ubiquitination, and sumoylation, alter the interaction between histones and DNA, which consequently affects nucleosome locations as well as higher-order chromatin folding. In doing so, these posttranslational modifications form the basis for the epigenetic regulation of chromatin structure and gene function (Strahl and Allis 2000; Jenuwein and Allis 2001). Modifications of distinct amino acid residues in histones have unique functions. For example, dimethylation of lysine (K) 4 in histone H3 is associated with transcriptional activation of surrounding DNA, whereas

TABLE 10.1
Modifications of Distinct Amino Acid Residues in Histones Have Unique Functions

Type of Modification	Sites and Functions of Histone Modifications				
Acetylation	H3K9 activation[a]	H3K14 activation[a]	H4K5 activation[b]	H4K8 activation[c]	H4K16 activation[c]
Biotinylation	H3K4 gene expression[d]	H3K9 gene expression[d]	H3K18 gene expression[d]	H4K8 gene expression[e]	H4K12 gene expression[e]
Methylation (mono)	H3K4 activation[f]	H3K9 activation[g]	H3K27 activation[g]	H3K79 activation[g,h]	H4K20 activation[g]
Methylation (di)				activation[h]	
Methylation (tri)	activation[a]	repression[g]	repression[g]	repression[g] activation[h]	
Phosphorylation	H3S10 activation[i]	H3S28 activation[j]			
Sumoylation	H4 N-terminal tail Repression[k]				

[a] Koch et al. (2007).
[b] Kawasaki et al. (2000).
[c] Kuo et al. (1996).
[d] Kobza et al. (2005).
[e] Camporeale et al. (2004).
[f] Benevolenskaya (2007).
[g] Barski et al. (2007).
[h] Steger et al. (2008).
[i] Saloaga et al. (2003).
[j] Zhong et al. (2001).
[k] Shiio and Eisenman (2003).

tri-methylation of K9 in histone H3 is associated with transcriptional repression (Table 10.1).

10.2 PLACENTAL DEVELOPMENT AND GENE EXPRESSION

Placenta is a structure formed during gestation. Its main function is to provide supplies for fetal growth. Because the placenta acts as the sole nutritional bridge between mother and fetus, its development has been a crucial focal point of studies interested in the programming of adult diseases. Placental development has been connected to maternal nutritional status as well as the mother's overall health. In turn, placental insufficiency, in either size or transfer capacity, has been strongly linked to fetal IUGR (intrauterine growth restriction). Initial structural studies began by showing that placentas of IUGR offspring had altered vasculature, and therefore altered oxygen and nutrient transport. Further studies, focusing on the molecular causes of these placental phenotypes, attribute the alterations to imprinted genes, a relatively small group of genes that have vital functions within the placenta. Because imprinted genes are expressed only from one parental allele, any epigenetic change in these genes has dire consequences for placental development and transport. Additionally, because the protein products of these genes can affect other placental transporters, such as the system A amino acid transporter (Cutfield et al. 2007), their epigenetic regulation is of great consequence to overall placental and fetal development. Several placental microarray studies have been performed to determine how imprinted genes from IUGR births are expressed in comparison to healthy patients. A study by McMinn et al. (2006) looking at mouse placentas of IUGR and healthy animals found that IUGR placentas demonstrated different expression patterns of important imprinted genes, including those responsible for placental development and function. However, a more recent human microarray study by Sitras et al. (2009) comparing placental gene expression between IUGR and healthy women showed that although MBD3 (methyl-CpG binding protein domain 3, an imprinted gene important for recruiting histone deacetylases and DNA methyltransferases for methylation-dependent gene silencing) was down-regulated in the microarray analysis but not after RT-PCR confirmation, none of the other microarray-analyzed imprinted genes showed altered expression in IUGR cases. The McMinn group discussed that although there may be epigenetic modifications in syndromic IUGR, their results suggest that there may be little epigenetic effects in the pathogenesis of placental insufficiency. They also suggest that not all epigenetically modified genes are imprinted, so their analysis is a small representation of IUGR-respondent genes. Despite the somewhat conflicting results between these studies, it is clear that imprinted genes are, at least in part, affected during IUGR, and that it is possible that these genes, because of their imprinted nature, can be easily regulated by the same physiological conditions that cause IUGR itself, such as maternal stress, disease, and potentially diet.

Once the relationship between placental and fetal development was established by a variety of human and animal studies, the primary question became whether maternal diet alone is capable of producing the changes seen within IUGR fetuses and the associated small placentas. Despite the multitude of studies linking maternal dietary interventions to altered fetal outcomes, so far few of these have been able to connect specific epigenetic events within the placenta or fetus that respond to maternal diet. The most commonly utilized interventions have included maternal low-protein (LP), high-fat (HF), calorie-restricted, and one-carbon (1-C)-supplemented diets. Of these, only diets high in or depleted of 1-C units have shown direct and clear effects on placental and fetal epigenetic markers. Placental and fetal genes do respond to other maternal dietary manipulations, since several studies have shown that maternal LP and HF diets can alter placental transport and therefore fetal health (Kwong et al. 2007; McArdle et al. 2006; Symonds et al. 2003), but many more analyses will need to be performed to show whether these genes are regulated by epigenetic mechanisms and whether these changes will program the fetus into a disease-prone adult.

10.3 MATERNAL NUTRITION AND PLACENTAL DEVELOPMENT

Several approaches have been taken in order to study the effects of maternal diet on placental development and fetal programming. As expected, human studies are complicated by the ethical responsibility to do no harm. Therefore most human studies have been observational. These focus on interviewing mothers about their habits before and during pregnancy, and doing a few blood and placental tissue analyses. However, human studies do not allow scientists to intervene and observe the outcomes, compelling us to turn to animal models of fetal programming. Animal studies have undoubtedly demonstrated that maternal macronutrient undernutrition, particularly during rapid placental growth, affects placental and fetal growth (Nafee et al. 2008). In such models, where the intention is often to induce IUGR, a definite set of genes has been related to placental underdevelopment. These genes, known as imprinted genes, have been suggested to act as the primary epigenetic mechanisms responding to maternal malnutrition or diet imbalance. The nutrient–gene interaction has been implied for these genes in particular because of their monoallelic nature, where only one allele is expressed, from either the mother or the father, and the other is silenced. The chief mechanism for silencing of the second allele is via hypermethylation of the imprinting control regions (ICR), a CG-rich chromosomal domain within imprinted genes. During normal development, the first methylation events occur at fertilization, when all paternal genes, except the paternally imprinted genes, heterochromatin, and few repetitive elements, become actively demethylated. During normal early embryogenesis, the inner cell mass, the precursor of all fetal tissues, becomes hypermethylated, but the trophectoderm, which develops into the placenta, is hypomethylated. During early gametogenesis, genomewide methylation stops, allowing for the development of the intended parent-of-origin and sex-specific methylation patterns. Because of the many methylation events that occur during early implantation and further gestation, any shifts in the availability or deposition of methyl groups could have dire consequences for both the placenta and fetus.

Some of the placental imprinted genes and the observed effects of their manipulations are listed in Table 10.2. Dysregulation of imprinted genes is an important factor for placental and fetal development. Because of their monoallelic nature and vital function within the placenta, the effect of epigenetic modifications on imprinted genes has become a widely studied phenomenon. Of particular interest is the reciprocally imprinted gene cluster of Igf2/H19, which has been shown to regulate both placental nutrient transport as well as matching maternal nutritional status to fetal supply. The link between these two genes is a result of their coordinated regulation, where the maternally expressed H19 silences the paternally expressed Igf2, and vice versa. One major difference between methylation of Igf2 and other imprinted genes is that Igf2 requires methylation upstream of its promoter in order to recruit transcription machinery, so increased methylation will increase gene expression of this gene. Because Igf2 is maternally imprinted and H19 is paternally imprinted, these genes are believed to be coordinately regulated via methylation of their ICRs. The function of these genes is indispensable to the proper functioning of the placenta. Gene knockouts of Igf2P0, a specific promoter region of Igf2, result in significant shrinking of the placenta and poor outcome in the fetus, while overexpression of Igf2

TABLE 10.2
Imprinted Genes in Placenta

Imprinted Gene	KO or Abnormality Phenotype	References
Peg10	Incomplete placenta formation, undeveloped labyrinth zone, absence of spongiotrophoblasts	Ono et al. 2006
Mest/Peg1	Embryonic and placental growth retardation	Lefebvre et al. 1998 M. Takahashi et al. 2005
Peg3	Embryonic and placental growth retardation	L. Li et al. 1999
H19	Increase in placental weight, fetal overgrowth, slightly decreased fetal:placental ratio	Leighton et al. 1995
Igf2	Igf2: decreased placental weight, fetal weight, and fetal:placental ratio Igf2-P0: decreased placental weight and fetal weight, and increased fetal:placental ratio Igf2R: increased placental weight and fetal weight, and normal fetal:placental ratio Disruption of both Igf2R and H19: increased placental weight and fetal weight and decreased fetal:placental ratio	Chiao et al. 2002 Constancia et al. 2002, 2005 Efstratiadis 1998 Moore et al. 1997
Ascl2/Mash2	Placental failure: absence of spongiotrophoblasts, disorganized labyrinth zone	Guillemot et al. 1994, 1995
Cdkn1c	Abnormal placental development (placentomegaly and trophoblast dysplasia)	K. Takahashi et al. 2000 Yan et al. 1997
Phlda2	Placental overgrowth, decreased fetal-to-placental weight ratio, decreased labyrinthine zone	Frank et al. 2002 Salas et al. 2004
Grb10	Embryo and placenta overgrowth	Charalambous et al. 2003, 2010
Rtl1	Placental abnormalities and functional deficiencies, placental growth retardation	Sekita et al. 2008
Slc38a4	Placental and fetal growth restriction	Angiolini et al. 2006

resulted in the overgrowth of the placenta and an overabundance of nutrients to the fetus. However, studies of maternal nutrient restriction in rats have not been able to clearly demonstrate that limiting nutrient availability directly affects Igf2 expression in the placenta (Kwong et al. 2007). Lillycrop et al. (2005) showed that manipulating both maternal folate and protein can generate changes in methylation patterns of both Igf2 and H19 within the livers of the offspring. Because DNA methylation occurs primarily in utero, these changes within the promoters of offspring livers suggest that maternal environment has dire consequences for the offspring. Because the Igf2/H19 cluster has a similar regulation within the placenta as in other organs (such as liver) (Fowden et al. 2006), it has been suggested that maternal nutrition should affect the methylation and expression of these genes in the placenta, but the clear connection has been elusive. The effects of maternal nutrition on other imprinted genes are unclear. Recent unpublished observations from our lab (Rita Strakovsky) show that a 9% low-protein diet during Sprague-Dawley (S-D) gestation did not affect maternal food intake or body weight compared with an 18% protein control diet. There was a significant decrease in Peg3, Dnmt1, Cd68, and Emr1 expression in the maternal low-protein placenta, and a significant increase in ATF3, Glut1, and Snat2 expression. We detected changes in placental DNA methylation in only some of these genes (increased methylation in the coding region of Peg3 gene), indicating that control of gene expression for the remaining genes is accomplished by means other than DNA methylation. However, the observed changes in DNA methylation suggest that a maternal low-protein diet has the potential to cause epigenetic programming, which will have significant consequences for the offspring. Interestingly, livers of newborn male offspring born to mothers fed the gestational LP diet showed an increase in Peg3, as well as an increase in Snat2 and Glut1. Livers of newborn female offspring showed a decrease in Snat2 and Dnmt1 as well as a slight but significant increase in Pon2. The exact consequences of these changes and physiological significance for both the placenta and fetus are still under investigation.

Oversupply or limitation of methyl groups has been an important dietary intervention in animal studies interested in epigenetic events within the placenta and fetus. Additionally, human observational and folate-intervention studies have been an important tool for establishing the importance of methyl groups in placental and fetal development. As a cofactor, folate acts as a carrier of methyl groups for the production of S-adenosyl methionine, which becomes the primary donor of methyl groups for gene methylation. Additionally, any dietary factors that directly or indirectly feed into the folate cycle, including exogenous choline, methionine, folic acid, vitamins B_6 and B_{12}, and zinc, will have an effect on the availability of methyl groups. Although human studies have suggested a strong link between folate deficiency and placental growth retardation and rupture, few genetic analyses have shown the direct effect of nutrition on imprinted genes in placenta. However, the relationship is still being considered because of the placental phenotypes observed when these genes are knocked out or overexpressed and because of the fetal outcomes that are observed when the methyl supply is affected. The effects of gestational folate manipulation on fetal outcomes are slightly clear. In mice, choline or methionine restriction during pregnancy led to a decrease in CpG methylation of genes important for fetal brain development and cell-cycling inhibition (Zeisel 2009). Additionally, in sheep gestational methyl donor restriction led to insulin resistance and blood pressure elevation in male offspring, which was accompanied by a 4% change in global CpG methylation in the fetal liver (Langley-Evans 2008). Because, like all nutrients, folate transport to the fetus is controlled via the placenta, folate availability for the fetus may be a direct result of the crosstalk between the placenta and the fetus. In this case, methylation or demethylation within the placenta sends a message of maternal nutrient status, thus resulting in epigenetic events within the fetus. This message is not only important for immediate fetal development, but also for the heritable attribute of epigenetic changes. Although the term "epigenetic changes" describes events that do not directly affect the genome, the methylation of DNA and associated histones can be maintained throughout the process of cell division by DNMT1, and can therefore be passed on to future generations, to affect their health and disease susceptibility.

The rising obesity rate in the United States is a risk to the health of both women and fetuses. About one-third of U.S. women of reproductive age are obese. Obesity has become a common complication of pregnancy. Obesity during pregnancy raises the risk of numerous problems for the mother, such as hypertension and diabetes. Babies born to obese mothers are one-third more likely to have a significant birth defect such as neural tube defects, heart defects, and hydrocephaly. Birth defects are responsible for about 20% of all infant deaths in the United States (Stothard et al. 2009). Maternal obesity during pregnancy may also increase the risk for metabolic compromise in the offspring that is already apparent at birth (Catalano et al. 2009). What it is about obesity that may lead to all these developmental defects is not completely understood. One possibility is that the maternal obesity may have some type of nutritional imbalance that modifies the development of placenta and fetus aberrantly. While placental response to obesity has been moderately investigated, data related to the effect of maternal nutrition during obese pregnancy on placental development are limited. Several studies have shown that high-fat feeding during gestation may affect placental size, efficiency, and nutrient transfer. A recent unpublished observation in our laboratory (Rita Strakovsky) demonstrated that maternal obesity and gestational high-fat diet affect fetal weight in combination with a dysregulation of the placental Wnt pathway. The Wnt pathway involves a number of proteins that can regulate the production of Wnt-signaling molecules, their interactions with receptors on target cells, and the physiological responses of target cells that result from the exposure of cells to the extracellular Wnt ligands. Selective deletion studies in placenta demonstrated the importance of Wnt components in placental development, including angiogenesis, vascularization, cell adhesion, and differentiation. In our observation, Wnt pathway components, including Axin2, Dkk1, Sfrp5, and Wnt3a genes, and nuclear β-catenin, are affected differently by gestational high-fat diet in obese animals. Our results indicate that obese and normal-weight pregnant animals have different responses to dietary modulation, and specifically to the switch between high-fat and balanced diets. The fact that birth weights of offspring of obese dams remained lighter than control regardless of maternal diet may suggest that for an obese animal, dietary modulations during gestation may not be enough to lead to an improvement in pregnancy outcomes.

Maternal Nutrition and Offspring Disease Risk: Many adult diseases have fetal origins, including obesity, diabetes, and cancers. Links between maternal nutrition, intrauterine environment of the fetus, and susceptibility to these adult diseases have attracted great attention. An unfavorable prenatal environment can trigger epigenetic changes that increase the risk of developing those diseases and reduce the chances of postnatal survival. Major epigenetic programming events take place during fetal development. This chapter describes investigations that target the relationship between maternal nutrition and offspring's obesity, diabetes, and cancer risks.

10.4 MATERNAL NUTRITION AND OBESITY

Obesity is among those adult diseases that have roots in the fetal programming and is studied extensively because of its prevalence in the modern world (Stocker et al.

TABLE 10.3
Maternal Nutrition and Offspring's Risk of Obesity

Subjects	Maternal Nutrition	Observations	References
Human	Malnutrition	Low birth weight but gaining weight rapidly in childhood	Yajnik et al. 2003
	Malnutrition in early stages of gestation	Increased obesity rate in young men and 50-year-old women	G. P. Ravelli et al. 1976 A. C. Ravelli et al. 1999
	Overexposure to glucocorticoids	Reduction in body size but an increase in central distribution of fat	Gillman et al. 2006
	Smoking during pregnancy	Low birth weight but elevated risk of obesity at age 33	Power and Jefferis 2002
Animal	Calorie restriction	Induce hyperphagic behavior and obesity of offspring	Vickers et al. 2005
	Protein restriction	Low birth weight but catch-up growth on obesity	Ozanne et al. 2004
	High fat	A marked obesity independent of postnatal nutrition	Howie et al. 2009
	Malnutrition	Leptin treatment of offspring reverses obesity risk resulting from relative fetal undernutrition	Vickers et al. 2005
	Iron restriction	Low birth weight	Lewis et al. 2001

2005; Fernandez-Twinn and Ozanne 2006). Maternal nutrition status is one of the most important causes of programming that influence the risk of obesity in later life. Both maternal undernutrition and overnutrition are linked to abnormalities of fetal growth and their postnatal obesity risk. The intrauterine environment for fetal growth can be altered by maternal calorie/protein restriction, high-fat feeding, iron intake, mother's smoking status, and other maternal factors. Fetal response and adaptation to the altered intrauterine environment during the critical period of development may lead to long-term changes and prevalence of many chronic diseases in postnatal life including obesity (Table 10.3).

Maternal undernutrition during the pregnancy has been demonstrated to result in IUGR and low birth weight. Malnutrition of pregnant women is still an important public health problem in the world, especially in the developing countries. In order to adapt to the maternal undernutrition environment and increase the chance of postnatal survival, the fetus responds with a number of strategies, such as changing its metabolic rate, altering the production of hormones, storing nutrients as fat, and redistribution of fetal blood flow to protect brain at the expenses of other tissues such as muscle. These factors lead to a slower fetal growth and low birth weight.

The term IUGR is often used and assigned to newborns with a birth weight and/or length below the 10th percentile for their gestational age and whose abdominal circumference is below the 2.5th percentile with pathologic restriction of fetal growth (Wollmann 1998). During recent years significant progress has been made in the understanding of IUGR-associated pathophysiology. Data have shown that IUGR is associated with a late-life increased prevalence of metabolic syndrome, like obesity. Large cohorts in epidemiological research programs have also studied IUGR phenomena. For example, an Indian cohort study showed that babies who are thin and lack muscle at birth gain weight rapidly in childhood, leading to a disproportionately high fat mass in later life (Yajnik et al. 2003). It is also shown that timing of maternal nutrient restriction has a major influence on the outcome in terms of predisposing the offspring to adult obesity. In a Dutch Hunger Winter Study, maternal undernutrition in the first and second trimesters of pregnancy is associated with higher obesity rates in young men (G. P. Ravelli et al. 1976) and higher BMI and waist circumference in 50-year-old women (A. C. Ravelli et al. 1999).

A "thrifty phenotype" hypothesis was proposed by Hales and Barker (1992) in order to link the fetal intrauterine environment to the susceptibility to chronic diseases in later life. It is believed that adaptations associated with fetal malnutrition become detrimental to the health of the offspring if they experience a period of adequate or plentiful nutrition leading to postnatal obesity.

Maternal calorie restriction to 50% of ad-lib in the last week of pregnancy retards beta-cell development. Continued restriction of the mother during the lactation period results in a permanent reduction of beta-cell mass and impaired glucose tolerance in the offspring. Calorie restriction in the pregnancy period was also found to induce hyperphagic behavior and obesity of offspring (Vickers et al. 2000). IGF-I treatment of 6-month-old undernourished offspring was found to alleviate hyperinsulinemia and obesity. Leptin treatment of neonatal rats normalized caloric intake, body weight, fat mass, and insulin concentrations in later life.

The maternal LP model is one of the most extensively studied models. Ozanne and coworkers (2004) tested the "thrifty phenotype hypothesis" and investigated the effects of fetal programming and postnatal catch-up growth on obesity and longevity. The offspring of LP-fed dams were switch to mothers receiving control diets for lactation. The body weights of those rats caught up and exceeded the weight of the control group by 7 days of age, and this pattern persisted in adulthood (Ozanne et al. 2004).

Impacts of Western-style diet, which contain a high percentage of saturated fats, on offspring risk have attracted the attention of researchers in developed countries. Animal studies have shown the adverse effects caused by high-fat feeding including fetal insulin resistance, gender-specific hypertension, and increased adiposity. A marked obesity in male and female offspring of Virgin Wistar rats can be induced by maternal high-fat intake during the pregnancy. These phenomena are independent of postnatal nutrition (Howie et al. 2009).

Data from human studies comparing breast-fed and bottle-fed infants also suggest that the lactation period is a critical time window for determination of obesity risk in humans. Breast-fed babies are at reduced risk of obesity compared to those who were formula fed, as bottle-fed infants have higher total and protein caloric intake (Locke 2002). Higher circulating leptin levels in breast-fed infants may contribute to the subsequent reduced obesity risk. Breast milk is also rich in long-chain polyunsaturated fatty acids, which are thought to be protective against the development of obesity.

Leptin plays a major role in the regulation of metabolism and neuroendocrine functions. It increases energy expenditure and modulates appetite by inhibition of hypothalamic arcuate nucleus neurons through the leptin receptor (Ashworth et al. 2000). Obese individuals have sustained elevated adipocyte-derived leptin levels, which may cause selective leptin resistance at the hypothalamic level. In rodents, leptin levels are very low at birth and then display a surge at the end of the second postnatal week. Injecting leptin into the offspring of undernourished mothers prevents hyperphagia and excessive body weight and fat mass gain of neonatal rat offspring (Vickers et al. 2005). In contrast, treating male offspring of normally nourished mothers caused an induced weight gain and increased total body adiposity (Vickers et al. 2008). Altering leptin levels during early postnatal key periods of hypothalamic development may induce long-lasting susceptibility to a postnatal obesity depending on the prenatal maternal nutrition status. In humans, fetuses in gestational age and known to be at risk of obesity in later life had lower levels of cord leptin (Ong 2006).

About 50% of pregnant women are iron deficient. The main causes of iron deficiency are poor absorption of iron due to insufficient vitamin C levels and inadequate daily intake of iron or high menstrual blood loss. Maternal anemia has been shown to associate with low birth weight. The rodent maternal iron restriction model resulted in low-birth-weight offspring and programs hypertension and obesity through adult life (Lewis et al. 2001).

In humans, glucocorticoids are administered during pregnancy for the treatment of neonatal respiratory morbidity and maternal asthma. But overexposure of glucocorticoids results in the fetus with reduced birth weight and program responses that lead to later adult disease. Glucocorticoid is regulated by its receptor in the cell. An isoform of 11β-hydroxysteroid dehydrogenase,11β-HSD-2, inactivates corticosol by converting it to corticosterone. Dietary protein restriction attenuates 11β-HSD-2 in the placenta and may provide a mechanism relating the maternal nutrition to fetal programming (Lindsay et al. 1996). In an epidemiology study, it was found that overexposure to maternal glucocorticoids in humans is associated with reduced birth weight and an increase in central adiposity (Gillman et al. 2006).

In a British cohort study, C and Jefferis (2002) studied the insult effect of maternal smoking during pregnancy on fetal growth and the influence of obesity risk through childhood to age 33. It was found the infants of mothers who smoked in pregnancy have lower birth weight than infants of nonsmokers and had an increased risk of obesity from adolescence to age 33.

10.5 MATERNAL NUTRITION AND TYPE 2 DIABETES

Diabetes influences more than 180 million people worldwide, and this number is most likely to double by 2030 (www.who.int/mediacentre/factsheets/fs312/en). Type 2 diabetes, a major prevalent form of diabetes, is suggested to have its origin during fetal development. Maternal nutrition during all the stages of gestation and lactation plays an important role in the control of type 2 diabetes risk. Although type 1 diabetes has been suggested to also have its origins in the fetal period, most epidemiological studies have confirmed the association between low birth weight and adult impairment of glucose metabolism and increased predisposition to type 2 diabetes

TABLE 10.4
Maternal Nutrition and Offspring's Risk of Type 2 Diabetes

Subjects	Maternal Nutrition	Observations or Conclusions	References
Human	Low birth weight	Smallest at birth (<2.5 kg) were more likely to have impaired glucose tolerance or type 2 diabetes	Barker et al. 1993
		Low birth weight is associated with the increased risk for the onset of type 2 diabetes	Harder et al. 2007
		An increased ratio of placental weight to birth weight exhibited impaired glucose tolerance or type 2 diabetes in adult life	Godfrey and Barker 2001
	IUGR	IUGR children have a specific impairment in insulin sensitivity	Hofman and Cutfeld 2006
	Obesity	Big babies had the increased risk of developing type 2 diabetes	Hadden 2008
Animal	Calorie restriction	A defect in β-cell function, reduced insulin content in β-cell, hypoinsulinemia, hyperglycemia, and onset of type 2 diabetes in later life	Jimenez-Chillaron et al. 2005
	Protein restriction	Adverse effect on pancreas development, β cell proliferation	Snoeck et al. 1990
		A reduction in both GLUT4 and PKC zeta in muscle in males	Ozanne et al. 2005
		Down-regulation of pdx-1 and IGF-II	Arantes et al. 2002 Petrik et al. 1998
		Up-regulation of PEPCK and down-regulation of glycolytic glucokinase	Desai et al. Rees et al. 2000
		Down-regulation of p110β subunit of PI3-kinase	Ozanne et al. 2005
		Lower methylation level in the promoter regions of PPAR and GR promoter and higher expression of these genes in F1 and F2 generations	Burdge et al. 2007

in adult life. It was also suggested that some physical parameters of the baby, such as abdominal circumference, body length, head circumference, low birth weight up to one year of age, and catch-up growth are associated with the development of type 2 diabetes risk in later life. As we discussed regarding the relation of maternal nutrition with obesity, it is because maternal undernutrition causes the fetus to adapt through endocrine and metabolic changes that such adaptations result in insulin resistance and a predisposition of offspring to developing type 2 diabetes (Table 10.4).

Barker et al. (1993) were the first to associate the birth size to the later development of metabolic syndrome in adult life. Over 20,000 newborns between 1911 and 1930 in a county were studied, and they demonstrated that men who were smallest at birth (<2.5 kg) were nearly seven times more likely to have impaired glucose tolerance or type 2 diabetes than those heaviest at birth (>4.3 kg). Hofman and Cutfield (2006) showed that short prepubertal IUGR children have a specific impairment in insulin sensitivity compared to their normal-birth-weight peers. A recent study conducted in Minnesota found that excess mortality for the adult-onset type 2 diabetes was concentrated in individuals who have abnormal birth weight. A recent meta-analysis conducted by Harder and coworkers (2007) involved a total of 132,280 people and concluded that low birth weight is associated with the increased risk for the onset of type 2 diabetes in later life. Another study conducted in Preston, UK, performed glucose tolerance tests on 226 men and women and found that an increased ratio of placental weight to birth weight exhibited impaired glucose tolerance or type 2 diabetes in adult life (Godfrey and Barker 2001).

Further evidence of the contribution of low birth weight to type 2 diabetes comes from some animal studies. Jimenez-Chillaron and coworkers (2005) performed a study to determine whether insulin resistance or insulin secretion dysfunction is associated with low birth weight and type 2 diabetes. In offspring of undernourished mothers, comparisons of offspring from mice fed with a control diet and offspring from mice fed with an undernourished diet (restricted to 50% of that of controls) were studied to explore whether undernutrition contributes to a defect in β-cell function resulting in abnormal glucose-stimulated insulin release and eventually in type 2 diabetes. Insulin content of β-cells was reduced by 25% in mice of the undernourished mothers, the cause of which was suggested to be that reduced nutrient delivery compromised insulin secretion. Others have also shown that nutrient restriction during pregnancy results in impaired glucose tolerance in low-birth-weight mice. A restriction in the maternal food intake by 50% during the last week of pregnancy led to the generation of IUGR pups with impaired β-cell development and a reduction in plasma glucose and insulin concentrations. Moreover, at approximately 8 months of age, the offspring demonstrated 40% decrease in pancreatic insulin content and an increase in nonfasting plasma glucose concentrations, which eventually led to fasting hypoinsulinemia, hyperglycemia, and insulin-to-glucose ratio, all of which contribute to the development of type 2 diabetes (Martin-Gronert and Ozanne 2005).

It is recommended that the mother increase her protein intake during pregnancy to provide the additional nitrogen source that is demanded by both the mother and the fetus because of an increase in protein metabolism due to the rapid growth demands of the fetus. In both early and late pregnancy periods, the amount of energy obtained from protein is positively associated with birth weight.

A great deal of evidence comes from the extensively studied LP animal models. Some mechanism insights of early growth restriction caused by insufficient protein were obtained from such a model. The model was established by Snoeck and coworkers (1990), where dams were fed a diet containing 8% protein throughout pregnancy and lactation and compared offspring to those of a control dam fed an isocaloric 20% protein diet. A low protein diet reflects the relevance to cultures, economies, and social-economic groups, where protein sources (meat, egg, etc.) are expensive. The effects on pregnancy outcome are therefore expected to have no effect on litter size. The model has firstly shown that protein restriction dams deliver litters with lower birth weight. There were also adverse effects on pancreas development and β-cell proliferation. Insulin content was reduced and its secretion was impaired. Giving offspring control diet improved these adverse effects in a time-dependent matter. After 6 weeks to 3 months, litters demonstrate improved glucose tolerance, but after 15 months they had an impaired glucose tolerance, and by 17 months, frank diabetes with insulin resistance was observed in male low-protein offspring. Insulin-stimulated glucose uptake was reduced in both muscle and adipose tissue of 15-month-old males. The observation was accompanied by a reduction in both GLUT4 and PKC zeta in muscle (Ozanne et al. 2003). Female offspring demonstrated only hyperglycemia and impaired glucose tolerance. Fernandez-Twinn et al. (2003) employed a maternal protein restriction rat model throughout gestation and lactation. Offspring were again born smaller than controls and developed diabetes, hyperinsulinemia, and tissue insulin resistance in adulthood.

In the mechanistic investigations, in pups delivered from LP-fed dams, expression of pancreatic duodenal homeobox-1 (pdx-1) and insulinlike growth factor II (IGF-II) in the islet cells were reduced (Arantes et al. 2002; Petrik et al. 1998). Pdx-1 is responsible for the regulation of a number of genes that are accountable for the proper development and maturation of the pancreas. IGF II functions in the prevention of apoptosis from development. When these two gene expressions were reduced, development of the pancreas and production of insulin are compromised, thus contributing to an increased predisposition to adult-onset type 2 diabetes. In other animal studies, an increase in hepatic gluconeogenic phosphoenolpyruvate carboxykinase (PEPCK) activity (Desai et al. 1997b) and reduced glycolytic glucokinase levels (Desai et al. 1997a; Rees et al. 2000) have been reported in response to a maternal low-protein diet. Of note is the decrease in glycolytic glucokinase activity seen in subjects diagnosed with type 2 diabetes (Vaxillaire and Froguel 2006). Overexpression of PEPCK has also been shown to cause the development of type 2 diabetes in mice (Franckhauser et al. 2006). Ozanne and coworkers (2005) reported that the offspring of LP diet-treated mothers had a reduction in epididymal adipocytes and an increase in basal and insulin-stimulated glucose uptake. However, both the muscle and adipocyte insulin receptor expression in the restricted-protein offspring were similar to that of the control group, and the molecular alteration resulting in insulin resistance must occur downstream of the insulin receptor. In this regard, a reduction in the expression of the p110β subunit of phosphatidylinositol (PI3-kinase) and a reduction of the activity of insulin-stimulated protein kinase B were found in the adipocytes of IUGR animals at 3 and 15 months of age.

Accumulating evidence has shown that epigenetic regulation of transcription is a mechanism for inducing changes in phenotype of fetal programming and the increased risk of diabetes in adulthood. The reversible changes that occur as a result of heritable modifications without involvement of primary DNA sequence alterations are defined as epigenetic changes. Histone modification and DNA methylation are two main molecular events known to initiate and sustain such epigenetic modifications.

Our laboratory also investigated the impact of a maternal low-protein diet on the expression of glucose transporter 4 (GLUT4) in offspring skeletal muscle (Zheng and Pan, submitted). A maternal LP diet during pregnancy and/or lactation was reported as affecting postnatal growth, appetite, triglyceride and cholesterol concentrations, as well as insulin resistance in male but not female offspring (Zambrano et al. 2006). However, the molecular mechanism underlying the sex-differentiated carbohydrate metabolism and insulin sensitivity is rarely explained. The major insulin-responsive protein involved in glucose absorption is glucose transporter 4 (GLUT4). We observed sex-dependent GLUT4 mRNA expression and increased GLUT4 protein content in female pup skeletal muscle by maternal low protein. Analysis of transcriptional and epigenetic regulation underlying increased skeletal muscle GLUT4 expression in offspring rats also revealed the regulatory mechanisms. Increases in regulatory genes associated with carbohydrate metabolism, C/EBPβ and Nur77, were observed in female pups whose mothers were fed a low-protein diet. Modifications of chromatin structure, including acetylated histone 3, acetylated histone 4, and dimethylated histone 3 at lysine 4, were detected at a significantly increased level at the GLUT4 promoter region in female pup muscle following maternal low-protein diet. Such modifications include increased levels of acetylated histone 3, acetylated histone 4, and di-methylated histone 3 at lysine 4 in female offspring rats. However, the restricted diet did not activate the amino acid response pathway or alter GLUT4 expression through DNA methylation. These results demonstrated that maternal protein restriction during pregnancy induces GLUT4 expression in female offspring skeletal muscle but not in males, which may indicate sex-dependent adaptation of insulin sensitivity and of glucose metabolism to maternal-protein diet (Figure 10.1).

In another report, Burdge et al. (2007) investigated whether the altered methylation of PPARγ and glucocorticoid receptor (GR) promoters are passed to the F2 generation in response to a maternal low-protein diet. It was observed that hepatic PPARα and GR promoter methylation was significantly lower in the protein restriction group in the F1 and F2 generations. There were also trends toward a higher expression of PPARγ, GR, acyl-CoA oxidase, and PEPCK in the F1 and F2 males, although this was significant only for PEPCK. These data showed that the altered methylation of gene promoters in the F1 generation by maternal protein restriction during pregnancy is transmitted to the F2 generation. This may represent a mechanism for the transmission of induced phenotypes between generations. The potential to affect subsequent generations by maternal undernutrition during pregnancy has also been reflected in other studies. The Dutch famine study showed that the pregnant women who experienced undernourishment during the famine period consisting of 5 months of food deprivation in the winter of 1944–1945 gave birth to offspring that were of average or normal birth weight. Those offspring went on to give birth to offspring of low birth weight, which illustrated again that maternal undernutrition can affect more than one generation (Stein and Lumey 2000).

Maternal undernutrition can result in the overexposure of the fetus to glucocorticoids. An animal model showed the adverse effects of overexposure to glucocorticoids continue to further generations. Offspring overexposed to dexamethansone (glucocorticoid) during late pregnancy have low birth weight and developed glucose intolerance and hypertension in adulthood (Seckl 2004). When these offspring gave birth to offspring of their own, their offspring also had low birth weights and impaired glucose intolerance, thus confirming that adverse events during pregnancy affect not only the first generation but also future generations. A recent study reveals that a low-protein diet during gestation had adverse effects on glucose, insulin, and

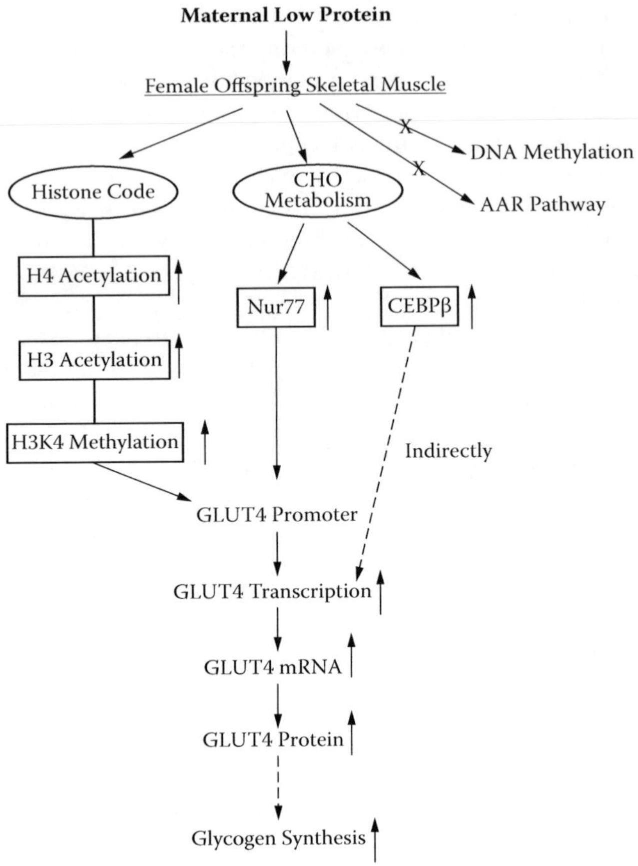

FIGURE 10.1 Schematic presentation of the combinatorial progression of upstream regulators and epigenetic modifications leading to the increased GLUT4 transcription in female offspring skeletal muscle by maternal low-protein diet.

leptin metabolism, resulting in insulin resistance in adult F2 offspring (Lindsay et al. 1996). Future prenatal and gestational nutritional recommendations may need to be formulated on the basis of region, culture, and risk assessment criteria.

Both clinical and experimental studies have shown that maternal obesity has been linked to the alteration in fetal development and increased risk of diabetes. Particularly, human studies have linked maternal diabetes with a higher incidence of type 2 diabetes in the offspring in adulthood (Hadden 2008). The concept of prediabetes has come to be recognized again with the worldwide epidemic of type 2 diabetes. Maternal hyperglycemia has been related to fetal macrosomia in diabetic pregnancy. The recent large studies about the outcome of pregnancy in type 2 diabetic mothers in England, Wales, and Northern Ireland in 2002–2003 shows that there is still an increase in the numbers of both big babies and small babies born to these diabetic women in comparison with the national population data (babies ≥4.5 kg: type 2 diabetes 5.1%, national population 1.7%) (Confidential Enquiry into

Maternal and Child Health Pregnancy in Women with Type 1 and Type 2 Diabetes in 2002–2003; England, Wales, and Northern Ireland).

In a small study of 99 babies with verified birth weight > 5 kg in 45 prediabetic women, 5 years prior to the diagnosis of diabetes, 36% of the mothers producing a big baby were obese, while none of the babies born to nonobese mothers weighed as much as 5 kg. A similar difference persisted at 1.5 years before the onset of diabetes but was reversed after the onset of diabetes, where non-obese mothers had a 28% prevalence of babies above 5 kg (Pedersen 1977).

Hypotheses exist to explain the relationship between intrauterine nutrition and risk of diabetes relevant to the big baby and the small baby. The Pedersen hypothesis, simply stated, is that maternal hyperglycemia causes fetal hyperglycemia by direct transplacental passage. After the fetal islet cells become functional, this leads to excess fetal insulin secretion, which facilitates excess fetal growth, particularly the fat component. The alternative but not exclusive concept, known as the Barker hypothesis, is that maternal malnutrition associated with placental insufficiency will lead to impaired fetal growth and small babies, which are associated with neonatal insulin resistance and long-term epidemiological evidence of type 2 diabetes and hypertension. It is possible to consider these two ends of the nutritional background to type 2 diabetes in a single concept of transgenerational diabetes. The studies of David McCance and coworkers (1994) on the Pima Indians demonstrated in that population, where type 2 diabetes is very prevalent, that there was a "U-shaped" curve where type 2 diabetes at age 30–34 years was more common among those who had been less than 2.5 kg or more than 4.5 kg at birth.

10.5.1 Recommendations

To prevent obesity and reduce diabetes risk, one immediate measure would be to ensure adequate maternal nutrition throughout pregnancy and lactation; however, it is important that any dietary enhancement in the mother should promote lean and not fat mass. Smoking is the lifestyle choice that affects the unborn child. Stress and infection are also factors that are not easily controllable but have adverse effect for the fetus. Suitable therapeutic approaches must be applied to address these factors. On the other hand, the risk of obesity could be reduced by controlling postnatal nutrition. Encouragement to breast-feed should remain a high priority given the benefits of immunity to certain diseases as well as protection against obesity.

10.6 MATERNAL NUTRITION AND CANCERS

Cancers are age-related diseases caused by the interaction of genetic susceptibility and environmental factors (Uauy and Solomons 2005). In recent years, cancer prevention has attracted great attention. Avoiding exposure to some environmental factors, such as radiation, carcinogenic compounds, infectious agents, some dietary factors, and contaminants in food, water, and air, has been used to prevent cancers from occurring. However, the knowledge of cancer risk management is still very limited, especially in early-life intervention. More recently, approaches to identify early metabolic syndrome and its underlying genetic susceptibility have shown promising prevention of cancers before any symptoms or laboratory indicators. For example, obesity has been recognized as a contributory risk factor for some types of cancers (Key et al. 2004). Relationships between maternal nutrition and obesity risk have been addressed in previous discussions. There is research evidence indicating that maternal nutrition should be considered as a factor to control postnatal cancer risk. The first human evidence came from a study of transient exposures to chemical substances during embryonic and fetal life on triggering cancer in offspring in 1970 (Herbst et al. 1999). Herbst found seven women who were treated with diethylstilbestrol (DES) during the first trimester of pregnancy, and they have a clear-cell vaginal adenocarcinoma. Diethylstilbestrol was a drug that was believed to reduce the incidence of premature births and neonatal deaths. However, the Herbst study demonstrated that the diethylstilbestrol treatment in the pregnancy period has the potential adverse consequence of exposing the fetus to a toxic substance and may increase fetal susceptibility to cancer during such a rapid growth and differentiation period. Along this line, maternal intake of the synthetic estrogen during the pregnancy period increases multigenerational uterine cancer risk in female offspring, and this effect is demonstrated in animals as well (Giusti et al. 1995). Thus recently natural estrogens have been classified as known human carcinogens (Table 10.5).

10.6.1 Breast Cancer

The link between the breast cancer susceptibility and the fetal environment has attracted extensive attention recently (Hilakivi-Clarke and de Assis, 2006). Maternal exposure to dietary factors during pregnancy or lactation alters the risk of offspring in developing breast cancer. Some exposures induce epigenetic changes and cause gene expression alteration in the mammary gland development. However, evidence is still lacking to link the epigenetic changes in the fetal development to the increased vulnerability for malignant transformation in postnatal life. Thus clarifying the target genes in the epigenetic modifications and the underlying mechanism may lead to useful approaches to prevent breast cancer in women.

Recent epidemiological studies have indicated that women with either high birth weight (>4kg) or very low birth weight are at increased risk of breast cancer (Troisi et al. 2003). The effect is more apparent in women who developed premenopausal rather than postmenopausal breast cancer. For instance, in a British cohort of 2547 girls followed from birth in 1946 to 1999, girls who were heavy at birth reached menarche earlier than others with similar growth in infancy. The findings suggested that women who grow faster in childhood and reach an adult height above the average for their menarche category are at a particularly increased risk of breast cancer (De Stavola et al. 2004). Elevated estrogenic environment is related to high birth weight in most studies except Asian women, who have a low risk of breast cancer but have significantly higher estrogen concentrations during pregnancy than Caucasian women (Shibata et al. 2002). This phenomenon might indicate that the increased risk of breast cancer in daughters caused by high concentration of estrogen during pregnancy could be potentially overridden by other factors during the pregnancy or postnatal life. On the other hand, a very low birth weight may also increase the risk of developing breast cancer (Ahlgren et al. 2003).

TABLE 10.5
Maternal Nutrition and Offspring's Risk of Cancer

Subjects	Maternal Nutrition	Observations	References
Human	Diethylstilbestrol exposure in pregnancy	Vaginal adenocarcinoma in mothers and increased fetal susceptibility to develop uterine cancers	Herbst et al. 1999 Giusti et al. 1995
	Elevated estrogenic environment	High birth weight and increased risk of breast cancer	De Stavola et al. 2004
	Maternal intake of fruits, vegetables, and fish and seafood	Reduced risk of developing acute lymphoblastic leukemia in offspring	Petridou et al. 2005
Animal	Exposure to synthetic estrogen	Increased carcinogen-induced mammary tumorigenesis in the offspring	Ottaviano et al. 1994
	In utero overexposure of estrogen	Increased estrogen receptor α level in the offspring's mammary gland and overexpression of estrogen-regulated genes such as cyclin D1, pSrc, and pAkt and down-regulate a tumor suppressor gene Caveolin-1	Hilakivi-Clarke et al. 2006
		Rats with high birth weight have reduced mammary ERα and develop mammary tumors earlier than control rats	De Assis et al. 2006
	High-fat diet	Increased E2 level in utero environment and increased susceptibility of female offspring to develop carcinogen-induced mammary tumors as adults	Hilakivi-Clarke et al. 1996
	Genistein exposure	Increased number of TEBs, which are targets for malignant transformation	Hilakivi-Clarke et al. 2006
	Soy intake	Did not increase offspring's risk of developing mammary tumors	Trock et al. 2006

In a rat model, exposing the mother to E2, the synthetic estrogen, during pregnancy increases carcinogen-induced mammary tumorigenesis in the offspring. Elevated maternal and/or cord-blood levels of estrogens, leptin, adiponectin, and IGF-1 might all increase the risk of a daughter developing breast cancer, but an increase in one of these does not necessary translate to an increase in later breast cancer risk. Rather, many alterations in utero could contribute to the later breast cancer risk modification.

At the cellular and molecular level, estrogen is known to act through two distinct pathways: (1) through intracellular signaling following activation of membrane bound estrogen receptors and (2) through more classical genomic routes in which estrogen binds to nuclear estrogen receptors leading to transcriptional activation.

Estrogens function through binding to estrogen receptors, ERα and ERβ. These estrogen receptors are nuclear transcription factors that regulate gene expressions. Genes encoding ERs are targets of epigenetic modifications and can be silenced by hypermethylation (Ottaviano et al. 1994). In normal mammary epithelial cells, ERα is not located in the cells that proliferate, but close to them, whereas in breast tumors, proliferating cells frequently express ERα (Potten and Morris 1988). It was suggested that ERβ might negatively control cell proliferation and may have a protective role in the normal breast (Cheng et al. 2004). In utero exposure of increased estrogen could increase or reduce the concentration of ERα in offspring (Newbold et al. 2004), and correspondingly trigger the ER-mediated pathways, including cell proliferation and expression of estrogen-regulated genes. It has been shown that in-utero exposure to elevated E2 increases ERα levels in the offspring mammary gland and up-regulates estrogen-regulated genes such as cyclin D1, pSrc, and pAkt, and down-regulates a suppressor gene, Caveolin-1. It is known that the expression of Caveolin-1 inhibits phosphorylation deactivation of the oncogenes Src and ras and prosurvival factors such as Akt (Acconcia et al. 2005). Therefore it was suggested that elevated exposure to estrogen in utero would alter postnatal susceptibility to malignant transformation in breast tissue by altering the expression of ERα and ERα-mediated downstream gene expression.

On the other hand, low concentrations of ERα in the mammary gland are also associated with increased risk of breast cancer. For example, rats with high birth weight have reduced mammary ERα and develop mammary tumors earlier than control rats (de Assis et al. 2006). Maternal leptin intake during pregnancy reduces ERα in the mammary gland and increases mammary tumorigenesis in offspring.

Maternal diet and nutrition are important factors because during pregnancy they may modify pregnancy hormone levels, especially estrogen levels in utero, and affect the later risk of breast cancer of the offspring. A high-fat diet increases E2 concentration during pregnancy. A high-fat diet, which is high in n-6 polyunsaturated fatty acids, also increases the susceptibility of female offspring to develop carcinogen-induced mammary tumors as adults (de Assis et al. 2006). It also shows mammary tumorigenesis in outbred mice and mice that overexpress the c-neu oncogene (Luijten et al. 2004). These phenomena may be caused by increased exposure of the fetus to estrogens because of a high-fat diet (Hilakivi-Clarke et al. 1996). The Ozanne group (Fernandez-Twinn et al. 2007) reported that compensatory mammary growth in the offspring following protein restriction during both pregnancy and lactation increased the number of early mammary tumors. The increased incidence of mammary tumors was accompanied by elevated expression of receptors to insulin, IGF-1, epidermal growth factor, and estrogen. This report became the first study on the effect of maternal low-protein diet on mammary cancer development and provided an extremely relevant model for further study of such processes, and ultimately the development of potential interventions. Our laboratory employed two well-established models (Lillycrop et al. 2005; Fernandez-Twinn et al. 2007) to measure the effect of maternal protein restriction during pregnancy on the promoter chromatin status and expression of p16 (CDKN2A) in the mammary gland of the offspring after weaning. The protein p16 (CDKN2A) is known to negatively control cell cycle and to retain pRb in a hypophosphorylated form in order to inhibit cellular growth (Voorhoeve and Agami, 2004). Down-regulated p16 (CDKN2A) expression has been related to aggressive cell growth, leading to many types of tumors in humans (Enders 2003). Our present study took advantage of the clinical features of the rat maternal low-protein model and addressed the mechanistic question by uncovering changes in epigenetic regulation related to p16 (CDKN2A). Our results show the histone modifications at the p16 (CDKN2A) promoter in the mammary gland of offspring rats. An 84.6% decrease in acetylated H4 and a 92.5% decrease of methylation at H3K4 were detected in the maternal LP pups, correlating to the 75.8% transcript repression. The data obtained from our analysis support that maternal low protein programs p16 (CDKN2A) gene expression in offspring mammary glands through histone modifications and transcriptional repression. A rapid compensatory mammary epithelial growth in LP offspring pups may occur after weaning with normal diet, which led us to hypothesize that maternal protein restriction resulted in the repression of some cell cycle regulators, including p16 (CDKN2A), which could contribute to increase the offspring's breast cancer risk later in life.

Genistein, a phytochemical in soybeans, induces epigenetic changes and influences estrogenic activity by binding to estrogen receptors (Uauy and Solomons 2005). Exposure in utero increases the number of TEBs, which are targets for malignant transformation (Hilakivi-Clarke et al. 2001). In contrast, prepubertal exposure to genistein reduces later mammary tumorigenesis. Maternal soy intake did not increase offspring's risk of developing mammary tumors, although the soy diet contained high levels of genistein (Trock et al. 2006). The soy diet also increased pregnancy estrogen levels. These results indicate that soy must contain some additional components, which reverse the effects of genistein on offspring's breast cancer risk when administered in utero.

10.6.2 Other Cancers

Maternal intake of the synthetic estrogen DES during pregnancy induces multigenerational uterine cancer risk in daughters. Investigations reveal that persistent expression of the proto-oncogene c-fos (S. Li et al. 2003) and the lactoferrin gene, and permanent repression of Hoxa-10 and Hoxa-11 in the female uterine tract (Block et al. 2000), correlate with increased uterine cancers and structural defects in the reproductive system. In particular, the CpG sites in the promoters and other regulatory regions of the c-fos and lactoferrin genes are hypomethylated. However, promoter CpG methylation of Hox genes is not altered by DES exposure, which indicates histone modifications might be involved (S. Li et al. 2001).

Causes of developing acute lymphoblastic leukemia (ALL) have not been revealed. Some evidence indicates that intrauterine environments may play an important etiologic role. At first, leukemia clone-specific chromosomal translocations are present at birth in children who have later developed leukemia (Hjalgrim et al. 2002). Second, birth weight has been suggested to be associated with risk of ALL (Hjalgrim et al. 2003). The relationship is particularly observed among children ages <5 years. But there are some null results being reported as well. Maternal diet and nutrition during pregnancy should be important factors in controlling the risk of ALL if the relationship between birth weight and ALL risk exists. A recent epidemiology study of ALL among children ages <5 years in Greece with a focus on maternal diet during the pregnancy indicates that the risk of ALL in the offspring was lower with increased maternal intake of fruits, vegetables, and fish and seafood. The risk of ALL is higher with increased maternal intake of sugars and syrups, and meat and meat products (Petridou et al. 2005).

10.6.3 Recommendations

Mothers should start pregnancy with a healthy weight for their height and avoid excessive or low weight gain during pregnancy. Key micronutrients such as folate, iron, copper, zinc, and vitamins are important for normal embryonic development and fetal growth. The present bigger-is-better model may increase cancer risk in later life. For women who are already overweight or obese, careful weight management in the prenatal period and a modest weight loss in the postpartum period can have important health benefits and can be undertaken safely for mother and infant.

10.7 SUMMARY AND FUTURE DIRECTIONS

Maternal nutrient supplementation or exposure to environmental agents may change metabolic processes and the accompanied alteration in epigenome, thereby predisposing offspring to the development of disease. However, the molecular mechanism remains relatively limited, which retards our further determining the physiological role of maternal nutrition in disease prevention. Experiments designed specifically to improve our understanding of the role of maternal dietary intake in nutritional signaling and regulation to placenta and fetus will provide novel insight into the key factors contributing to the current increasing offspring diseases. Using pathway-specific analysis, one should be able to show that maternal diets are associated with gene expression changes not only to those nutrient transporters, but also to genes that may affect tissue functions broadly. New knowledge generated may ultimately enable the development of nutritional intervention to promote prenatal intrauterine nutritional regulation that would be beneficial to fetus development and reduce offspring diseases.

ACKNOWLEDGMENTS

This project was supported by the USDA Cooperative State Research, Education and Extension Service, hatch project number #ILLU-698-374, and by the NIH-NCI Grant 1R03CA139557-01A1.

REFERENCES

Acconcia, F., P. Ascenzi, A. Bocedi, E. Spisni, V. Tomasi, A. Trentalance, P. Visca, and M. Marino. 2005. Palmitoylation-dependent estrogen receptor alpha membrane localization: regulation by 17beta-estradiol. *Mol Biol Cell* 16:231–37.

Ahlgren, M., T. Sorensen, J. Wohlfahrt, A. Haflidadottir, C. Holst, and M. Melbye. 2003. Birth weight and risk of breast cancer in a cohort of 106,504 women. *Int J Cancer* 107:997–1000.

Angiolini, E., A. Fowden, P. Coan, I. Sandovici, P. Smith, W. Dean, G. Burton, B. Tycko, W. Reik, C. Sibley, and M. Constancia. 2006. Regulation of placental efficiency for nutrient transport by imprinted genes. *Placenta* 27 Suppl A:S98–102.

Arantes, V. C., V. P. Teixeira, M. A. Reis, M. Q. Latorraca, A. R. Leite, E. M. Carneiro, A. T. Yamada, and A. C. Boschero. 2002. Expression of PDX-1 is reduced in pancreatic islets from pups of rat dams fed a low protein diet during gestation and lactation. *J Nutr* 132:3030–35.

Ashworth, C. J., N. Hoggard, L. Thomas, J. G. Mercer, J. M. Wallace, and R. G. Lea. 2000. Placental leptin. *Rev Reprod* 5:18–24.

Barker, D. J., C. N. Hales, C. H. Fall, C. Osmond, K. Phipps, and P. M. Clark. 1993. Type 2 (non-insulin-dependent) diabetes mellitus, hypertension and hyperlipidaemia (syndrome X): relation to reduced fetal growth. *Diabetologia* 36:62–67.

Barski, A., S. Cuddapah, K. Cui, T. Y. Roh, D. E. Schones, Z. Wang, G. Wei, I. Chepelev, and K. Zhao. 2007. High-resolution profiling of histone methylations in the human genome. *Cell* 129:823–37.

Benevolenskaya, E. V. 2007. Histone H3K4 demethylases are essential in development and differentiation. *Biochem Cell Biol* 85:435–43.

Block, K., A. Kardana, P. Igarashi, and H. S. Taylor. 2000. In utero diethylstilbestrol (DES) exposure alters Hox gene expression in the developing mullerian system. *FASEB J* 14:1101–8.

Burdge, G. C., J. Slater-Jefferies, C. Torrens, E. S. Phillips, M. A. Hanson, and K. A. Lillycrop. 2007. Dietary protein restriction of pregnant rats in the F0 generation induces altered methylation of hepatic gene promoters in the adult male offspring in the F1 and F2 generations. *Br J Nutr* 97:435–39.

Camporeale, G., E. E. Shubert, G. Sarath, R. Cerny, and J. Zempleni. 2004. K8 and K12 are biotinylated in human histone H4. *Eur J Biochem* 271:2257–63.

Catalano, P. M., L. Presley, J. Minium, and M. S. Hauguel-de. 2009. Fetuses of obese mothers develop insulin resistance in utero. *Diabetes Care* 32:1076–80.

Charalambous, M., M. Cowley, F. Geoghegan, F. M. Smith, E. J. Radford, B. P. Marlow, C. F. Graham, L. D. Hurst, and A. Ward. 2010. Maternally-inherited Grb10 reduces placental size and efficiency. *Dev Biol* 337:1–8.

Charalambous, M., F. M. Smith, W. R. Bennett, T. E. Crew, F. Mackenzie, and A. Ward. 2003. Disruption of the imprinted Grb10 gene leads to disproportionate overgrowth by an Igf2-independent mechanism. *Proc Natl Acad Sci U S A* 100:8292–97.

Cheng, G., Z. Weihua, M. Warner, and J. A. Gustafsson. 2004. Estrogen receptors ER alpha and ER beta in proliferation in the rodent mammary gland. *Proc Natl Acad Sci U S A* 101:3739–46.

Chiao, E., P. Fisher, L. Crisponi, M. Deiana, I. Dragatsis, D. Schlessinger, G. Pilia, and A. Efstratiadis. 2002. Overgrowth of a mouse model of the Simpson-Golabi-Behmel syndrome is independent of IGF signaling. *Dev Biol* 243:185–206.

Constancia, M., E. Angiolini, I. Sandovici, P. Smith, R. Smith, G. Kelsey, W. Dean, A. Ferguson-Smith, C. P. Sibley, W. Reik, and A. Fowden. 2005. Adaptation of nutrient supply to fetal demand in the mouse involves interaction between the Igf2 gene and placental transporter systems. *Proc Natl Acad Sci U S A* 102:19219–24.

Constancia, M., M. Hemberger M, J. Hughes, W. Dean, A. Ferguson-Smith, R. Fundele, F. Stewart, G. Kelsey, A. Fowden, C. Sibley, and W. Reik. 2002. Placental-specific IGF-II is a major modulator of placental and fetal growth. *Nature* 417:945–48.

Cutfield, W. S., P. L. Hofman, M. Mitchell, and I. M. Morison. 2007. Could epigenetics play a role in the developmental origins of health and disease? *Pediatr Res* 61:68R–75R.

de Assis, S., G. Khan, and L. Hilakivi-Clarke. 2006. High birth weight increases mammary tumorigenesis in rats. *Int J Cancer* 119:1537–46.

Desai, M., C. D. Byrne, K. Meeran, N. D. Martenz, S. R. Bloom, and C. N. Hales. 1997a. Regulation of hepatic enzymes and insulin levels in offspring of rat dams fed a reduced-protein diet. *Am J Physiol* 273:G899–G904.

Desai, M., C. D. Byrne, J. Zhang, C. J. Petry, A. Lucas, and C. N. Hales. 1997b. Programming of hepatic insulin-sensitive enzymes in offspring of rat dams fed a protein-restricted diet. *Am J Physiol* 272:G1083–G1090.

De Stavola, B. L., I. dos Santos Silva, V. McCormack, R. J. Hardy, D. J. Kuh, and M. E. Wadsworth. 2004. Childhood growth and breast cancer. *Am J Epidemiol* 159:671–82.

Efstratiadis, A. 1998. Genetics of mouse growth. *Int J Dev Biol* 42:955–76.

Enders, G. H. 2003. The INK4a/ARF locus and human cancer. *Methods Mol Biol* 222:197–209.

Fernandez-Twinn, D. S., S. Ekizoglou, B. A. Gusterson, J. Luan, and S. E. Ozanne. 2007. Compensatory mammary growth following protein restriction during pregnancy and lactation increases early-onset mammary tumor incidence in rats. *Carcinogenesis* 28:545–52.

Fernandez-Twinn, D. S., and S. E. Ozanne. 2006. Mechanisms by which poor early growth programs type-2 diabetes, obesity and the metabolic syndrome. *Physiol Behav* 88:234–43.

Fernandez-Twinn, D. S., S. E. Ozanne, S. Ekizoglou, C. Doherty, L. James, B. Gusterson, and C. N. Hales. 2003. The maternal endocrine environment in the low-protein model of intra-uterine growth restriction. *Br J Nutr* 90:815–22.

Fowden, A. L., J. W. Ward, F. P. Wooding, A. J. Forhead, and M. Constancia. 2006. Programming placental nutrient transport capacity. *J Physiol* 572:5–15.

Franckhauser, S., S. Munoz, I. Elias, T. Ferre, and F. Bosch. 2006. Adipose overexpression of phosphoenolpyruvate carboxykinase leads to high susceptibility to diet-induced insulin resistance and obesity. *Diabetes* 55:273–80.

Frank, D., W. Fortino, L. Clark, R. Musalo, W. Wang, A. Saxena, C. M. Li, W. Reik, T. Ludwig, and B. Tycko. 2002. Placental overgrowth in mice lacking the imprinted gene Ipl. *Proc Natl Acad Sci U S A* 99:7490–95.

Gillman, M. W., J. W. Rich-Edwards, S. Huh, J. A. Majzoub, E. Oken, E. M. Taveras, and S. L. Rifas-Shiman. 2006. Maternal corticotropin-releasing hormone levels during pregnancy and offspring adiposity. *Obesity (Silver Spring)* 14:1647–53.

Giusti, R. M., K. Iwamoto, and E. E. Hatch. 1995. Diethylstilbestrol revisited: a review of the long-term health effects. *Ann Intern Med* 122:778–88.

Godfrey, K. M., and D. J. Barker. 2001. Fetal programming and adult health. *Public Health Nutr* 4:611–24.

Guillemot, F., T. Caspary, S. M. Tilghman, N. G. Copeland, D. J. Gilbert, N. A. Jenkins, D. J. Anderson, A. L. Joyner, J. Rossant, and A. Nagy. 1995. Genomic imprinting of Mash2, a mouse gene required for trophoblast development. *Nat Genet* 9:235–42.

Guillemot, F., A. Nagy, A. Auerbach, J. Rossant J, and A. L. Joyner. 1994. Essential role of Mash-2 in extraembryonic development. *Nature* 371:333–36.

Hadden, D. R. 2008. Prediabetes and the big baby. *Diabet Med* 25:1–10.

Hales, C. N., and D. J. Barker. 1992. Type 2 (non-insulin-dependent) diabetes mellitus: the thrifty phenotype hypothesis. *Diabetologia* 35:595–601.

Harder, T., E. Rodekamp, K. Schellong, J. W. Dudenhausen, and A. Plagemann. 2007. Birth weight and subsequent risk of type 2 diabetes: a meta-analysis. *Am J Epidemiol* 165:849–57.

Herbst, A.L., H. Ulfelder, D. C. Poskanzer, and L. D. Longo. 1999. Adenocarcinoma of the vagina. Association of maternal stilbestrol therapy with tumor appearance in young women. 1971. *Am J Obstet Gynecol* 181:1574–75.

Hilakivi-Clarke, L., E. Cho, S. deAssis, S. Olivo, E. Ealley, K. B. Bouker, J. N. Welch, G. Khan, R. Clarke R, and A. Cabanes A. 2001. Maternal and prepubertal diet, mammary development and breast cancer risk. *J Nutr* 131:154S–157S.

Hilakivi-Clarke, L., and S. de Assis. 2006. Fetal origins of breast cancer. *Trends Endocrinol Metab* 17:340–48.

Hilakivi-Clarke, L., I. Onojafe, M. Raygada, E. Cho, R. Clarke, and M. E. Lippman. 1996. Breast cancer risk in rats fed a diet high in n-6 polyunsaturated fatty acids during pregnancy. *J Natl Cancer Inst* 88:1821–27.

Hjalgrim, L. L., H. O. Madsen, M. Melbye, Jorgensen, M. Christiansen, M. T. Andersen, N. Pallisgaard, P. Hokland, N. Clausen, L. P. Ryder, K. Schmiegelow, and H. Hjalgrim. 2002. Presence of clone-specific markers at birth in children with acute lymphoblastic leukaemia. *Br J Cancer* 87:994–99.

Hjalgrim, L. L., T. Westergaard, K. Rostgaard, K. Schmiegelow, M. Melbye, H. Hjalgrim, and E. A. Engels. 2003. Birth weight as a risk factor for childhood leukemia: a meta-analysis of 18 epidemiologic studies. *Am J Epidemiol* 158:724–35.

Hofman, P. L., and W. S. Cutfield. 2006. Insulin sensitivity in people born pre-term, with low or very low birth weight and small for gestational age. *J Endocrinol Invest* 29:2–8.

Howie, G. J., D. M. Sloboda, T. Kamal, and M. H. Vickers. 2009. Maternal nutritional history predicts obesity in adult offspring independent of postnatal diet. *J Physiol* 587:905–15.

Jenuwein, T., and C. D. Allis. 2001. Translating the histone code. *Science* 293:1074–80.

Jimenez-Chillaron, J. C., M. Hernandez-Valencia, C. Reamer, S. Fisher, A. Joszi, M. Hirshman, A. Oge A, S. Walrond, R. Przybyla, C. Boozer, L. J. Goodyear, and M. E. Patti. 2005. Beta-cell secretory dysfunction in the pathogenesis of low birth weight-associated diabetes: a murine model. *Diabetes* 54:702–11.

Kawasaki, H., L. Schiltz, R. Chiu, K. Itakura, K. Taira, Y. Nakatani, and K. K. Yokoyama. 2000. ATF-2 has intrinsic histone acetyltransferase activity which is modulated by phosphorylation. *Nature* 405:195–200.

Key, T. J., A. Schatzkin, W. C. Willett, N. E. Allen, E. A. Spencer, and R. C. Travis. 2004. Diet, nutrition and the prevention of cancer. *Public Health Nutr* 7:187–200.

Kobza, K., G. Camporeale, B. Rueckert, A. Kueh, J. B. Griffin, B. Sarath, and J. Zempleni. 2005. K4, K9 and K18 in human histone H3 are targets for biotinylation by biotinidase. *FEBS J* 272:4249–59.

Koch, C. M., R. M. Andrews, P. Flicek, S. C. Dillon, U. Karaoz, G. K. Clelland, S. Wilcox, D. M. Beare DM, Fowler, Couttet, K. D. James, G. C. Lefebvre, A. W. Bruce, O. M. Dovey, P. D. Ellis, P. Dhami, C. F. Langford, Z. Weng, E. Birney, N. P. Carter, D. Vetrie, and I. Dunham. 2007. The landscape of histone modifications across 1% of the human genome in five human cell lines. *Genome Res* 17:691–707.

Kuo, M. H., J. E. Brownell, R. E. Sobel, T. A. Ranalli, R. G. Cook, D. G. Edmondson, S. Y. Roth, and C. D. Allis CD. 1996. Transcription-linked acetylation by Gcn5p of histones H3 and H4 at specific lysines. *Nature* 383:269–72.

Kwong, W. Y., D. J. Miller, A. P. Wilkins, M. S. Dear, J. N. Wright, C. Osmond, J. Zhang, and T. P. Fleming. 2007. Maternal low protein diet restricted to the preimplantation period induces a gender-specific change on hepatic gene expression in rat fetuses. *Mol Reprod Dev* 74:48–56.

Langley-Evans, S. C. 2008. Nutritional programming of disease: unravelling the mechanism. *J Anat* (e-pub).

Lefebvre, L., S. Viville, S. C. Barton, F. Ishino, E. B. Keverne, and M. A. Surani. 1998. Abnormal maternal behaviour and growth retardation associated with loss of the imprinted gene Mest. *Nat Genet* 20:163–69.

Leighton, P. A., R. S. Ingram, J. Eggenschwiler, A. Efstratiadis, and S. M. Tilghman. 1995. Disruption of imprinting caused by deletion of the H19 gene region in mice. *Nature* 375:34–39.

Lewis, R. M., L. A. James, J. Zhang, C. D. Byrne, and C. N. Hales. 2001. Effects of maternal iron restriction in the rat on hypoxia-induced gene expression and fetal metabolite levels. *Br J Nutr* 85:193–201.

Li, L., E. B. Keverne, A. A. Aparicio, F. Ishino, S. C. Barton, and M. A. Surani. 1999. Regulation of maternal behavior and offspring growth by paternally expressed Peg3. *Science* 284:330–33.

Li, S., R. Hansman, R. Newbold R, B. Davis, J. A. McLachlan, and J. C. Barrett. 2003. Neonatal diethylstilbestrol exposure induces persistent elevation of c-fos expression and hypomethylation in its exon-4 in mouse uterus. *Mol Carcinog* 38:78–84.

Li, S., L. Ma L, T. Chiang, M. Burow, R. R. Newbold, M. Negishi, J. C. Barrett, and J. A. McLachlan. 2001. Promoter CpG methylation of Hox-a10 and Hox-a11 in mouse uterus not altered upon neonatal diethylstilbestrol exposure. *Mol Carcinog* 32:213–19.

Lillycrop, K. A., E. S. Phillips, A. A. Jackson, M. A. Hanson, and G. C. Burdge. 2005. Dietary protein restriction of pregnant rats induces and folic acid supplementation prevents epigenetic modification of hepatic gene expression in the offspring. *J Nutr* 135:1382–86.

Lindsay, R. S., R. M. Lindsay, B. J. Waddell, and J. R. Seckl. 1996. Prenatal glucocorticoid exposure leads to offspring hyperglycaemia in the rat: studies with the 11 beta-hydroxysteroid dehydrogenase inhibitor carbenoxolone. *Diabetologia* 39:1299–1305.

Locke, R. 2002. Preventing obesity: the breast milk-leptin connection. *Acta Paediatr* 91:891–94.

Luijten, M., A. R. Thomsen, J. A. van den Berg, P. W. Wester, A. Verhoef, N. J. Nagelkerke, H. Adlercreutz, H. J. van Kranen, A. H. Piersma, I. K. Sorensen, G. N. Rao, and C. F. van Kreijl. 2004. Effects of soy-derived isoflavones and a high-fat diet on spontaneous mammary tumor development in Tg.NK (MMTV/c-neu) mice. *Nutr Cancer* 50:46–54.

Martin-Gronert, M. S., and S. E. Ozanne. 2005. Programming of appetite and type 2 diabetes. *Early Hum Dev* 81:981–88.

McArdle, H. J., H. S. Andersen, H. Jones, and L. Gambling. 2006. Fetal programming: causes and consequences as revealed by studies of dietary manipulation in rats—a review. *Placenta* 27 Suppl A:S56–S60.

McCance, D.R., D. J. Pettitt, R. L. Hanson, L. T. Jacobsson, W. C. Knowler, and P. H. Bennett. 1994. Birth weight and non-insulin dependent diabetes: thrifty genotype, thrifty phenotype, or surviving small baby genotype? *BMJ* 308:942–45.

McMinn, J., M. Wei, Y. Sadovsky, H. M. Thaker, and B. Tycko. 2006. Imprinting of PEG1/MEST isoform 2 in human placenta. *Placenta* 27:119–26.

Moore, T., M. Constancia, M. Zubair, B. Bailleul, R. Feil, H. Sasaki, and W. Reik. 1997. Multiple imprinted sense and antisense transcripts, differential methylation and tandem repeats in a putative imprinting control region upstream of mouse Igf2. *Proc Natl Acad Sci U S A* 94:12509–14.

Nafee, T. M., W. E. Farrell, W. D. Carroll, A. A. Fryer, and K. M. Ismail. 2008. Epigenetic control of fetal gene expression. *BJOG* 115:158–68.

Newbold, R. R., W. N. Jefferson, E. Padilla-Banks, and J. Haseman. 2004. Developmental exposure to diethylstilbestrol (DES) alters uterine response to estrogens in prepubescent mice: low versus high dose effects. *Reprod Toxicol* 18:399–406.

Ong, K. K. 2006. Size at birth, postnatal growth and risk of obesity. *Horm Res* 65 (Suppl 3):65–69.

Ono, R., K. Nakamura, K. Inoue, M. Naruse, T. Usami, N. Wakisaka-Saito, T. Hino, R. Suzuki-Migishima, N. Ogonuki, Miki, T. Kohda, A. Ogura, M. Yokoyama, T. Kaneko-Ishino, and F. Ishino. 2006. Deletion of Peg10, an imprinted gene acquired from a retrotransposon, causes early embryonic lethality. *Nat Genet* 38:101–6.

Ottaviano, Y. L., J. P. Issa, F. F. Parl, H. S. Smith, S. B. Baylin, and N. E. Davidson. 1994. Methylation of the estrogen receptor gene CpG island marks loss of estrogen receptor expression in human breast cancer cells. *Cancer Res* 54:2552–55.

Ozanne, S. E., C. B. Jensen, K. J. Tingey, H. Storgaard, S. Madsbad, and A. A. Vaag. 2005. Low birthweight is associated with specific changes in muscle insulin-signalling protein expression. *Diabetologia* 48:547–52.

Ozanne, S. E., R. Lewis, B. J. Jennings, and C. N. Hales. 2004. Early programming of weight gain in mice prevents the induction of obesity by a highly palatable diet. *Clin Sci (Lond)* 106:141–45.

Ozanne, S. E., G. S. Olsen, L. L. Hansen, K. J. Tingey, B. T. Nave, C. L. Wang, K. Hartil, C. J. Petry, A. J. Buckley, and L. Mosthaf-Seedorf. 2003. Early growth restriction leads to down regulation of protein kinase C zeta and insulin resistance in skeletal muscle. *J Endocrinol* 177:235–41.

Pedersen, J. 1977. *The pregnant diabetic and her newborn: problems and management* (2nd ed.). Baltimore: Williams & Wilkins.

Petridou, E., E. Ntouvelis, N. Dessypris, A. Terzidis, and D. Trichopoulos. 2005. Maternal diet and acute lymphoblastic leukemia in young children. *Cancer Epidemiol Biomarkers Prev* 14:1935–39.

Petrik, J., E. Arany, T. J. McDonald, and D. J. Hill. 1998. Apoptosis in the pancreatic islet cells of the neonatal rat is associated with a reduced expression of insulin-like growth factor II that may act as a survival factor. *Endocrinology* 139:2994–3004.

Potten, C. S., and R. J. Morris. 1988. Epithelial stem cells in vivo. *J Cell Sci* Suppl 10:45–62.

Power, C., and B. J. Jefferis. 2002. Fetal environment and subsequent obesity: a study of maternal smoking. *Int J Epidemiol* 31:413–19.

Ravelli, A. C., J. H. van der Meulen, C. Osmond, D. J. Barker, and O. P. Bleker. 1999. Obesity at the age of 50 y in men and women exposed to famine prenatally. *Am J Clin Nutr* 70:811–16.

Ravelli, G. P., Z. A. Stein, and M. W. Susser. 1976. Obesity in young men after famine exposure in utero and early infancy. *N Engl J Med* 295:349–53.

Rees, W. D., S. M. Hay, D. S. Brown, C. Antipatis, and R. M. Palmer. 2000. Maternal protein deficiency causes hypermethylation of DNA in the livers of rat fetuses. *J Nutr* 130:1821–26.

Salas, M., R. John, A. Saxena, S. Barton, D. Frank, G. Fitzpatrick, M. J. Higgins, and B. Tycko. 2004. Placental growth retardation due to loss of imprinting of Phlda2. *Mech Dev* 121:1199–1210.

Seckl, J. R.. 2004. Prenatal glucocorticoids and long-term programming. *Eur J Endocrinol* 151 (Suppl 3):U49–U62.

Sekita, Y., H. Wagatsuma, K. Nakamura, R. Ono, M. Kagami, N. Wakisaka, T. Hino, R. Suzuki-R. Migishima, T. Kohda, A. Ogura, T. Ogata, M. Yokoyama, T. Kaneko-Ishino, and F. Ishino. 2008. Role of retrotransposon-derived imprinted gene, Rtl1, in the feto-maternal interface of mouse placenta. *Nat Genet* 40:243–48.

Shibata, A., D. T. Harris, and P. R. Billings. 2002. Concentrations of estrogens and IGFs in umbilical cord blood plasma: a comparison among Caucasian, Hispanic, and Asian-American females. *J Clin Endocrinol Metab* 87:810–15.

Shiio, Y., and R. N. Eisenman. 2003. Histone sumoylation is associated with transcriptional repression. *Proc Natl Acad Sci U S A* 100:13225–30.

Sitras, V., R. Paulssen, J. Leirvik, A. Vartun, and G. Acharya. 2009. Placental gene expression profile in intrauterine growth restriction due to placental insufficiency. *Reprod Sci* 16:701–11.

Snoeck, A., C. Remacle, B. Reusens, and J. J. Hoet. 1990. Effect of a low protein diet during pregnancy on the fetal rat endocrine pancreas. *Biol Neonate* 57:107–18.

Soloaga, A., S. Thomson, G. R. Wiggin, N. Rampersaud, M. H. Dyson, C. A. Hazzalin, L. C. Mahadevan, and J. S. Arthur. 2003. MSK2 and MSK1 mediate the mitogen- and stress-induced phosphorylation of histone H3 and HMG-14. *EMBO J* 22:2788–97.

Steger, D. J., M. I. Lefterova, L. Ying, A. J. Stonestrom, M. Schupp, D. Zhuo, A. L. Vakoc, J. E. Kim, J. Chen, M. A. Lazar, G. A. Blobel, and C. R. Vakoc. 2008. DOT1L/KMT4 recruitment and H3K79 methylation are ubiquitously coupled with gene transcription in mammalian cells. *Mol Cell Biol* 28:2825–39.

Stein, A. D., and L. H. Lumey. 2000. The relationship between maternal and offspring birth weights after maternal prenatal famine exposure: the Dutch Famine Birth Cohort Study. *Hum Biol* 72:641–54.

Stocker, C. J., J. R. Arch, and M. A. Cawthorne. 2005. Fetal origins of insulin resistance and obesity. *Proc Nutr Soc* 64:143–51.

Stothard, K. J., P. W. Tennant, R. Bell, and J. Rankin. 2009. Maternal overweight and obesity and the risk of congenital anomalies: a systematic review and meta-analysis. *JAMA* 301:636–50.

Strahl, B. D., and C. D. Allis. 2000. The language of covalent histone modifications. *Nature* 403:41–45.

Symonds, M. E., G. Gopalakrishnan, J. Bispham, S. Pearce, J. Dandrea, A. Mostyn, M. M. Ramsay, and T. Stephenson. 2003. Maternal nutrient restriction during placental growth, programming of fetal adiposity and juvenile blood pressure control. *Arch Physiol Biochem* 111:45–52.

Takahashi, K., T. Kobayashi, and N. Kanayama. 2000. p57(Kip2) regulates the proper development of labyrinthine and spongiotrophoblasts. *Mol Hum Reprod* 6:1019–25.

Takahashi, M., Y. Kamei, and O. Ezaki. 2005. Mest/Peg1 imprinted gene enlarges adipocytes and is a marker of adipocyte size. *Am J Physiol Endocrinol Metab* 288:E117–E124.

Trock, B. J., L. Hilakivi-Clarke, and R. Clarke. 2006. Meta-analysis of soy intake and breast cancer risk. *J Natl Cancer Inst* 98:459–71.

Troisi, R., N. Potischman, J. Roberts, P. Siiteri, A. Daftary, C. Sims, and R. N. Hoover. 2003. Associations of maternal and umbilical cord hormone concentrations with maternal, gestational and neonatal factors (United States). *Cancer Causes Control* 14:347–55.

Uauy, R., and N. Solomons. 2005. Diet, nutrition, and the life-course approach to cancer prevention. *J Nutr* 135:2934S–2945S.

Vaxillaire, M., and P. Froguel. 2006. Genetic basis of maturity-onset diabetes of the young. *Endocrinol Metab Clin North Am* 35:371–84, x.

Vickers, M. H., B. H. Breier, W. S. Cutfield, P. L. Hofman, and P. D. Gluckman. 2000. Fetal origins of hyperphagia, obesity, and hypertension and postnatal amplification by hypercaloric nutrition. *Am J Physiol Endocrinol Metab* 279:E83–E87.

Vickers, M. H., P. D. Gluckman, A. H. Coveny, Hofman, W. S. Cutfield, A. Gertler, B. H. Breier, and M. Harris. 2005. Neonatal leptin treatment reverses developmental programming. *Endocrinology* 146:4211–16.

———. 2008. The effect of neonatal leptin treatment on postnatal weight gain in male rats is dependent on maternal nutritional status during pregnancy. *Endocrinology* 149:1906–13.

Voorhoeve, P. M, and R. Agami. 2004. Unraveling human tumor suppressor pathways: a tale of the INK4A locus. *Cell Cycle* 3:616–20.

Wollmann, H. A. 1998. Intrauterine growth restriction: definition and etiology. *Horm Res* 49 (Suppl 2):1–6.

Yajnik, C. S., C. H. Fall, K. J. Coyaji, S. S. Hirve, S. Rao, D. J. Barker, C. Joglekar, and S. Kellingray. 2003. Neonatal anthropometry: the thin-fat Indian baby. The Pune Maternal Nutrition Study. *Int J Obes Relat Metab Disord* 27:173–80.

Yan, Y., J. Frisen, M. H. Lee, J. Massague, and M. Barbacid. 1997. Ablation of the CDK inhibitor p57Kip2 results in increased apoptosis and delayed differentiation during mouse development. *Genes Dev* 11:973–83.

Zambrano, E., C. J. Bautista, M. Deas, P. M. Martinez-Samayoa, M. Gonzalez-Zamorano, H. Ledesma, Morales, F. Larrea, and P. W. Nathanielsz. 2006. A low maternal protein diet during pregnancy and lactation has sex- and window of exposure-specific effects on offspring growth and food intake, glucose metabolism and serum leptin in the rat. *J Physiol* 571:221–30.

Zeisel, S. H. 2009. Epigenetic mechanisms for nutrition determinants of later health outcomes. *Am J Clin Nutr* 89:1488S–1493S.

Zheng, S., and Y.-X. Pan. Manuscript submitted.

Zhong, S., C. Jansen, Q. B. She, H. Goto, M. Inagaki, A. M. Bode, W. Y. Ma, and Z. Dong. 2001. Ultraviolet B-induced phosphorylation of histone H3 at serine 28 is mediated by MSK1. *J Biol Chem* 276:33213–19.

11 Nutritional Epigenetics
Impact on Metabolic Syndrome

Lakshminarasimhan Pavithra, Sreenivas Chavali, and Samit Chattopadhyay

CONTENTS

11.1 Metabolic Syndrome: Definition and Pathogenesis 259
11.2 Metabolic Syndrome Components Underlying T2D and CVD 260
11.2.1 Insulin Resistance 260
11.2.2 Glucose Intolerance 262
11.2.3 Abdominal Obesity 262
11.2.4 Dyslipidemia 263
11.2.5 Hypertension 263
11.3 Heritability of Metabolic Syndrome 264
11.4 Epigenetics in Metabolic Syndrome 265
11.5 Epigenetic Processes Underlying Metabolic Syndrome 265
11.5.1 Type 2 Diabetes 266
11.5.2 Hypertension 267
11.5.3 Obesity 267
11.5.4 Cardiovascular Diseases 268
11.6 Diet as an Epigenetic Modulator in Metabolic Processes 268
11.7 Dietary Components Influencing Epigenetic Modifications 270
11.8 Nutrients and Nutrient Sensors in Gene Regulation 270
11.9 Epigenetic Reprogramming by Nutrients During Different Phases 272
11.9.1 Gender-Dependent Nutri-Epigenome Inheritance 274
11.10 Nutrients in Epigenetic Therapy 275
11.11 Future Directions 276
Acknowledgments 277
References 277

11.1 METABOLIC SYNDROME: DEFINITION AND PATHOGENESIS

Metabolic syndrome, one the fastest-growing health problems worldwide, is characterized by a cluster of metabolic abnormalities. Classically, it can be defined as a condition pertaining to a combination of abdominal obesity, dyslipidemia, hypertension, disturbances in glucose, and insulin metabolism (insulin resistance and glucose tolerance) that subsequently leads to type 2 diabetes (T2D) and/or cardiovascular disease (CVD). This syndrome is often synonymously referred as Syndrome X, the deadly quartet, and insulin resistance syndrome. Until 1998, it was called the insulin resistance syndrome, when the World Health Organization proposed a more unifying definition for the syndrome and chose to call it metabolic syndrome.

Metabolic syndrome is a result of multifactorial aberrations, with visceral obesity being one of the major driving factors. Notably, visceral obesity is associated with many pathophysiological changes like increase in sympathetic nervous system activity, volume expansion due to sodium retention, and increased secretion of adipokines that in turn affects vasculature and metabolism (Bogaert and Linas 2009). Moreover, the proinflammatory state associated with obesity often appears to mediate the progression to T2D and CVD. Dyslipidemia is characterized by abnormalities in the lipoprotein levels, resulting from an elevated triglyceride, low density lipoprotein (LDL), and low high density lipoprotein (HDL) levels. These abnormalities result in an atherogenic state. Elevated blood pressure or hypertension frequently is associated with obesity and atherogenic state in elders. Though several nonmetabolic contributors are known, hypertension is considered a primary risk factor for metabolic syndrome. Insulin resistance strongly associates with other metabolic risk factors and correlates to incidence of CVD. This also is correlated to manifestation of glucose tolerance, which, when evolved into diabetes-level hyperglycemia, predisposes the individual to CVD. A proinflammatory state characterized by elevated C-reactive protein (CRP) levels is often contributed to by the adipokines released due to obesity and is a common feature of individuals with metabolic syndrome. Similarly, in prothrombic state, fibrinogen, an acute-phase protein, is elevated, pointing out the possible metabolic connection between prothrombic and proinflammatory states.

11.2 METABOLIC SYNDROME COMPONENTS UNDERLYING T2D AND CVD

Overall, the risk for T2D in patients with the metabolic syndrome is 3- to 5-fold higher, and for new incidences of CVD the risk is 1.5- to 3-fold higher (Eckel et al. 2005); however, follow-up studies in middle-age men and women suggest a much higher percentage of incidences of CVD and T2D (Rutter et al. 2005; Wilson et al. 2005). The current section will elaborate on the individual components and their contribution to the disease manifestation (depicted in Figure 11.1).

11.2.1 Insulin Resistance

Insulin resistance is best defined as a defect in insulin secretion and/or action that results in fasting hyperinsulinemia to maintain euglycemia. Circulating free fatty acids are one of the major reasons for development of insulin resistance. This is primarily due to the antilipolytic activity of insulin by stimulation of lipoprotein lipases in adipocytes. However, when the fat accumulation in adipocytes increases, causing monocyte infiltration and release of proinflammatory cytokines, insulin sensitivity

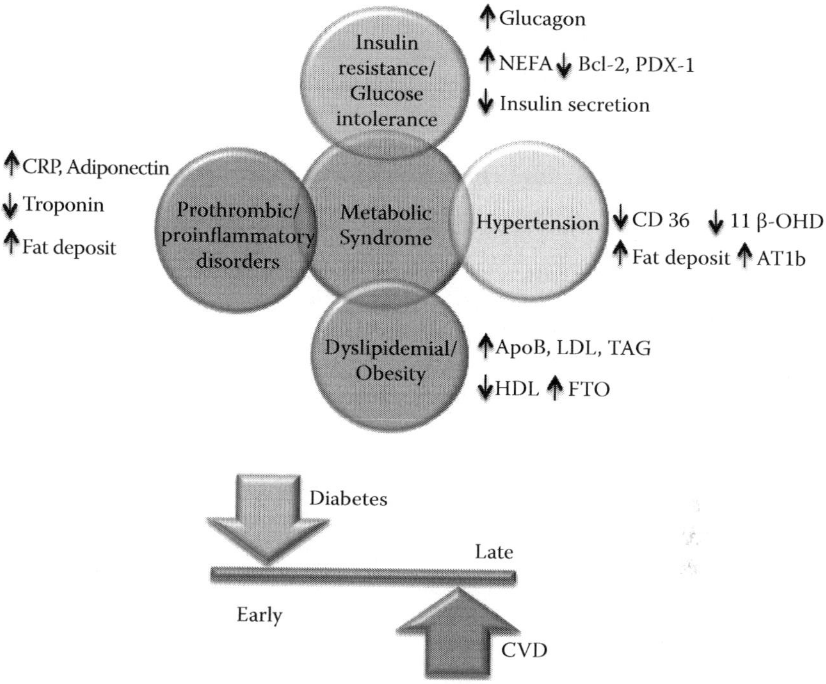

FIGURE 11.1 The metabolic syndrome represents a summation of risk factors for atherosclerotic CVD and type 2 diabetes. Insulin resistance appears to explain much of the pathophysiology of the syndrome. Increased fatty acid flux resulting in obesity and hypertension is another major contributor. Although not included in the diagnostic criteria, increases in proinflammatory cytokines and prothrombotic factors may also contribute to the increased incidence of atherosclerotic CVD and diabetes. The onset of diabetes could be an early-stage manifestation of metabolic syndrome; CVD is often considered a late-stage complication.

is impaired, promoting lipolysis. This causes an elevated level of Acetyl coA that unfavorably modifies a number of downstream pathways including insulin signaling, insulin-dependent glucose transport and phosphorylation, insulin-stimulated glycogen synthesis, insulin-stimulated oxidative phosphorylation (ATP synthesis), accumulation of triglycerides, expression of PPARγ coactivator-1 and PPARγ coactivator-1-controlled genes involved in mitochondrial biogenesis and oxidative phosphorylation, and potentially the initiation of inflammatory processes by activation of protein kinase C and NF-kB, and expression of matrix metalloproteinases (Roden 2005).

Detailed studies of cellular and molecular basis of insulin resistance show a defect in mitochondrial oxidative phosphorylation in insulin-resistant subjects with obesity and/or T2D (Petersen et al. 2003, 2004; Barbato et al. 2004). Murine models have established that the endoplasmic reticulum X-box-binding protein-1 and hyperactivation of c-Jun N-terminal kinase increase serine phosphorylation of insulin receptor substrate-1 (IRS1) and insulin resistance (Ozcan et al. 2004). Mice lacking the fatty acid-binding proteins aP2 and mal1 exhibit a striking phenotype with strong protection from diet-induced obesity, insulin resistance, T2D, and fatty liver disease (Maeda et al. 2005). Studies by Shuldiner and McLenithan (2004) demonstrate that PPARγ coactivator-1a and -1b mRNA levels decrease with age in individuals with a genetic variant in PPARγ coactivator-1a, and that these decreases correlate with alterations in whole-body glucose and fatty acid oxidation, linking changes in oxidative phosphorya-tion to T2D. Oxidative stress has also been viewed as a cellular mechanism for insulin resistance. For example, the knockout of Neil 1, a purine pyrimidine base excision repair enzyme, promotes severe obesity, dyslipidemia, and hepatic steatosis with more modest hyperinsulinemia (Vartanian et al. 2006). Therefore these biochemical changes in insulin-mediated signaling pathways result in decreases in insulin-mediated glucose transport and metabolism in the metabolic syndrome as well.

11.2.2 Glucose Intolerance

The relationship between impaired fasting glucose or glucose intolerance and insulin resistance is well supported by human studies (Abdul-Ghani et al. 2006). Insulin resistance leads to glucose intolerance, both prediabetes (largely IGT) and type 2 diabetes, if insulin secretion from pancreatic beta cells is insufficient to overcome the insulin resistance. Oxidative stress is a major contributor to the development of glucose intolerance. Underlying mechanisms include increased apoptosis due to decrease in the antiapoptotic protein Bcl-2, decreased insulin secretion due to decrease in the transcription factor PDX-1, and decreased glucose-stimulated insulin secretion due to increases in uncoupling protein-2 (Kawamori et al. 2003; Krauss 1995). Interestingly, the tissue-specific autoregulation of insulin also plays a role in manifestation of glucose intolerance. This has been demonstrated in mice, where the deletion of insulin receptor in skeletal muscle does not result in hyperglycemia; however, the β-cell-specific knockout of the insulin receptor produces progressive glucose intolerance and diabetes (Kulkarni et al. 1999).

11.2.3 Abdominal Obesity

The worldwide obesity epidemic has been one of the major contributors for the increasing incidence of metabolic syndrome. The increase in waist circumference has routinely been considered as an indicator for obesity that is a result of increased nonessential fatty acid (NEFA) flux and fat deposits. The impact of body fat distribution on insulin resistance and the metabolic syndrome have been evaluated in the mouse model where the overexpression of 11-hydroxysteroid dehydrogenase 1 results in central obesity, hypertension, impaired glucose tolerance, hypertriacylglyceroemia, and elevated intra-adipose corticosterone levels (Masuzaki et al. 2001). Conversely, mice that fail to express 11-hydroxysteroid dehydrogenase 1 are resistant to the development of metabolic syndrome (Kotelevstev et al. 1997). More recent studies show that uncoupling protein 1 expression in epididymal fat improves glucose tolerance and decreases food intake in both diet-induced and genetically obese mouse models (Yamada et al. 2006). Moreover, in the case of lipoatrophy, it is well documented that insulin resistance and the metabolic syndrome typically coexist (Garg and Misra 2004). The genetic basis of the syndrome has been supported by studies demonstrating the role of single-gene defects in PPAR-l, lamin

A/C, 1-acylglycerol-3-phosphate, O-acyltransferase, Seipin, β-2 adrenergic receptor, and adiponectin (Hegele 2003; Dallongeville et al. 2003; Fumeron et al. 2004).

11.2.4 Dyslipidemia

In the case of insulin resistance, increased flux of NEFA to the liver increases hepatic triacylglycerol (TAG) synthesis, while under physiological conditions insulin inhibits the secretion of VLDL into the systemic circulation (Lewis and Steiner 1996). This two-prong effect of insulin can be attributed to the degradation of ApoB (Taghibiglou et al. 2002) along with increasing the transcription and enzymatic activity of many genes that relate to TAG biosynthesis (Foufelle and Ferré 2002). Recent evidence suggests that mutations in genes that affect LPL expression are associated with both dyslipidemia and insulin resistance (Hölzl et al. 2002; Ma et al. 2003; Goodarzi et al. 2004). The levels of lipoproteinlipase (LPL) mass in preheparin plasma may also be a quantitative indicator of whole-body insulin resistance (Miyashita and Shirai 2005). Alterations in the HDL metabolism are another major change in dyslipidemia. The ability of HDL to inhibit or to enhance vascular inflammation, lipid oxidation, plaque growth, and thrombosis reflects changes in specific enzyme and protein components, e.g., increases in serum amyloid A, an acute-phase reactant and proinflammatory molecule (Kontush and Chapman 2006). This change in lipoprotein composition also results in an increased clearance of HDL from the circulation (Brinton et al. 1991; Chan et al. 2006). The relationship between these changes in HDL and insulin resistance is indirect, occurring in concert with the changes in TAG-rich lipoprotein metabolism. In addition to HDL, LDL is also modified in composition in a similar way by depletion of nonesterified cholesterol, esterified cholesterol, and phospholipid with either no change or an increase in LDL-TAG (Halle et al. 1999; Kwiterovich 2002). The small, dense LDL is more atherogenic because it is more likely to permeate the basement endothelial membrane, adhere to arterial wall glycosaminoglycans, be susceptible to oxidation, and get bound to scavenger receptors on monocyte-derived macrophages (Packard, 1996).

11.2.5 Hypertension

The relationship between insulin resistance and hypertension is well established (Ferrannini et al. 1987), and the spontaneous-hypertensive rat is a widely studied model of hypertension that exhibits metabolic abnormalities that share features with the human metabolic syndrome. Insulin-resistant subjects lack the vasodilatory effect of insulin (Tooke and Hannemann 2000), and the vasoconstriction due to fat deposition may also contribute to this. Interestingly, a defective CD36 gene encoding a membrane fatty acid transporter results in hyperinsulinemia in the spontaneous-hypertensive rat because of uncoupling of glucose oxidation from its uptake and enhanced protein O-linked N-acetylglucosaminylation, suggesting increased glucose shunt through the hexose monophosphate pathway (Tanaka et al. 2007). In addition to mild hypertension and dyslipidemia, whole-body insulin-mediated body glucose uptake is reduced in CD36-deficient patients, indicating the presence of insulin resistance. The frequency of CD36 deficiency is also higher in patients with CVD than in control subjects (Yamashita et al. 2007).

11.3 HERITABILITY OF METABOLIC SYNDROME

Although the heritability of metabolic syndrome has not been estimated, individual components of the syndrome are known to be strongly inherited (Groop and Orho-Melander 2001). For example, in the case of hypertension, estimates of heritability vary between 22% and 62% for systolic blood pressure and 38% and 63% for diastolic BP. Similarly, family aggregation studies have shown that 45% of first-degree relatives of T2D patients are insulin resistant compared to only 20% of individuals without a family history (Groop et al. 1996).

The Lamarckian theory of "inheritance of acquired characters," though lampooned in the nineteenth and early twentieth centuries, has now been accepted as the origin of the concept of "epigenetics." Waddington in 1942 introduced the term epigenetics to best explain the causal mechanisms by which genotype brings about the phenotypic effects. Nanney in 1958 further refined this term and described it as mechanisms of cellular heredity that were not based on template-replicating mechanisms. Since then, epigenetics has come to refer causes of heritable differences that are not dependent on changes in DNA sequence.

In higher organisms, the interplay between finely tuned genetic and epigenetic programs by means of proliferation, differentiation, and apoptosis tailors the tissue and organ modalities. Like the genetic component, the epigenetic mechanisms can also arise in mature organisms by random changes or by the influence of environment. It has now been widely acknowledged that epigenetic code comprises several layers of complex yet coordinated codes like the DNA methylation, the histones, modifications, and various other coregulators. This in turn is dependent on the myriad of chromatin remodelers that govern the accessibility of transcription factors to the specific regions in the genome, thereby activating or silencing genes either transiently or permanently. This requires that the epigenetic marks be initially set up in an establishment phase that then subsequently gets transmitted from the mother to daughter cell and potentially from generation to generation (transgenerational inheritance). Therefore these epigenetic marks serve as a memory bank to maintain the genome function for cell identity, viability, and proliferation after differentiation. From the aforementioned facts, it is quite fathomable that disturbances in the balance of epigenetic networks may lead to manifestation of diseases.

The first clue about the involvement of epigenetic changes in manifestation of human diseases came from the research by Feinberg and Vogelstein (1983), who showed that cancer cells exhibit an unusual pattern of DNA methylation. Since then a number of studies have documented correlation between aberrant DNA methylation and manifestation of cancer. However, the relevance of epigenetics to physiopathological mechanisms in common diseases such as metabolic syndrome was less clear. Although the individual system components of metabolic syndrome can be clustered, the underlying molecular and pathological mechanisms can be explained only by taking into account the interplay between environment and genetic determinants. Recently, there has been an immense interest in this field, and data are now available to support the hypothesis that individuals with metabolic syndrome have undergone aberrant "epigenetic programming" due to nutrition at postnatal/neonatal states or due to adult-life metabolic disturbances.

11.4 EPIGENETICS IN METABOLIC SYNDROME

Steep temporal trends in the incidence rates of metabolic syndrome across various populations suggest that these epidemiological associations are unlikely to have arisen exclusively through the pleiotropic effects of genes. Accordingly, the effects are now viewed as phenotypes established by the interaction between genes and the developmental environment using the mechanisms of developmental plasticity. Developmental plasticity attempts to "tune" gene expression to produce a phenotype best suited to the predicted later environment (Godfrey et al. 2007). A match between the resulting phenotype and its environment ensures a healthy life. However, a mismatch might trigger an inappropriate response to environmental challenges, increasing the risk for disease manifestation. Thus the degree of the mismatch determines the degree of an individual's susceptibility to chronic disease. The processes of phenotypic induction through developmental plasticity produce integrated changes in a range of organs via epigenetic processes. This was first established by Waddington (1952) in *Drosophila melanogaster* where he used heat shock to alter the wing vein pattern, which eventually led to a stable population exhibiting this phenotype even without the environmental stimulus. This demonstrated a dynamic interaction between the genome and the environment during the plastic phase of development, producing effects that can be heritable. Though it is difficult to dissect the relative contribution of genetic, epigenetic, and environmental or behavioral factors in humans, a great deal of evidence suggests transgenerational nongenomic inheritance.

It is now quite conceivable that the epigenetic modifications establish a life course strategy for meeting the demands of the predicted later environment. This explains why an impaired early environment produces a range of effects causing alterations in cardiovascular and metabolic homeostasis, growth and body composition, cognitive and behavioral development, reproductive function, repair processes, and longevity. Some of these are associated with increased risk of cardiovascular and metabolic disease, "precocious" puberty, osteoporosis, and some forms of cancer (Gluckman et al. 2007, 2008). Understanding the underlying epigenetic processes thus holds the key to determining pathophysiology and to developing approaches for early diagnosis, prevention, and therapeutic intervention. What remains elusive, though, is the knowledge of the underlying mechanisms detailing the contributions from environmental and genetic components and their interplay.

11.5 EPIGENETIC PROCESSES UNDERLYING METABOLIC SYNDROME

The major epigenetic changes accomplishing gene regulation include DNA methylation, histone tail modifications, chromatin remodeling, and RNA interference (depicted in Figure 11.2). In the following sections, specific cases of epigenetic

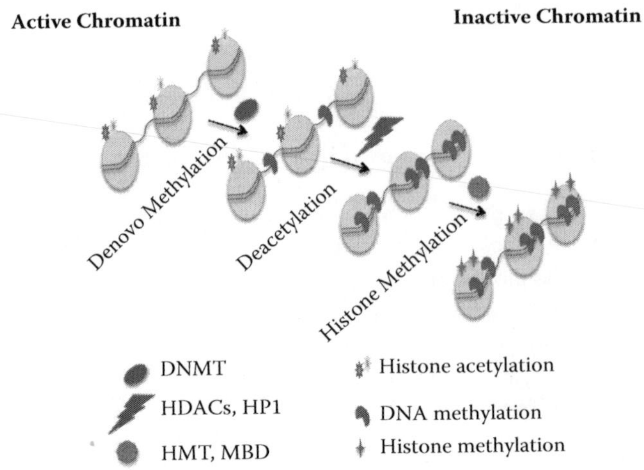

FIGURE 11.2 Epigenetic modifications tune gene expression. The major factors contributing to epigenetic programming are the interdependent DNA methylation and histone modifications. DNA de novo methylation is carried out by DNA methyltransferases, with SAM as the cofactor. Once DNA methylation marks are set, it acts as cue for heterochromatin protein 1 and histone deacetylases to remove the active chromatin acetyl marks. This is followed by the action of histine methyltransferases that efficiently methylate the lysine/arginine residues, resulting in a silenced chromatin.

modifications of candidate genes and their ramification in the onset of various components of metabolic syndrome shall be discussed.

11.5.1 Type 2 Diabetes

Type 2 diabetes is typically characterized by a combination of peripheral insulin resistance, mostly accompanied by defects in insulin secretion that varies in severity. Moreover, the production of insulin appears to be governed by constraints imposed at the level of transcription of the gene encoding insulin (*INS*), by an intricate interplay between transcription factors. Intriguingly, the proximal promoter of *INS* is selectively hyperacetylated at histone H3 only in β–islet cells and is highly correlated with recruitment of the histone acetyltransferase p300 (Chakrabarti et al. 2003). This indicates that epigenetic variables relating to the chromatin structure of *INS* might preclude transcription factor-mediated chromatin remodeling under unfavorable conditions. The known mechanisms for disrupted insulin secretion in T2D include accumulated damage caused by hyperglycemia, hyperlipidemia, and oxidative stress. In combination these directly affect DNA methylation pattern and histone organization, therefore tuning the expression of multiple genes like the glucocorticoid receptor (GR) and peroxisome proliferator-activated receptor gamma (PPARγ) (Simmons 2007). PPARγ coactivator 1 α (PGC-1 α encoded by *PPARGC1A*) is a well-established node in the pathogenesis of T2D and a master transcriptional coactivator of mitochondrial genes (Scarpulla 2006). A reduced expression of this gene is related to impaired ATP production caused by a reduced oxidative phosphorylation.

The increased DNA methylation of the *PPARGC1A* promoter in diabetic islets has been demonstrated as a plausible mechanism for reduced *PPARGC1A* mRNA expression and insulin secretion in pancreatic islets of patients with T2D (Ling et al. 2008). Similarly, the developmental regulator Pax2, which is associated with H3K4 methyltransferase complex, has been shown to specifically transactivate glucagon promoter (Patel et al. 2007).

An additional layer of transcriptional control is provided by microRNAs (miRs). For instance, the pancreatic islet-specific miR-375 has been shown to inhibit insulin secretion in mouse pancreatic β-cells by inhibiting the expression of the protein myotrophin (Poy et al. 2009). Also, levels of miR-192, which controls TGF-β induced extracellular matrix proteins collagen 1-α 1 and 2 expression levels, have been shown to be increased in glomeruli isolated from streptozotocin-injected diabetic mice as well as diabetic mice (db/db) when compared to nondiabetic mice, suggesting a potential role of miR-192 in kidney diseases associated with T2D (Kato et al. 2007). Recent studies indicate that miR-30d up-regulated by glucose increased insulin gene expression, while its inhibition abolished glucose-stimulated insulin gene transcription (Tang et al. 2009). miR-143 regulates genes that are crucial for adipocyte differentiation, including GLUT4, HSL, fatty acid-binding protein aP2, and PPAR-γ 2, demonstrating a role for miRNAs in fat metabolism and in endocrine function in humans (Esau et al. 2004).

11.5.2 Hypertension

Although hypertension is recognized as one of the major contributing factors to cerebrovascular and cardiovascular diseases, its pathogenesis is not completely understood. Several studies suggest a significant role of the renin-angiotensin system (RAS) in hypertension pathogenesis. Studies by Bogdarina et al. (2007) show that the development of hypertension in rat offspring was associated with increased expression of the AT1b angiotensin receptor (AT1b) and reduced methylation of its promoter in the adrenal gland. This provides a link between fetal insults to epigenetic modification of genes and the resultant alteration of gene expression in adult life leading to the development of hypertension. Also, studies indicate that the Dot1a-Af9 pathway may also be involved in the control of genes implicated in renal fibrosis and hypertension by methylation at the Histone H3 Lysine 79 (W. Zhang et al. 2004). Recent studies also highlight the role of promoter methylation of 11 β-hydroxysteroid dehydrogenase 2 (11β-HSD2), a major epigenetic modification of DNA, in regulation of *HSD11B2* expression, which classically induces hypertension (Friso et al. 2008).

11.5.3 Obesity

The protein product of *FTO* (fat mass and obesity associated) contains a 2-oxoglutarate oxygenase fold (2OG), and this enzyme has been found to be involved in diverse cellular processes, ranging from DNA repair to histone demethylation (Gerken et al. 2007). Interestingly, variants in the FTO gene are associated with increased body mass index in humans. The histone demethylase activity of this protein would therefore present an interesting study if there is an actual causal relationship between obesity and epigenetics.

11.5.4 Cardiovascular Diseases

The essential role of a histone acetyltransferase (HAT) protein in cardiac muscle was first proven by deletion of the coactivator p300, which perturbed heart development and cell proliferation (Gusterson et al. 2003). Class II histone deacetylases (HDACs) can act as signal-responsive repressors of cardiac hypertrophy, inhibiting gene expression that is dependent on myocyte enhancer factor-2 (C. L. Zhang et al. 2002). In addition, overexpression of the transcriptional corepressor homeodomain-only protein (Hop) causes cardiac hypertrophy by the recruitment of a class I HDAC. Also, different HDACs can act, in some contexts, as repressors of cardiac hypertrophy by inhibiting the gene expression of prohypertrophic genes, while in other contexts they effectively contribute to cardiac hypertrophy by inhibiting the expression of growth-suppressing antihypertrophic transcriptional pathways (e.g., recruitment by HOP that inhibits transcriptional activity of serum response factor) (Hammamori and Schneider 2003; Kook et al. 2003).

The stability and expression of the cardiac troponin gene associated with cardiac contractility function (and disease) is affected by cytosine methylation (D'Cruz et al. 2000). Recently, it has also been demonstrated that the expression of genes known to be essential in maintaining homeostatic cardiac physiology can be modulated by targeted DNA methylation; e.g., the *KVLQT1* gene involved in cardiac membrane transport is subject to regulation by DNA methylation, which alters its expression (Smilinich et al. 1999; Cerrato et al. 2002).

11.6 DIET AS AN EPIGENETIC MODULATOR IN METABOLIC PROCESSES

The past decade has seen a series of seminal studies, notably by Hales and Barker, whose studies on "fetal programming" led to the "thrifty phenotype" hypothesis, proposing an early initiation of these common traits that laid the foundation for the epigenetic basis for metabolic syndrome (1992). According to the thrifty phenotype hypothesis, poor fetal and infant growth resulting from poor nutrition in early life, in part due to maternal malnourishment, produces permanent changes in insulin secretion by pancreatic β-cells, and hence glucose metabolism. These changes, when combined with the effects of obesity, aging, and physical inactivity, strongly predispose to T2D and metabolic syndrome. Several studies have reiterated these findings in different populations and ethnic groups, laying the foundation for what is popularly known as "developmental origins of health and disease" (Figure 11.3). A thorough introspection of the thrifty phenotype hypothesis carried out by several groups worldwide demonstrated that diet is a critical modulator of epigenetic programming in metabolic syndrome.

Several recent studies demonstrate the effect of diet and dietary components at the molecular level on both DNA methylation and posttranslational modifications of histones. This is exerted at the levels of methyl donors such as folate, choline, and methionine. Alcohol and zinc are the other factors that influence the CpG methylation by influencing the availability of S-adenosyl methionine (SAM), essential for DNA methyltransferase activity (Ross 2003; Davis and Uthus 2004; Pogribny et al. 2006). It is postulated that diet may also exert influence on the response to bioactive compounds in food (Niitsu et al. 2001). For example, dietary methyl supplement

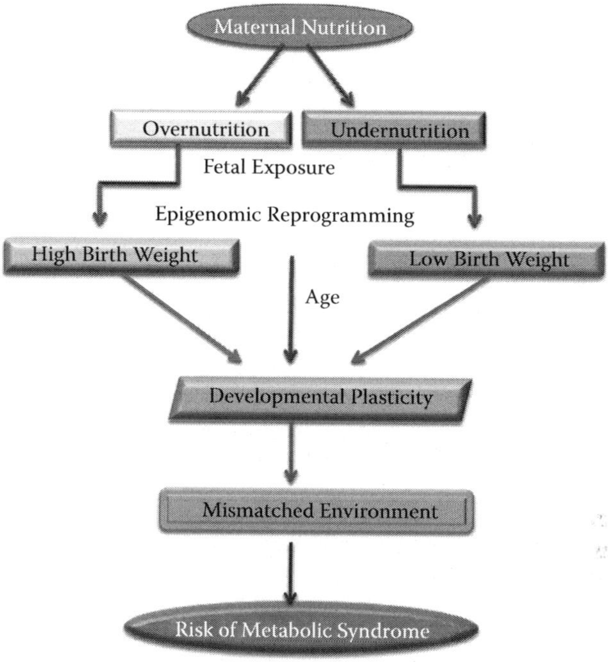

FIGURE 11.3 The epigenetic reprogramming at the fetal stage is mainly attributed to the nutritional status of the mother. The maternal nutrition can modulate the fetal epigenetic programming mediated by specific dietary components, which play a role during a specific window period. Once born, the child is subjected to environmental cues to be fine-tuned by the external system. If such a fine-tuning fails to occur, the individual is subjected to a mismatched environment that increases the incidence of risk factors contributing to the metabolic syndrome.

deficiency led to hepatocarcinogenesis even in the absence of an external carcinogen, and this can be reversed by refeeding of methyl-sufficient diet in early weeks of the disease onset in rat model (Poirier 1994). Thus epigenetic regulation by diet appears to be critical for regulation of cellular proliferation.

Studies by Dashwood et al. (2006) suggested that butyrate, diallyl sulfide, and sulforapane, some of the essential components of a normal diet, can lead to conformational changes in the histone deacetylase pocket, leading to its inactivation. Moreover, acute methyl deficiency has also been shown to be involved in histone 3 lysine 9 and histone 4 lysine 20 trimethylation by affecting the Suv 4, Suv 39 methyltransferases (Schotta et al. 2006). The role of biotin has also now been established in regulating the chromatin modifications, causing an epigenetic reprogramming (Hassan and Zempleni 2006). In the case of some chronic noncommunicable diseases, certain histone modifications have been identified as markers along with the increased of proinflammatory genes. For example, decreased H3 methylation associated with transcriptional repression has been proposed to underlie the sustained proinflammatory phenotype of vascular smooth muscles seen in diabetic animals even after control of glycemia (Villeneuve et al. 2008). Similarly, transient hyperglycemia causes changes in histone 3 methylation in vascular endothelial cells, causing changes in proatherogenic genes (El-Osta et al. 2008). Moreover, a single nucleotide polymorphism in the promoter of respiratory chain component gene NDUFB6 introduces a CpG site, the methylation of which negatively correlates with gene expression and insulin sensitivity (Ling et al. 2007).

Noncoding RNAs have also been shown to be affected by methyl deficiency (Datta et al. 2008). In the case of methyl deficiency-induced hepatocarcinogenesis, some of the microRNA genes were up-regulated while some were down-regulated when compared with normal tissue in mouse. Interestingly, miR 122, a liver-specific microRNA, was down-regulated in hepatocellular carcinoma in both rodent and human tissues (Kutay et al. 2006). Similarly, miR 155 has been shown to have a role in early hepatocarcinogensis (B. Wang et al. 2009). The subsequent effects of RNA-mediated transcriptional or posttranscriptional gene silencing by dietary components and its role in metabolic diseases is yet to be investigated.

11.7 DIETARY COMPONENTS INFLUENCING EPIGENETIC MODIFICATIONS

Methyl donors have been the most well-characterized factors in diet that affect epigenetic mechanisms (Mariman 2007). For example, SAM is an important component of folic acid metabolism, and low intake by the mother during pregnancy may alter the activity and status of genes that may in later life contribute to risk for diet-related disorders. Apart from methylation of DNA, other histone modifications like acetylation/deacetylation, phosphorylation, and methylation are also influenced by dietary components (Yang and Sauve 2006). Histone acetylation has been associated with gene activation and deacetylation with gene silencing. The specific family of acetyltransferases/deacetylases are involved in this histone tail-modifying activity. Therefore inhibition or activation of any of these enzymes by bioactive components in the food could change the epigenetic programming in the cell. Calorie restriction studies in many organisms have been shown to influence metabolic changes such as decreasing blood glucose, insulin, glycogen, fat, and body weight, and improving insulin sensitivity. In this regard the gene that has been found to be overexpressed during calorie restriction in yeast is the silent information regulator (Sir2), a class III histone deacetlyase dependent on NAD+ for its enzymatic activity. In humans, the most well-studied class III histone deacetylase is SIRT1. This enzyme acetylates not only histones but also other transcription factors like p53, FOXO, and NFκB. Interestingly, NAD levels determine the deacetylase activity, providing a link between the activity of glycolysis and TCA cycle as sensors of food availability and chromatin remodeling with altered gene expression (Meijer and Codogno 2008).

11.8 NUTRIENTS AND NUTRIENT SENSORS IN GENE REGULATION

Though environment and genes have an absolute impact on the risk, a general disease risk is the result of their interaction. An important finding supporting this aspect is that nutrients can act as signaling molecules for gene regulation (Van Sluijters et al. 2000; Müller and Kersten 2003). These nutrients are recognized by the cellular sensor systems that influence gene and protein expression and subsequently metabolite production. These "dietary signatures" of nutrients/nutrient regimes can affect specific cells and tissues and regulate the homeostasis of the organism (Corthésy-Theulaz et al. 2005). Similar to normal gene regulation, transcription factors are the major players in regulating the nutrient-mediated gene expression (Afman and Muller 2006). The nuclear hormone receptors and superfamily members are the prime nutrient sensors, which bind to nutrients and metabolites. Some of these include retinoic acid receptor (RAR), retinoic X receptor (RXR), liver X receptor (LXR), oxysterols, bile salts (farnesoid X receptor), estrogen receptor, and carbohydrate-responsive element binding protein (ChREBP) (Glass 1994). Nuclear receptors bind to specific target sequences in promoter regions of a number of genes, and when bound by respective ligands undergo conformational changes that result in a coordinated recruitment or dissociation of corepressor and coactivators, thereby affecting the transcription machinery (Bocher et al. 2002). In metabolically active organs, these factors act as nutrient sensors and affect the rate of gene transcription in response to change in the nutrient influx.

For instance, the nuclear hormone receptor FXR mediates the response to high bile acids by activating the expression of genes involved in lipid metabolism like PPARα, apolipoprotein E, and APOC11. This inhibits further synthesis of bile acids and facilitates the export of bile acids from the cell. Similarly, PPARα acts as nutrient sensors for fatty acids and is important during food deprivation and starvation. These fatty acids bind to PPARα, which then activates a cascade of genes involved in ketogenesis, amino acid metabolism, cellular proliferation, and acute phase response (Jeter et al. 1975). This is an elegant pathway in which the initial signal that originates from adipocytes acts through a receptor whose expression is up-regulated by fatty acids during fasting and affects the metabolism in liver.

Several studies detail the relation between the nutrient availability and the chromatin compaction. Macronuclear chromatin normally condenses upon starvation, coincident with decreased global gene expression and cellular metabolic activity. These events require linker histone H1 for the regulation of specific genes and an HP1-like protein (Shen et al. 1995; Parker et al. 2007).

Studies by Moraes et al. (1995) also show that in adult mice, starvation increased the chromatin-packing states, especially in areas of noncondensed chromatin, and induced drastic decreases in concanavalin. This reactivity results due to nuclear matrix glycoproteins and the frequency of nuclei with chromatin extensibility under gravity. Changes in the liver cell nuclei associated with starvation and refeeding of adult mice involved chromatin supraorganization, hepatocyte proliferation (refeeding), and the loss, regain, and redistribution of nuclear proteins, especially nuclear matrix components, related to chromatin organization and extensibility. These changes are suggested as favoring the silencing and reactivation of transcriptional activities, depending on the organism's nutritional state. Moreover, the chromatin from the livers of rats fed with varying nutritional diets—i.e., a high-carbohydrate, fat-free diet (diet 1) or a low-carbohydrate, protein-free diet (diet 2)—revealed an increase in ratios of RNA:DNA and nonhistone:DNA, relative to control ratios. The nonhistone:DNA ratios in liver of rats fed diet 1 or diet 2 were 2.4-fold and 3.5-fold, respectively, larger than control ratios (Castro and Sevall 1982). Rozovski (1984) and others have demonstrated that during the first 3 weeks of malnutrition in adult male rats, there is an increase in RNA polymerase activity and increased incorporation of orotic acid into nuclear RNA. Further refeeding of proteins to animals deprived of all essential amino acids led to an increase of Pol I activity and an increase in the nucleolar material in the nucleus.

11.9 EPIGENETIC REPROGRAMMING BY NUTRIENTS DURING DIFFERENT PHASES

The epigenetic programming by nutrients is critical during early life development. The Dutch Hunger Winter (1944–1945), during which there was a lack of food supply that led to severe malnutrition in northern Holland, serves as a classic model for this proposal. Detailed studies by Roseboom et al. (2006) suggest that the first trimester could be the most critical period that decides the risk for later-life metabolic syndrome. This could be because of the lack of certain nutrients or metabolites that alters the expression activity of particular genes to a level during the critical developmental period that is fixed and remains so throughout life (de Rooij et al. 2007). It could also be explained on the basis of the thrifty phenotype hypothesis where the nutritional stress in the prenatal state changes its metabolism to a thrifty phenotype by resetting the activity of certain genes; the individual will be able to survive better under those conditions. But if in the postnatal state food is readily available, then those with thrifty phenotype will catch up with normalcy during childhood and early adulthood (catch-up growth); but in later life, this may lead to disease. For instance, low birth weight is reported as a risk condition for later-life obesity (Monteiro and Victora 2005).

Changes in methylation patterns during early embryo and development have been extensively studied. For instance, apolipoproteins encoded by the gene cluster *APOA1-C3-A4* at a given stage in development show dynamic changes in methylation pattern during the entire phase of development, which can lead to synchronous locking of transcription (Shemer et al. 1991). Similarly, the hormone leptin, which regulates the electrophysiological modulation of orexigenic neurons and anorexigenic neurons, has a tropic effect on some of the neurons in hypothalamus in response to nutrients intake that affects their plasticity. However, this effect of leptin is exerted only during a narrow window period of neonatal period, after which it is involved in regulation of food intake in adults (Melzner et al. 2002). Therefore failure of leptin intake (from breast milk) during that particular period may have long-standing consequences due to incorrect epigenetic remodeling of chromatin. At the chromatin level, a loss of DNA methylation and H3K9me1/2, replaced by increased levels of H3K27me3, characterizes the reprogramming events occurring in primordial germ cells (Seki et al. 2005). At the differentiated oocyte stage, prophase I displays a high level of acetylation at histones H3 and H4, whereas the level dramatically drops at metaphase I, and a complete deacetylation affecting H3K14, H4K12, H4K8, H4K16, and H4K5 occurs later at metaphase II (Akiyama et al. 2006; Adenot et al. 1997). The global histone deacetylation during the progression of meiosis may be required for an appropriate recruitment of heterochromatin proteins, and consumption of deacetylase inhibitors in diet during these phases may affect the development of the fetus (Delage and Dashwood 2008).

Animal models have been extensively used to evaluate the effect of dietary restriction during gestation and development of adult disease. In murine model, the offspring of rat fed with a low-protein diet during pregnancy have increased susceptibility to diabetes, insulin resistance, and hypertension when fed with a high-fat diet, promoting obesity (Galler and Tonkiss 1991). In fact, changes in methionine metabolism result in increased homocysteine production, which lead to an increased endogenous methylation of DNA in liver alone but not in kidney or heart tissue (Rees et al. 2006). The epigenetic reprogramming in one-carbon metabolism pathways and increased methyl donor supply are suggested to be the causative mechanisms for this increased methylation, which in turn silence the growth-promoting genes during an influential growth period (Waterland and Garza 1999). Similarly, a low-protein maternal diet has been shown to affect the promoter methylation status of glucocorticoid receptor and PPAR genes in the liver of offspring. However, when this diet was supplemented with folic acid, these changes could be prevented, suggesting that a one-carbon metabolism defect was involved. That the hypomethylation effect also persisted after weaning when the maternal effect ceased to exist demonstrates the influence of maternal diet in persistent changes to the phenotype of the offspring (Lillycrop et al. 2007). Early-life exposure to glucocorticoid receptor has a role in hypertension, and PPAR activity is associated with induction of dyslipidemia. Further, increase in PPAR activity may promote a program of lipid-induced activation of genes regulating fatty acid uptake, β-oxidation, fatty acid transport into peroxisomes, and β-oxidation of unsaturated fatty acids. Studies by Anway et al. (2006) demonstrated that developmental exposure to fungicide vinclozolin affects male rat fertility, which is transmitted through four generations without further exposure to the chemical. Moreover, the adult offspring of dams fed with vinclozolin during pregnancy exhibited a significant increase in the abnormalities of prostate, kidneys, testes, and immune system. These effects at similar proportions were seen also in F2-F4 generation, by inheritance through the male germline (Nilsson et al. 2008).

Studies in the agouti model of mice showed that dietary methylation in utero can have profound effects on the phenotype of offspring by inducing changes in DNA methylation. In these studies, a retotransposon with a cryptic promoter normally silenced was inserted close to the gene-encoding agouti gene, enabling a tissue-specific and regulated agouti expression. Hypomethylation of this site leads to constitutive expression of agouti gene, leading to yellow coat color, obesity, and other symptoms for metabolic syndrome. Dietary supplementation with folic acid, betaine, zinc, and vitamin B_{12} in maternal diet lead to a change in the coat color from yellow to normal agouti coat, which is associated with lower risks of obesity, cancer, diabetes, and a prolonged life compared with yellow mice. This was associated with the hypomethylation of long terminal repeats of 5′ agouti gene in yellow mice compared to the degree of hypermethylation of this site in normal/brown mice. Another study found that the changes in the pup hair or pigmentation are directly associated with the supplementation of methyl groups in the mother's diet (Wolff et al. 1998). Studies by Dolinoy et al. (2006) demonstrated that the ingestion of genistein, a major phytoestrogen in soy, protected the offspring against obesity in adulthood, which suggests that dietary supplementation is also involved in methylation-dependent susceptibility to disease. One example by which dietary exposure may affect genomic imprinting involves imprinted gene insulin like growth factor-2 (IGF2). The maternally inherited allele is normally silenced, and the phenotype is exclusively affected by the paternal allele. The loss of heterozygosity at IGF2 locus has been implicated in biallelic expression of this mitogenic growth factor in human adults as well as in various cancers and Beckwith-Weiderman syndrome (phenotypic and metabolic abnormalities). Though the reason for loss of imprinting in mice is yet unclear, mice weaned on synthetic or methyl donor cofactor-deficient diet displayed hypomethylation of IGF2, and this persisted during a subsequent 100-day recuperation period even after mice were switched to the control diet (Waterland et al. 2006). These tantalizing findings suggest that early nutrition may influence adult obesity, diabetes, CVD, and cancers.

11.9.1 Gender-Dependent Nutri-Epigenome Inheritance

Several lines of evidence indicate that changes in the epigenome can be established during prenatal life by a variety of environmental factors affecting the mother (nutrition, stress, xenibiotics, nursing behavior) and to some extent the father. There have also been instances when epigenetic influence is stronger than the environmental influence as is the case with the "parent of origin effect," a strong indicator for epigenetic involvement. This is especially seen in multiple case of families with multifactorial disorders like spina bifida and anencephaly where the second patient from the index is more often found among the maternal relatives (Mariman and Hamel 1992; Chatkupt et al. 1992). This might be because of the risk factors carried as an imprint of the transmitting parent. Genomic imprinting in which gene activity depends on the sex of the parent of origin has been demonstrated for several cases like the Prader-Willi syndrome (where the gene in maternal chromosome is inactive while the paternal one is active), type 1 diabetes, and other autosomal diseases (Mascari et al. 1992; Bennett et al. 1996). Apart from the transgenerational phenomena explained by the fetal and parent-of-origin effects, grandparental effect has also been established. Longevity was associated with the food availability to the paternal grandfather during his slow growth period before the prepubertal growth peak. Moreover, the food supply to the paternal grandparent had an effect only on the risk of grandson, and the paternal grandmother's food supply was linked with the mortality risk of her granddaughter (Bygren et al. 2001; Kaati et al. 2007). For example, patterns of smoking, diet, and exercise can affect risk across more than one generation by several mechanisms (Brook et al. 1999). Records from Overkalix in northern Sweden for individuals born in 1890, 1905, and 1920 have shown that diabetes mortality increased in men if the paternal grandfather was exposed to abundant nutrition during his prepubertal growth period, an effect later extended to paternal grandmother/granddaughter pairs and transmitted in a gender-specific fashion. Intriguingly, the nutrition-related circumstances of social environment have transgenerational effects in diabetes- and cardiovascular-related mortality down the

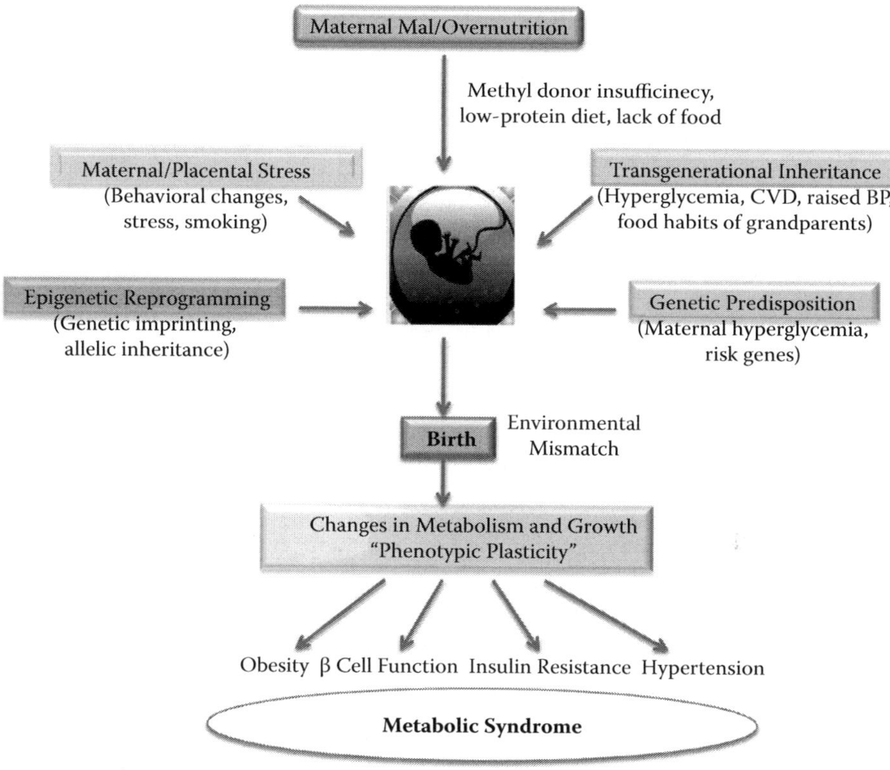

FIGURE 11.4 External factors affecting the onset of metabolic syndrome include the maternal malnutrition that leads to epigenetic reprogramming and maternal stress. These then jointly with an individual's predisposition to the disease, namely transgenerational inheritance (from grandparents), and genetic predisposition contribute to an altered metabolism and growth resulting in phenotypic plasticity where the individual is unable to match to the environmental cues. This results in various components contributing to metabolic syndrome.

male line (Kaati et al. 2002). Nutrition seems to have an effect on the regulation in developing oocytes, and there might be an involvement of X and Y chromosomes. The complex sex-specific male line transgenerational responses to environmental cues like nutrition are more complex in humans and resemble Lamarckism (Pembrey et al. 2006). Under this perspective, the epigenome can also be described as the "cellular memory" of nutritional exposure of the parents and/or even grandparents (Figure 11.4). The precise mechanisms by which this information (posed by environmentally induced methylation/genomic imprinting) survives the reprogramming and is transmitted is yet to be unraveled.

11.10 NUTRIENTS IN EPIGENETIC THERAPY

The preceding sections obviate the effect of alteration in epigenetic program in manifestation of disease. Intuitively, correcting these epigenetic defects may provide an attractive alternative to ameliorate the effects of deleterious genes. The fact that epigenetic marks can be readily reversed and nutrition can have influence on these generates an interesting possibility that these epigenetic processes can be influenced by diet.

This is supported by the observation that several drugs are able to modify the activity of enzymes like methyltransferases and histone deacetylases (Allen 2007). Currently, some of them are under clinical and preclinical trials for cancer therapy (solid tumors) and hematological malignancies. As mentioned earlier, treatment with folic acid alleviates neural tube defects by removing the epigenetic block. A population-based study has revealed that up to 65% occurrence of neural tube defects and 50% recurrence can be prevented by periconceptional folic acid supplementation (Van der Put 1996). The effectiveness of folic acid has also been demonstrated in activation of PPARα and glucocorticoid receptor in liver of offspring even in a protein-restricted maternal diet (Lillycrop et al. 2008). Both type 1 and 2 diabetes have been identified as pathological factors resulting from modulation of methyl metabolism in rat models (Williams and Schalinske 2007). The metabolic disturbances in diabetic state lead to dysregulation of hepatic methyl group methylation, characterized by elevation in glycine-N-methylase expression and activity leading to hypomethylation. Therefore interventions in these processes by supplementation of methyl group donors would be an effective therapy.

Inhibiting the activity of DNMT has been shown to have effects on cellular differentiation (Vogiatzi et al. 2007). The key polyphenol in green tea called epigallocatechin-3-gallate has been shown to have a strong DNMT inhibitory activity. Similarly, a bioactive component from *Artemisia dracunculus L.* has been identified as a promising candidate for development of a nutritional supplement for diabetes owing to its strong hypoglycemic activity. Data are also suggestive of this flavanoid being effective in increasing the molecular events of insulin action in skeletal muscle, thereby improving carbohydrate metabolism (Z. Q. Wang et al. 2008). This is thought to be brought about by the down-regulation of DNMT 1 and 3B activity while up-regulating the NANOG expression. Similarly, the fruit extracts of *Ligustrum lucidum* have been reported to have antidiabetic, anticancer, antioxidant, and neuroprotective activities (Li et al. 1994). This is attributed to the up-regulation of glycine-N-methyltransferase gene activity by NZ-01, a bioactive component of the fruit. A list of some bioactive components in natural diet with their enzyme-modifying activity is shown in Table 11.1. Taken together, it is evident that the modulators of epigenetic patterns are found in nutrients/bioactive components in diet and may be relevant to modulate/maintain the epigenome for maintaining the homeostasis.

11.11 FUTURE DIRECTIONS

The possibility of intervening and correcting the epigenetic processes that precipitate metabolic syndrome and its symptoms has generated a lot of interest and speculation in the field of nutritional (epi)genomics. Figure 11.4 shows a schematic depiction of the molecular events through the life span of an individual affected by nutrition. Novel techniques to unravel the information about "epigenome" would enhance our understanding of the relation between food availability and diet and their role on

TABLE 11.1
List of Certain Bioactive Components in Daily Diet with Source and the Epigenetic Modulation in Which It Plays a Role

Compound Name	Source	Activity
Curcumin	Turmeric	HAT inhibitor (p300) DNA hypomethylation
Garcinol	*Garcinia indica* fruit	HAT inhibitor (p300)
Anacardic acid	Cashew nut shell liquid	HAT inhibitor (p300 and PCAF)
Annurca polyphenol extract	Annurca apples	DNMT inhibitor
EGCG-polyphenol derivative	Green tea	DNMT inhibitor
Sulforaphane	Broccoli	HDAC inhibitor
Pomiferin	Maclura pomifera	HDAC inhibitor

gene expression and phenotype in later life or even for future generations. Further such studies will provide knowledge for the use of specific nutrients or diets to treat nutrition-related disorders with epigenetics as part of their etiology. However, several important issues need to be addressed before use of these nutrient-derived compounds as inhibitors/activators of epigenetic machinery:

1. As to specificity of the treatment, will the nutrients/bioactive components be safe and not affect the nontarget cells?
2. Since these treatments will not be exclusively restricted to fetal origin, will they be relevant in maintaining good health throughout the life span?
3. What is the best way to design effective means of delivery to targets?
4. Considering that the epigenome is more flexible than the genome per se, will it be possible to optimize nutritional interventions in a personalized manner by a detailed study of nutrient-induced epigenetic changes?

ACKNOWLEDGMENTS

Supported by a grant from the Department of Biotechnology, Government of India. LP acknowledges Swedish Institute for scholarship and AG Fonden Sweden for Research Grant FB 09-09-38. We thank all the authors whose work has contributed directly or indirectly to making this chapter possible.

REFERENCES

Abdul-Ghani, M. A., D. Tripathy, and R. A. DeFronzo. 2006. Contributions of beta-cell dysfunction and insulin resistance to the pathogenesis of impaired glucose tolerance and impaired fasting glucose. *Diabetes Care* 29:1130–39.

Adenot, P. G., Y. Mercier, J. P. Renard, and E. M. Thompson. 1997. Differential H4 acetylation of paternal and maternal chromatin precedes DNA replication and differential transcriptional activity in pronuclei of 1-cell mouse embryos. *Development* 124:4615–25.

Afman, L., and M. Müller. 2006. Nutrigenomics: from molecular nutrition to prevention of disease. *J Am Diet Assoc* 106:569–76.

Akiyama, T., M. Nagata, and F. Aoki. 2006. Inadequate histone deacetylation during oocyte meiosis causes aneuploidy and embryo death in mice. *Proc Natl Acad Sci U S A* 103:7339–44.

Allen, A. 2007. Epigenetic alterations and cancer: new targets for therapy. *Idrugs* 10:709–12.

Anway, M. D., C. Leathers, and M. K. Skinner. 2006. Endocrine disruptor vinclozolin induced epigenetic transgenerational adult-onset disease. *Endocrinology* 147:5515–23.

Barbato, A., F. P. Cappuccio, E. J. Folkerd, P. Strazzullo, B. Sampson, D. G. Cook, and K. G. Alberti. 2004. Metabolic syndrome and renal sodium handling in three ethnic groups living in England. *Diabetologia* 47:40–46.

Bennett, S. T., A. J. Wilson, F. Cucca, J. Nerup, F. Pociot, P. A. McKinney, A. H. Barnett, S. C. Bain, and J. A. Todd. 1996. IDDM2-VNTR-encoded susceptibility to type 1 diabetes: dominant protection and parental transmission of alleles of the insulin gene-linked minisatellite locus. *J Autoimmun* 9:415–21.

Bocher, V., I. Pineda-Torra, J. C. Fruchart, and B. Staels. 2002. PPARs: transcription factors controlling lipid and lipoprotein metabolism. *Ann N Y Acad Sci* 967:7–18.

Bogaert ,Y. E., and S. Linas. 2009. The role of obesity in the pathogenesis of hypertension. *Nat Clin Pract Nephrol* 5:101–11.

Bogdarina, I., S. Welham, P. J. King, S. P. Burns, and A. J. Clark. 2007. Epigenetic modification of the renin-angiotensin system in the fetal programming of hypertension. *Circ Res* 100:520–26.

Brinton, E. A., S. Eisenberg, and J. L. Breslow. 1991. Increased apo A-I and apo A-II fractional catabolic rate in patients with low high density lipoprotein-cholesterol levels with or without hypertriglyceridemia. *J Clin Invest* 87:536–44.

Brook, J. S., M. Whiteman, and D. W. Brook. 1999. Transmission of risk factors across three generations. *Psychol Rep* 85:227–41.

Bygren, L. O., G. Kaati, and S. Edvinsson. 2001. Longevity determined by paternal ancestors' nutrition during their slow growth period. *Acta Biotheor* 49:53–59.

Castro, C. E., and J. S. Sevall. 1982. Analysis of rat liver chromatin and nuclear proteins after nutritional variation 1,2. *J Nutr* 112:1203–11.

Cerrato, F., M. Vernucci, P. V. Pedone, L. Chiariotti, G. Sebastio, C. B. Bruni, and A. Riccio. 2002. The 5′ end of the KCNQ1OT1 gene is hypomethylated in the Beckwith-Wiedemann syndrome. *Hum Genet* 111:105–7.

Chakrabarti, S. K., J. Francis, S. M. Ziesmann, J. C. Garmey, and R. G. Mirmira. 2003. Covalent histone modifications underlie the developmental regulation of insulin gene transcription in pancreatic beta cells. *J Biol Chem* 278:23617–23.

Chan, D. C., P. H. Barrett, and G. F. Watts. 2006. Recent studies of lipoprotein kinetics in the metabolic syndrome and related disorders. *Curr Opin Lipidol* 17:28–36.

Chatkupt, S., P. R. Lucek , M. R. Koenigsberger, and W. G. Johnson. 1992. Parental sex effect in spina bifida: a role for genomic imprinting? *Am J Med Genet* 44:508–12.

Corthésy-Theulaz, I., J. T. den Dunnen, P. Ferré, J. M. Geurts, M. Müller, N. van Belzen, and B. van Ommen. 2005. Nutrigenomics: the impact of biomics technology on nutrition research. *Ann Nutr Metab* 49:355–65.

Dallongeville, J., N. Helbecque, D. Cottel, P. Amouyel, and A. Meirhaeghe. 2003. The Gly16→Arg16 and Gln27→Glu27 polymorphisms of beta2-adrenergic receptor are associated with metabolic syndrome in men. *J Clin Endocrinol Metab* 88:4862–66.

Dashwood, R. H., M. C. Myzak, and E. Ho. 2006. Dietary HDAC inhibitors: time to rethink weak ligands in cancer chemoprevention? *Carcinogenesis* 27:344–49.

Datta, J., H. Kutay, M. W. Nasser, G. J. Nuovo, B.Wang, S. Majumder, C. G. Liu, S, C. M. Volinia, T. D. Schmittgen, K. Ghoshal, and S. T. Jacob. 2008. Methylation mediated silencing of MicroRNA-1 gene and its role in hepatocellular carcinogenesis. *Cancer Res* 68:5049–58.

Davis, C. D., and E. O. Uthus. 2004. DNA methylation, cancer susceptibility, and nutrient interactions. *Exp Biol Med* 229:988–95.

D'Cruz, L. G., C. Baboonian, H. E. Phillimore, R. Taylor , P. M. Elliott, A. Varnava, F. Davison, W. J. McKenna, and N. D. Carter. 2000. Cytosine methylation confers instability on the cardiac troponin T gene in hypertrophic cardiomyopathy. *J Med Genet* 37:E18.

Delage, B., and R. H. Dashwood. 2008. Dietary manipulation of histone structure and function. *Annu Rev Nutr* 28:347–66.

de Rooij, S. R., R. C. Painter, F. Holleman, P. M. Bossuytand, and T. J. Roseboom. 2007. The metabolic syndrome in adults prenatally exposed to the Dutch famine. *Am J Clin Nutr* 86:1219–24.

Dolinoy, D. C., J. R. Weidman, P. A. Waterland, and R. L. Jirtle. 2006. Maternal genistein alters coat color and protects Avy mouse offspring from obesity by modifying the fetal epigenome. *Environ Health Perspect* 114:567–72.

Eckel, R. H., S. M. Grundy, and P. Z. Zimmet. 2005. The metabolic syndrome. *Lancet* 365:1415–28.

El-Osta, A., D. Brasacchio, D. Yao, A. Pocai, P. L. Jones, R. G. Roeder, M. E. Cooper, and M. Brownlee. 2008. Transient high glucose causes persistent epigenetic changes and altered gene expression during subsequent normoglycemia. *J Exp Med* 205:2409–17.

Esau, C., X. Kang, E. Peralta, E. Hanson, E. G. Marcusson, L. V. Ravichandran, Y. Sun, S. Koo, R. J. Perera, R. Jain, N. M., Dean, S. M. Freier, C. F. Bennett, B. Lollo, and R. Griffey. 2004. MicroRNA-143 regulates adipocyte differentiation. *J Biol Chem* 279:52361–65.

Feinberg, A. P., and B. Vogelstein. 1983. Hypomethylation distinguishes genes of some human cancers from their normal counterparts. *Nature* 301:89–92.

Ferrannini, E., G. Buzzigoli, R. Bonadonna, M. A. Giorico, M. Oleggini, L. Graziadei, R. Pedrinelli, L. Brandi, and S. Bevilacqua. 1987. Insulin resistance in essential hypertension. *N Engl J Med* 317:350–57.

Foufelle, F., and P. Ferré. 2002. New perspectives in the regulation of hepatic glycolytic and lipogenic genes by insulin and glucose: a role for the transcription factor sterol regulatory element binding protein-1c. *Biochem J* 366:377–91.

Friso, S., F. Pizzolo, S. W. Choi, P. Guarini, A. Castagna, V. Ravagnani, A. Carletto, P. Pattini, R. Corrocher, and O. Olivieri. 2008. Epigenetic control of 11 beta-hydroxysteroid dehydrogenase 2 gene promoter is related to human hypertension. *Atherosclerosis* 199:323–27.

Fumeron, F., R. Aubert, A. Siddiq, D. Betoulle, F. Péan, S. Hadjadj, J. Tichet, E. Wilpart, M. C. Chesnier, B. Balkau, P. Froguel, and M. Marre. 2004. Epidemiologic data on the Insulin Resistance Syndrome (DESIR) Study Group. Adiponectin gene polymorphisms and adiponectin levels are independently associated with the development of hyperglycemia during a 3-year period: the epidemiologic data on the insulin resistance syndrome prospective study. *Diabetes* 53:1150–57.

Galler, J. R., and J. Tonkiss. 1991. Prenatal protein malnutrition and maternal behavior in Sprague-Dawley rats. *J Nutr* 121:762–69.

Garg, A., and A. Misra. 2004. Lipodystrophies: rare disorders causing metabolic syndrome. *Endocrinol Metab Clin North Am* 33:305–31.

Gerken, T., C. A. Girard, Y. C. Tung, C. J. Webby, V. Saudek , K. S. Hewitson, G. S. Yeo, M. A. McDonough, S. Cunliffe, L. A. McNeill, J. Galvanovskis, P. Rorsman, P. Robins, X. Prieur, A. P. Coll, M. Ma, Z. Jovanovic, I. S. Farooqi, B. Sedgwick, I. Barroso,

T. Lindahl, C. P. Ponting, F. M. Ashcroft, S. O'Rahilly, and C. J. Schofield. 2007. The obesity-associated FTO gene encodes a 2-oxoglutarate-dependent nucleic acid demethylase. *Science* 318:1469–72.

Glass, C. K. 1994. Differential recognition of target genes by nuclear receptor monomers, dimers, and heterodimers. *Endocr Rev*15:391–407.

Gluckman, P. D., M. A. Hanson, and A. S. Beedle. 2007. Early life events and their consequences for later disease: a life history and evolutionary perspective. *Am J Hum Biol* 19:1–19.

Gluckman, P. D., M. A. Hanson, C. Cooper, and Thornburg, K. L. 2008. Effect of in utero and early-life conditions on adult health and disease. *N Engl J Med* 359:61–73.

Godfrey, K. M., K. A. Lillycrop, G. C. Burdge, P. D. Gluckman, and M. A. Hanson. 2007. Epigenetic mechanisms and the mismatch concept of the developmental origins of health and disease. *Pediatr Res* 61:5R–10R.

Goodarzi, M. O., X. Guo, K. D. Taylor, M. J. Quiñones, M. F. Saad, H. Yang, W. A. Hsueh, and J. I. Rotter. 2004. Lipoprotein lipase is a gene for insulin resistance in Mexican Americans. *Diabetes* 53:214–20.

Groop, L., C. Forsblom, M. Lehtovirta, T. Tuomi, S. Karanko, M. Nissén, B. O. Ehrnström, B. Forsén, B. Isomaa, B. Snickars, and M. R. Taskinen. 1996. Metabolic consequences of a family history of NIDDM (the Botnia study): evidence for sex-specific parental effects. *Diabetes* 45:1585, 1593.

Groop, L., and M. Orho-Melander. 2001. The dysmetabolic syndrome. *J Intern Med* 250:105–20.

Gusterson, R. J., E. Jazrawi, I. M. Adcock, and D. S. Latchman. 2003. The transcriptional co-activators CREB-binding protein (CBP) and p300 play a critical role in cardiac hypertrophy that is dependent on their histone acetyltransferase activity. *J Biol Chem* 278:6838–47.

Hales, C. N., and D. J. Barker. 1992. Type 2 (non-insulin-dependent) diabetes mellitus: the thrifty phenotype hypothesis. *Diabetologia* 35:595–601.

Halle, M., A. Berg, M. W. Baumstark, D. König, M. Huonker, and Keul Joseph. 1999. Influence of mild to moderately elevated triglycerides on low density lipoprotein subfraction concentration and composition in healthy men with low high density lipoprotein cholesterol levels. *Atherosclerosis* 143:185–92.

Hamamori, Y., and M. D. Schneider. 2003. HATs off to Hop: recruitment of a class I histone deacetylase incriminates a novel transcriptional pathway that opposes cardiac hypertrophy. *J Clin Invest* 112:824–26.

Hassan, Y. I., and J. Zempleni. 2006. Epigenetic regulation of chromatin structure and gene function by biotin. *J Nutr* 136:1763–65.

Hegele, R. A. 2003. Monogenic forms of insulin resistance: apertures that expose the common metabolic syndrome. *Trends Endocrinol Metab* 14:371-7.

Hölzl, B., B. Iglseder, A. Sandhofer, L. Malaimare, J. Lang, B. Paulweber, and F. Sandhofer. 2002. Insulin sensitivity is impaired in heterozygous carriers of lipoprotein lipase deficiency. *Diabetologia* 45:378–84.

Jeter, J. R, Jr., W. A. Pavlat, and I. L. Cameron. 1975. Changes in the nuclear acidic proteins and chromatin structure in starved and refed tetrahymena. *Exp Cell Res* 93:79–88.

Kaati, G., L. O. Bygren, and S. Edvinsson. 2002. Cardiovascular and diabetes mortality determined by nutrition during parents' and grandparents' slow growth period. *Eur J Hum Genet* 10:682–88.

Kaati, G., L. O. Bygren, M. Pembrey, and M. Sjöström. 2007. Transgenerational response to nutrition, early life circumstances and longevity. *Eur J Hum Genet* 15:784–90.

Kato, M., J. Zhang, M. Wang, L. Lanting, H. Yuan, H. J. J. Rossi, and R. Natarajan. 2007. MicroRNA-192 in diabetic kidney glomeruli and its function in TGF-beta-induced collagen expression via inhibition of E-box repressors. *Proc Natl Acad Sci U S A* 104:3432–37.

Kawamori, D., Y. Kajimoto, H. Kaneto, Y. Umayahara, Y. Fujitani, T. Miyatsuka, H. Watada, I. B. Leibiger, Y. Yamasaki and M. Hori. 2003. Oxidative stress induces nucleo-cytoplasmic translocation of pancreatic transcription factor PDX-1 through activation of c-Jun NH(2)-terminal kinase. *Diabetes* 52:2896–2904.

Kontush, A., and M. J. Chapman. 2006. Antiatherogenic small, dense HDL—guardian angel of the arterial wall? *Nat Clin Pract Cardiovasc Med* 3:144–53.

Kook, H., J. J. Lepore, A. D. Gitler, M. M. Lu, W. Wing-Man Yung, J. Mackay, R. Zhou, V. Ferrari, P. Gruber, and J. A. Epstein. 2003. Cardiac hypertrophy and histone deacetylase-dependent transcriptional repression mediated by the atypical homeodomain protein Hop. *J Clin Invest* 112:863–71.

Kotelevtsev, Y., M. C. Holmes, A. Burchell, P. M. Houston, D. Schmoll, P. Jamieson, R. Best, R. Brown, C. R. Edwards, J. R. Seckl and J. J. Mullins. 1997. 11beta-hydroxysteroid dehydrogenase type 1 knockout mice show attenuated glucocorticoid-inducible responses and resist hyperglycemia on obesity or stress. *Proc Natl Acad Sci U S A* 94:14924–29.

Krauss, R. M. 1995. Dense low density lipoproteins and coronary artery disease. *Am J Cardiol* 75:53B–57B.

Kulkarni, R. N., J. C. Brüning, J. N. Winnay, C. Postic, M. A. Magnuson, and C. R. Kahn. 1999. Tissue-specific knockout of the insulin receptor in pancreatic beta cells creates an insulin secretory defect similar to that in type 2 diabetes. *Cell* 96:329–39.

Kutay, H., S. Bai, J. Datta, T. Motiwala, I. Pogribny, W. Frankel, S. T. Jacob, and K. Ghoshal. 2006. Downregulation of miR-122 in the rodent and human hepatocellular carcinomas. *J Cell Biochem* 99:671–78.

Kwiterovich, P. O., Jr. 2002. Clinical relevance of the biochemical, metabolic, and genetic factors that influence low-density lipoprotein heterogeneity. *Am J Cardiol* 90:30i–47i.

Lewis, G. F., and G. Steiner. 1996. Acute effects of insulin in the control of VLDL production in humans. Implications for the insulin-resistant state. *Diabetes Care* 19:390–93.

Li, M. L., M. L. Lui, and W. H. Feng. 1994. Advances in the research on the fruits of *Ligustrum lucidum*. *Zhongguo Zhongyao Zhazhi* 198:504.

Lillycrop, K. A., E. S. Phillips, C. Torrens Hanson, A. A. Jackson, and G. C. Burdge. 2008. Feeding pregnant rats a protein-restricted diet persistently alters the methylation of specific cytosines in the hepatic PPAR alpha promoter of the offspring. *Br J Nutr* 100:278–82.

Lillycrop, K. A., J. L. Slater-Jefferies, M. A. Hanson, K. M. Godfrey, A. A. Jackson, and G. C. Burdge. 2007. Induction of altered epigenetic regulation of the hepatic glucocorticoid receptor in the offspring of rats fed a protein-restricted diet during pregnancy suggests that reduced DNA methyltransferase-1 expression is involved in impaired DNA methylation and changes in histone modifications. *Br J Nutr* 97:1064–73.

Ling, C., S. Del Guerra, R. Lupi, T. Rönn, C. Granhall, H. Luthman, P. Masiello, P. Marchetti, L. Groop, and S. Del Prato. 2008. Epigenetic regulation of PPARGC1A in human type 2 diabetic islets and effect on insulin secretion. *Diabetologia* 51:615–22.

Ling, C., P. Poulsen, S. Simonsson, T. Rönn, J. Holmkvist, P. Almgren, P. Hagert, E. Nilsson, A. G. Mabey, P. Nilsson, A. Vaag, and L. Groop. 2007. Genetic and epigenetic factors are associated with expression of respiratory chain component NDUFB6 in human skeletal muscle. *J Clin Invest* 117:3427–35.

Ma, Y. Q., G. N. Thomas, M. C. Ng, J. A. Critchley, J. C. Chan, and B. Tomlinson. 2003. The lipoprotein lipase gene HindIII polymorphism is associated with lipid levels in early-onset type 2 diabetic patients. *Metabolism* 52:338–43.

Maeda, K., H. Cao, K. Kono, C. Z. Gorgun, M. Furuhashi, K. T. Uysal, Q. Cao, G. Atsumi, H. Malone, B. Krishnan, Y. Minokoshi, B. B. Kahn, R. A. Parker, and G. S. Hotamisligil. 2005. Adipocyte/macrophage fatty acid binding proteins control integrated metabolic responses in obesity and diabetes. *Cell Metab* 1:107–19.

Mariman, E. C. 2007. Nutrigenomics in perspective. *Pharmacogenomics* 8:421–24.

Mariman, E. C., and B. C. Hamel. 1992. Sex ratios of affected and transmitting members of multiple case families with neural tube defects. *J Med Genet* 29:695–98.

Mascari, M. J., W. Gottlieb, P. K. Rogan, M. G. Butler, D. A. Waller, J. A. Armour, A. J. Jeffreys, R. L. Ladda, and R. D. Nicholls. 1992. The frequency of uniparental disomy in Prader-Willi syndrome. Implications for molecular diagnosis. *N Engl J Med* 1326:1599–1607.

Masuzaki, H., J. Paterson, H. Shinyama, N. M. Morton, J. J. Mullins, J. R. Seckl, and J. S. Flier. 2001. A transgenic model of visceral obesity and the metabolic syndrome. *Science* 294:2166–70.

Meijer, A. J., and P. Codogno. 2008. Nutrient sensing: TOR's Ragtime. *Nat Cell Biol* 10:881–83.

Melzner, I., V. Scott, K. Dorsch, P. Fischer, M. Wabitsch, S. Brüderlein, C. Hasel, and P. Möller. 2002. Leptin gene expression in human preadipocytes is switched on by maturation-induced demethylation of distinct CpGs in its proximal promoter. *J Biol Chem* 277:45420–27.

Miyashita, Y., and K. Shirai. 2005. Clinical determination of the severity of metabolic syndrome: preheparin lipoprotein lipase mass as a new marker of metabolic syndrome. *Curr Med Chem Cardiovasc Hematol Agents* 3:377–81.

Monteiro, P. O., and C. G. Victora. 2005. Rapid growth in infancy and childhood and obesity in later life—a systematic review. *Obes Rev* 6:143–54.

Moraes, A.S., B. de Campos Vidal, A. M. Guaraldo, and M. L. Mello. 2005. Chromatin supraorganization and extensibility in mouse hepatocytes following starvation and refeeding. *Cytometry A* 63:94–107.

Müller, M., and S. Kersten. 2003. Nutrigenomics: goals and strategies. *Nat Rev Genet* 4:315–22.

Nanney, D. L. 1958. Epigenetic control systems. *Proc Natl Acad Sci U S A* 44:712–17.

Niitsu, N., Y. Hayashi, K. Sugita, and Y. Honma. 2001. Sensitization by 5-aza-2'-deoxycytidine of leukaemia cells with MLL abnormalities to induction of differentiation by all-trans retinoic acid and 1alpha,25-dihydroxyvitamin D3. *Br J Haematol* 112:315–26.

Nilsson, E. E., M. D. Anway, J. Stanfield, and M. K. Skinner. 2008. Transgenerational epigenetic effects of the endocrine disruptor vinclozolin on pregnancies and female adult onset disease. *Reproduction* 135:713–21.

Ozcan, U., Q. Cao, E. Yilmaz, A. H. Lee, N. N. Iwakoshi, E. Ozdelen, G. Tuncman, C. Görgün, L. H. Glimcher and G. S. Hotamisligil. 2004. Endoplasmic reticulum stress links obesity, insulin action, and type 2 diabetes. *Science* 306:457–61.

Packard, C. J. 1996. LDL subfractions and atherogenicity: an hypothesis from the University of Glasgow. *Curr Med Res Opin* 13:379–90.

Parker, K., J. Maxson, A. Mooney, and E. A. Wiley. 2007. Class I histone deacetylase Thd1p promotes global chromatin condensation in Tetrahymena thermophila. *Eukaryot Cell* 6:1913–24.

Patel, S. R., D. Kim, I. Levitan, and G. R. Dressler. 2007. The BRCT-domain containing protein PTIP links PAX2 to a histone H3, lysine 4 methyltransferase complex. *Dev Cell* 13:580–92.

Pembrey, M. E., L. O. Bygren, G. Kaati, S. Edvinsson, K. Northstone, M. Sjöström, and J. Golding. 2006. ALSPAC Study Team. Sex-specific, male-line transgenerational responses in humans. *Eur J Hum Genet* 14:159–66.

Petersen, K. F., D. Befroy, S. Dufour, J. Dziura, C. Ariyan, D. L. Rothman, L. DiPietro, G. W. Cline, and G. I. Shulman. 2003. Mitochondrial dysfunction in the elderly: possible role in insulin resistance. *Science* 300:1140–42.

Petersen, K. F., S. Dufour, D. Befroy, R. Garcia, and G. I. Shulman. 2004. Impaired mitochondrial activity in the insulin-resistant offspring of patients with type 2 diabetes. *N Engl J Med* 350:664–71.

Pogribny, I. P., S. A. Ross, V. P. Tryndyak, M. Pogribna, L. A. Poirier, and T. V. Karpinets. 2006. Histone H3 lysine 9 and H4 lysine 20 trimethylation and the expression of Suv4-20h2 and Suv-39h1 histone methyltransferases in hepatocarcinogenesis induced by methyl deficiency in rats. *Carcinogenesis* 27:1180–86.

Poirier, L. A. 1994. Methyl group deficiency in hepatocarcinogenesis. *Drug Metab Rev* 26:185–99.

Poy, M. N., J. Hausser, M. Trajkovski, M. Braun, S. Collins, P. Rorsman, M. Zavolan, and M. Stoffel. 2009. miR-375 maintains normal pancreatic alpha- and beta-cell mass. *Proc Natl Acad Sci U S A* 106:5813–18.

Rees, W. D., S. M. Hay, and M. Cruickshank. 2006. An imbalance in the methionine content of the maternal diet reduces postnatal growth in the rat. *Metabolism* 55:763–70.

Roden, M. 2005. Muscle triglycerides and mitochondrial function: possible mechanisms for the development of type 2 diabetes. *Int J Obes (Lond)* 29 Suppl 2:S111–15.

Roseboom, T., S. de Rooij, and R. Painter. 2006. The Dutch famine and its long-term consequences for adult health. *Early Hum Dev* 82:485–91.

Ross, S. A. 2003. Diet and DNA methylation interactions in cancer prevention. *Ann N Y Acad Sci* 983:197–207.

Rozovski, S.J. 1984. Nutrition and aging. *Curr Concepts Nutr* 13:137–69.

Rutter, M. K., J. B. Meigs, L. M. Sullivan, R. B. D'Agostino, and P. W. Wilson. 2005. Insulin resistance, the metabolic syndrome, and incident cardiovascular events in the Framingham Offspring Study. *Diabetes* 54:3252–57.

Scarpulla, R. C. 2006. Nuclear control of respiratory gene expression in mammalian cells. *J Cell Biochem* 97:673–83.

Schotta, G., M. Lachner, K. Sarma, A. Ebert, R. Sengupta, G. Reuter, D. Reinberg, and T. A. Jenuwein. 2004. Silencing pathway to induce H3-K9 and H4-K20 trimethylation at constitutive heterochromatin. *Genes Dev* 18:1251–62.

Seki, Y., K. Hayashi, K. Itoh, M. Mizugaki, M. Saitou, and Y. Matsui. 2005. Extensive and orderly reprogramming of genome-wide chromatin modifications associated with specification and early development of germ cells in mice. *Dev Biol* 278:440–58.

Shemer, R., S. Eisenberg, J. L. Breslow, and A. Razin. 1991. Methylation patterns of the human apoA-I/C-III/A-IV gene cluster in adult and embryonic tissues suggest dynamic changes in methylation during development. *J Biol Chem* 266:23676–81.

Shen, X., L. Yu, J. W. Weir and M. A. Gorovsky. 1995. Linker histones are not essential and affect chromatin condensation in vivo. *Cell* 82:47–56.

Shuldiner, A. R, and J. C. McLenithan. 2004. Genes and pathophysiology of type 2 diabetes: more than just the Randle cycle all over again. *J Clin Invest* 114:1414–17.

Simmons, R. A. 2007. Developmental origins of diabetes: the role of epigenetic mechanisms. *Curr Opin Endocrinol Diabetes Obes* 14:13–16.

Smilinich, N. J., C. D. Day, G. V. Fitzpatrick, G. M. Caldwell, A. C. Lossie, P. R. Cooper, A. C. Smallwood, J. A. Joyce, P. N. Schofield, W. Reik, R. D. Nicholls, R. Weksberg, D. J. Driscoll, E. R. Maher, T. B. Shows, and M. J. Higgins. 1999. A maternally methylated CpG island in KvLQT1 is associated with an antisense paternal transcript and loss of imprinting in Beckwith-Wiedemann syndrome. *Proc Natl Acad Sci U S A* 96:8064–69.

Taghibiglou, C., F. Rashid-Kolvear, S. C. van Iderstine, H. Le-Tien, I. G. Fantus, G. F. Lewis, and K. Adeli. 2002. Hepatic very low density lipoprotein-ApoB overproduction is associated with attenuated hepatic insulin signaling and overexpression of protein-tyrosine phosphatase 1B in a fructose-fed hamster model of insulin resistance. *J Biol Chem* 277:793–803.

Tanaka, T., K. Sohmiya, T. Kono, F. Terasaki, R. Horie, Y. Ohkaru, M. Muramatsu, S. Takai, M. Miyazakiand, and Y. Kitaura. 2007. Thiamine attenuates the hypertension and metabolic abnormalities in CD36-defective SHR: uncoupling of glucose oxidation from cellular entry accompanied with enhanced protein O-GlcNAcylation in CD36 deficiency. *Mol Cell Biochem* 299:23–35.

Tang, X., L. Muniappan, G. Tang, and S. Ozcan. 2009. Identification of glucose-regulated miRNAs from pancreatic {beta} cells reveals a role for miR-30d in insulin transcription. *RNA* 15:287–93.

Tooke, J. E., and M. M. Hannemann. 2000. Adverse endothelial function and the insulin resistance syndrome. *J Intern Med* 247:425–31.

van der Put, N. M., L. P. van den Heuvel, R. P. Steegers-Theunissen, F. J. Trijbels,T. K. Eskes, E. C. Mariman, M. den Heyer and H. J. Blom. 1996. Decreased methylene tetrahydrofolate reductase activity due to the 677C→T mutation in families with spina bifida offspring. *J Mol Med* 74:691–94.

van Sluijters, D. A., P. F. Dubbelhuis, E. F. Blommaart, and A. J. Meijer. 2000. Amino-acid-dependent signal transduction. *Biochem J* 351:545–50.

Vartanian, V., B. Lowell, I. G. MinkoWood, J. D. Ceci, S. George, S. W. Ballinger, C. L. Corless, A. K. McCullough and R. S. Lloyd. 2006. The metabolic syndrome resulting from a knockout of the NEIL1 DNA glycosylase. *Proc Natl Acad Sci U S A* 103:1864–69.

Villeneuve, L. M., M. A. Reddy, L. L. Lanting, M. Wang, L. Meng, and R. Natarajan. 2008. Epigenetic histone H3 lysine 9 methylation in metabolic memory and inflammatory phenotype of vascular smooth muscle cells in diabetes. *Proc Natl Acad Sci U S A* 105:9047–52.

Vogiatzi, P., P. Aimola, M. I. Scarano, and P. P. Claudio. 2007. Epigenome-derived drugs: recent advances and future perspectives. *Drug News Perspect* 20:627–33.

Waddington, C. H. 1952. Selection of the genetic basis for an acquired character. *Nature* 169:625–26.

Wang, B., S. Majumder, G. Nuovo, H. Kutay, S. Volinia, T. Patel, T. D. Schmittgen, C. Croce, K. Ghoshal, and S. T. Jacob. 2009. Role of microRNA-155 at early stages of hepatocarcinogenesis induced by choline-deficient and amino acid-defined diet in C57BL/6 mice. *Hepatology* 50:1152–61.

Wang, Z. Q., D. Ribnicky, X. H. Zhang, I. Raskin, Y. Yu, and W. T. Cefalu. 2008. Bioactives of *Artemisia dracunculus L* enhance cellular insulin signaling in primary human skeletal muscle culture. *Metabolism* 57:S58–64.

Waterland, R. A., and C. Garza. 1999. Potential mechanisms of metabolic imprinting that lead to chronic disease. *Am J Clin Nutr* 69:179–97.

Waterland, R. A., J. R. Lin, C. A. Smith, and R. L. Jirtle. 2006. Post-weaning diet affects genomic imprinting at the insulin-like growth factor 2 (Igf2) locus. *Hum Mol Genet* 15:705–16.

Williams, K. T., and K. L. Schalinske. 2007. New insights into the regulation of methyl group and homocysteine metabolism. *J Nutr* 137:311–14.

Wilson, P. W., R. B. D'Agostino, H. Parise, L. Sullivan, and J. B. Meigs. 2005. Metabolic syndrome as a precursor of cardiovascular disease and type 2 diabetes mellitus. *Circulation* 112:3066–72.

Wolff, G. L., R. L. Kodell, S. R. Moore, and C. A. Cooney. 1998. Maternal epigenetics and methyl supplements affect agouti gene expression in Avy/a mice. *FASEB J* 12:949–57.

Yamada, T., H. Katagiri, Y. Ishigaki, T. Ogihara, J. Imai, K. Uno, Y. Hasegawa, J. Gao, H. Ishihara, A. Niijima, H. Mano, H. Aburatani, T. Asano, and Y. Oka. 2006. Signals from intra-abdominal fat modulate insulin and leptin sensitivity through different mechanisms: neuronal involvement in food-intake regulation. *Cell Metab* 3:223–29.

Yamashita, S., K. Hirano, T. Kuwasako, M. Janabi, Y. Toyama, M. Ishigami, and N. Sakai. 2007. Physiological and pathological roles of a multi-ligand receptor CD36 in atherogenesis; insights from CD36-deficient patients. *Mol Cell Biochem* 299:19–22.

Yang, T., and A. A. Sauve. 2006. NAD metabolism and sirtuins: metabolic regulation of protein deacetylation in stress and toxicity. *AAPS J* 8:E632–43.

Zhang, C. L, T. A. McKinsey, S. Chang, C. L. Antos, J. A. Hill, and E. N. Olson. 2002. Class II histone deacetylases act as signal-responsive repressors of cardiac hypertrophy. *Cell* 110:479–88.

Zhang, W., Y. Hayashizaki, and B. C. Kone. 2004. Structure and regulation of the mDot1 gene, a mouse histone H3 methyltransferase. *Biochem J* 377:641–51.

12 Nutrition and the Emerging Epigenetic Paradigm

Lessons from Neurobehavioral Disorders

Axel Schumacher

CONTENTS

12.1 Epigenetics and Complex Disorders 288
12.1.1 Epigenetic Mechanisms 289
12.1.2 DNA Methylation 290
12.2 Epigenetic Alteration in Major Psychosis 291
12.3 Nutritional Deficiency and One-Carbon Metabolism 295
12.3.1 Aberrant One-Carbon Metabolism in Major Psychosis 296
12.4 Diet and Major Psychosis 298
12.4.1 Methyl Donors and Vitamins 299
12.4.2 Natural Dietary Factors with Impact on the Epigenome 301
12.4.3 Transgenerational Effects 303
12.4.4 Malnutrition and Major Psychosis 304
12.4.5 Anticipation in Major Psychosis—an Epigenetic Component? 306
12.5 Conclusions 308
Acknowledgments 309
References 309

Schizophrenia (SCZ) (MIM 181500) and bipolar disorder (BD) (MIM 125480) are etiologically related psychiatric disorders, collectively termed "major psychosis," a disorder that affects approximately 2% of the population (Craddock et al. 2006; Mill et al. 2008). SCZ is primarily characterized by abnormal perceptions of reality, social apathy, disorganized thought, delusions, or hallucinations and is usually accompanied in varying degrees by other emotional, behavioral, or intellectual disturbances. BD, also known as manic depressive disorder or bipolar affective disorder, is an illness that characteristically involves alternating periods of elevated mood, called manic episodes, and depression. SCZ is often classified as a "thought" disorder, while BD is sometimes classified as a "mood" disorder. Nevertheless, there is increasing evidence of substantial overlap between the two, indicating that these diseases may share a common etiology (Craddock et al. 2006; Owen et al. 2007). For example, extreme manic episodes in BD can sometimes lead to psychotic symptoms, such as delusions and hallucinations, that are also present in SCZ, thereby challenging the validity of the dichotomous classification of these disorders (Mill and Petronis 2009).

Although numerous genetic linkage and association studies have been conducted in order to explain the heritable predisposition to major psychosis, these studies suffer from small effect sizes, and the resulting data are quite often inconsistent. In addition to the immense complexity of major psychosis, the slow progress in understanding the molecular origins of major psychosis could be due to limitations in classical hypotheses concerning the origin of complex disorders as well as in traditional research strategies, which are primarily focused on DNA sequence variation (e.g., mutations, single-nucleotide polymorphisms) and the exposure to environmental risk factors during critical periods of brain development (Tandon et al. 2008). However, DNA sequence variation and environment-based mechanisms do not fully explain many of the epidemiological, clinical, and molecular peculiarities associated with these disorders. As a result, a new interpretation of the classical paradigm of "genes plus environment" emerged in recent years, whereby the emphasis has been shifted to epigenetic dysregulation as a major etiopathogenic factor (Petronis 2004; Petronis et al. 2003; Schumacher and Petronis 2006).

12.1 EPIGENETICS AND COMPLEX DISORDERS

Epigenetics refers to regulation of various genomic functions that are brought about by heritable but reversible changes in DNA modification (particularly methylation of cytosines) and chromatin organization, i.e., histone modifications such as acetylation, methylation, and phosphorylation, among others. Growing epidemiological and experimental evidence supports a role for epigenetic dysfunction in SCZ and BD. Indeed, consistent with the epigenetic theory of major psychosis (Schumacher and Petronis 2006), a number of loci were found to be epigenetically altered in the brain of SCZ and BD patients relative to unaffected controls (Mill et al. 2008). Abnormal epigenetic patterns in the human brain could be caused by a chain of aberrant epigenetic events that begins with a pre-epimutation, a primary epigenetic problem that takes place during the maturation of the germline (Mill and Petronis 2009; Flanagan et al. 2006; Schumacher and Petronis 2006). Generally, the epigenetic status of genes and genomes is far more dynamic in comparison to the DNA sequence and can be changed under the influence of developmental programs and the internal and external environment of the organism (Cooney et al. 2002; Sutherland and Costa 2003; Waterland and Jirtle 2003; Weaver et al. 2004). Many of the environmental risk factors that have been consistently proposed to interact with genetic factors in major psychosis exert their influence in very early developmental stages, particularly in prenatal stages. The most significant risk factors during the prenatal period have been reported to be paternal age, hypoxia, nutritional deficiency, impaired homocysteine metabolism, maternal infection, and maternal stress (reviewed in Rutten and Mill 2009). Other environmental risk factors that become increasingly important during childhood and early adolescence include chronic stress/victimization, urban environment, migration, and drug abuse. However, the mechanism by which environmental factors act upon the epigenetic machinery in the human brain and ultimately give rise to psychosis-related phenotypes and pathology remains poorly understood.

Of all of these environmental risk factors, epigenetic changes induced by diet have been of particular interest. Diet is one of the most important environmental factors that humans are exposed to during their lifetime, and it is able to modulate epigenetic blueprints in various cell types, thereby affecting critical patterns of gene expression. For example, intake of folic acid and other nutrients affects both the global methylation level in the genome and the regulation of imprinted genes (Ingrosso et al. 2003; Wolff et al. 1998). In this way, dietary lifestyle choices affect physiologic and pathologic processes in our body. It can be speculated that a certain nutrient deficiency or overexposure may induce an abnormality in the epigenetic condition that may in turn induce psychotic symptoms, and conversely a proper nutritional intervention program may reverse abnormal epigenetic patterns. In this chapter, we summarize current evidence on how nutrition may affect epigenetic processes in SCZ and BD, and review the emerging epigenetic paradigm that has led to the reinterpretation of a series of clinical and molecular findings in major psychosis.

12.1.1 Epigenetic Mechanisms

Change in the composition and higher-order structure of chromatin with age is a major causative mechanism for the deterioration of cellular and tissue functions. The major components of chromatin are DNA and histones, although numerous other chromosomal proteins also have important functions. The smallest unit of chromatin is the nucleosome, consisting of a histone octamer core that contains two copies of histones H2A, H2B, H3, and H4, wrapped by a 147 base pair DNA segment in a superhelical turn. To lock the DNA in place, the linker histone H1 binds to the nucleosome at the entry and exit sites of the DNA, therefore allowing the formation of higher-order chromatin structures. Long N-terminal tails protrude from the nucleosome at histones H3 and H4, and several locations on these tails can undergo a number of covalent posttranslational modifications, such as acetylation, methylation, sumoylation, phosphorylation, ubiquitination, ribosylation, biotinylation, deamination (citrullination), proline isomerisation, and the lesser known carbonylation (Kouzarides 2007; Sharma et al. 2006). Different combinations of modifications are thought to constitute a histone code, which directs the binding of transcription factors to the DNA and changes the chromatin structure on a genomewide scale. A 2005 study examining different age categories in a large cohort of monozygotic (MZ) twins found major differences in DNA methylation and histone modifications, signifying that there is a significant epigenetic drift between siblings during aging (Fraga et al. 2005). At the time, these changes were associated with phenotypic discordance, which was attributed to an unshared environment, although several of the observed epigenetic changes could also be the result of stochastic or other influences, rather than environmental effects. Overall, it appears that histone modifications drastically change during aging and frequently occur in concert with changes in DNA methylation.

12.1.2 DNA Methylation

Not much is known about histone modifications in complex neurobehavioral disorders, partly due to the instability of histone modification patterns in postmortem brain samples. In contrast, DNA methylation patterns remain relatively stable in postmortem tissues and therefore provide a better target to study epigenetic mechanisms in complex disorders such as major psychosis. DNA methylation is a universal phenomenon observed in bacteria, plants, and animals. In mammals, DNA methylation usually refers to methylation at the 5-position of cytosine bases (^{m}C; Figure 12.1), and nearly all CpG dinucleotides are symmetrically methylated. Recently, 5-hydroxymethylcytosine (^{hm}C) was discovered to be an additional DNA modification in humans (Kriaucionis and Heintz 2009; Loenarz and Schofield 2009; Tahiliani et al. 2009). 5-hydroxymethylcytosine is a stable base that is formed through conversion of ^{m}C by the oxygenase TET1 (Figure12.1). It seems that ^{hm}C is overrepresented in regulatory regions and is highly tissue specific. For example, ^{hm}C was identified as significant component of human embryonic stem (ES) cells (Tahiliani et al. 2009) as well as Purkinje cells in the brain (Kriaucionis and Heintz 2009), where it represents around 40% of the abundance of ^{m}C, but it was not detected in dendritic cells or T cells. Expression patterns in Purkinje cells of SCZ patients were reported to be abnormal (Bernstein et al. 2001); however, since ^{hm}C is only a recent discovery, no information exists on the molecular, cellular, and physiological roles of the hmC modification and it is unknown if this DNA modification plays an essential role in brain regulatory processes.

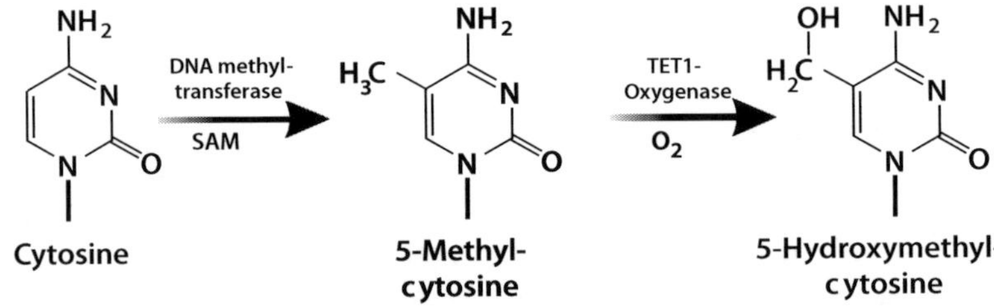

FIGURE 12.1 DNA methylation in humans. DNA methyltransferases catalyze the transfer of a methyl group (-CH3) from S-adenosylmethionine (SAM) to the five-carbon position of cytosine. In humans, the resulting methylcytosine can be oxygenized to 5-hydroxymethylcytosine. It is possible that further modifications at the cytosine 5-position occur, either to enable active demethylation of cytosine or to reverse the function of cytosine methylation by modulating the affinity of proteins that bind to the methylated CpG dinucleotides (Kriaucionis and Heintz 2009; Loenarz and Schofield 2009). This concept is supported by the discovery that the methyl-CpG binding protein (MeCP2) has a lower affinity toward sequences containing 5-hydroxymethylcytosine (Valinluck et al., *Nucleic Acids Res.* 32(14) 4100–8, 2004).

DNA methylation is mediated by a family of conserved DNA methyltransferases (Dnmts). Maintenance methylation, which is necessary after each round of DNA replication, is carried out by Dnmt1, the most abundant DNA methyltransferase in mammalian cells (Bestor et al. 1988). Accurate transmission of epigenetic information to the next cellular generation critically depends on the degree of specificity of Dnmt1 for hemimethylated DNA. In contrast, DNMT3a and DNMT3b are responsible for de novo DNA methylation reactions of unmethylated DNA. DNA methylation is associated with the silencing of gene expression. Consequently, DNA methyltransferases are able to influence numerous biological processes by modulating gene repression in a heritable manner, allowing DNA methylation to act as a form of information or cellular memory.

12.2 EPIGENETIC ALTERATION IN MAJOR PSYCHOSIS

Although the idea of epigenetic dysregulation in major psychosis is not new, so far only a limited number of studies have empirically investigated the role of epigenetic factors in major psychotic disorders. The majority of studies examine global changes in methylation or opt for a relatively more biased approach of selecting candidate genes in the vicinity of which methylation differences are assessed. In a study by Kuratomi et al. (2008), differences in DNA methylation were observed in the lymphoblastoid cells of healthy and bipolar monozygotic twins discordant for the disorder. Particularly, two regions upstream of the spermine synthase (SMS) gene and the peptidylprolyl isomerase E-like gene (PPIEL) showed aberrant DNA methylation status in BD patients. While DNA methylation at the SMS locus was not correlated with expression of the SMS gene, a strong inverse correlation between expression of the PPIEL gene and DNA methylation was observed. In contrast, Bromberg and colleagues (2009) did not find any methylation differences in leukocytes of BD patients compared to age-matched controls. Similarly, the same group did not find any changes in leukocyte global DNA methylation in SCZ patients, although a correlation between leukocyte methylation levels and the smoking habits of the probands was observed (Bromberg et al. 2008). In another study that controlled for effects of sex, a tendency of global hypomethylation of peripheral leukocyte DNA was observed in male SCZ patients but not females (Shimabukuro et al. 2007).

DNA methylation patterns were also investigated in the 5′-regulatory region of the dopamine D2 receptor gene (DRD2) in lymphocytes of two pairs of monozygotic twins (MZ), one concordant and one discordant for SCZ. Numerous DNA methylation differences were identified in the analyzed regions of DRD2, both within and between the pairs of MZ twins (Petronis et al. 2003). “Epigenetic distances” between MZ twins were calculated and used for the comparison of twin DRD2 methylation profiles. It was detected that the affected twin from the pair discordant for SCZ was epigenetically “closer” to the affected concordant twins than to his unaffected MZ cotwin. This result suggested that even between genetically identical individuals the expression and function of the genome is not equivalent. Such phenotypic discordance between MZ twins is often attributed to nonshared environmental factors. However, the empirical evidence for such an influential environmental contribution to major psychosis is still lacking, and no specific environmental risk factor has been conclusively linked to etiology (Schumacher and Petronis 2006; Rutten and Mill 2009). DRD2 seems to be, generally, up-regulated in peripheral blood lymphocytes of SCZ patients. Yet no statistically significant differences in frequency of site-specific cytosine methylation in DRD2 were found between patients and normal controls (Zhang et al. 2007). Furthermore, DRD2 methylation patterns did not show any significant association between sex, age on admission, or age at onset of SCZ, indicating that the methylation status of DRD2 has no significant role in the etiology of SCZ. This would suggest that the observed methylation difference in the MZ twin pair is not directly linked with SCZ. However, an alternative explanation would be that the epigenetic and phenotypic discordance between MZ twins is related to the partial instability of epigenetic signals that change progressively during life (Petronis et al. 2003; Schumacher and Petronis 2006; Rutten and Mill 2009). Indeed, one study demonstrated that although twins are fairly indistinguishable on an epigenetic level during the early years of life, older MZ twins exhibit remarkable differences in their overall content and genomic distribution of DNA methylation and histone acetylation, consequently affecting their gene-expression portrait and highlighting the dynamic nature of epigenetic processes (Fraga et al. 2005). MZ twin methylation differences in non-disease-related tissues, such as buccal swabs, have previously been reported at CpG sites located at several genes associated with major psychosis, including the catechol-O-methyltransferase (COMT) gene (Mill et al. 2006).

Aberrant epigenetic patterns are most likely to be found in disease-related tissues, e.g., the parietal lobe or prefrontal cortex in major psychosis. Early studies on brain samples reported DNA methylation differences associated with SCZ in the vicinity of both the COMT gene (Abdolmaleky et al. 2006) and the reelin gene (RELN) (Abdolmaleky et al. 2005; Grayson et al. 2005), which control processes of neuronal migration and positioning in the developing brain (reviewed in Connor and Akbarian 2008). Unfortunately, these findings were not confirmed by other groups using more sophisticated methylation-profiling methods (Dempster et al. 2006; Murphy et al. 2005; Tamura et al. 2007; Tochigi et al. 2008; Mill et al. 2008; Schumacher et al. 2006). Another study found that a CpG island in the sex-determining region Y-box containing gene 10 (SOX10), an oligodendrocyte-specific transcription factor, tended to be highly methylated in brains of patients with SCZ, correlating with a reduced expression of SOX10 (Iwamoto et al. 2005). One interesting and very thorough study by Mill and coworkers (2008) utilized 105 frontal cortex brain tissue samples from SCZ and BD patients. Following enrichment of the unmethylated fraction of genomic DNA, samples were hybridized on high-density CpG island microarrays to identify DNA methylation changes associated with SCZ and BD. This study found evidence for psychosis-associated DNA methylation differences in numerous loci, including several genes that have been functionally linked to disease etiology such as dysbindin, which is known to confer a genetic risk for psychosis. Consistent with increasing evidence for altered glutamatergic and GABAergic neurotransmission in the pathogenesis of major psychosis (Coyle 2004), the Mill et al. study identified epigenetic changes at a number of loci associated with both these neurotransmitter pathways. Glutamate is the most abundant fast excitatory neurotransmitter in the mammalian nervous system and has a critical role in synaptic plasticity. Dysfunction of glutamatergic neurotransmission may play a significant role in the pathophysiology of major psychosis. Indeed, a number of studies have revealed alterations in pre- and postsynaptic markers for glutamatergic neurons in several brain regions in SCZ and BD (Harrison and Weinberger 2005; Millan 2005). The major subtype of glutamate receptors are the N-methyl-D-aspartate (NMDA) receptors, whose function is considered critical for the proper expression of numerous complex behaviors, such as associative learning, working memory, behavioral flexibility, and attention, many of which are impaired in psychosis (Coyle 2006). In the last decades, basic and clinical evidence has been accumulating to support the idea that abnormal NMDA receptor functions elicit many aspects of molecular, cellular, and behavioral abnormalities associated with SCZ.

Intriguingly, the Mill et al. (2008) study demonstrated that the methylome in the brain and other tissues possesses a highly modular network structure, indicating that the human epigenome can be split into distinct groups of correlated loci, which potentially correspond to distinct functional pathways and/or physical regions. A network comprises distinct clusters of elements, termed "modules," which are highly connected within themselves but have fewer connections with the rest of the network (Newman 2006; Mill et al. 2008). Importantly, even though DNA methylation in both affected and unaffected groups was clearly modular, the number of interconnections between specific genomic regions was clearly higher in the psychosis group compared to the unaffected control group, demonstrating more between-module interference in both brain and germline DNA of patients with major psychosis (Figure 12.2). Given that modules within such biological networks are likely to have

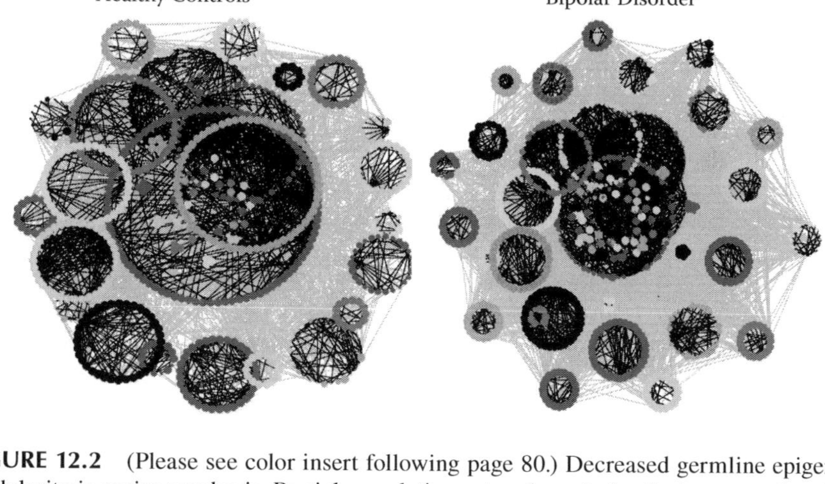

FIGURE 12.2 (Please see color insert following page 80.) Decreased germline epigenetic modularity in major psychosis. Partial-correlation network analysis of microarray data illustrating significant connections between nodes, demonstrating strong hierarchical modularity for male healthy controls and BD samples. (Data from Mill, J. et al., *Am J Hum Genet,* 141B, 4, 421–5, 2008). Although modularity is apparent in both sample groups, it is significantly lower in the BD group, resulting in increased between-module interference, reflected by the higher number of interconnections between specific genomic regions in the psychosis group. A similar pattern of modularity was seen in the comparison of methylation between brain SCZ samples and unaffected controls.

specific functional tasks separate from those of other modules, the lower degree of methylation modularity observed in the major psychosis samples indicates a certain level of systemic epigenetic dysfunction associated with major psychosis.

Another potentially important epigenetic phenomenon in major psychosis is skewed X chromosome inactivation. MZ female twin pairs are sometimes discordant for monogenic X-linked disorders because of differential X inactivation. Hence it was postulated that similar mechanisms may also occur in disorders with more complex inheritance patterns, including major psychosis. Examination of X chromosome inactivation patterns in DNA samples from blood and/or buccal swabs in a series of 63 female MZ twin pairs concordant or discordant for BD or SCZ and healthy MZ controls suggested that X-linked loci may contribute to discordance within twin pairs for BD (Rosa et al. 2008). Discordant female BD twins showed greater differences in the methylation of the maternal and paternal X alleles than concordant twin pairs. This suggests that differential skewing of X chromosome inactivation may contribute to the discordance observed for BD in female MZ twin pairs.

Although significant differences in methylation were observed in major psychosis, the absolute methylation differences between psychosis and control samples were relatively small in most studies. Nevertheless, because of its enormous complexity, the human brain is likely to be susceptible to even mild epigenetic changes, which may lead to a wide variety of small morphological and functional changes in major psychosis. Overall, DNA methylation changes in major psychosis appear to be more subtle compared to those observed in other complex disorders such as cancer. One of the central obstacles hampering progress in the field of epigenetics related to neurobehavioral disease is the inherent heterogeneity of brain cells. To date, most methylation studies in the brain have utilized DNA extracted from pooled cell lysates, without taking into account the multiple different cell types such as neurons, astrocytes, and satellite cells, among others, which are known to possess unique methylation patterns. Moreover, specific subpopulations of cells within distinct brain regions are thought to be affected in psychotic disorders, such as parvalbumin-immunoreactive neurons, oligodendrocytes, and astrocytes (reviewed in Connor and Akbarian 2008). However, standard population measurement techniques merely describe average behavior and are insufficient to investigate variability among cells (Le et al. 2005). Epigenetic analyses traditionally probe cell ensembles, thereby completely averaging the relevant individual cell responses, such as differences in cell proliferation, lack of synchrony of cells in a culture, responses to external stimuli, disease onset, or stochastic events (Sims and Allbritton 2007). As a consequence, it is presently impossible to understand whether a small increase in methylation determined in the ensemble of cells results from a small, homogeneous increase across all cells or a large increase in a subset of cells. It is likely that differences in the cellular composition of the analyzed brain tissue samples account for some of the controversial results observed in the aforementioned studies. In cancer studies, methylation patterns in CpG islands that are important in gene regulation could be different from cell to cell even in a single tumor tissue. Indeed, cancer cells with distinct epigenetic profiles display distinct phenotypic behavior and drug response (Cluzel et al. 2000; Wilkinson 2005).

In contrast to DNA methylation, histone modifications are far less stable in postmortem brain; hence it is significantly more difficult to derive accurate measurements from such samples. Nevertheless, using postmortem prefrontal cortex samples, one study reported decreased GAD1 expression and H3K4-trimethylation (H3K4me3), a type of histone modification linked to transcriptional mechanisms and RNA polymerase II activity, predominantly in female SCZ patients (Huang et al. 2007). Altered expression of GAD1, a key enzyme for GABA synthesis that is regulated by neuronal activity, has been implicated in prefrontal dysfunction in major psychosis. The same study demonstrated that SCZ subjects biallelic for GAD1 haplotypes previously associated with major psychosis were affected by a deficit in prefrontal GAD1 mRNA, in conjunction with a shift from open (H3K4me3) toward repressive (H3K27me3) chromatin-associated histone methylation. Additional support for the proposal that epigenetic factors are operative in the pathophysiology of SCZ came from a study that measured the histone deacetylase 1 expression in the prefrontal cortex of SCZ patients (Sharma et al. 2008). The authors reported that HDAC1 expression levels were significantly higher in SCZ versus normal subjects and that the mRNA expression level of one of the SCZ candidate genes, GAD67, was strongly and negatively correlated with the expression levels of HDAC1, HDAC3, and HDAC4.

Previously, it has been suggested that major psychosis may result from a primary epigenetic defect (preepimutation) occurring already in the germline and in addition to further changes to an individual's chromatin state as a result of hormones, stochastic factors, and various environmental influences such as diet or overall lifestyle (Schumacher and Petronis 2006; Petronis 2004). If this hypothesis is correct, one would expect to observe epigenetic modification also in tissues that are not primarily associated with the disease. Using lymphocytes from 19 healthy controls and 25 patients with SCZ, Gavin and coworkers (2009) examined mean baseline levels of dimethylated lysine 9 of histone 3 (H3K9me2), a histone modification that likely represents a repressive mark and is associated with a decreased probability and intensity of genomewide promoter activity. Gavin et al. found that SCZ patients had significantly higher basal levels of H3K9me2 compared to healthy controls. Moreover, there was a significant negative correlation between age at onset of illness and levels of H3K9me2. A restrictive chromatin state has been thought to be operant in the pathophysiology of SCZ. Accordingly, patients with SCZ have been found to have significantly lower levels of acetylated histone 3, an indicator of open chromatin, compared with healthy controls (Gavin et al. 2008). Alternatively, these results may indicate that individuals with major psychosis are afflicted with an overall dysfunction and lack of organization in the epigenetic machinery, rather than simply an overly restrictive chromatin state (Gavin and Sharma 2009). Furthermore, aberrant histone modification in the non-disease-related tissues may be an indicator of the severity of the individual's preepimutation.

12.3 NUTRITIONAL DEFICIENCY AND ONE-CARBON METABOLISM

Converging evidence suggests that disruption of the one-carbon metabolism may play a central role in major psychosis. The one-carbon metabolism comprises chemical reactions and pathways involving the transfer of one-carbon units in various oxidation states and influences a variety of epigenetic processes such as DNA or histone methylation. An in-depth description of its components is beyond the scope of this chapter; however, the reader is directed to several excellent reviews on the role of one-carbon metabolism in psychiatric disorders (Rutten and Mill 2009; Mill and Petronis 2009; McGowan et al. 2008; Frankenburg 2007; Krebs et al. 2009).

12.3.1 Aberrant One-Carbon Metabolism in Major Psychosis

One of the main components of the one-carbon metabolism is homocysteine (Hcy), a sulfur-containing amino acid that has been widely investigated for its putative role in neuropsychiatric and neurodegenerative disorders (Figure 12.3). Normal serum level is maintained by enzymes that require B vitamins such as folate (B_9), cobalamin (B_{12}), and pyridoxine (B_6). High levels of Hcy were suggested to contribute to the pathogenesis of major psychosis, as mean plasma Hcy levels are usually significantly higher in patients than in the control individuals (Applebaum et al. 2004; Adler Nevo et al. 2006; Levine, Sela et al. 2005; Dittmann et al. 2007).

It seems that compared to individuals with average plasma levels, BD patients with hyperhomocysteinemia display greater functional deterioration and cognitive impairment, particularly related to verbal learning, delayed memory tasks, and executive function (Dittmann et al. 2007, 2008). These observations suggest that elevated Hcy levels may play a role in the pathophysiology of cognitive deficits, specifically with executive dysfunction. Executive function processes are highly dependent on the prefrontal cortex, a brain region that is centrally involved in complex cognitive behaviors, personality expression, decision making, and social behavior. Both SCZ and BD have been repeatedly linked to dysfunction of the prefrontal cortex. Interestingly, plasma levels of Hcy tend to increase with age, which is correlated with the finding that homocysteinemia may have a higher impact in the pathophysiology of neurocognitive deficits among older patients or those who have a delayed onset of illness (Dias et al. 2009). Variation in the age at onset and rate of progression is a hallmark of complex disorders with non-Mendelian anomalies (Wang et al. 2008).

Some of the observed phenotypes may derive from neurotoxic properties of Hcy, e.g., through stimulation of the N-methyl-D-aspartate receptors (Lipton et al. 1997). Nevertheless, it is important to note that increased Hcy levels in patients with major psychosis are not present in all tissues. For example, no difference was found in cerebrospinal fluid (CSF) Hcy levels between SCZ patients and controls (Levine, Sela et al. 2005; Levine, Agam et al. 2005). The nature of the Hcy imbalance in major-psychosis patients is still poorly understood, and clinical parameters such as folate or vitamin B_{12} concentration, age, duration of illness, age at onset of disease, and family history of mental illness, as well as smoking or coffee consumption habits do not appear to be related to serum Hcy levels. Nevertheless, in SCZ patients Hcy levels seem to correlate with the disease state; i.e., levels vary between the exacerbation phase (immediately upon admission to the hospital) and later stages (remission phase) (Petronijevic et al. 2008). Folate and B_{12} levels do not seem to vary between the exacerbation and remission phases of the illness. The significant decrease of plasma Hcy levels, without changes in folate and vitamin B_{12} concentrations in the remission phase of SCZ, could indicate that dietary deficiencies are less likely to

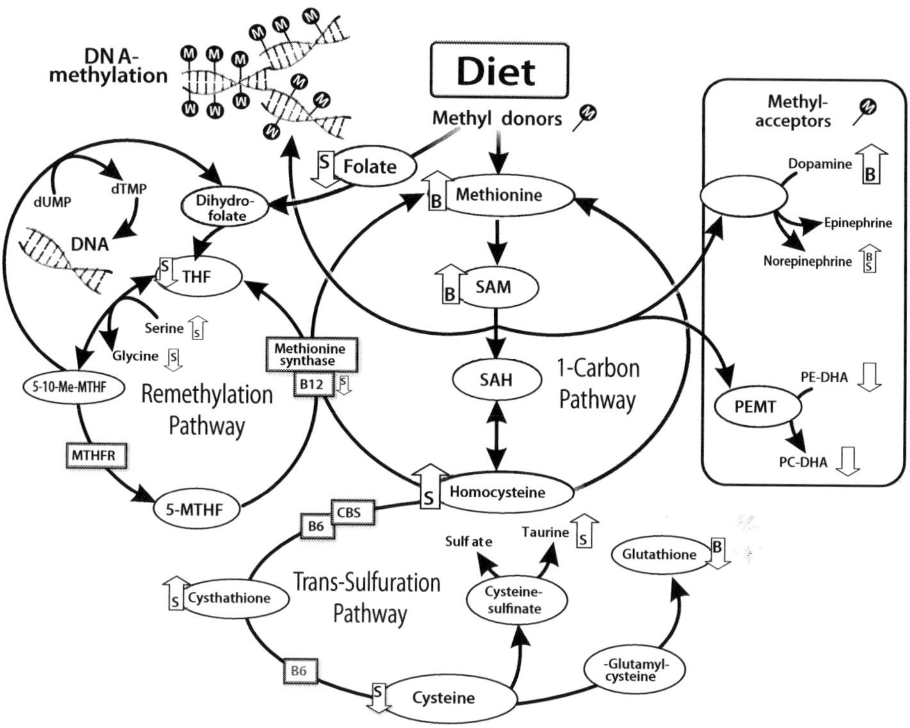

FIGURE 12.3 Aberrant one-carbon metabolism in major psychosis. The transmethylation (one-carbon) pathway is present in most mammalian tissues and centrally involves S-adenosylmethionine (SAM) as the universal methyl group donor for a variety of methyltransferases, resulting in the methylation of substrates such as nucleic acids, histones, and lipids, among others. SAM is generated via the activation of methionine by methionine adenosyltransferase. During DNA methylation and other methyl group transfers, the DNA methyltransferase (DNMT) reaction converts SAM to S-adenosylhomocysteine, which can subsequently be converted to homocysteine by SAH hydrolase. Hcy levels as well as other components of the one-carbon metabolisms are typically abnormal in MD patients (see arrows). Folate and vitamin B_{12} are essential cofactors for the methionine/homocysteine cycle in the brain and mediate the remethylation of Hcy in the remethylation pathway. Saturating one of the two key carbon-carbon double bonds in folate yields dihydrofolate, and adding another hydrogen to the second carbon yields tetrahydrofolate (THF). Alternatively, in the liver and kidney, Hcy can be metabolized by the irreversible trans-sulfuration pathway, which degrades homocysteine via cystathionine to cysteine, beginning with the irreversible conversion to cystathionine by cystathionine β-synthase (CBS). Cysteine can be further catabolized into other important biological compounds such as the antioxidant glutathione. Cellular concentrations of cysteine influence Hcy metabolism by a feedback mechanism: high concentrations of cysteine decrease homocysteine catabolism to cysteine and may stimulate homocysteine remethylation. 5,10-MTHF 5,10-methylene tetrahydrofolate; 5-MTHF, 5-methylene tetrahydrofolate; methionine; PEDHA, phosphatidylethanolamine with docosahexaenoic acid attached to position 2; PC-DHA, phosphatidylcholine with docosahexaenoic acid attached to position 2; NE, nor-epinephrine; EP, epinephrine; DNA, deoxyribonucleic acid; RNA, ribonucleic acid, histone. Key enzymes are shown in rounded rectangles: MTHFR, methylene, tetrahydrofolate reductase; PEMT, phosphatidyl ethanolamine methyl transferase; COMT, catecholamine-O-methyl transferase; MT, methyl transferase; CBS, cystathionine beta synthase. B = Brain, S = Serum.

play a role in the disease. The influence of pathogenetic processes involved in SCZ that affects Hcy metabolism and various epigenetic processes is a more plausible causative factor for Hcy abnormalities in major psychosis.

The major methyl donor in the one-carbon metabolism is S-adenosylmethionine (SAM). The methyl group (CH3) attached to the methionine sulfur atom in SAM is chemically highly reactive. SAM is the sole methyl donor in many important methylation reactions in the nervous system, involving neurotransmitters, amines and polyamines, membrane phospholipids, proteins (e.g., histones), and DNA (Figure 12.1). Hence the abundance of key components of the one-carbon metabolic pathway directly influences the epigenetic machinery of the cell. The process of methylation generates S-adenosyl-homocysteine (SAH), which is a potent and competitive inhibitor of methylation reactions. The maintenance of healthy SAM levels in various cell types is a complex process and seems to be disturbed for specific tissues in major psychosis. For example, prefrontal cortex levels of SAM were reported to be increased by about twofold in SCZ and BD patients (Guidotti et al. 2007). Because the DNA methylation is dependent on the supply of SAM and removal of SAH, the [SAM]:[SAH] ratio has been proposed as a "methylation index" to indicate the likelihood of hyper- or hypomethylation of DNA (Waterland 2006). In summary, it is becoming evident that the availability of methyl donors and cofactors is a major external influence on DNA methylation and the development of neuropsychiatric disorders. Yet an estimation of the impact deriving from an aberrant one-carbon metabolism is difficult and not straightforward. It was postulated that altered one-carbon metabolism with an increase in brain SAM may be associated with DNMT1 overexpression and may become a crucial factor mediating DNA hypermethylation and transcriptional down-regulation of putative candidate genes in cortical GABAergic neurons of SCZ or BP patients (Guidotti et al. 2007). In contrast, a negative correlation between plasma Hcy and global DNA methylation has been reported (Castro et al. 2003; Yi et al. 2000), although other reports failed to demonstrate such a correlation (Bonsch et al. 2004; Fux et al. 2005; Bromberg et al. 2009).

12.4 DIET AND MAJOR PSYCHOSIS

Nutrition during diverse stages of life can influence epigenetic gene regulation (Waterland 2006), and the evidence that nutrition plays a role at the interface between the environment and the genome in complex disorders is slowly beginning to be recognized. It is entirely possible that sustained exposure to specific dietary components, especially during a developmental period, benefits mental health and may lower the risk for psychopathology.

An obvious target for dietary intervention is nutritional deficiency, one of the main risk factors for major psychosis during the prenatal period (Brown and Susser 2008; Xu et al. 2009). Indeed, epidemiological data suggest that dietary changes in methyl contents affect DNA methylation and gene expression programming (McGowan et al. 2008). Potentially, nutritional sufficiency in methyl donors can therefore help to minimize some of the unpleasant symptoms or completely resolve them. Generally, a balanced diet and regular exercise are probably the best measures to protect the brain and ward off mental disorders. In general, the proper maintenance of the epigenomic landscape in an otherwise disease-free brain appears to depend on the adequate supply of essential nutrients involved in the metabolism of methyl groups. Long-term feeding of a folate- or methyl-deficient diet to Fisher 344 rats resulted in genomewide DNA hypermethylation of brain cells; crucially, this increase extended to normally unmethylated CpG-rich DNA (Pogribny et al. 2008). In contrast, the level of SAM, SAH, and SAM:SAH ratio in the brain of folate/methyl-deficient rats remained mostly unaffected, whereas the level of Hcy steadily increased, with differences being significant after 36 weeks of deficiency (Pogribny et al. 2006). Dietary supplementation may be able to target specific epigenetically metastable loci in the genome as demonstrated for the *Avy* locus (expressing the pigmentation protein Agouti) in mice (Wolff et al. 1998). Viable yellow (*Avy/a*) mice are epigenetic mosaics, ranging from a yellow fur phenotype due to maximum ectopic agouti overexpression to a pseudoagouti phenotype due to minimal ectopic expression. Pseudoagouti Avy/a mice are leaner, healthier, and longer-lived than their yellow littermates. It was shown that feeding pregnant black *a/a* dams methyl-supplemented diets altered the epigenetic regulation of Agouti expression in their offspring, as indicated by increased agouti/black mottling in the direction of the pseudoagouti phenotype (Wolff et al. 1998). Moreover, this epigenetic phenotype was maternally heritable, suggesting that maternal dietary supplementation may positively affect the health and longevity of multiple generations. A similar developmental effect has been observed when genistein (the major phytoestrogen in soy) is added to the diet of pregnant dams at levels comparable with humans consuming high-soy diets (Dolinoy et al. 2006). The genistein supplementation shifted the coat color of heterozygous viable yellow agouti (*Avy/a*) offspring toward the pseudoagouti phenotype associated with increased methylation upstream of the *Avy* locus. Interestingly, this genistein-induced hypermethylation persisted into adulthood, decreasing ectopic Agouti expression and protecting offspring from obesity.

12.4.1 Methyl Donors and Vitamins

A plausible target for nutritional intervention in major psychosis may involve components of one-carbon metabolism, because several of the involved methylation pathways are highly dependent on diet. Unfortunately, our current knowledge in the matter is inadequate, making it difficult to give specific dietary advice. Consuming excessive quantities of one-carbon metabolism components may affect DNA methylation, thereby causing misregulation of gene expression. For example, it has been shown that nutritional supplementation during early development can increase DNA methylation at specific loci, resulting in permanent changes in gene expression (Waterland and Jirtle 2003). However, it cannot be excluded there is a possibility that for certain individuals, an increased intake of dietary methyl-donors such as methionine or folate may be an effective component of a therapeutic dietary regimen to restore appropriate levels of locus-specific DNA methylation. In general, deficiencies in important components of methylation homeostasis, such as folate or Hcy, should be corrected. Hcy levels can be lowered by oral folic acid and B_6 (Levine et al. 2006). B_{12} may also lower Hcy levels; however, supplementation is probably not helpful in some cases. It was shown that Hcy levels decrease significantly with vitamin therapy relative to placebo treatment, a decrease that in turn correlates with a decline in clinical symptoms of SCZ. Overall, neuropsychological test results were significantly better after vitamin treatment, indicating that a subgroup of patients with hyperhomocysteinemia might benefit from the simple addition of B vitamins. Hence it is probably important to check a patient's folate, B_{12}, homocysteine, and methylmalonic levels in certain situations such as alcoholism, malnutrition, malabsorption of vitamins, and the concurrent use of some medications, because B vitamin supplementation may help alleviate symptoms in such cases.

Folate supplementation is routinely recommended to pregnant women, but it may exacerbate the effects of vitamin B_{12} deficiency. Natural folate (also known as pteroylmonoglutamic acid) is found in animal products, particularly liver, yeast, and many green vegetables such as spinach (reviewed in Frankenburg 2007). As with many vitamins, folate is highly temperature-sensitive and thus can be destroyed by heat during cooking. Folate is not stored in large amounts in the body, so a regular intake of this vitamin is essential. A common cause of folate deficiency is alcoholism, as alcohol possibly exerts toxic effects on hepatic parenchymal cells, thereby inhibiting folate secretion into bile, which then inhibits reabsorbtion of the folate into the gut and delivery to other tissues. Additionally, some medications that inhibit dihydrofolate reductase such as antineoplastic or antimicrobial agents, as well as medications that interfere with absorption and storage of folate (e.g., certain anticonvulsants and oral contraceptives) interfere with folate metabolism and can lead to folate deficiency (Frankenburg 2007). Following absorption in the small intestine, ingested folates undergo hydrolysis to methyltetrahydrofolate, the predominant form of folate in plasma; however, this process becomes saturated at doses of ~270 μg, so that additional folate is transported into the plasma unchanged (Sweeney et al. 2007). Hence a daily intake of ~400 μg, a dose commonly used in supplements, produces a sustained level of plasma folic acid. Although folic acid seems to exhibits a higher bioavailability than natural folate, dietary foods naturally rich in folate can also be effective in increasing serum folate levels. A reduced dose of folate in supplements may help to restore parts of the one-carbon metabolism, in turn helping alleviating symptoms of major psychosis.

Like folate, methionine is considered essential because it cannot be manufactured in the body and must be obtained through diet. This particular amino acid is found primarily in meat, fish, eggs, and dairy products. Methionine is quickly metabolized and directly influences the [SAM]:[SAH] ratio in brain cells toward SAH (Finkelstein 1998), which presumably causes global DNA hypomethylation. The connection between methionine and DNA methylation status appears not to be straightforward; subcutaneous injections of methionine in mice paradoxically hypermethylate specific CpG sites in the reelin promoter of cortical cells and were also associated with silencing of the reelin promoter (Dong et al. 2005). However, high doses of dietary methionine may be a risk factor for psychiatric disorders; there was a report showing that treatment with methionine actually exacerbates psychotic symptoms in patients with SCZ (Cohen et al. 1974). Similarly, prolonged methionine treatment in mice induces behavior patterns that mimic several phenotypic aspects of SCZ (Tremolizzo et al. 2002). Since the effect of methionine supplementation seems to be age and tissue specific, some have suggested that high dietary intake of methionine may induce DNA hypermethylation in some circumstances and hypomethylation in others (Waterland 2006). Hence despite the potential risks in dietary methionine uptake, intake of additional dietary methionine may be an effective element of therapeutic dietary regimens for some individuals to restore appropriate locus-specific DNA methylation. However, more data on interindividual methylation pattern in patients are needed to understand the relationship between abnormal epigenetic patterns and course of disease.

Many studies have found SAM to be an effective antidepressant, although questions remain about the mechanism of action, bioavailability, and absorption of oral SAM (Williams et al. 2005). The fact that SAM is effective in the treatment of depression is apparently contradictory to the effects of methionine. On the other hand, SAM is a methyl donor not only for DNA methylation but also for other enzymatic reactions (McGowan et al. 2008). For example, SAM treatment increases phosphocreatine levels in the brain, which may also contribute to the antidepressive effect of SAM, as decreased phosphocreatine levels have been reported in BD (Kato et al. 1994). However, despite its antidepressant effect, those suffering from BD should not take SAM, as it was shown that it may induce or heighten the manic phase of this condition (Carney et al. 1989).

One of the most promising approaches in treating symptoms in SCZ and BD is to lower serum Hcy levels. Some studies have demonstrated that vitamin B_6 supplementation in low doses may lead to the lowering of plasma Hcy levels in individuals with hyperhomocysteinemia depending on their folate status (McKinley et al. 2001). Another study analyzed the efficacy of vitamin B_6 and folic acid administered either alone or in combination and demonstrated a reduction in Hcy serum level induced by B_6 administered alone during a 5-week period in a moderate dose (120 mg/d) (Mansoor et al. 1999). It was further demonstrated that after vitamin B_6 treatment, Hcy serum levels significantly decreased in patients diagnosed with SCZ or with schizoaffective disorders; this decrease was however only statistically significant in men and not in women (Miodownik et al. 2007). Additional research efforts in a larger sample are necessary to substantiate such gender-specific effects, and to determine if the decrease in Hcy influences the course of the disease.

In summary, despite the recent progress in the field, the medical sciences still need a better understanding of how dietary supplements impact the metabolism of methyl groups, and in turn the epigenetic machinery for optimizing health and minimizing adverse consequences.

12.4.2 Natural Dietary Factors with Impact on the Epigenome

It is generally acknowledged in medical circles that a balanced diet has the potential to alter our overall health and mental function. However, in contrast to methylation supplements, it is difficult to estimate the impact of specific natural dietary sources on cellular epigenomic patterns, since dietary nutrients are normally ingested in complex combinations, and they interact with each other in their normal metabolic and physiological functions; consequently, identifying effective components is difficult. Nevertheless, in recent years a number of natural substances were reported to affect epigenetic patterns and health; these include taurin, all-trans-retinoic acid, the flavonoid baicalein, creatine, theophyllin, curcumin, pomegranate extracts, quercetin, selenium, alpha- and gamma-tocopherols, the antioxidant carotenoid lycopene, and the isoflavone genistein as well as various isothiocyanates from plant foods. For example, the polyphenolic plant compound curcumin (diferuloymethane), which is found in the rhizome of the Indian curry spice turmeric (*Curcuma longa L.*), may alter several cellular pathways involved in DNA and histone modification (Moiseeva et al. 2007). Similarly, genistein, one of the phytoestrogens contained in soy, was positively associated with changes in DNA methylation at CpG islands of mouse genes and is able to dose-dependently inhibit the activity of DNA methyltransferases thereby reactivating methylation-silenced genes (Day et a!. 2002). Indeed, maternal dietary supplementation of genistein was shown to reduce the DNA methylation status of the agouti locus, thereby changing the coat color of heterozygous yellow agouti (Avy/a) pups toward the black phenotype (Dolinoy et al. 2006). Also, the main polyphenol from green tea, Epigallocatechin-3-gallate (EGCG), was found to be an inhibitor of nuclear DNA methytransferase activity and reactivate methylation-silenced genes in cancer cells (Fang et al. 2003). These findings raise the possibility that changes in diet are a viable strategy for enhancing cognitive abilities, protecting the brain from damage and counteracting the effects of psychiatric phenotypes.

Various clinical studies have shown that omega-3 fatty acids could potentially be useful in the prevention and treatment of SCZ. Omega-3 fatty acids, also known as polyunsaturated fatty acids (PUFAs), are considered essential fatty acids and play a crucial role in synaptic plasticity, affecting the expression of several molecules related to, and enhancing, learning and memory. There are three major types of omega-3 fatty acids: alpha-linolenic acid (ALA), eicosapentaenoic acid (EPA), and docosahexaenoic acid (DHA). They are essential to human health but cannot be manufactured by the human body. Consequently, omega-3 fatty acids must be obtained from food, particularly fish (i.e., salmon or halibut) or other marine life such as algae and krill, nut oils, and various plants, e.g., the leaf vegetable purslane (*Portulaca oleracea*) and kiwis. Dietary omega-3 fatty acids have effects on diverse physiological processes that affect gene expression and hence are likely to influence several epigenetic pathways (Benatti et al. 2004), although the effect on DNA methylation or histone modification patterns is still unknown. Nevertheless, omega-3 fatty acids, particularly EPA, are increasingly being used by psychiatric patients, and many studies have shown reduced levels of omega-3 fatty acids in the cell membranes of red blood cells from SCZ patients (reviewed in Peet 2008). Several lines of evidence suggest that people with SCZ experience an improvement in symptoms when given omega-3 fatty acids. However, results of omega-3 fatty acid supplementation in chronic and acutely relapsing SCZ patients have been mixed (Peet 2008). One advantage of omega-3 fatty acid DHA may be its effect on the signaling molecule brain-derived neurotrophic factor (BDNF). For example, DHA dietary supplementation has been found to elevate levels of hippocampal BDNF and enhance cognitive function in rat models of brain trauma (Wu et al. 2004). In patients with major depression and SCZ, levels of BDNF are often reduced in serum, hippocampus, and various cortical areas (Durany et al. 2001). In humans, a mutation in a BDNF receptor has been linked to impairments in learning and memory (reviewed in Monteggia 2007). BDNF is one of the most promising targets for pharmacological intervention, and it could be shown that antidepressant can elevate BDNF levels. The conflicting results suggest that more research is needed before conclusions can be drawn about the benefit of omega-3 fatty acids on reducing the cognitive deficits that are associated with psychiatric disorders.

Many other substances, especially medications such as histone deacetylase inhibitors, ibuprofen, or valproic acid, are known for their effect on epigenetic processes. Given the dynamic nature of the epigenome, it is not surprising that methylation patterns are also highly susceptible to medication. For example, methylation of a CpG island located upstream of the MEK1 gene was found to be strongly correlated with antipsychotic use in the frontal cortex of SCZ patients (Mill et al. 2008). The link between MEK1 and antipsychotic exposure in SCZ is remarkable given the involvement of mitogen-activated protein-kinase (MAPK) signaling pathways in mediation of intraneuronal signaling and the observation that clozapine, a widely used medication in the treatment of SCZ, selectively activates this pathway via an interaction with MEK1.

Overall, a wealth of epidemiological data and genetic studies highlight a crucial role for environmental factors, such as Western-style diet, in the etiology of major psychotic disorders (reviewed in Rutten and Mill 2009). Intriguingly, due to epigenome-genome interaction, individuals from different genetic backgrounds may show significantly different responses to such environmental influences. For example, it could be shown that frontal-cortex DNA methylation in BDNF genes is associated with the genotype at a nearby nonsynonymous SNP (rs6265–val66met) (Mill et al. 2008), which has been previously associated with major psychosis. The observation of a correlation between genotype and DNA methylation adds to the increasing evidence that DNA sequences can influence epigenetic profiles (Flanagan et al. 2006). The notion that epigenetic changes might be associated with DNA sequence variation is relevant in light of many inconsistent genetic-association studies in complex diseases and suggests that a comprehensive epigenetic analysis of candidate SNPs and haplotypes in neurobehavioral disorders may be necessary.

12.4.3 Transgenerational Effects

In addition to environmental influences, accumulating evidence from epidemiologic studies suggests that epigenetic marks can be inherited across multiple generations and may affect the likelihood of complex diseases in the offspring. Several mechanisms, such as genetic anticipation and nutritional imbalances during fetal development, or a combination of these factors, are likely to influence epigenetic marks passed through the germline.

One of the best documented transgenerational phenomena is the correlation between advanced paternal age with an increased risk of SCZ in the offspring (Lopez-Castroman et al. 2009; Malaspina et al. 2001). It was shown that the incidence of SCZ increased progressively with increasing paternal age, the risk being up to threefold for offspring of fathers aged 45 or more years, compared with those of fathers aged less than 25 years (reviewed in Perrin et al. 2007). Classical theories of advanced paternal age are primarily focused on a potential higher frequency of point mutations in males, the frequency of which increases with age. However, studies in a number of other disorders that show clear effects of paternal age suggest that mutations in sperm DNA cannot fully explain the association (Tiemann-Boege et al. 2002). Instead, advanced paternal age may be associated with epigenetic dysregulation, leading to abnormalities in genomic imprinting and other epigenetic processes. Genomic imprinting is a molecular mechanism by which a subset of genes is expressed in certain tissues in a parent-of-origin-specific manner.

According to the model of evolutionary epigenetics of aging (Schumacher 2010), human aging results from progressive accumulation of epigenetic damage as a direct consequence of evolved limitations in the genetic and epigenetic settings of maintenance and repair functions. The theory suggests that, along with other causes, age-dependent epigenetic drift is a natural phenomenon that is present in all healthy individuals, but may become hazardous with age, thereby potentially playing a central role in many complex disorders. Indeed, epigenetic drift was reported to be present in tissues from healthy individuals, but was also notably increased in tissues derived from patients with late-onset Alzheimer's disease (Fraga et al. 2005; Wang et al. 2008). Supporting the model of evolutionary epigenetics of aging, a microarray analysis by Flanagan et al. (2006) identified numerous intra- and interindividual DNA methylation-variable positions in the human germ cell genome. The largest degree of variation was detected within the promoter CpG islands and pericentromeric satellites among the single-copy DNA fragments and repetitive elements, respectively. A number of genes, such as EED, CTNNA2, CALM1, CDH13, and STMN2, exhibited age-related DNA methylation changes. Age-dependent germline epigenotypes may arise from a variety of risk factors, such as multiple cell divisions, malnutrition, or stochastic fluctuations, among others, that directly act on the epigenomic machinery, thereby increasing epigenetic variability with age, the consequence of which may be a greater susceptibility to neuropsychiatric disorders in the offspring (Perrin et al. 2007).

12.4.4 Malnutrition and Major Psychosis

In a landmark study, David Barker and colleagues (1989) demonstrated an inverse relationship between birth weight and the incidence of cardiovascular disease. The Barker hypothesis states that many adult chronic conditions might have a developmental origin, resulting from adaptations made by the fetus as a result of limited (or excessive) supply of specific nutrients (Barker 1997). A large body of subsequent discoveries has supported this hypothesis and has shown that diseases of the cardiovascular system, hypothalamic-pituitary-adrenal (HPA) axis, and diabetes, among other complex disorders, can also be affected by nutritional imbalances during periconceptional, embryonic, fetal, and infantile phases of life. Study of the Barker hypothesis and recent findings demonstrating epigenetic drift have prompted a new line of thinking that accounts for the shortcomings and paradigms of earlier hypotheses, resulting in theories such as the "thrifty epigenotype" (Stoger 2008), the "Latent Early-life Associated Regulation" (LEARn) model (Lahiri et al. 2009), the hypothesis of age-dependent epigenetic drift (Wang et al. 2008), and the epigenetic paradigm of major psychosis (Schumacher and Petronis 2006; Wang et al. 2008; Petronis 2004). Thus the epigenome is able to unify a wide variety of biological and psychological theories as well as empirical findings that pertain to major psychosis.

The main evidence demonstrating that prenatal nutritional deficiency may increase risk of major psychosis comes from studies that analyzed some of the most severe human-made famines: the 1944–1945 Dutch Hunger Winter and the 1959–1961 Chinese famine. The Dutch famine was the result of a blockade imposed by the Nazi occupation, in which the Nazi regime banned all transport of food to the western part of the Netherlands, resulting in a sharp but time-limited decline in food intake (Heijmans et al. 2008). A total of 18,000 people died during the famine, which ended abruptly when the Allied forces liberated the nation in May 1945. The Dutch famine had several phenotypic consequences in the offspring. First of all, as expected, the children of women who were pregnant during the famine were smaller than average. However, surprisingly, when these children grew up and had children themselves, those children were also smaller than average, suggesting the involvement of transgenerational epigenetic processes. Furthermore, a twofold increase in the cumulative risk of SCZ was found among children conceived during the famine (Susser et al. 1996), as well as a significant increased risk of developing schizoid personality disorder (Hoek et al. 1996). Importantly, caloric rations of <1000 kcal were associated with an increased risk of SCZ, while exposure to a lesser severity of famine (1000–1500 kcal) was not related to an increased SCZ risk, indicating that the degree of nutritional deficiency may also be relevant (Brown and Susser 2008; Brown et al. 1995).

Similarly, a correlation between prenatal famine and risk of SCZ was observed in Chinese cohorts from the Wuhu region of the Anhui Province (St. Clair et al. 2005) and the Liuzhou prefecture of Guangxi autonomous region (Xu et al. 2009). The 1959–1961 Chinese famine affected all provinces of China and was one of the twentieth century's tragic horrors. It was partially caused by political and social upheaval in a period termed the Great Leap Forward. In addition, bad weather, the collectivization of agriculture, the adoption of unsound agricultural practices, and a reduction of cultivated land led to the interruption food supplies to the population (St. Clair et al. 2005). By spring 1959, the provinces were starving and people were dying in huge numbers (over 30 million deaths in China, with a mortality rate of over 3.0%). Among the births that occurred during the famine years in the Anhui Province, the risk of developing SCZ in later life increased significantly, from 0.84% in 1959 to over 2% in 1960 (St. Clair et al. 2005). Similarly, the mortality-adjusted relative risk for SCZ in the Liuzhou prefecture was 1.5% in 1960 and 2.05% in 1961 (Xu et al. 2009). Interestingly, epidemiologic studies conducted nationwide revealed a significant difference in the way the Great Leap Forward Famine influenced the risk of SCZ in urban and rural populations. In rural populations, the post-famine cohort had a significantly higher SCZ risk than either the famine or the pre-famine cohort, while SCZ risk in the famine cohort was only marginally higher than the pre-famine cohort (Song et al. 2009). In contrast, the urban famine cohort had higher SCZ risk than both the pre-famine and the post-famine cohort. A possible explanation for the urban/rural gap may be the significantly higher mortality level caused by the famine among the rural populations (Song et al. 2009).

Epigenetic mechanisms have been proposed to underlie the associations between famine and SCZ. In support of this hypothesis, individuals who received prenatal exposure to famine in the Dutch Hunger Winter had six decades later less DNA methylation in the imprinted IGF2 gene compared to their unexposed, same-sex siblings (Heijmans et al. 2008). The association was specific to periconceptional exposure, reinforcing that very early mammalian development is a crucial period for establishing and maintaining epigenetic marks. Furthermore, it was reported that methylation in the INSIGF gene was lower in individuals who were periconceptionally exposed to famine in comparison to their unexposed same-sex siblings, whereas methylation in the IL10, LEP, ABCA1, GNASAS, and MEG3 genes was higher (Tobi et al. 2009). The methylation changes in the INSIGF, LEP, and GNASAS genes were also sex dependent. Furthermore, exposure to famine late in gestation produced different methylation patterns in the GNASAS gene and, in men, the LEP gene compared to unexposed siblings. These data indicate that persistent changes in DNA methylation may be a common consequence of prenatal famine exposure and that these changes depend on the sex of the exposed individual and the gestational timing of the exposure (Tobi et al. 2009).

12.4.5 Anticipation in Major Psychosis—an Epigenetic Component?

Anticipation refers to the increase in disease severity or decrease in age of onset as it is transmitted down through successive family generations. There have been several reports of the phenomenon of anticipation existing in cases of major psychosis that date back even to the 19th century. However, these incidents of anticipation were largely ignored because no molecular mechanism to explain these types of inheritance patterns was envisioned at the time. The putative role of epigenetic factors in genetic anticipation comes from several experimental findings describing intergenerational changes in the degree of DNA methylation (Petronis et al. 1999). For example, experiments with transgenic mouse strains revealed gradual intergenerational changes in DNA methylation in the mouse offspring (Allen et al. 1990; Schumacher et al. 2000). The level of DNA methylation increased or decreased in the offspring, depending on the genetic background of the nontransgenic parent. This suggests that differences in the genome of the unaffected parent may determine whether an epigenetic condition progresses or regresses in the subsequent generation, hence contributing to phenotypic variance (Figure 12.4). So far, the only other known mechanism for anticipation is a class of mutations containing unstable repetitive sequences, as exemplified by pathogenic trinucleotide repeats that have been observed in Huntington's disease, myotonic dystrophy, pancreatic cancer, and other diseases. Many of the non-Mendelian genetic features of major psychosis have the potential to be explained by the behavior of unstable sequences; however, despite promising findings in the mid-1990s, no trinucleotide repeat expansions have yet to be identified as a causative factor of idiopathic SCZ or BD (Fortune et al. 2003). In contrast, epimutations may play a significant role in determining the age of disease onset, challenging the traditional trinucleotide repeat expansion-based mechanism of genetic anticipation (Petronis et al. 1999).

In this context it is noteworthy that epidemiologic analysis has revealed several social factors as having a significant influence on the age of disease onset in

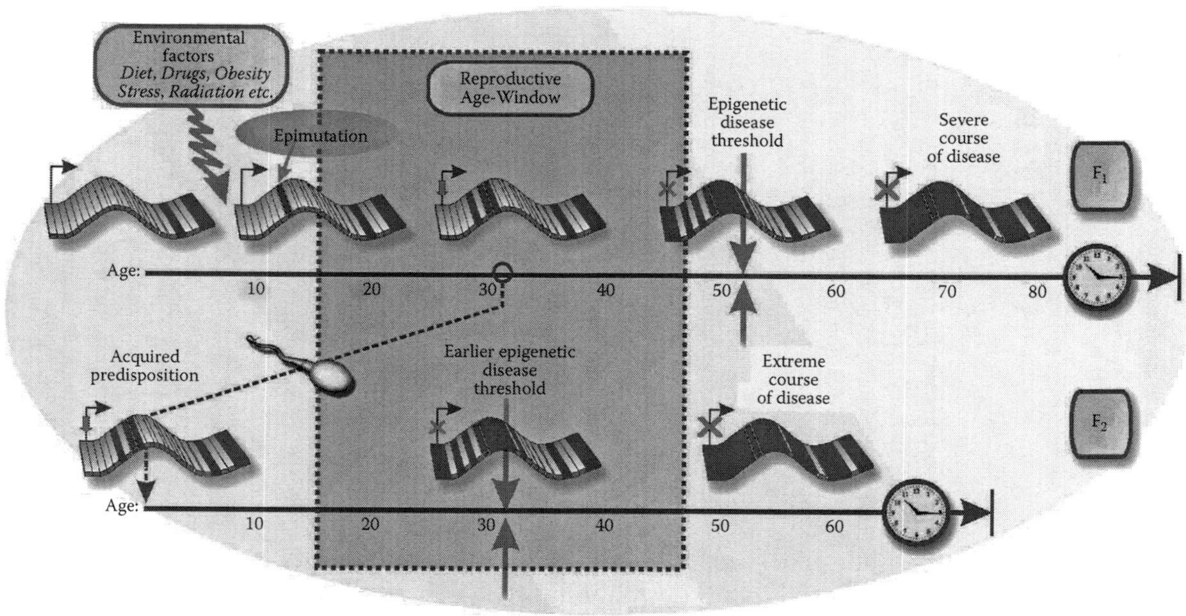

FIGURE 12.4 (Please see color insert.) Anticipation and heritable epigenetic variation. Shown is an example of a gene promoter, which is primarily unmethylated (white boxes) and active in early life, but may become aberrantly methylated (black boxes) during aging and therefore increasingly inactivated. The relatively high frequency of de novo epimutations suggests that epigenetic alterations accumulate during aging. In order to have a deleterious effect, the number of de novo methylation events must cross a certain threshold. Depending on the overall effect of the series of pre- and postnatal impacts on the preepimutation, only some predisposed individuals will reach the "threshold" of epigenetic deregulation that causes the phenotypic changes that meet the diagnostic criteria for a clinical disorder. Some individuals may already be epigenetically predisposed at birth, due to transgenerational epigenetic effects. For example, if epimutations occur in the germline and are not cleared following fertilization, meiotically heritable epigenetic modifications may spread from the inherited epimutation. As a result, the symptoms of the disorder become apparent at an earlier age as it is passed on to the next generation, potentially accompanied with an increase of severity of symptoms or even premature death. Deleterious epigenetic drift occurring after the reproductive phase is relatively neutral to selection, because their bearers have already transmitted their genes (and potentially epigenetic information) to the next generation.

neuropsychiatric disorders, including year of birth and residence (urban dwellers show psychotic symptoms sooner than rural ones) (Stompe et al. 2000). Environmental risk factors such as urban environment (Krabbendam and Van Os 2005; Crow 2007) as well as migration (Cantor-Graae and Selten 2005) are known to increase the rate of major psychosis significantly. Intriguingly, this risk of developing psychosis as well as its severity is further increased in the second generation (Dealberto 2007). This phenomenon could suggest that, among other causes, a change in nutritional exposure may also influence anticipation (e.g., the introduction of immigrants to a North American diet). This hypothesis could answer some puzzling issues in the epidemiology of major psychosis, including the association between short birth interval and SCZ in the offspring (Smits et al. 2004), which has been related to folic acid deficiency but could also be related to vitamin D insufficiency (Dealberto 2007). A nutrition component in anticipation could also clarify the role of the protective factor of ethnic density for ethnic minorities (Boydell et al. 2001), which could be explained by easier availability of ethnic food in these areas.

Overall, the impact of nutrition on the epigenome and its relation to the development of major psychosis is still poorly understood. To assess the influence of dietary factors, particularly methyl donors, cohort studies were suggested to monitor changes in SCZ and psychosis rates with systematic folic acid supplementation during pregnancy (Dealberto 2007). Folic acid supplementation is often recommended for pregnant women to prevent neural tube defects, making it relatively straightforward to monitor changes in the prospective rate of major psychosis. Other translational research efforts across many disciplines are also warranted because there are currently more questions than answers in the field of nutritional epigenetics.

12.5 CONCLUSIONS

It is now increasingly accepted that the human epigenetic machinery interacts with dietary factors in mediating susceptibility to various complex diseases such as neuropsychiatric disease or age-related disorders. Although investigations are still at a very early stage, studies on the interaction of specific dietary compounds and the human epigenome have the potential to lead us to safe and effective nutritional and pharmaceutical intervention strategies against major psychosis and various nonpsychiatric disorders. To understand these interactions, it will be beneficial not only to focus on research of neurobehavioral disorders, but also to incorporate studies in healthy cohorts, e.g., observing the epigenetic patterns in centenarians to see if health-promoting factors exist. One of the main obstacles in nutritional epigenetics is interindividual variance, which is partly avoided by large-scale epidemiologic studies that are able to test models that describe the establishment and spread of epigenetic variations in human populations. The new molecular methods that allow the epigenetic profile of millions of sequences to be simultaneously observed and studied will greatly broaden and refine the ability to assess disease phenotypes and relate them to present and past environmental factors, as well as to genotypes. Such studies may focus on critical periods of prenatal and postnatal mammalian development, to establish how nutrition and other environmental stimuli influence developmental pathways and thereby induce susceptibility to complex diseases. Another line of research should focus on long-term changes in epigenetic regulation and disease, which may ultimately enable specific early-life interventions to improve human health by finding effective treatments for complex diseases and for reducing their incidence. Although it would be premature to conclude that epigenetics will lead to ground-breaking discoveries, research on the epigenome-environment interface may one day contribute to the growing awareness of the importance of the environment, in particular, the Western-style diet, in the neurobiology of severe mental illness.

ACKNOWLEDGMENTS

I would like to thank Shradda Pai and Viviane Labrie for their invaluable comments and for proofreading the manuscript, and Sun-Chong Wang for preparing the modularity plots.

REFERENCES

Abdolmaleky, H. M., K. H. Cheng, S. V. Faraone, M. Wilcox, S. J. Glatt, F. Gao, C. L. Smith, R. Shafa, B. Aeali, J. Carnevale, H. Pan, P. Papageorgis, J. F. Ponte, V. Sivaraman, M. T. Tsuang, and S. Thiagalingam. 2006. Hypomethylation of MB-COMT promoter is a major risk factor for schizophrenia and bipolar disorder. *Hum Mol Genet* 15 (21):3132–45.

Abdolmaleky, H. M., K. H. Cheng, A. Russo, C. L. Smith, S. V. Faraone, M. Wilcox, R. Shafa, S. J. Glatt, G. Nguyen, J. F. Ponte, S. Thiagalingam, and M. T. Tsuang. 2005. Hypermethylation of the reelin (RELN) promoter in the brain of schizophrenic patients: a preliminary report. *Am J Med Genet B Neuropsychiatr Genet* 134B (1):60–66.

Adler Nevo, G., S. Meged, B. A. Sela, A. Hanoch-Levi, R. Hershko, and A. Weizman. 2006. Homocysteine levels in adolescent schizophrenia patients. *Eur Neuropsychopharmacol* 16 (8):588–91.

Allen, N. D., M. L. Norris, and M. A. Surani. 1990. Epigenetic control of transgene expression and imprinting by genotype-specific modifiers. *Cell* 61 (5):853–61.

Applebaum, J., H. Shimon, B. A. Sela, R. H. Belmaker, and J. Levine. 2004. Homocysteine levels in newly admitted schizophrenic patients. *J Psychiatr Res* 38 (4):413–16.

Barker, D. J. 1997. Maternal nutrition, fetal nutrition, and disease in later life. *Nutrition* 13 (9):807–13.

Barker, D. J., C. Osmond, J. Golding, D. Kuh, and M. E. Wadsworth. 1989. Growth in utero, blood pressure in childhood and adult life, and mortality from cardiovascular disease. *BMJ* 298 (6673):564–67.

Benatti, P., G. Peluso, R. Nicolai, and M. Calvani. 2004. Polyunsaturated fatty acids: biochemical, nutritional and epigenetic properties. *J Am Coll Nutr* 23 (4):281–302.

Bernstein, H. G., D. Krell, K. H. Braunewell, B. Baumann, E. D. Gundelfinger, S. Diekmann, P. Danos, and B. Bogerts. 2001. Increased number of nitric oxide synthase immunoreactive Purkinje cells and dentate nucleus neurons in schizophrenia. *J Neurocytol* 30 (8):661–70.

Bestor, T., A. Laudano, R. Mattaliano, and V. Ingram. 1988. Cloning and sequencing of a cDNA encoding DNA methyltransferase of mouse cells. The carboxyl-terminal domain of the mammalian enzymes is related to bacterial restriction methyltransferases. *J Mol Biol* 203 (4):971–83.

Bonsch, D., B. Lenz, U. Reulbach, J. Kornhuber, and S. Bleich. 2004. Homocysteine associated genomic DNA hypermethylation in patients with chronic alcoholism. *J Neural Transm* 111 (12):1611–16.

Boydell, J., J. van Os, K. McKenzie, J. Allardyce, R. Goel, R. G. McCreadie, and R. M. Murray. 2001. Incidence of schizophrenia in ethnic minorities in London: ecological study into interactions with environment. *BMJ* 323 (7325):1336–38.

Bromberg, A., Y. Bersudsky, J. Levine, and G. Agam. 2009. Global leukocyte DNA methylation is not altered in euthymic bipolar patients. *J Affect Disord* 118 (1–3):234–39.

Bromberg, A., J. Levine, B. Nemetz, R. H. Belmaker, and G. Agam. 2008. No association between global leukocyte DNA methylation and homocysteine levels in schizophrenia patients. *Schizophr Res* 101 (1–3):50–57.

Brown, A. S., and E. S. Susser. 2008. Prenatal nutritional deficiency and risk of adult schizophrenia. *Schizophr Bull* 34 (6):1054–63.

Brown, A. S., E. S. Susser, S. P. Lin, R. Neugebauer, and J. M. Gorman. 1995. Increased risk of affective disorders in males after second trimester prenatal exposure to the Dutch hunger winter of 1944–45. *Br J Psychiatry* 166 (5):601–6.

Cantor-Graae, E., and J. P. Selten. 2005. Schizophrenia and migration: a meta-analysis and review. *Am J Psychiatry* 162 (1):12–24.

Carney, M. W., T. K. Chary, T. Bottiglieri, and E. H. Reynolds. 1989. The switch mechanism and the bipolar/unipolar dichotomy. *Br J Psychiatry* 154:48–51.

Castro, R., I. Rivera, E. A. Struys, E. E. Jansen, P. Ravasco, M. E. Camilo, H. J. Blom, C. Jakobs, and I. Tavares de Almeida. 2003. Increased homocysteine and S-adenosylhomocysteine concentrations and DNA hypomethylation in vascular disease. *Clin Chem* 49 (8):1292–96.

Cluzel, P., M. Surette, and S. Leibler. 2000. An ultrasensitive bacterial motor revealed by monitoring signaling proteins in single cells. *Science* 287 (5458):1652–55.

Cohen, S. M., A. Nichols, R. Wyatt, and W. Pollin. 1974. The administration of methionine to chronic schizophrenic patients: a review of ten studies. *Biol Psychiatry* 8 (2):209–25.

Connor, C. M., and S. Akbarian. 2008. DNA methylation changes in schizophrenia and bipolar disorder. *Epigenetics* 3 (2):55–58.

Cooney, C. A., A. A. Dave, and G. L. Wolff. 2002. Maternal methyl supplements in mice affect epigenetic variation and DNA methylation of offspring. *J Nutr* 132 (8 Suppl):2393S–2400S.

Coyle, J. T. 2004. The GABA-glutamate connection in schizophrenia: which is the proximate cause? *Biochem Pharmacol* 68 (8):1507–14.

———. 2006. Glutamate and schizophrenia: beyond the dopamine hypothesis. *Cell Mol Neurobiol* 26 (4–6):365–84.

Craddock, N., M. C. O'Donovan, and M. J. Owen. 2006. Genes for schizophrenia and bipolar disorder? Implications for psychiatric nosology. *Schizophr Bull* 32 (1):9–16.

Crow, T. J. 2007. How and why genetic linkage has not solved the problem of psychosis: review and hypothesis. *Am J Psychiatry* 164 (1):13–21.

Day, J. K., A. M. Bauer, C. DesBordes, Y. Zhuang, B. E. Kim, L. G. Newton, V. Nehra, K. M. Forsee, R. S. MacDonald, C. Besch-Williford, T. H. Huang, and D. B. Lubahn. 2002. Genistein alters methylation patterns in mice. *J Nutr* 132 (8 Suppl):2419S–2423S.

Dealberto, M. J. 2007. Why are immigrants at increased risk for psychosis? Vitamin D insufficiency, epigenetic mechanisms, or both? *Med Hypotheses* 68 (2):259–67.

Dempster, E. L., J. Mill, I. W. Craig, and D. A. Collier. 2006. The quantification of COMT mRNA in post mortem cerebellum tissue: diagnosis, genotype, methylation and expression. *BMC Med Genet* 7:10.

Dias, V. V., S. Brissos, C. Cardoso, A. C. Andreazza, and F. Kapczinski. 2009. Serum homocysteine levels and cognitive functioning in euthymic bipolar patients. *J Affect Disord* 113 (3):285–90.

Dittmann, S., F. Seemuller, H. C. Grunze, M. J. Schwarz, J. Zach, K. Fast, C. Born, S. Dargel, R. R. Engel, B. Bernhard, H. J. Moller, M. Riedel, and W. E. Severus. 2008. The impact of homocysteine levels on cognition in euthymic bipolar patients: a cross-sectional study. *J Clin Psychiatry* 69 (6):899–906.

Dittmann, S., F. Seemuller, M. J. Schwarz, N. Kleindienst, R. Stampfer, J. Zach, C. Born, B. Bernhard, K. Fast, H. Grunze, R. R. Engel, and E. Severus. 2007. Association of cognitive deficits with elevated homocysteine levels in euthymic bipolar patients and its impact on psychosocial functioning: preliminary results. *Bipolar Disord* 9 (1–2):63–70.

Dolinoy, D. C., J. R. Weidman, R. A. Waterland, and R. L. Jirtle. 2006. Maternal genistein alters coat color and protects Avy mouse offspring from obesity by modifying the fetal epigenome. *Environ Health Perspect* 114 (4):567–72.

Dong, E., R. C. Agis-Balboa, M. V. Simonini, D. R. Grayson, E. Costa, and A. Guidotti. 2005. Reelin and glutamic acid decarboxylase67 promoter remodeling in an epigenetic methionine-induced mouse model of schizophrenia. *Proc Natl Acad Sci U S A* 102 (35):12578–83.

Durany, N., T. Michel, R. Zochling, K. W. Boissl, F. F. Cruz-Sanchez, P. Riederer, and J. Thome. 2001. Brain-derived neurotrophic factor and neurotrophin 3 in schizophrenic psychoses. *Schizophr Res* 52 (1–2):79–86.

Fang, M. Z., Y. Wang, N. Ai, Z. Hou, Y. Sun, H. Lu, W. Welsh, and C. S. Yang. 2003. Tea polyphenol (-)-epigallocatechin-3-gallate inhibits DNA methyltransferase and reactivates methylation-silenced genes in cancer cell lines. *Cancer Res* 63 (22):7563–70.

Finkelstein, J. D. 1998. The metabolism of homocysteine: pathways and regulation. *Eur J Pediatr* 157 Suppl 2:S40–44.

Flanagan, J. M., V. Popendikyte, N. Pozdniakovaite, M. Sobolev, A. Assadzadeh, A. Schumacher, M. Zangeneh, L. Lau, C. Virtanen, S. C. Wang, and A. Petronis. 2006. Intra- and interindividual epigenetic variation in human germ cells. *Am J Hum Genet* 79 (1):67–84.

Fortune, M. T., J. L. Kennedy, and J. B. Vincent. 2003. Anticipation and CAG*CTG repeat expansion in schizophrenia and bipolar affective disorder. *Curr Psychiatry Rep* 5 (2):145–54.

Fraga, M. F., E. Ballestar, M. F. Paz, S. Ropero, F. Setien, M. L. Ballestar, D. Heine-Suner, J. C. Cigudosa, M. Urioste, J. Benitez, M. Boix-Chornet, A. Sanchez-Aguilera, C. Ling, E. Carlsson, P. Poulsen, A. Vaag, Z. Stephan, T. D. Spector, Y. Z. Wu, C. Plass, and M. Esteller. 2005. Epigenetic differences arise during the lifetime of monozygotic twins. *Proc Natl Acad Sci U S A* 102 (30):10604–9.

Frankenburg, F. R. 2007. The role of one-carbon metabolism in schizophrenia and depression. *Harv Rev Psychiatry* 15 (4):146–60.

Fux, R., D. Kloor, M. Hermes, T. Rock, B. Proksch, A. Grenz, U. Delabar, R. Bucheler, S. Igel, K. Morike, C. H. Gleiter, and H. Osswald. 2005. Effect of acute hyperhomocysteinemia on methylation potential of erythrocytes and on DNA methylation of lymphocytes in healthy male volunteers. *Am J Physiol Renal Physiol* 289 (4):F786–92.

Gavin, D. P., S. Kartan, K. Chase, D. R. Grayson, and R. P. Sharma. 2008. Reduced baseline acetylated histone 3 levels, and a blunted response to HDAC inhibition in lymphocyte cultures from schizophrenia subjects. *Schizophr Res* 103 (1–3):330–32.

Gavin, D. P., C. Rosen, K. Chase, D. R. Grayson, N. Tun, and R. P. Sharma. 2009. Dimethylated lysine 9 of histone 3 is elevated in schizophrenia and exhibits a divergent response to histone deacetylase inhibitors in lymphocyte cultures. *J Psychiatry Neurosci* 34 (3):232–37.

Gavin, D. P., and R. P. Sharma. 2009. Histone modifications, DNA methylation, and schizophrenia. *Neurosci Biobehav Rev* (e-pub).

Grayson, D. R., X. Jia, Y. Chen, R. P. Sharma, C. P. Mitchell, A. Guidotti, and E. Costa. 2005. Reelin promoter hypermethylation in schizophrenia. *Proc Natl Acad Sci U S A* 102 (26):9341–46.

Guidotti, A., W. Ruzicka, D. R. Grayson, M. Veldic, G. Pinna, J. M. Davis, and E. Costa. 2007. S-adenosyl methionine and DNA methyltransferase-1 mRNA overexpression in psychosis. *Neuroreport* 18 (1):57–60.

Harrison, P. J. and D. R. Weinberger. 2005. Schizophrenia genes, gene expression, and neuropathology: On the matter of their convergence. *Mol Psychiatry* 10 (1):40–68; image 45.

Heijmans, B. T., E. W. Tobi, A. D. Stein, H. Putter, G. J. Blauw, E. S. Susser, P. E. Slagboom, and L. H. Lumey. 2008. Persistent epigenetic differences associated with prenatal exposure to famine in humans. *Proc Natl Acad Sci U S A* 105 (44):17046–49.

Hoek, H. W., E. Susser, K. A. Buck, L. H. Lumey, S. P. Lin, and J. M. Gorman. 1996. Schizoid personality disorder after prenatal exposure to famine. *Am J Psychiatry* 153 (12):1637–39.

Huang, H. S., A. Matevossian, C. Whittle, S. Y. Kim, A. Schumacher, S. P. Baker, and S. Akbarian. 2007. Prefrontal dysfunction in schizophrenia involves mixed-lineage leukemia 1-regulated histone methylation at GABAergic gene promoters. *J Neurosci* 27 (42):11254–62.

Ingrosso, D., A. Cimmino, A. F. Perna, L. Masella, N. G. De Santo, M. L. De Bonis, M. Vacca, M. D'Esposito, M. D'Urso, P. Galletti, and V. Zappia. 2003. Folate treatment and unbalanced methylation and changes of allelic expression induced by hyperhomocysteinaemia in patients with uraemia. *Lancet* 361 (9370):1693–99.

Iwamoto, K., M. Bundo, K. Yamada, H. Takao, Y. Iwayama-Shigeno, T. Yoshikawa, and T. Kato. 2005. DNA methylation status of SOX10 correlates with its downregulation and oligodendrocyte dysfunction in schizophrenia. *J Neurosci* 25 (22):5376–81.

Kato, T., S. Takahashi, T. Shioiri, J. Murashita, H. Hamakawa, and T. Inubushi. 1994. Reduction of brain phosphocreatine in bipolar II disorder detected by phosphorus-31 magnetic resonance spectroscopy. *J Affect Disord* 31 (2):125–33.

Kouzarides, T. 2007. Chromatin modifications and their function. *Cell* 128 (4):693–705.

Krabbendam, L., and J. van Os. 2005. Schizophrenia and urbanicity: a major environmental influence—conditional on genetic risk. *Schizophr Bull* 31 (4):795–99.

Krebs, M. O., A. Bellon, G. Mainguy, T. M. Jay, and H. Frieling. 2009. One-carbon metabolism and schizophrenia: current challenges and future directions. *Trends Mol Med* 15 (12):562–70.

Kriaucionis, S., and N. Heintz. 2009. The nuclear DNA base 5-hydroxymethylcytosine is present in Purkinje neurons and the brain. *Science* 324 (5929):929–30.

Kuratomi, G., K. Iwamoto, M. Bundo, I. Kusumi, N. Kato, N. Iwata, N. Ozaki, and T. Kato. 2008. Aberrant DNA methylation associated with bipolar disorder identified from discordant monozygotic twins. *Mol Psychiatry* 13 (4):429–41.

Lahiri, D. K., B. Maloney, and N. H. Zawia. 2009. The LEARn model: an epigenetic explanation for idiopathic neurobiological diseases. *Mol Psychiatry* 14 (11):992–1003.

Le, T. T., S. Harlepp, C. C. Guet, K. Dittmar, T. Emonet, T. Pan, and P. Cluzel. 2005. Real-time RNA profiling within a single bacterium. *Proc Natl Acad Sci U S A* 102 (26):9160–64.

Levine, J., G. Agam, B. A. Sela, D. L. Garver, E. F. Torrey, and R. H. Belmaker. 2005. CSF homocysteine is not elevated in schizophrenia. *J Neural Transm* 112 (2):297–302.

Levine, J., B. A. Sela, Y. Osher, and R. H. Belmaker. 2005. High homocysteine serum levels in young male schizophrenia and bipolar patients and in an animal model. *Prog Neuropsychopharmacol Biol Psychiatry* 29 (7):1181–91.

Levine, J., Z. Stahl, B. A. Sela, V. Ruderman, O. Shumaico, I. Babushkin, Y. Osher, Y. Bersudsky, and R. H. Belmaker. 2006. Homocysteine-reducing strategies improve symptoms in chronic schizophrenic patients with hyperhomocysteinemia. *Biol Psychiatry* 60 (3):265–69.

Lipton, S. A., W. K. Kim, Y. B. Choi, S. Kumar, D. M. D'Emilia, P. V. Rayudu, D. R. Arnelle, and J. S. Stamler. 1997. Neurotoxicity associated with dual actions of homocysteine at the N-methyl-D-aspartate receptor. *Proc Natl Acad Sci U S A* 94 (11):5923–28.

Loenarz, C., and C. J. Schofield. 2009. Oxygenase catalyzed 5-methylcytosine hydroxylation. *Chem Biol* 16 (6):580–83.

Lopez-Castroman, J., D. D. Gomez, J. J. Belloso, P. Fernandez-Navarro, M. M. Perez-Rodriguez, I. B. Villamor, F. F. Navarrete, C. M. Ginestar, D. Currier, M. R. Torres, M. Navio-Acosta, J. Saiz-Ruiz, M. A. Jimenez-Arriero, and E. Baca-Garcia. 2009. Differences in maternal and paternal age between schizophrenia and other psychiatric disorders. *Schizophr Res* (e-pub).

Malaspina, D., S. Harlap, S. Fennig, D. Heiman, D. Nahon, D. Feldman, and E. S. Susser. 2001. Advancing paternal age and the risk of schizophrenia. *Arch Gen Psychiatry* 58 (4):361–67.

Mansoor, M. A., O. Kristensen, T. Hervig, C. J. Bates, K. Pentieva, H. Vefring, A. Osland, T. Berge, P. A. Drablos, O. Hetland, and S. Rolfsen. 1999. Plasma total homocysteine response to oral doses of folic acid and pyridoxine hydrochloride (vitamin B6) in healthy individuals. Oral doses of vitamin B6 reduce concentrations of serum folate. *Scand J Clin Lab Invest* 59 (2):139–46.

McGowan, P. O., M. J. Meaney, and M. Szyf. 2008. Diet and the epigenetic (re)programming of phenotypic differences in behavior. *Brain Res* 1237:12–24.

McKinley, M. C., H. McNulty, J. McPartlin, J. J. Strain, K. Pentieva, M. Ward, D. G. Weir, and J. M. Scott. 2001. Low-dose vitamin B-6 effectively lowers fasting plasma homocysteine in healthy elderly persons who are folate and riboflavin replete. *Am J Clin Nutr* 73 (4):759–64.

Mill, J., E. Dempster, A. Caspi, B. Williams, T. Moffitt, and I. Craig. 2006. Evidence for monozygotic twin (MZ) discordance in methylation level at two CpG sites in the promoter region of the catechol-O-methyltransferase (COMT) gene. *Am J Med Genet B Neuropsychiatr Genet* 141B (4):421–25.

Mill, J., and A. Petronis. 2009. The relevance of epigenetics to major psychosis. In *Epigenomics*, ed. A. Ferguson-Smith, J. Greally, and R. Martienssen. New York: Springer, 411–34.

Mill, J., T. Tang, Z. Kaminsky, T. Khare, S. Yazdanpanah, L. Bouchard, P. Jia, A. Assadzadeh, J. Flanagan, A. Schumacher, S. C. Wang, and A. Petronis. 2008. Epigenomic profiling reveals DNA-methylation changes associated with major psychosis. *Am J Hum Genet* 82 (3):696–711.

Millan, M. J. 2005. [Dopamine D3 receptors as a novel target for improving the treatment of schizophrenia]. *Med Sci (Paris)* 21 (4):434–442.

Miodownik, C., V. Lerner, T. Vishne, B. A. Sela, and J. Levine. 2007. High-dose vitamin B6 decreases homocysteine serum levels in patients with schizophrenia and schizoaffective disorders: a preliminary study. *Clin Neuropharmacol* 30 (1):13–17.

Moiseeva, E. P., G. M. Almeida, G. D. Jones, and M. M. Manson. 2007. Extended treatment with physiologic concentrations of dietary phytochemicals results in altered gene expression, reduced growth, and apoptosis of cancer cells. *Mol Cancer Ther* 6 (11):3071–79.

Monteggia, L. M. 2007. Elucidating the role of brain-derived neurotrophic factor in the brain. *Am J Psychiatry* 164 (12):1790.

Murphy, B. C., R. L. O'Reilly, and S. M. Singh. 2005. Site-specific cytosine methylation in S-COMT promoter in 31 brain regions with implications for studies involving schizophrenia. *Am J Med Genet B Neuropsychiatr Genet* 133B (1):37–42.

Newman, M. E. 2006. Modularity and community structure in networks. *Proc Natl Acad Sci U S A* 103 (23):8577–82.

Owen, M. J., N. Craddock, and A. Jablensky. 2007. The genetic deconstruction of psychosis. *Schizophr Bull* 33 (4):905–11.

Peet, M. 2008. Omega-3 polyunsaturated fatty acids in the treatment of schizophrenia. *Isr J Psychiatry Relat Sci* 45 (1):19–25.

Perrin, M. C., A. S. Brown, and D. Malaspina. 2007. Aberrant epigenetic regulation could explain the relationship of paternal age to schizophrenia. *Schizophr Bull* 33 (6):1270–73.

Petronijevic, N. D., N. V. Radonjic, M. D. Ivkovic, D. Marinkovic, V. D. Piperski, B. M. Duricic, and V. R. Paunovic. 2008. Plasma homocysteine levels in young male patients in the exacerbation and remission phase of schizophrenia. *Prog Neuropsychopharmacol Biol Psychiatry* 32 (8):1921–26.

Petronis, A. 2004. The origin of schizophrenia: genetic thesis, epigenetic antithesis, and resolving synthesis. *Biol Psychiatry* 55 (10):965–70.

Petronis, A., Gottesman, II, P. Kan, J. L. Kennedy, V. S. Basile, A. D. Paterson, and V. Popendikyte. 2003. Monozygotic twins exhibit numerous epigenetic differences: clues to twin discordance? *Schizophr Bull* 29 (1):169–78.

Petronis, A., A. D. Paterson, and J. L. Kennedy. 1999. Schizophrenia: an epigenetic puzzle? *Schizophr Bull* 25 (4):639–55.

Pogribny, I. P., A. R. Karpf, S. R. James, S. Melnyk, T. Han, and V. P. Tryndyak. 2008. Epigenetic alterations in the brains of Fisher 344 rats induced by long-term administration of folate/methyl-deficient diet. *Brain Res* 1237:25–34.

Pogribny, I. P., S. A. Ross, C. Wise, M. Pogribna, E. A. Jones, V. P. Tryndyak, S. J. James, Y. P. Dragan, and L. A. Poirier. 2006. Irreversible global DNA hypomethylation as a key step in hepatocarcinogenesis induced by dietary methyl deficiency. *Mutat Res* 593 (1–2):80–87.

Rosa, A., M. M. Picchioni, S. Kalidindi, C. S. Loat, J. Knight, T. Toulopoulou, R. Vonk, A. C. van der Schot, W. Nolen, R. S. Kahn, P. McGuffin, R. M. Murray, and I. W. Craig. 2008. Differential methylation of the X-chromosome is a possible source of discordance for bipolar disorder female monozygotic twins. *Am J Med Genet B Neuropsychiatr Genet* 147B (4):459–62.

Rutten, B. P., and J. Mill. 2009. Epigenetic mediation of environmental influences in major psychotic disorders. *Schizophr Bull* 35 (6):1045–56.

Schumacher, A. 2010. Aging epigenetics. In *Handbook of epigenetics: the new molecular and medical genetics,* ed. T. Tollefsbol. New York: Elsevier (in press).

Schumacher, A., P. Kapranov, Z. Kaminsky, J. Flanagan, A. Assadzadeh, P. Yau, C. Virtanen, N. Winegarden, J. Cheng, T. Gingeras, and A. Petronis. 2006. Microarray-based DNA methylation profiling: technology and applications. *Nucleic Acids Res* 34 (2):528–42.

Schumacher, A., P. A. Koetsier, J. Hertz, and W. Doerfler. 2000. Epigenetic and genotype-specific effects on the stability of de novo imposed methylation patterns in transgenic mice. *J Biol Chem* 275 (48):37915–21.

Schumacher, A., and A. Petronis. 2006. Epigenetics of complex diseases: from general theory to laboratory experiments. *Curr Top Microbiol Immunol* 310:81–115.

Sharma, R. P., D. R. Grayson, and D. P. Gavin. 2008. Histone deactylase 1 expression is increased in the prefrontal cortex of schizophrenia subjects: analysis of the National Brain Databank microarray collection. *Schizophr Res* 98 (1–3):111–17.

Sharma, R., A. Nakamura, R. Takahashi, H. Nakamoto, and S. Goto. 2006. Carbonyl modification in rat liver histones: decrease with age and increase by dietary restriction. *Free Radic Biol Med* 40 (7):1179–84.

Shimabukuro, M., T. Sasaki, A. Imamura, T. Tsujita, C. Fuke, T. Umekage, M. Tochigi, K. Hiramatsu, T. Miyazaki, T. Oda, J. Sugimoto, Y. Jinno, and Y. Okazaki. 2007. Global hypomethylation of peripheral leukocyte DNA in male patients with schizophrenia: a potential link between epigenetics and schizophrenia. *J Psychiatr Res* 41 (12):1042–46.

Sims, C. E., and N. L. Allbritton. 2007. Analysis of single mammalian cells on-chip. *Lab Chip* 7 (4):423–40.

Smits, L., C. Pedersen, P. Mortensen, and J. van Os. 2004. Association between short birth intervals and schizophrenia in the offspring. *Schizophr Res* 70 (1):49–56.

Song, S., W. Wang, and P. Hu. 2009. Famine, death, and madness: schizophrenia in early adulthood after prenatal exposure to the Chinese Great Leap Forward Famine. *Soc Sci Med* 68 (7):1315–21.

St. Clair, D., M. Xu, P. Wang, Y. Yu, Y. Fang, F. Zhang, X. Zheng, N. Gu, G. Feng, P. Sham, and L. He. 2005. Rates of adult schizophrenia following prenatal exposure to the Chinese famine of 1959-1961. *Jama* 294 (5):557–62.

Stoger, R. 2008. The thrifty epigenotype: an acquired and heritable predisposition for obesity and diabetes? *Bioessays* 30 (2):156–66.

Stompe, T., G. Ortwein-Swoboda, R. Strobl, and A. Friedmann. 2000. The age of onset of schizophrenia and the theory of anticipation. *Psychiatry Res* 93 (2):125–34.

Susser, E., R. Neugebauer, H. W. Hoek, A. S. Brown, S. Lin, D. Labovitz, and J. M. Gorman. 1996. Schizophrenia after prenatal famine. Further evidence. *Arch Gen Psychiatry* 53 (1):25–31.

Sutherland, J. E., and M. Costa. 2003. Epigenetics and the environment. *Ann N Y Acad Sci* 983:151–60.

Sweeney, M. R., J. McPartlin, and J. Scott. 2007. Folic acid fortification and public health: report on threshold doses above which unmetabolised folic acid appear in serum. *BMC Public Health* 7:41.

Tahiliani, M., K. P. Koh, Y. Shen, W. A. Pastor, H. Bandukwala, Y. Brudno, S. Agarwal, L. M. Iyer, D. R. Liu, L. Aravind, and A. Rao. 2009. Conversion of 5-methylcytosine to 5-hydroxymethylcytosine in mammalian DNA by MLL partner TET1. *Science* 324 (5929):930–35.

Tamura, Y., H. Kunugi, J. Ohashi, and H. Hohjoh. 2007. Epigenetic aberration of the human REELIN gene in psychiatric disorders. *Mol Psychiatry* 12 (6):519, 593–600.

Tandon, R., M. S. Keshavan, and H. A. Nasrallah. 2008. Schizophrenia, "just the facts" what we know in 2008. 2. Epidemiology and etiology. *Schizophr Res* 102 (1–3):1–18.

Tiemann-Boege, I., W. Navidi, R. Grewal, D. Cohn, B. Eskenazi, A. J. Wyrobek, and N. Arnheim. 2002. The observed human sperm mutation frequency cannot explain the achondroplasia paternal age effect. *Proc Natl Acad Sci U S A* 99 (23):14952–57.

Tobi, E. W., L. H. Lumey, R. P. Talens, D. Kremer, H. Putter, A. D. Stein, P. E. Slagboom, and B. T. Heijmans. 2009. DNA methylation differences after exposure to prenatal famine are common and timing- and sex-specific. *Hum Mol Genet* 18 (21):4046–53.

Tochigi, M., K. Iwamoto, M. Bundo, A. Komori, T. Sasaki, N. Kato, and T. Kato. 2008. Methylation status of the reelin promoter region in the brain of schizophrenic patients. *Biol Psychiatry* 63 (5):530–33.

Tremolizzo, L., G. Carboni, W. B. Ruzicka, C. P. Mitchell, I. Sugaya, P. Tueting, R. Sharma, D. R. Grayson, E. Costa, and A. Guidotti. 2002. An epigenetic mouse model for molecular and behavioral neuropathologies related to schizophrenia vulnerability. *Proc Natl Acad Sci U S A* 99 (26):17095–100.

Valinluck, V., H. H. Tsai, D. K. Rogstad, A. Burdzy, A. Bird, and L. C. Sowers. 2004. Oxidative damage to methyl-CpG sequences inhibits the binding of the methyl-CpG binding domain (MBD) of methyl-CpG binding protein 2 (MeCP2). *Nucleic Acids Res* 32 (14):4100–4108.

Wang, S-C., B. Oelze, and A. Schumacher. 2008. Age-specific epigenetic drift in late-onset Alzheimer's disease. *PLoS ONE* 3 (7):e2698.

Waterland, R. A. 2006. Assessing the effects of high methionine intake on DNA methylation. *J Nutr* 136 (6 Suppl):1706S–1710S.

Waterland, R. A., and R. L. Jirtle. 2003. Transposable elements: targets for early nutritional effects on epigenetic gene regulation. *Mol Cell Biol* 23 (15):5293–5300.

Weaver, I. C., N. Cervoni, F. A. Champagne, A. C. D'Alessio, S. Sharma, J. R. Seckl, S. Dymov, M. Szyf, and M. J. Meaney. 2004. Epigenetic programming by maternal behavior. *Nat Neurosci* 7:847–54.

Wilkinson, G. R. 2005. Drug metabolism and variability among patients in drug response. *N Engl J Med* 352 (21):2211–21.

Williams, A. L., C. Girard, D. Jui, A. Sabina, and D. L. Katz. 2005. S-adenosylmethionine (SAMe) as treatment for depression: a systematic review. *Clin Invest Med* 28 (3):132–39.

Wolff, G. L., R. L. Kodell, S. R. Moore, and C. A. Cooney. 1998. Maternal epigenetics and methyl supplements affect agouti gene expression in Avy/a mice. *FASEB J* 12 (11):949–57.

Wu, A., Z. Ying, and F. Gomez-Pinilla. 2004. Dietary omega-3 fatty acids normalize BDNF levels, reduce oxidative damage, and counteract learning disability after traumatic brain injury in rats. *J Neurotrauma* 21 (10):1457–67.

Xu, M. Q., W. S. Sun, B. X. Liu, G. Y. Feng, L. Yu, L. Yang, G. He, P. Sham, E. Susser, D. St. Clair, and L. He. 2009. Prenatal malnutrition and adult schizophrenia: further evidence from the 1959–1961 Chinese famine. *Schizophr Bull* 35 (3):568–76.

Yi, P., S. Melnyk, M. Pogribna, I. P. Pogribny, R. J. Hine, and S. J. James. 2000. Increase in plasma homocysteine associated with parallel increases in plasma S-adenosylhomocysteine and lymphocyte DNA hypomethylation. *J Biol Chem* 275 (38):29318–23.

Zhang, A. P., J. Yu, J. X. Liu, H. Y. Zhang, Y. Y. Du, J. D. Zhu, G. He, X. W. Li, N. F. Gu, G. Y. Feng, and L. He. 2007. The DNA methylation profile within the 5′-regulatory region of DRD2 in discordant sib pairs with schizophrenia. *Schizophr Res* 90 (1–3):97–103.

13 Interactions between Folate, Other B Vitamins, DNA Methylation, and Neurodevelopmental Disorders

Rebecca J. Schmidt and Janine M. LaSalle

CONTENTS

13.1 Role of Folate and Other B Vitamins in Neurodevelopment 318
13.1.1 Description of the Role of Folate and Other B Vitamins in Neurodevelopment 318
13.1.1.1 Folate/Folic Acid 318
13.1.1.2 Vitamin B_{12} 319
13.1.1.3 Choline and Other B Vitamins 321
13.1.2 B Vitamin-Dependent Metabolic Pathways and Enzymes Relevant to Neurodevelopment 322
13.2 Folate, B Vitamins, and Neurodevelopmental Disorders 323
13.2.1 Neural Tube Defects 323
13.2.2 Autism Spectrum Disorders 328
13.2.2.1 Metabolites 329
13.2.2.2 Folate Pathway Genes Implicated in Autism 329
13.2.2.3 Uptake to the Central Nervous System 330
13.2.2.4 Treatment 330
13.2.2.5 Rett Syndrome 331
13.2.2.6 Fragile X Syndrome 332
13.2.3 Schizophrenia 332
13.2.3.1 Genes 333
13.2.3.2 Treatment 333
13.2.4 Down Syndrome (Trisomy 21) 333
13.2.5 Other Neurodevelopmental Outcomes 334
13.2.5.1 Cognitive Functioning 334
13.2.5.2 Behaviors 335
13.3 Role of Folate and Other B Vitamins in Methylation 336
13.3.1 Relationship between Folate Levels and DNA Methylation Status 336

13.3.2 Overlapping Timing of Neurodevelopmental Need for Folic Acid and DNA Methylation in the Periconceptional Period 337
13.3.3 B Vitamins and Methylation of Products Other Than DNA 338
13.4 Epigenetics and Neurodevelopmental Disorders 339
13.5 Chapter Summary and Conclusions 341
Acknowledgments 342
References 342

Maternal nutrition plays an important role in the proper neurodevelopment of mammals, particularly humans. Of all the important nutrients in the maternal diet, folate and other B vitamins are the most implicated dietary factors for preventing neural tube defects and perhaps other neurodevelopmental disorders, but their mechanisms of action are likely complex. This chapter reviews the biochemistry and genetics of folate and B vitamin metabolism, as well as the neurodevelopmental disorders with known or suspected genetic causes in which folate supplementation or administration has been tested. The emerging role of DNA methylation as an epigenetic mechanism important for normal neurodevelopment and the development of multiple neurodevelopmental disorders are discussed. Folate and other B vitamins may impact the essential pathways required for establishing and maintaining high global levels of DNA methylation and histone methylation during the critical periconception window of dynamic epigenetic changes in the mammalian genome.

13.1 ROLE OF FOLATE AND OTHER B VITAMINS IN NEURODEVELOPMENT

Most of what we know about the role of B vitamins in neurodevelopment is derived from examining the consequences of dietary or metabolic deficiencies. While discovering associations between low levels of B vitamins and neurodevelopmental outcomes has been important for determining B vitamin involvement in neurodevelopmental aspects, often the mechanisms behind their action remain unclear.

13.1.1 Description of the Role of Folate and Other B Vitamins in Neurodevelopment

13.1.1.1 Folate/Folic Acid

Folate is the natural form of the water-soluble vitamin B_9, obtained from the diet from green, leafy vegetables such as spinach, Brussels sprouts, broccoli, asparagus, turnip greens, and lettuces. Other dietary sources of folate include beans, peas, sunflower seeds, oranges, baker's yeast, and liver. Dietary folate absorption involves conversion of polyglutamates to monoglutamates in the jejunum, which then enter the intestinal cells. Folic acid, or pteroyl-L-glutamic acid, is the synthetic form of folate found in supplements and fortified foods. Folic acid must be reduced to tetrahydrofolate (THF) and converted to methyl or formyl forms before entering portal circulation. The predominant form of folate in serum and tissues is 5-methyl THF monoglutamate. Storage locations include the liver, pancreas, kidneys, brain, and red cells. Folate is necessary for the production and maintenance of new cells, making it especially important during periods of rapid cell division and growth. More specifically, folate acts as a donor and acceptor of one-carbon units, important for the biosynthesis of nucleic acids, proteins, and methyl groups. Given folate's essential role in nucleic acid synthesis, folate deficiency hinders DNA synthesis and cell division. Hematopoietic cells appear to be the most affected cell type because of their frequent cell divisions compared to other cell types. Folate deficiency results in a limited production of red blood cells, leading to megaloblastic anemia, characterized by large immature red blood cells called megaloblasts, showing clumping and fragmentation of nuclear chromatin (Zittoun 1993). In addition to the essential roles for folate in hematopoesis and nucleic acid synthesis, dietary folate provides the methyl donors for methylation of proteins involved in the formation and maintenance of neuronal and glial membrane lipids (Hirata and Axelrod 1980), and numerous other methylation reactions of DNA, RNA, histones, neurotransmitters, membrane phospholipids, and proteins, as discussed in more detail later. As discussed in the sections that follow, adequate levels of folate are required during the first weeks of pregnancy for proper closure of the neural tube, which later forms the brain and spinal cord.

13.1.1.2 Vitamin B_{12}

Vitamin B_{12} is another water-soluble vitamin with important roles in brain and nervous system function and the formation of blood. Cyanocobalamin is the common synthetic form of the vitamin. Intrinsic factor, a glycoprotein produced by the parietal cells of the stomach, is necessary for the absorption of vitamin B_{12}. As discussed more later, vitamin B_{12} is necessary for synthesis of neurotransmitters and catecholamines and is a required cofactor for the conversion of methylTHF to THF, and homocysteine to methionine, a substrate for production of S-adenosylmethionine (SAM), a key enzyme for methylation reactions including DNA methylation (see later discussion of Figure 13.1) and methylation of myelin sheath phospholipids.

The effects of vitamin B_{12} deficiency were first discovered in association with pernicious anemia, an autoimmune disease that destroys parietal cells in the stomach that release intrinsic factor, and consequently vitamin B_{12} cannot be absorbed. If left untreated, pernicious anemia results in neurological complications including neuropathy, intellectual disability, and ultimately death. Because vitamin B_{12} stores can last years, nutritional deficiencies of vitamin B_{12} are uncommon in adults. However, deficiencies of vitamin B_{12} can occur more rapidly in infants. Vitamin B_{12} deficiencies rarely occur before about 4 months of age, when vitamin B_{12} stores established in utero are typically depleted. Infants of vitamin B_{12}-deficient, often vegan, vegetarian, or lacto-ovo vegetarian, breastfeeding mothers and infants receiving low amounts of animal-source foods are vulnerable to vitamin B_{12} deficiency and neurological symptoms. Case reports observe central nervous system effects, developmental delays, and in severe cases, brain atrophy (Wighton et al. 1979; Lovblad et al. 1997).

The function of vitamin B_{12} in DNA synthesis overlaps with folate and can be compensated for with sufficient quantities of folic acid. However, even in the presence of adequate folate and absence of anemia, vitamin B_{12} deficiencies cause neuropathies.

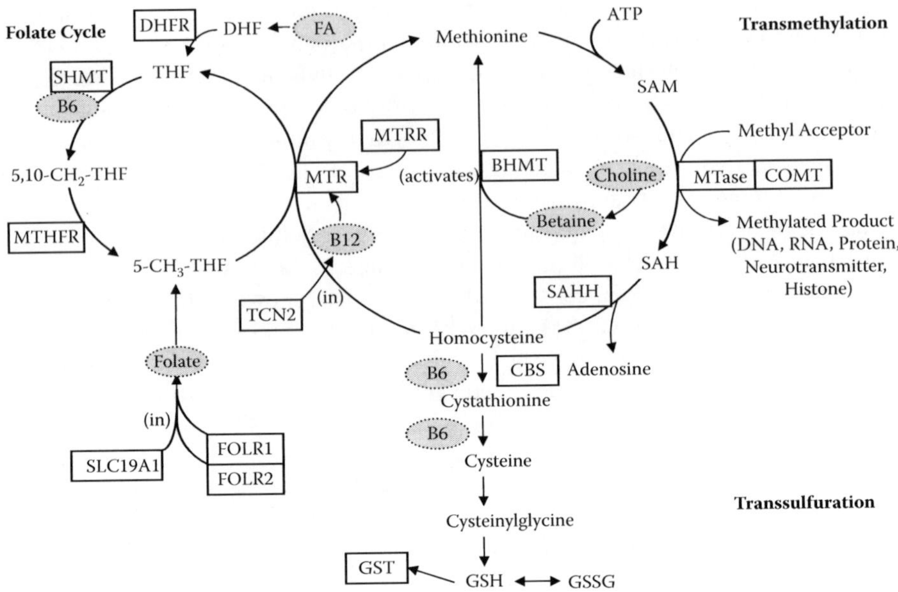

FIGURE 13.1 B vitamin-dependent folate cycle and transmethylation and transsulfuration pathways. Enzymes are in boxes; vitamin substrates or cofactors are in ellipses. Abbreviations: DHFR, dihydrofolate reductase; DHF, dihydrofolate; FA, Folic acid; THF, tetrahydrofolate; SHMT, serine hydromethyltransferase; 5,10-CH2-THF, 5,10-methylenetetrahydrofolate; MTHFR, methylenetetrahydrofolate reductase; 5-CH3-THF, 5-methyltetrahydrofolate; SLC19A1, solute carrier family 19 member 1; FOLR1, folate receptor 1; FOLR2, folate receptor 2; MTR, 5-methyltetrahydrofolate-homocysteine methyltransferase; MTRR, 5-methyltetrahydrofolate-homocysteine methyltransferase reductase; TCN2, transcobalamin II; BHMT, betaine-homocysteine methyltransferase; SAM, S-adenosylmethionine; MTase, methyltransferase; COMT, catechol-O-methyltransferase; SAH, S-adenosylhomocysteine; SAHH, SAH hydrolase; CBS, cystathionine beta synthase; GSH, reduced active glutathione; GSSG, oxidized disulfide form of glutathione; GST, glutathione-S-transferase. (Adapted from S. J. James et al., *Am J Genet B Neuropsychiatr Genet,* 141, 947–56, 2006.)

A function uniquely lost in vitamin B_{12} deficiency (unaffected by folate supplementation) is the ability of methylmalonyl coenzyme A mutase (MUT) to use the 5′-deoxyadenosylcobalamin form to catalyze the conversion of methylmalonyl coenzyme A to succinyl coenzyme A. Failure of this reaction to occur results in elevated methylmalonic acid (MMA) levels. Some suggest that MMA is a myelin destabilizer, that excessive MMA prevents normal fatty acid synthesis, and that abnormal fatty acids are then incorporated into myelin, which results in fragile myelin, demyelinatination, and subacute combined degeneration of the central nervous system and spinal cord (Naidich and Ho 2005). Others believe that subacute combined degeneration of the spinal cord results from methyl group deficiency and deficient methylation (Scott et al. 1981). This hypothesis is supported by evidence that prolonged exposure in mammals to nitrous oxide, which inactivates methionine synthetase and impairs methionine synthesis, results in a myelopathy indistinguishable from subacute combined degeneration of the spinal cord.

13.1.1.3 Choline and Other B Vitamins

Choline is an essential water-soluble nutrient, usually grouped with the B complex vitamins. It is a natural amine found in lipids of cell membranes and in the neurotransmitter acetylcholine. A variety of foods provide both free and esterified forms of choline; excellent sources include liver, eggs, and wheat germ. Choline can also be synthesized de novo primarily in the liver when phosphatidylcholine is formed through a methylation reaction by phosphatidylethanolamine N-methyltransferase (PEMT) (Zeisel 2006). As reviewed by Zeisel, this PEMT activity tends to be higher in females than in males (Noga and Vance 2003), especially for premenopausal women, and could be mediated by estrogen (Drouva et al. 1986; Zeisel 2006). Demand for choline is especially high during mammalian pregnancy, as large amounts of choline are delivered to the fetus across the placenta, resulting in a much greater concentration of choline in the amniotic fluid and fetus than in the maternal blood (Zeisel and Niculescu 2006). In rodents, perinatal choline supplementation increases choline metabolite concentrations in the blood and fetal brain (Garner et al. 1995) as it is transported across the blood–brain barrier (Pardridge 1986). Research on associations between choline levels and neurodevelopment and neurologic function is growing. Similar to folate, evidence suggests that choline is essential for normal neural tube closure in early pregnancy in both animal and human studies (Fisher et al. 2001, 2002; Shaw et al. 2004). Choline also has effects similar to folate on cell proliferation and apoptosis in the fetal rat brain (Craciunescu et al. 2003, 2004). Studies show that choline supplementation during days 11–18 of gestation in rats (corresponding to day 56 of pregnancy through the first few months of pregnancy in humans) can increase proliferation and decrease apoptosis of hippocampal progenitor cells (Albright, Friedrich et al. 1999; Albright, Tsai et al. 1999), with long-term effects on memory capacity (Meck and Williams 1997a, 1997b, 1997c, 1999, 2003; Meck et al. 1988; Williams et al. 1998). Maternal choline deficiency during this period has the opposite effects on offspring (Albright, Friedrich et al. 1999; Albright, Tsai et al. 1999; Meck and Williams 1997b, 1999). The timing of this period of choline sensitivity correlates with periods of neurogenesis and synaptogenesis in the hippocampus and basal forebrain (Zeisel 2006). Though mechanisms for neurodevelopmental effects of altered choline levels are not completely understood, evidence for decreased methylation and increased expression of certain genes, like CDKN3, which inhibits cell proliferation (Niculescu et al. 2004), suggest that DNA methylation might play a role. Protective effects of perinatal choline supplementation have also been shown against a few different neurotoxic exposures (Guo-Ross et al. 2002, 2003; Holmes et al. 2002; Riley and McGee 2005; Thomas et al. 2000, 2004, 2007, 2009).

Choline's metabolite, betaine (trimethylglycine), is a methyl donor that participates in the synthesis of SAM through conversion of homocysteine to methionine. Methionine is another essential amino acid that needs to be obtained from the diet in addition to being regenerated from homocysteine (discussed as part of the methionine cycle in a later section).

13.1.2 B Vitamin-Dependent Metabolic Pathways and Enzymes Relevant to Neurodevelopment

The folate cycle is intimately intertwined with the methionine/transmethylation cycle, as shown in Figure 13.1. The folate cycle and the role of the participatory enzymes have been reviewed extensively, often in context of the candidate genes for neural tube defect research (van der Linden et al. 2006; van der Put and Blom 2000). Dietary folates are primarily polyglutamates (Tamura and Stokstad 1973), and must be deconjugated to monoglutamates by folypoly-γ-glutamate carboxypeptidase, encoded by the glutamate carboxypeptidase II (GCPII) gene, before uptake and transport in the body (van der Linden et al. 2006). The solute carrier family 19, member 1 (SLC19A1), also known as the reduced folate carrier (RFC1) enzyme, participates in the absorption of folate monoglutamates in the proximal small intestine. Synthetic folic acid is converted to dihydrofolate and is reduced into tetrahydrofolate (THF) by dihydrofolate reductase (DHFR). Folic acid itself is not biologically active; all the biological functions are performed by THF and other derivatives, the availability of which depends on DHFR. A deletion in the DHFR gene affecting gene expression (Johnson et al. 2004) has been associated with decreased serum and red blood cell folate (Stanislawska-Sachadyn et al. 2008; Kalmbach et al. 2008). THF can be metabolized into 5,10-CH_2-THF by serine hydroxymethyltransferase (SHMT) transferring a methylene group from serine using the active form of vitamin B_6 (pyridoxal phosphate) as a cofactor, or through other pathways. Methylenetetrahydrofolate reductase (MTHFR) enzyme then reduces the methylene group in 5,10-CH_2-THF with flavin adenine dinucleotide (FAD) and nicotinamide adenine dinucleotide phosphate (NADPH) serving as cofactors. MTHFR is a key regulatory enzyme for folate availability and the remethylation of homocysteine that is allosterically inhibited by SAM. As discussed previously, the C677T and perhaps the C1298A polymorphism in the MTHFR gene are associated with reduced MTHFR enzyme activity and higher plasma homocysteine levels (van der Put et al. 1998).

Once in the bloodstream, folate, mainly present as 5-CH_3-THF, can enter cells through receptor- or carrier-mediated transport. Folate receptor α (FR-α), encoded by FOLR1, has a high affinity for 5-CH_3-THF but is expressed in a limited number of epithelial cells, predominantly in the proximal tubules of the kidney, the choroid plexus, and the placenta (Kamen and Smith 2004). The other folate receptors, FR-β (encoded by FOLR2) and FR-γ, and the ubiquitously expressed SLC19A1 responsible for carrier-mediated transport, possess a lower affinity for 5-CH_3-THF than does FR-α.

In the cell, the function of 5-CH_3-THF as a methyl donor with subsequent formation of THF, during the remethylation of homocysteine to methionine, is the key connection between the folate cycle and the methionine cycle. The methyl-group transfer is mediated by 5-methyltetrahydrofolate-homocysteine methyltransferase (MTR or methionine synthase) and requires activation of MTR by methionine synthase reductase (MTRR) and vitamin B_{12} (methylcobalamin, present in the cytosol) as a cofactor. It is important to note that vitamin B_{12} is the only acceptor of methyl groups from 5-CH_3-THF and that homocysteine is the only acceptor of methyl groups from methylcobalamin. Thus a defective methionine synthase enzyme or a vitamin

B_{12} deficiency can lead to a "methyl trap," generating a pool of 5-CH_3-THF that is unable to undergo reactions, creating the equivalent of a folate deficiency (Scott 1992). Transcobalamin (encoded by TCN2) is a plasma globulin that serves as the primary transport protein for vitamin B_{12} cellular uptake (Hakami et al. 1971).

The conversion of homocysteine to methionine can also occur by a secondary pathway involving choline and its metabolite, betaine, which can provide a methyl group for betaine-homocysteine methyltransferase (BHMT). Although the MTR enzyme is expressed in almost every cell and BHMT expression is primarily limited to the liver and kidneys, this pathway is responsible for about 50% of the homocysteine remethylation (van der Linden et al. 2006). Notably, because of the close metabolic relationship in the one-carbon transfer cycle between folate and choline, deficiencies in folate are compensated for by choline, leading to its depletion in tissues (Zeisel 2009; Kim et al. 1994). Dietary deficiencies in either choline or folate are associated with elevated plasma homocysteine concentrations (da Costa et al. 2005; Chiuve et al. 2007) and reduced methylation capacity.

Methionine and ATP are biosynthesized into SAM in reactions catalyzed by methionine adenosyltransferases. SAM is the universal donor of methyl groups for various methylation reactions, including DNA methylation. The transfer of a methyl group from SAM to a methyl receptor involves a methyltransferase and results in the formation of S-adenosylhomocysteine (SAH). Different methyltransferases act on DNA (DNMT1, DNMT3A, and DNMT3B), proteins such as histones (G9a) and catecholamines such as the neurotransmitter dopamine (catechol-O-methyltransferase, COMT). SAH is then hydrolyzed into adenosine and homocysteine by S-adenosylhomocysteine hydrolase (SAHH). The equilibrium of this reversible reaction favors SAH formation, which is an allosteric inhibitor of methylation. Thus homocysteine and adenosine need to be metabolized rapidly in order to maintain low SAH levels. The resultant homocysteine either begins the methionine cycle again or is permanently removed from the cycle through degradation into cysteine by cystathionine-beta-synthase (CBS) with vitamin B_6 as a cofactor in the transsulfuration cycle.

13.2 FOLATE, B VITAMINS, AND NEURODEVELOPMENTAL DISORDERS

13.2.1 Neural Tube Defects

The essential role of folate in the proper development of the brain and spinal cord was recognized through the observation that the incidence of neural tube defects (NTDs) was notably higher in children of women who were folate deficient during the first trimester of pregnancy (Smithells et al. 1976). This observation was supported by predictions of a dietary role in NTD pathology based on the seasonal patterns observed in NTD prevalence (Elwood 1975; Sandahl 1977; Barry et al. 1983; Fraser et al. 1986), and increased prevalence of NTDs in cohorts of children conceived during periods of famine (Susser et al. 1996). There was also an increased risk of NTDs in women of lower social classes (Fedrick 1976), who tended to have lower nutrient intake in the first trimester of pregnancy (Smithells et al. 1977). Clinical trials in the early 1990s confirmed the ability of folic acid supplementation prior to and early in pregnancy to prevent or rescue 50–70% of NTDs (MRC Vitamin Study Research Group 1991; Czeizel and Dudas 1992). A randomized, double-blind trial conducted by the Medical Research Council (MRC) Vitamin Study Research Group demonstrated that administration of 4000 μg folic acid per day before conception was associated with a 72% reduction in NTD occurrence for women at high risk of having a pregnancy with a neural tube defect, because of a previous affected pregnancy (MRC Vitamin Study Research Group 1991). Another randomized clinical trial demonstrated the ability of multivitamins containing 800 μg folic acid taken for at least 1 month before conception and during the first months of pregnancy to reduce the risk of first-occurrence NTDs (Czeizel and Dudas 1992). Based on the findings of these studies and several retrospective case-control studies (Milunsky et al. 1989; Mulinare et al. 1988; Shaw et al. 1995; Bower and Stanley 1989), recommendations were implemented in the United States for women of childbearing age to consume at least 400 μg of folic acid per day to reduce their risk for an NTD-affected pregnancy (Centers for Disease Control and Prevention 1992). All women of childbearing age were targeted due to the timing of neural tube closure, occurring in the third and fourth week of gestation, when many women are unaware that they are pregnant. Because about half of all pregnancies in the United States are unplanned (Henshaw 1998) and women might not comply with recommendations, the U.S. Food and Drug Administration (FDA) mandated the addition of folic acid to enriched breads, cereals, flours, corn meals, pastas, rice, and other grain products by January 1, 1998 (Daly et al. 1997). Several studies have shown a declining prevalence of NTDs in the United States (Honein et al. 2001; Boulet et al. 2008) after manufacturers were mandated to fortify cereals with folic acid, though some suggest that food fortification levels of folic acid are too low to reduce the risk of NTDs to the fullest extent possible (Oakley 1999; McNulty et al.; Cuskelly et al. 1999). Several other countries have also initiated folic acid fortification, while other countries still debate the issue.

The molecular mechanisms underlying the reduction in risk of NTDs associated with periconceptional folic acid supplementation are unknown (Finnell et al. 2003). The most intuitive possibility is that additional folic acid compensates for insufficient maternal folate levels. However, most mothers of affected fetuses have either normal folate status or are, at most, mildly folate-deficient, arguing against this mechanism for the majority of cases (van der Put et al. 2001; Mills et al. 1996; Scott 1999).

Alternatively, supplemental folic acid may overcome insufficient folate uptake or metabolism. Defects in folate uptake, transport, and metabolism that could result from maternal or fetal genetic variations have been thoroughly investigated as risk factors for NTDs. The first discovered and most studied single-nucleotide polymorphism (SNP) in association with NTDs is the C677T (Ala222Val) SNP in the MTHFR gene that results in a less efficiently functioning "thermolabile" MTHFR enzyme. In 1995, Frosst et al. demonstrated that the MTHFR 677 TT genotype was associated with elevated plasma homocysteine levels, and van der Put et al. (1995) reported increased risk of spina bifida risk associated with both maternal and child MTHFR 677 TT genotypes. Although it was well established that NTDs had genetic contributions based on skewed sex ratios and findings from heritability and twin studies (Bassuk and Kibar 2009), this was the first genetic risk factor identified for spina bifida. Following this study, many additional studies evaluated MTHFR C677T as a risk factor for NTDs. Although findings differed among populations, a meta-analysis observed an overall twofold increase in the risk of having NTD-affected children for mothers with the MTHFR 677 TT genotype (OR 2.0; 95% CI 1.5, 2.8), and an 80% higher risk for children with the MTHFR 677 TT genotype (OR 1.8; 95% CI 1.4, 2.2) (Botto and Yang 2000). It should be pointed out that, at most, 25% of NTDs can be explained exclusively by the MTHFR C677T variant (Posey et al. 1996). Excluding individuals with the MTHFR 677 TT genotype, children with NTDs and their parents still have decreased folate and increased homocysteine levels (van der Put et al. 1997).

As reviewed by van der Linden et al. (2006), several other genetic variants related to the folate cycle and related homocysteine and methionine metabolism have since been studied for their association with altered risk of NTDs. Only one variant other than MTHFR C677T has been somewhat consistently associated with increased NTD risk. Mothers with the MTRR A66G (Ile22Met) variant are associated with about a 48% increased risk of NTDs in their offspring (van der Linden et al. 2006). The substitution of a methionine for an isoleucine at position 22 of the MTRR enzyme results in less efficient repair of the methionine synthase enzyme (Olteanu et al. 2002, 2004). Although an association between the MTRR A66G variant and elevated plasma homocysteine levels has been observed (Gaughan et al. 2001), most studies have not observed an effect of MTRR A66G on homocysteine levels (Jacques et al. 2003; Kluijtmans et al. 2003; Feix et al. 2004; Wilson et al. 1999).

Other gene variants examined in association with NTD risk include: SLC19A1 A80G (rs1051266), maternal methylenetetrahydrofolate dehydrogenase (MTHFD) G1958A (rs2236225) (Brody et al. 2002), thymidylate synthase (TS or TYMS) 5′-untranslated region (UTR) 28 base pair (bp) tandem repeat (2 versus 3 repeat) and 3′-UTR 6 bp deletion (del) (Volcik et al. 2003), DHFR 19 bp del (Johnson et al. 2004), MTR A2756G (rs1805087) (Christensen et al. 1999), and BHMT rs3733890 (Boyles et al. 2006). Although maternal and fetal genetic variations leading to deficiencies in folate uptake, transport, and metabolism likely play a role in NTD etiology, few genetic variations have been found to be independently and consistently associated with NTD risk. The lack of consistent findings across studies evaluating genetic risk factors could result from different allele frequencies, diverse genetic backgrounds, or varied dietary folate or supplemental folic acid intake among populations. Notably, maternal nutrition, primarily supplemental B vitamin intake, has been shown to influence the association between several gene variants and the child's risk for NTDs.

Gene-by-environment interaction effects have been demonstrated for the MTHFR C677T variant such that the TT genotype is more strongly associated with NTD risk in children whose mothers did not consume a periconceptional folic acid supplement (Boyles et al. 2006; Volcik et al. 2003; Shaw et al. 1998) and those with low red blood cell folate (Christensen et al. 1999). Several other candidate genes related to the folate cycle demonstrate an association with increased risk for NTDs in children whose mothers did not consume vitamin supplements. The SLC19A1 A80G variant has been reported as a risk factor for NTDs primarily in children whose mothers did not use folic acid periconceptionally (Shaw et al. 2002; Pei et al. 2005). Conversely, the BHMT rs3733890 variant was significantly associated with NTD risk in children whose mothers were supplemented in a study by Boyles et al. (2006). In the same study, there were associations between CBS and MTR gene variants and NTDs only in children whose mothers did not supplement with folic acid before conception. Another study (Wilson et al. 1999) reported an association between the MTRR 66 GG genotype and NTDs only when plasma vitamin B_{12} concentration was low.

Gene-by-gene interaction effects within folate-related genes also contribute to NTD risk. If both the mother and child of a mother-child pair have gene variants associated with less efficient metabolism, the risk of the child having an NTD is increased. For example, if both the mother and the child possess the MTHFR C677T genotype, estimated risk of an NTD was three times greater than if either the mother or the child had the genotype alone in a study by Christensen et al. (1999). Relton et al. (2004) demonstrated that the combination of a maternal MTRR A66G variant with a child MTHFR C677T variant was associated with higher risk for NTDs than for either variant separately.

Similarly, the more variants associated with less efficient folate metabolism the child or the child's parent has, the higher his/her risk for NTDs tends to be. Multiple variants in the same gene can lead to higher risk through less efficient enzyme activity. For example, the combination of both the C677T and C1298A variants within the MTHFR gene is associated with reduced MTHFR specific activity, higher plasma homocysteine levels (van der Put et al. 1998), lower vitamin B_{12} concentrations (Cunha et al. 2002), and higher risk for NTDs (van der Put et al. 1998; Richter et al. 2001) than for either variant by itself. Notably, if an individual is homozygous for the variant allele at either MTHFR locus, he or she is almost always homozygous for the wild-type allele at the other locus and compound homozygosity for the two MTHFR SNPs appears to be embryonic lethal (van der Put et al. 1998; Yates and Lucock 2003). Variants across multiple genes can also increase risk by blocking compensation by genes in alternate pathways. Several studies have found interaction effects between the MTHFR C677T variant and other candidate gene variants related to one-carbon metabolism. The combination of the child CBS 844ins68 and MTHFR C677T variants are associated with both higher homocysteine levels (Afman et al. 2003) and NTD risk (Relton et al. 2004; Speer et al. 1999; Botto and Mastroiacovo 1998) than each single variant. Interaction effects have also been observed between child MTHFR C677T and BHMT (Boyles et al. 2006), MTR (Boyles et al. 2006), and MTRR (Relton et al. 2004), between child MTRR and GCPII, and between maternal CBS and SLC19A1 (Relton et al. 2004).

Although maternal and child genetic variations leading to deficiencies in folate uptake, transport, and metabolism are risk factors for at least a subset of NTDs, either independently, or in combination with each other, and/or nutritional factors, other disruptions leading to deficiencies in folate uptake and utilization have also been proposed. Rothenberg et al. (2004) reported the presence of autoantibodies directed against the folate receptor in the serum of women whose pregnancy is or was complicated by NTDs. These autoantibodies were shown to block the binding of folic acid to the folate receptors and to inhibit cellular folic acid uptake. Further studies are needed to replicate the finding of this small study, to determine whether this association is causal and to investigate how the folate receptors become self-antigens. One proposed mechanism involves homocysteinylation of the folate receptor (Taparia et al. 2007).

Folic acid might also protect against NTDs through its potentially toxic effects of homocysteine. Studies report that mothers of children with NTDs have increased postnatal homocysteine levels (Steegers-Theunissen et al. 1991, 1994; van der Put et al. 1995, 1997; Mills et al. 1995), and administration of folic acid can reduce homocysteine levels (Brouwer et al. 1999). It has been proposed that homocysteine is in itself a teratogen, and that the protective effect of folic acid works through the reduction of homocysteine levels (Ubbink 1995; van der Put and Blom 2000). Although animal studies have shown inconsistent associations between homocysteine administration and NTDs (Rosenquist et al. 1996; Afman et al. 2003; Bennett et al. 2006), genetic findings seem compatible with this mechanism (van der Put and Blom 2000).

Several lines of evidence are consistent with folic acid's ability to compensate for decreased DNA methylation capacity, as reviewed by Blom et al. (2006). Interestingly, all of the previously proposed mechanisms for the protective effect of folic acid on neural tube development can also be linked to DNA methylation. The genetic variants implicated as risk factors for NTDs, including MTHFR C677T, are involved in methylating homocysteine into methionine, needed to create SAM for carrying out methylation reactions (as discussed later). In addition, one proposed explanation behind folate receptor antibody production could be decreased T cell DNA methylation (Blom 2009), because decreased DNA methylation could result in the overexpression of genes and cause autoreactivity in vitro and autoimmunity in vivo (Richardson 2002, 2003). Finally, high levels of homocysteine negatively impact DNA methylation, as discussed further later. Increased levels or inefficient removal of homocysteine lead to an accumulation of SAH (Hoffman et al. 1980), and excess SAH inhibits DNA methyltransferases (Hoffman et al. 1979; Cox et al. 1977; De Cabo et al. 1995; Yi et al. 2000). This could be especially important for preimplantation embryos as methylation patterns are erased and reprogrammed (Reik et al. 2001), as discussed further later. Because DNA methylation patterns established during early embryogenesis determine tissue-specific gene expression during development, it seems plausible that any number of pathways influencing DNA methylation could alter neurodevelopmental trajectories. Indeed, disrupting de novo DNA methylation by inactivating the DNA methyltransferase DNMT3B results in multiple developmental defects, including NTDs, in mice (Okano et al. 1999).

Another convincing finding in favor of a DNA methylation mechanism for the protection against NTDs by folic acid supplementation is that other B vitamins that are methyl donors or cofactors in the methylation cycle also are associated with decreased risk for NTDs. Many studies have observed that deficient or inadequate maternal vitamin B_{12} status is a risk factor for NTDs independent of folate status (Molloy et al. 2009; Thompson et al. 2009; Zhang et al. 2009; Ray et al. 2007; Gaber et al. 2007; Groenen et al. 2004; Gardiki-Kouidou and Seller 1988; Kirke et al. 1993; Afman et al. 2001), though a few studies have not found evidence for an association (Molloy et al. 1985; Economides et al. 1992; Wald et al. 1996). Recent studies have also shown associations between low maternal dietary and serum levels of choline, and increased risk for NTDs (Shaw et al. 2004, 2009). Higher periconceptional levels of dietary betaine (choline metabolite) and methionine also have shown associations with lower risk of neural tube defects (Shaw et al. 2004). Direct evidence for global DNA hypomethylation has recently been published by Chen et al. (2010), showing significantly reduced global DNA methylation in fetal brain tissue of NTD cases compared to controls, especially for those with the MTHFR 677 TT genotype.

13.2.2 Autism Spectrum Disorders

Studies of maternal folic acid or other B vitamin supplementation and the risk for autism spectrum disorders (ASDs) are nonexistent to date, though several individuals have hypothesized a link. Autism prevalence in the United States has progressively increased during the period since recommendations for periconception folic acid supplements were released and mandatory folic acid fortification of cereal grains was implemented in 1998 (Hertz-Picciotto and Delwiche 2009). Some suggest that folic acid fortification could be responsible for this trend toward increased prevalence of autism (Leeming and Lucock 2009; Rogers 2008; Currenti 2009). This type of time trend analysis is crude, as comparisons over time must consider all other factors. For example, increases in environmental exposures to toxins, changes in patterns of medication use, growing prevalence of obesity and diabetes, or shifts in other factors could be contributing to the rising incidence of autism, which might be superimposed on contributions from changes in diagnostic criteria and practices (Hertz-Picciotto and Delwiche 2009). It is nearly impossible to disentangle the contribution of folic acid fortification to either increases or decreases in autism across this time span from other potential contributors, especially given the current complexity of genetic and environmental factors contributing to autism risk.

Despite the hypothetical nature of the correlation between folic acid fortification and autism prevalence, mechanisms behind such a connection have been speculated. Rogers (2008) suggested that enhanced folate status during pregnancy could have altered natural selection by increasing the survival rates of fetuses possessing genetic polymorphisms associated with hyperhomocysteinemia and subsequently requiring higher levels of folic acid for proper neurodevelopment. A study by Haggarty et al. (2008) failed to find evidence that the frequencies of variants in relevant folate-pathway genes have increased between generations since the initiation of folic acid fortification, as would be predicted under this "rescue" hypothesis. However, as discussed later, researchers have found a higher prevalence of variants in folate-dependent genes in children with autism (Boris et al. 2004; M. Adams et al. 2007; James et al. 2006). Similarly, it is possible that folate fortification has increased survival of fetuses with gene variants in other pathways affected by hyperhomocysteinemia and folate status, possibly including some of the known genetic risk factors for ASD. Finally, for women who do not consume prenatal vitamins periconceptionally, current levels of folic acid fortification may have raised blood folate levels enough to facilitate neural tube closure, but not enough to prevent more subtle anomalies of brain development in certain fetuses that later display neurodevelopmental conditions like autism. Despite few studies on folate and B vitamin supplementation and risk for autism, many studies on folate and methionine metabolites, genes, and treatments in children with autism have been conducted, without definitive results.

This research is reviewed systematically by Main and colleagues (2010), and a brief review follows.

13.2.2.1 Metabolites

Altered metabolic profiles relating to the folate and one-carbon metabolism cycles, including atypical homocysteine levels, suboptimal vitamin B_{12} status, reduced methionine levels, and impaired methylation capacity, have been observed for children with autism and their parents compared to children who are typically developed and their parents (James et al. 2004, 2006, 2008; Pasca et al. 2006; Wakefield et al. 1998). One small study has examined serum and red blood cell folate levels in children with autism compared to typically developed children (M. Adams et al. 2007). Although no significant differences were found, levels of red cell folate were higher in children with autism. To our knowledge, no studies have evaluated maternal folate status during pregnancy and risk for autism. No study to date has examined choline blood or intake levels in children with autism compared to typically developing children, but one study has reported lower levels of choline-containing metabolites in the left side of the thalamus in children with autism (Hardan et al. 2008). Pasca et al. (2006) found increased homocysteine levels in children with autism compared to typically developed controls, while studies by James et al. found reduced homocysteine levels for children with autism (2004), but increased homocysteine levels for mothers of children with autism (2008). Wakefield et al. (1998) reported raised urinary concentrations of methylmalonic-acid (MMA), a marker of functional B_{12} deficiency, in all eight of the children with a pervasive developmental disorder (PDD) examined compared with age-matched controls. In another study, by Pasca et al. (2006), suboptimal or low levels of plasma vitamin B_{12} were found for the majority (9 of 12) of children with autism. In contrast, findings from a small study found non-significantly higher mean levels of vitamin B_{12} for children with autism compared to typical controls (M. Adams et al. 2007). Individuals with autism have shown significantly higher total vitamin B_6, and lower pyridoxal kinase activity compared to typically developing (TD) children, suggesting that low conversion of pyridoxal and pyridoxine to pyridoxal 5 phosphate (PLP) results in low levels of PLP, which is the active cofactor for 113 known enzymatic reactions, including the formation of many key neurotransmitters (J. B. Adams et al. 2006). No peer-reviewed studies have measured the active form of vitamin B_6 (PLP) in children with autism. Although these studies have several limitations including small sample sizes and some inconsistent findings, and the retrospective nature of these studies precludes determining whether these differences are causal, they present some evidence for altered one-carbon metabolism and methylation capacity in families of children with autism.

13.2.2.2 Folate Pathway Genes Implicated in Autism

Several studies have identified some of the same folate- and vitamin B_{12}-dependent gene variants associated with NTDs (Whitehead et al. 1995; Ou et al. 1996; van der Put et al. 1998; De Marco et al. 2001; Johnson et al. 2004) as risk factors for autism, including variants in MTHFR, DHFR, SLC19A1, MTRR, and TCN2 in addition to variants of other genes connected to the one-carbon metabolism cycle, including glutathione-S-transferase M1 (GSTM1) and COMT (Boris et al. 2004; M. Adams et al.

2007; James et al. 2006). The study that investigated a 19 base-pair deletion within DHFR was small (M. Adams et al. 2007). Only one study examined associations with parental genotypes, which could be as or more pertinent to fetal health than infant genotypes because fetal homocysteine levels appear to be regulated maternally rather than intrinsically by the fetus (Molloy et al. 2002). In this study, James et al (2010) found an association between the maternal, but not the child SLC19A (reduced folate carrier) A80G variant and autism, suggesting that more studies on maternal genotypes are warranted. Despite their limitations, these studies suggest that families affected by autism and ASD might be genetically predisposed to less efficient folate metabolism and function. As described previously, studies of NTDs have shown significant effect modification between folic acid intake and common genetic variants in folate-dependent pathways, such that increased folic acid intake can resolve differences in risk across genotypes (Christensen et al. 1999). The intake of folic acid and other vitamins was not collected or examined as an effect modifier of genotypes in previous studies of autism. An additional study has identified correlations between the MTHFR C677T polymorphism and the frequency and severity of specific behaviors in children with autism (Goin-Kochel et al. 2009).

13.2.2.3 Uptake to the Central Nervous System

Impaired folate delivery to the central nervous system and maternal autoantibodies to folate receptors during pregnancy are potentially risk factors for ASDs (Ramaekers et al. 2007). Several case reports and small studies have shown that a subset of children with developmental delay, regression, seizures, and autistic symptoms suffer from cerebral folate deficiency, regardless of blood levels of folate in peripheral tissues (Moretti et al. 2005, 2008; Ramaekers et al. 2007). One of these studies showed that serum folate receptor autoantibodies might block the folate binding site on receptors and impede transport across the blood-CSF (cerebrospinal fluid) barrier (Ramaekers et al. 2007). Treatment with folinic acid in these children corrected CSF abnormalities and improved motor symptoms, but cognitive delays remained (Moretti et al. 2005; Ramaekers et al. 2007). One can speculate that if this subset of children had received supplemental folic acid during early neurodevelopment, there might have been a greater recovery or even prevention of developmental symptoms.

13.2.2.4 Treatment

Studies have also investigated the effects of interventions involving folic acid supplementation on symptoms of ASDs, but the evidence for folic acid supplementation as treatment for those symptoms is insufficient. Interventions including the administration of folinic acid or folic acid-containing supplements have been reported to help normalize metabolites and improve symptoms of ASD in some instances (Moretti et al. 2005; James et al. 2004, 2009; Ramaekers et al. 2007), but these studies were small and did not include rigorous measures of symptom improvement. Importantly, folic acid could potentially alleviate most of the differences described earlier; supplemental folic acid is able to resolve differences in risk across folate-related genotypes (Christensen et al. 1999), reduce homocysteine levels (Daly et al. 2002), increase global DNA methylation levels (Pufulete et al.

2005; Dolinoy et al. 2007), and overcome problems of central nervous system uptake. These preliminary findings have implications for folic acid as a potential treatment in children with ASDs, but as findings for NTDs suggest, there could exist a critical window during periconceptional development when folate supplementation could prevent the occurrence of ASDs.

Vitamin B_6 has also been used to treat children with autism, sometimes in combination with magnesium and/or other vitamins. Research on the treatment of children with autism with vitamin B_6 began in the 1960s with most studies reporting positive findings, as reviewed by Rimland (1988) and Pfeiffer et al. (1995). However, as reported by Pfeiffer et al., many of these studies had methodological limitations such as a small number of participants, no long-term follow-up, and lack of precise outcome measures. These studies included those by Rimland et al. (1973, 1974) that deemed vitamin B_6 the most effective of several vitamin treatments among over 200 children with autism after a 4-month trial for autism, and that reported positive findings for a double-blind placebo-controlled crossover experiment of 16 children with autism who were responsive to vitamin B_6 treatment in the first study (Rimland et al. 1978). Other researchers followed up these initial findings with studies of behavior and evoked potentials in addition to metabolite measures, and most reported improvement in at least a portion of the children with autism (Barthelemy et al., 1981, 1983; Jonas et al. 1984; Lelord et al. 1981; Martineau et al. 1985, 1986). In a survey of parents of children with autism, vitamin B_6 (combined with magnesium) was the biomedical treatment with the highest rating for efficacy and safety (Rimland 1988). However, a more recent 10-week double-blind, placebo-controlled trial found that an average dose of 638.9 mg of pyridoxine and 216.3 mg of magnesium oxide was ineffective in ameliorating autistic behaviors as assessed by the Children's Psychiatric Rating Scale (CPRS), the Clinical Global Impression Scale, and the NIMH Global Obsessive Compulsive Scale (Findling et al. 1997).

13.2.2.5 Rett Syndrome

Rett syndrome is an ASD primarily affecting females due to X-linked dominant mutation in the MECP2 gene (Amir et al. 1999). The protein product of MECP2 binds to methylated DNA and is an essential transcriptional modulator for postnatal neuronal maturation. Notably, some children with Rett syndrome have low CSF levels of the biologically active form of folate (cerebral folate deficiency) (Ormazabal et al. 2005; Temudo et al. 2009; Ramaekers et al. 2003). Treatment of Rett patients with folinic acid normalized CSF levels of the active folate metabolite, and was more effective for recovery of clinical symptoms if it was delivered early in development, before 6 years of age (Ramaekers and Blau 2004). A 12-month double-blind placebo-controlled trail of folate and betaine supplementation of Rett patients did not find significant improvement in objective measures, regardless of age group, although subjective improvement was reported by parents with children less than 5 years old (Glaze et al. 2009). Another study showed no clinical improvement for the 32% of Rett patients that were found to have low CSF folate levels and were subsequently treated with folinic acid (Temudo et al. 2009). Choline supplementation resulted in improved motor coordination, striatal nerve growth factor expression, and N-acetyl aspartate levels in a mouse model of Rett syndrome (Nag and Berger-Sweeney 2007;

Nag et al. 2008; Ward et al. 2009), but choline supplementation has not yet been evaluated in human Rett patients. Interestingly, MTHFR deficiency in humans can clinically mimic Rett, with overlapping symptoms including mental retardation, the presence of seizures, ataxia, and absent speech (Arn et al. 1998).

13.2.2.6 Fragile X Syndrome

Fragile X syndrome is another X-linked ASD caused by triplet expansion of CGG in the promoter of FMR1. DNA methylation of the full expansion mutation results in silencing of FMR1, encoding a protein required for translation at neuronal synapses (Oostra and Willemsen 2009). Folic acid has been used to treat children with fragile X syndrome (Rosenblatt et al. 1985; Strom et al. 1992; Madison et al. 1986; Lejeune et al. 1984; Hagerman et al. 1986; Gillberg et al. 1986; Fisch et al. 1988; W. T. Brown et al. 1986; Gustavson et al. 1985; Froster-Iskenius et al. 1986). Although study results are mixed and tend to not report significant benefits (Rueda et al. 2009), evidence indicates that fragile X-affected children might be more likely to benefit from folic acid treatment when it is administered early in development (prepuberty) (Froster-Iskenius et al. 1986; Greenblatt et al. 1994; Hagerman et al. 1986). Studies have also found evidence for cholinergic dysfunction in fragile X syndrome (D'Antuono et al. 2003), with significantly reduced levels of choline in the right prefrontal cortex compared to controls (Kesler et al. 2009). In controls, left and right prefrontal cortex choline was positively correlated with intelligence measures (Kesler et al. 2009).

13.2.3 Schizophrenia

There is now general acceptance that schizophrenia should be considered a neurodevelopmental disorder. Despite the late onset of schizophrenia in young adulthood, evidence points to prenatal or perinatal origins and abnormalities of brain function much earlier in life (Lewis and Levitt 2002; Rapoport et al. 2005). Evidence for effects of maternal nutrition also fit a neurodevelopmental model for schizophrenia. Susser et al. (1996) observed that individuals conceived at the height of the Dutch Hunger Winter of 1944–45 showed a twofold increase in the risk for schizophrenia in both male and female offspring. A similar study found that schizophrenia was increased for offspring exposed prenatally to the Chinese famine of 1959–61 (St. Clair et al. 2005). However, it remains unclear whether factors other than nutrition during periods of famine could have predisposed offspring to schizophrenia. A more recent study has compared prenatal maternal serum homocysteine levels from individuals diagnosed with schizophrenia and age- and sex-matched controls without any major affective disorders and found elevated third-trimester homocysteine levels for mothers of children who later developed schizophrenia (A. S. Brown et al. 2007). Elevated homocysteine levels have also been described for individuals with schizophrenia, with and without low plasma folate and vitamin B_{12} levels, and tend to correlate positively with negative symptom scores (Petronijevic et al. 2008; Feng et al. 2009; Muntjewerff et al. 2003; Goff et al. 2004; Regland Jet al. 1994; Kemperman et al. 2006). In addition, plasma levels of the active form of vitamin B_6 are lower in schizophrenic patients with movement disorders such as tardive dyskinesia (Miodownik et al. 2008).

13.2.3.1 Genes

Studies have identified gene variants associated with the folate-dependent one-carbon metabolism/methylation cycle as risk factors for schizophrenia. MTHFR C677T has been associated with increased risk of developing schizophrenia in several studies (Mavros et al. 2008; Kempisty et al. 2006; Feng et al. 2009; Muntjewerff et al. 2005; Sazci et al. 2003; Arinami et al. 1997). Other studies have not found evidence for an association between increased risk of schizophrenia and either maternal or patient MTHFR C677T (Muntjewerff et al. 2007; Philibert et al. 2006; Vilella et al. 2005; Virgos et al. 1999; Yu et al. 2004; Jonsson et al. 2008). Meta-analyses of maternal MTHFR C677T and schizophrenia risk show a small pooled effect, if any (Zintzaras 2006; Muntjewerff et al. 2006). MTHFR C677T has been associated with earlier age at onset (Vares et al. 2009), more negative symptoms (Roffman, Weiss, Purcell et al. 2008), and deficits in executive function (Roffman et al. 2007).

COMT Val158Met has also been identified as a genetic risk factor for schizophrenia (Egan et al. 2001), but a recent meta-analysis failed to find evidence for a significant contribution of COMT val158met to schizophrenia susceptibility (Okochi et al. 2009). One study suggests that both the Val allele and hypomethylation of the membrane-bound COMT promoter region associated with COMT overexpression contribute to schizophrenia pathology (Abdolmaleky et al. 2006). Interaction effects between the MTHFR C677T and the COMT variants have also been described as risk factors for susceptibility to schizophrenia (Muntjewerff et al. 2008) and for poor prefrontal executive function in schizophrenia (Roffman, Gollub et al. 2008; Roffman, Weiss, Deckersbach et al. 2008). Interestingly, COMT is located on chromosome 22q11.2, a common region of deletion and duplication associated with autism and mental retardation (Portnoi 2009).

13.2.3.2 Treatment

In addition to finding low levels of folate and vitamin B_{12} and high levels of homocysteine, some studies in psychiatric patients (including those with schizophrenia) showed improvements after administration of methylfolate (Godfrey et al. 1990; Procter 1991) or a combination of vitamin B_6, vitamin B_{12}, and folic acid (Levine et al. 2006). Although the research is limited, it has reinforced a hypothesis that methylation deficiencies contribute to schizophrenia pathology (Smythies et al. 1997; Regland et al. 1994). Preliminary trials of vitamin B_6 supplementation have been initiated in schizophrenic patients, both alone and in combination with existing treatments, with limited success at reducing certain motor symptoms (Miodownik et al. 2006, 2007; Lerner et al. 2004; Lerner et al. 2002, 2007). Individuals with schizophrenia often present with lower activity of choline acetyltransferase, the enzyme synthesized within neurons that forms the neurotransmitter acetylcholine from acetyl-CoA and choline (Bird et al. 1977; Karson et al. 1993, 1996). Consequently, choline chloride has been tried as a treatment for schizophrenia, but with no significant effects (Davis et al. 1979).

13.2.4 Down Syndrome (Trisomy 21)

Down syndrome is a genetic disorder that results from trisomy of chromosome 21. Down syndrome is the most common genetic cause of human intellectual disability, with a prevalence of 1 in 600 live births (Hobbs et al. 2000). In the majority of Down syndrome cases, the extra chromosome is a result of a failure of maternal chromosomal segregation (nondisjunction) during meiosis I in the maturing oocyte, prior to conception (Antonarakis et al. 1992). In 1999, James and colleagues (1999) found that mothers of children with Down syndrome had significantly higher plasma homocysteine levels and were more likely to possess the MTHFR 677T allele. More than 25 studies within various populations have gone on to investigate whether genetic polymorphisms within one-carbon metabolism pathways increase a mother's risk of delivering a child with Down syndrome. As reviewed by Patterson (2008), subsequent findings have been inconclusive as to whether there are differences between homocysteine levels and the prevalence of MTHFR alleles between mothers of children with or without Down syndrome. Some researchers report increased risk for having a child with Down syndrome in mothers with combinations of more than one variant within the folate metabolism pathway (Brandalize et al. 2009; Fintelman-Rodrigues et al. 2009; Hobbs et al. 2000; O'Leary et al. 2002; Grillo et al. 2002; Acacio et al. 2005; Coppede et al. 2006; Scala et al. 2006), and nutritional backgrounds might play a role. Additional studies have examined how alterations in one-carbon metabolism, including those due to extra copies of several genes known to be involved in folate metabolism located on chromosome 21, affect the development of intellectual disabilities in children with Down syndrome. One such study by Gueant et al. (2005) demonstrated correlations between the highest quartile of total homocysteine, the MTHFR 677 T and TCN2 776 G alleles, and low IQ in individuals with Down syndrome. Future investigations with mouse models of Down syndrome and folate deficiency as well as experiments designed to investigate potential epigenetic mechanisms of trisomy risk appear to be needed to sort out the important unresolved understanding of folate metabolism and risk for Down syndrome.

13.2.5 Other Neurodevelopmental Outcomes

13.2.5.1 Cognitive Functioning

Rodent studies show that the effects on hippocampal progenitor cells from choline supplementation during days 11–18 of gestation and postnatal days 16–30 (discussed earlier in Section 13.1.1.3) have substantial and irreversible changes in hippocampal function in the adult rodent, including altered long-term potentiation (J. P. Jones et al. 1999; Montoya et al. 2000; Pyapali et al. 1998) and visuospatial and auditory memory (Meck and Williams 1997a, 1997b, 1997c, 1999, 2003; Meck et al. 1988; Williams et al. 1998). These effects have not been tested in humans, though at least one pilot study is under way (Zeisel 2006).

Craciunescu et al. (2003) followed up the choline studies with a study of folic acid and observed that offspring of pregnant rats fed a folic acid-deficient diet during gestation had a reduction of 54% of progenitor cells in the fetal neocortex, 47% in the septum, and 43% in the caudate and putamen. They did not find differences between pups of folic acid-supplemented mice and pups of control mice. Though the neocortex is the part of the brain responsible for complex behaviors such as cognition, attention, and social competence, behavioral deficits were not tested.

In humans, case studies have found that inadequate B vitamin intake, uptake, or metabolism increases risk for deficits in cognitive functioning (Tallur et al. 2005; Gueant et al. 2005; Black 2003, 2008; Trimble et al. 1980). An early study demonstrated delayed maturation patterns in brain function for infants breast-fed by folate-deficient mothers compared to folate replete (6+ mγ/ml) mothers (Arakawa, Mizuno, Honda et al. 1969), which corroborated their previous study of treatment with a folic acid analog in rats (Arakawa, Mizuno, Sakai et al. 1969). Mechanisms proposed for the association between low vitamin B_{12} status during early development and reduced cognitive ability include disruptions to myelination (Black 2008). The functional CBS 844ins68 allele related to homocysteine metabolism has been linked to cognitive function in children, while variants in other genes, including MTHFR and MTR, have not (Barbaux et al. 2000). A couple of small studies showed that administration of folic acid-containing supplements to anemic children increased cognitive performance, though the effects could have been attributed to the iron contained in the supplements (Seshadri et al. 1982; Seshadri and Gopaldas 1989). There is limited research examining effects of maternal perinatal intake of folic acid and other B vitamins on cognitive development and intelligence, but most studies to date have reported a lack of correlations. A randomized clinical trial of multivitamin supplementation during pregnancy did not find significantly improved neurodevelopment in children at 2 years and 6 years of age, but found nonsignificantly higher IQ scores among 6-year-old children of mothers with periconceptional intake of multivitamins containing 800 μg folic acid and other B vitamins (Dobo and Czeizel 1998).

Another study found that folate status of mothers during pregnancy (after the first trimester) was not associated with later mental and psychomotor development of their children at 5 years of age (Tamura et al. 2005). Bhate and colleagues (2008) found that maternal plasma vitamin B_{12} concentration at 28 weeks gestation was associated with attention and short-term memory tasks, but not intelligence of the child at 9 years of age. In addition, Bhate and colleagues found that maternal concentrations of total homocysteine, methylmalonic acid, and folate were not associated with the child's cognitive performance. A study of gestational maternal serum at four gestational age intervals (16–18 wk, 24–26 wk, 30–32 wk, and 36–38 wk) and cord blood concentrations of free and total choline found no associations with overall or selected scales of childhood intelligence at age 5.5±0.5 years (Signore et al. 2008). Although these studies have not demonstrated consistent evidence for the importance of maternal and fetal B vitamin status for intelligence, these studies did not examine B vitamin levels during the very earliest stages of brain development in the first trimester of pregnancy when levels could be the most critical for neurodevelopment.

13.2.5.2 Behaviors

Animal studies have reported impaired learning of avoidance behaviors associated with folate deficiency (Bachevalier and Botez 1978), diminished exploratory behaviors with perinatal vitamin B_6 deficiency (Krishna and Ramakrishna 1984), and the preservation of exploratory behaviors and certain features of hippocampal plasticity by prenatal choline supplementation that last long into adulthood (Glenn et al. 2008;

Wong-Goodrich et al. 2008). Several studies in humans have reported findings on behavioral outcomes as they relate to B vitamin intake, most of which focused on folic acid. A recent population-based cohort study has found evidence for a higher risk of behavioral problems in the offspring of mothers with inadequate use of folic acid supplements during early pregnancy (before and during the first 10 weeks) (Roza et al. 2009). These behaviors, assessed using the Child Behavior Checklist at 18 months, included both internalizing and externalizing problems. Another recent population-based cohort study supplements these findings with improved neurodevelopment of children at 4 years of age associated with maternal use of folic acid supplements during early pregnancy (first trimester), including higher verbal (b = 3.98, SE = 1.69), motor (b = 4.54, SE = 1.66), and verbal-executive function (b = 3.97, SE = 1.68) scores, higher social competence ratings (b = 3.97, SE = 1.61), and fewer inattention symptoms (OR = 0.46; 95% CI 0.22, 0.95) (Julvez et al. 2009). Psychological outcomes were assessed by two psychologists and teachers, and associations were adjusted for a number of sociodemographic and behavioral factors. Another recent prospective cohort study by Schlotz et al. (2010) provides further evidence for reduced folate status in early pregnancy impairing fetal brain development and affecting hyperactivity/inattention and peer problems in childhood.

Earlier studies had results that are not easily interpretable. A study by Wehby and Murray (2008) reported improved gross-motor development and overall development, but poorer performance for the personal-social domain in 3-year-old children of mothers with periconceptional folic acid use (at least 3 days per week, 3 months before and/or after conception).

A double-blind study found that vitamin B_6 levels in the mothers' milk were associated with neonatal behaviors and maternal-infant interactions (McCullough et al. 1990). Finally, a large population-based study found that higher choline concentrations were associated with fewer anxiety symptoms in older adults (Bjelland et al. 2009).

13.3 ROLE OF FOLATE AND OTHER B VITAMINS IN METHYLATION

13.3.1 Relationship between Folate Levels and DNA Methylation Status

As described previously, the folate cycle and the methylation cycle are intimately connected. Importantly, early nutrition is able to influence DNA methylation, as dietary methyl donors and cofactors are needed for one-carbon metabolism, which provides the methyl groups for all biological methylation reactions (van den Veyver 2002). Dietary methyl donors include methionine and choline in addition to folate, and vitamin B_{12} is a critical cofactor in methyl metabolism (Waterland and Jirtle 2004). Low levels of folate and vitamin B_{12} lead to decreased methylation reactions of proteins, phospholipids, DNA, and neurotransmitters (Moretti et al. 2004).

In the mouse, the best example of an effect on dietary folate change on a specific phenotype through DNA methylation changes is at the IAP retrotransposon in the agouti viable yellow (A^{vy}) mouse model (Morgan et al. 1999). Increasing dietary methyl donors through perinatal maternal supplementation of folic acid, vitamin B_{12}, choline, and betaine reduced the frequency of pups born with the yellow obese phenotype through hypermethylation of the IAP allele (Waterland and Jirtle 2003). In addition, increasing maternal dietary methyl donors could negate the effect of hypomethylation associated with environmental exposure to bisphenol A (Dolinoy et al. 2007).

In humans, the first examples of the effects of maternal nutrition on DNA methylation patterns come from studies of humans who had been exposed to famine preconceptually during the Dutch Hunger Winter in 1944–1945 (Heijmans et al. 2008). There was a specific association between periconceptional exposure to famine and reduced methylation of the insulinlike growth factor 2 gene (IGF2) promoter, and a later study found altered DNA methylation patterns in an additional six genes, with some of the loci being specific to male offspring (Tobi et al. 2009). A recent study demonstrated a 4.5% higher methylation of the IGF2 differentially methylated region (DMR) for children of mothers who used periconceptional folic acid (400+ μg/day) than children of mothers who did not, providing the first direct evidence that maternal periconceptional folic acid use relates to DNA methylation in the child (Steegers-Theunissen et al. 2009). IGF2 DMR methylation was also inversely associated with birth weight, implying that developmental growth, birth outcome, and perhaps long-term health could be affected by these changes. These studies suggest that nutrition effects of DNA methylation patterns in the genome are widespread and specific in timing to the periconception period.

Human genetic variation at the MTHFR locus has also been shown to influence DNA methylation patterns. Individuals with a MTHFR 677 T/T genotype had diminished levels of global DNA methylation compared to those with the C/C genotype, but only the T/T subjects with low serum folate levels accounted for the reduced DNA methylation (Friso et al. 2002). Another study took a different approach of examining by an enzymatic assay the capacity of DNA to accept methyl groups and also found evidence in favor of global DNA hypomethylation associated with the MTHFR 677 T/T genotype and low folate (Stern et al. 2000). These combined results suggest that individuals with the risk allele at MTHFR and low dietary folate have detectable deficiencies in global measures of DNA methylation.

13.3.2 Overlapping Timing of Neurodevelopmental Need for Folic Acid and DNA Methylation in the Periconceptional Period

As evidence grows in support of a link between dietary methyl donors and detectable changes to DNA methylation pattern, the critical question is why is the perinatal period of human development the critical window for protection by folate and other B vitamins? Studies investigating the reprogramming of DNA methylation patterns during mouse embryonic development have shed light on the global waves of demethylation and remethylation occurring with the first week of embryonic life (reviewed in Guibert et al. 2009). Immediately after fertilization of the zygote, the paternal genome is actively demethylated, while the maternal genome is passively and more gradually demethylated, resulting in a diminished methylation of the embryo by the preimplantation blastocyst stage (Santos et al. 2002). After implantation, DNA methylation patterns are reset in the developing embryo and restored to the global DNA methylation levels observed in adulthood. The DNA methyltransferases DNMT3a and DNMT3b are primarily responsible for the de novo methylation marks occurring in the early embryo, while the maintenance DNA methyltransferase DNMT1 is responsible for maintaining DNA methylation patterns following each somatic cell division in the embryo and adult.

In addition to CpG methylation, which is the primary and virtually exclusive site of methylation in somatic tissues of mammals, non-CpG methylation at CpA or CpT sites is present at high levels in embryonic stem cells and likely mediated by DNMT3a (Ramsahoye et al. 2000). The specific heavy non-CpG methylation of human embryonic stem cells was especially apparent in the first description of the human DNA methylome at base resolution compared to that between a pluripotent embryonic stem cell line and a fetal fibroblast line (Lister et al. 2009). The pluripotent embryonic stem cells had a greater total methylcytosine count than differentiated fetal fibroblasts largely because 25% of the methylC sites were non-CpG compared to 0.02% in the somatic cells. Remarkably, non-CpG methylation was primarily found in gene bodies and appeared to be an important mark of pluripotency because it was restored in the fibroblasts induced to pluripotency in culture. In addition, there were multiple regions showing differentiation-specific changes in CpG methylation patterns between embryonic stem cells and fetal fibroblast, providing the first large-scale glimpse of dynamic epigenetic changes in DNA methylation patterns occurring in the preimplantation stages of human development.

Together, these studies suggest that the mammalian periconceptional period is characterized as having the most critical need for dietary methyl donors for DNA methylation compared to any other stage of development. Pluripotent stem cells of the preimplantation blastocyst appear to have additional sites requiring DNA methylation at non-CpG sites and overall increased global methylC. The global wave of demethylation and reestablishment of DNA methylation patterns that occurs around implantation is likely another critical periconception period where dietary methyl donors are likely to be critical. A limiting dietary supply of methyl donors in a genetically susceptible individual (MTHFR 677 T/T) could impact the epigenetic switch between pluripotency and cellular identity as well as the establishment of DNA methylation patterns for tissue differentiation, X chromosome inactivation, and parental imprinting.

13.3.3 B Vitamins and Methylation of Products Other Than DNA

Exclusive of DNA methylation, methylation of other products could also play a role in neurodevelopmental outcomes, and can be influenced by B vitamin levels. SAM-dependent methyltransferase activity is essential for hundreds of other cellular methylation reactions, including methylation of phospholipids, proteins, RNA, and other molecules, many of which are potentially relevant for neurodevelopment and nervous system functioning (Chiang et al. 1996). Of particular relevance is the involvement of SAM as a cofactor in methylation reactions in catecholamine synthesis and metabolism (A. J. Turner 1977), and requirement of 5-methylhydrofolate, vitamin B_{12}, and SAM for biosynthesis and maintenance of adequate amounts of tetrahydrobiopterin (Hamon et al. 1986), a key cofactor in the synthesis of serotonin and the catecholamine neurotransmitters (A. J. Turner 1977). Serotonin deficits during synaptogenesis lead to decreased synaptic density and learning deficits in adult rats (Mazer et al. 1997), and evidence suggests that early abnormalities in the stimulation of dopamine or serotonin receptor subtypes during brain development can lead to the types of neuroanatomical changes observed in individuals with schizophrenia, bipolar affective disorder, and autism (Todd 1992). As reviewed by Yi et al. (2000), SAH inhibits cellular methyltransferases including COMT (Schatz et al. 1981), phosphatidylethanolamine methyltransferase (Hoffman et al. 1981), histone methyltransferase (Hoffman et al. 1979), tRNA and mRNA methyltransferases (Glick et al. 1975; Pugh and Borchardt 1982), acetylserotonin methyltransferase (Deguchi and Barchas 1971), and histamine N-methyltransferase (Schatz et al. 1981; Borchardt et al. 1978), which in addition to reduced neurotransmitter synthesis (Deguchi and Barchas 1971; Schatz et al. 1981) can lead to central nervous system demyelination (Scott et al. 1994; Molloy et al. 1992), decreased chemotaxis and macrophage phagocytosis (Garcia-Castro et al. 1983; Leonard et al. 1978), altered membrane phospholipid composition (Chiang et al. 1980), and cell differentiation (Chiang 1981; Aarbakke et al. 1986). All of these potential functional consequences of SAH inhibition of methyltransferases could put a developing nervous system at increased risk for abnormal development. Hypomethylation specific for proteins located in neural tubes that failed to close and the ability to rescue neural tube closure with the addition of methionine suggests that protein methylation also might play a role in proper neural tube closure (Moephuli et al. 1997; Coelho and Klein 1990).

Histone methylation is another essential epigenetic mark of the developing embryo for the specification and maintenance of cellular identity. The histone H3 subunit has multiple sites of lysine methylation that can be either activating (H3K4me3) or repressing (H3K27me3) to the DNA sequences coiled by the modified nucleosomes (reviewed in Hublitz et al (2009). Mitotic cell divisions as well as tissue differentiation occurring in the early embryo undergo dynamic changes in histone methylation patterns, so the need for dietary methyl donors for the establishment and maintenance of histone methylation patterns is expected to be greatest in the early embryo.

13.4 EPIGENETICS AND NEURODEVELOPMENTAL DISORDERS

Epigenetic mechanisms, including DNA methylation, create pathways through which the environment can make its mark on the human genome. DNA methylation (addition of a carbon group) occurs at specific cytosine-phosphorous-guanosine (5′-CpG-3′ sites) sites (Holliday and Grigg 1993), and in mammals, about 60–80% of CpG sites in DNA are methylated, except for within CpG islands, where most CpGs are not methylated (Jeltsch 2002). This addition of a methyl group influences multiple cellular events, including gene transcription, genomic imprinting, X chromosome inactivation, and genomic stability (Jaenisch 1997; P. A. Jones and Gonzalgo 1997; Robertson and Wolffe 2000). Therefore DNA methylation analysis is an essential addition to a repertoire for understanding genetic and environmental risk factors in the etiology of neurodevelopmental disorders.

The involvement of epigenetic regulatory mechanisms in the pathway of ASDs has been implicated by the co-occurrence of ASD in disorders such as fragile X syndrome (FXS), Rett syndrome (RTT), Angelman syndrome (AS), and Prader-Willi syndrome (PWS), all of which have known epigenetic contributions (Samaco et al. 2005; Schanen 2006; Hogart et al. 2007). FXS arises through a combination of genetic and epigenetic mutations, where the expansion of a CGG repeat in the 5′-untranslated region of the FMR1 gene renders the region susceptible to methylation and epigenetic silencing, resulting in the loss of expression of the gene (Hagerman et al. 2005). Rett syndrome is a complex neurological disorder included among the pervasive developmental disorders in the *DSM-IV* (*Diagnostic and Statistical Manual of Mental Disorders,* 4th edition) that arises from mutations in the X-linked gene MECP2, encoding methyl-CpG-binding protein 2 (MeCP2) (Amir et al. 1999). MeCP2 is a key mediator of epigenetic regulation of gene expression, by binding with methylated cytosine residues and interacting with chromatin remodeling complexes to modulate gene expression. In addition to females with RTT, males with duplication of MECP2 have an increased risk of ASDs (Meins et al. 2005; Friez et al. 2006). Moreover, aberrant DNA methylation at MECP2 in male autism brain samples correlated with reduced MeCP2 in postmortem brain (Nagarajan et al. 2008). The X chromosome is probably the most vulnerable sector of the male genome, and mutations of brain-expressed genes on the X chromosome often cause mental retardation. This vulnerability is also true for epigenetic influences of the X chromosome, and thus the epigenetic alterations in X chromosome genes such as FMR1 and MECP2 are plausible mechanisms for males' increased vulnerability to ASDs. X chromosome DNA methylation patterns may depend on maternal capacity for methylation or demethylation, during precise developmental stages of XCI erasure and reestablishment.

Also relevant is an observation made by Schanen (2006) that many gene loci linked to autism are near or overlap with regions that are subject to genomic imprinting, with the most common linkage to chromosome 15q11-13. Several studies have found an association between autism and duplication of 15q11-13, a region subject to methylation-dependent genomic imprinting (Thatcher et al. 2005). The regulatory region of the small nuclear riboprotein N gene (SNRPN) is an important imprinting control region (ICR) for establishing and maintaining imprinting at the 15q11-13 locus associated with Prader-Willi and Angelman syndromes. PWS is caused by paternal deletion of 15q11-13, maternal disomy, or small paternal deletions in the ICR. In contrast, AS is caused by maternal 15q11-13 deletion, paternal disomy, maternal UBE3A mutations, or lack of maternal methylation at the ICR (Carrel et al. 1999). Imprinted genes within 15q11-13 include UBE3A, encoding a type 3 ubiquitin ligase, that is mutated in AS and shows reduced expression in RTT. In addition, three GABA receptor subunit genes (GABRB3, GABRA5, GABRG3) are located in this region, and GABAergic inhibition is suggested to predispose individuals to autism. Reduced expression levels of UBE3A and GABRB3 (encoding for the β3 subunit of the $GABA_A$ receptor) in Rett, Angelman, and autism brains are also associated with MeCP2 deficiency (Samaco et al. 2005). Hogart et al. (2007) showed that epigenetic dysregulation was the mechanism for the reduced expression of GABRB3 in neurons of ASD brains, or more specifically, by MeCP2 binding to methylated CpG sites within GABRB3. Decreased function of another methyl-CpG binding protein, MBD1, is also associated with autismlike behavioral phenotypes, potentially through the elevated expression of Htr2c, a serotonin receptor (Allan et al. 2008).

Other candidate genes for autism, including reelin (RELN), MET, and ROAA genes, are either epigenetically regulated or alter transcription factor binding (Campbell et al. 2006; Schanen 2006; Nguyen et al. 2010).

Most evidence for involvement of epigenetic mechanisms and studies of methylation in the pathway of autism has focused on specific genetic loci of imprinted genes, with findings for aberrant methylation. In a case-control study that included measurement of global DNA methylation levels, James and colleagues (2008) found evidence for genomewide DNA hypomethylaton in a subset of parents of children with autism with elevated SAH. However, because their measure for global methylation might have underestimated levels of DNA methylation at 6–7% compared to 30–90% expected from other studies (Tucker 2001; S. E. Brown et al. 2007), the level of differences between groups might have been obscured. Thus a more encompassing measurement of global DNA methylation in mothers and children in families who are and are not affected by autism is warranted. Epigenetic mechanisms have also been implicated for schizophrenia, as summarized in recent reviews (Roth et al. 2009; Akbarian 2009). Schizophrenia cortex is characterized by reduced expression due to increased promoter methylation of RELN and GAD67, encoding proteins essential for GABAergic neurons (Grayson et al. 2005; Huang and Akbarian 2007). Histone methylation modifications have also been observed in schizophrenia (Akbarian et al. 2005; Stadler et al. 2005), further implicating a potential epigenetic etiology of this common neuropsychiatric disorder. The evidence for maternal nutrition effects in the etiology of schizophrenia is also consistent with epigenetic involvement in schizophrenia development.

13.5 CHAPTER SUMMARY AND CONCLUSIONS

Folate, vitamin B_{12}, and choline are essential dietary ingredients for the transmethylation pathway that provide methyl donors for DNA, RNA, proteins, and neurotransmitters during brain development. Dietary deficiencies in methyl donors have been implicated in neural tube defects as well as more subtle neurodevelopmental outcomes of reduced IQ, behavioral problems, and schizophrenia. Dietary methyl-donor supplementation has been attempted in autism spectrum disorders including fragile X and Rett syndromes, but the critical timing of supplementation appears to be much earlier than tested in these studies. The early mammalian embryo is characterized by an additional need for methyl donors for DNA methylation patterns marking pluripotency and tissue differentiation, and the periconceptional period may be especially vulnerable to dietary deficiency in folate and other methyl donors. Prenatal vitamins have been recommended for women considering pregnancy since the early 1990s, and the increased usage of prenatal vitamins and folate supplementation of cereal have been credited with a significant reduction in neural tube defects. For neurodevelopmental disorders with a postnatal onset, including autism, schizophrenia, fragile X, and Rett syndromes, the protective nature of periconception folate and other B vitamins has not been investigated. Considering the emerging insight into the role of epigenetics in these disorders, investigations into the combined role of genetic and dietary factors appears warranted. Since dietary supplementation of B vitamins is a relatively low-cost and simple preventative measure, the protective nature of these important nutrients should be investigated in a broader spectrum of neurodevelopmental disorders.

ACKNOWLEDGMENTS

The authors thank Dr. Joshua Miller for reading and providing feedback on the manuscript. Dr. Schmidt's effort is supported by the Interdisciplinary Training for Autism Researchers grant program funded by the NIH, Grant T32 MH073124. Dr. LaSalle's research and effort has been supported by the NIH (Grants 2R01HD041462, 1R01HD048799, and 1R01ES015171), Autism Speaks, and the International Rett Syndrome Foundation.

REFERENCES

Aarbakke, J., G. A. Miura, P. S. Prytz, A. Bessesen, L. Slordal, R. K. Gordon, and P. K. Chiang. 1986. Induction of HL-60 cell differentiation by 3-deaza-(+/–)-aristeromycin, an inhibitor of S-adenosylhomocysteine hydrolase. *Cancer Res* 46 (11):5469–72.

Abdolmaleky, H. M., K. H. Cheng, S. V. Faraone, M. Wilcox, S. J. Glatt, F. Gao, C. L. Smith, R. Shafa, B. Aeali, J. Carnevale, H. Pan, P. Papageorgis, J. F. Ponte, V. Sivaraman, M. T. Tsuang, and S. Thiagalingam. 2006. Hypomethylation of MB-COMT promoter is a major risk factor for schizophrenia and bipolar disorder. *Hum Mol Genet* 15 (21):3132–45.

Acacio, G. L., R. Barini, C. S. Bertuzzo, E. C. Couto, J. M. Annichino-Bizzacchi, and W. P. Junior. 2005. Methylenetetrahydrofolate reductase gene polymorphisms and their association with trisomy 21. *Prenat Diagn* 25 (13):1196–99.

Adams, J. B., F. George, and T. Audhya. 2006. Abnormally high plasma levels of vitamin B6 in children with autism not taking supplements compared to controls not taking supplements. *J Altern Complement Med* 12 (1):59–63.

Adams, M., M. Lucock, J. Stuart, S. Fardell, K. Baker, and X. Ng. 2007. Preliminary evidence for involvement of the folate gene polymorphism 19bp deletion-DHFR in occurrence of autism. *Neurosci Lett* 422 (1):24–29.

Afman, L. A., H. J. Blom, N. M. van der Put, and H. W. van Straaten. 2003. Homocysteine interference in neurulation: a chick embryo model. *Birth Defects Res A Clin Mol Teratol* 67 (6):421–28.

Afman, L. A., N. M. van der Put, C. M. Thomas, J. M. Trijbels, and H. J. Blom. 2001. Reduced vitamin B12 binding by transcobalamin II increases the risk of neural tube defects. *QJM* 94 (3):159–66.

Akbarian, S. 2009. The molecular pathology of schizophrenia—focus on histone and DNA modifications. *Brain Res Bull* (in press).

Akbarian, S., M. G. Ruehl, E. Bliven, L. A. Luiz, A. C. Peranelli, S. P. Baker, R. C. Roberts, W. E. Bunney Jr., R. C. Conley, E. G. Jones, C. A. Tamminga, and Y. Guo. 2005. Chromatin alterations associated with down-regulated metabolic gene expression in the prefrontal cortex of subjects with schizophrenia. *Arch Gen Psychiatry* 62 (8):829–40.

Albright, C. D., C. B. Friedrich, E. C. Brown, M. H. Mar, and S. H. Zeisel. 1999. Maternal dietary choline availability alters mitosis, apoptosis and the localization of TOAD-64 protein in the developing fetal rat septum. *Brain Res Dev Brain Res* 115 (2):123–29.

Albright, C. D., A. Y. Tsai, C. B. Friedrich, M. H. Mar, and S. H. Zeisel. 1999. Choline availability alters embryonic development of the hippocampus and septum in the rat. *Brain Res Dev Brain Res* 113 (1–2):13–20.

Allan, A. M., X. Liang, Y. Luo, C. Pak, X. Li, K. E. Szulwach, D. Chen, P. Jin, and X. Zhao. 2008. The loss of methyl-CpG binding protein 1 leads to autism-like behavioral deficits. *Hum Mol Genet* 17 (13):2047–57.

Amir, R. E., I. B. van den Veyver, M. Wan, C. Q. Tran, U. Francke, and H. Y. Zoghbi. 1999. Rett syndrome is caused by mutations in X-linked *MECP2,* encoding methyl- CpG-binding protein 2. *Nat Genet* 23 (2):185–88.

Antonarakis, S. E., M. B. Petersen, M. G. McInnis, P. A. Adelsberger, A. A. Schinzel, F. Binkert, C. Pangalos, O. Raoul, S. A. Slaugenhaupt, M. Hafez et al. 1992. The meiotic stage of nondisjunction in trisomy 21: determination by using DNA polymorphisms. *Am J Hum Genet* 50 (3):544–50.

Arakawa, T., T. Mizuno, Y. Honda, T. Tamura, and K. Sakai. 1969. Brain function of infants fed on milk from mothers with low serum folate levels. *Tohoku J Exp Med* 97 (4):391–97.

Arakawa, T., T. Mizuno, K. Sakai, K. Chida, and A. Watanabe. 1969. Electroencephalographic frequency patterns of rats treated with aminopterin in early infancy. *Tohoku J Exp Med* 97 (4):385–90.

Arinami, T., N. Yamada, K. Yamakawa-Kobayashi, H. Hamaguchi, and M. Toru. 1997. Methylenetetrahydrofolate reductase variant and schizophrenia/depression. *Am J Med Genet* 74 (5):526–28.

Arn, P. H., C. A. Williams, R. T. Zori, D. J. Driscoll, and D. S. Rosenblatt. 1998. Methylenetetrahydrofolate reductase deficiency in a patient with phenotypic findings of Angelman syndrome. *Am J Med Genet* 77 (3):198–200.

Bachevalier, J., and M. I. Botez. 1978. Avoidance behavior in folate-deficient rats. *Tohoku J Exp Med* 126 (2):111–16.

Barbaux, S., R. Plomin, and A. S. Whitehead. 2000. Polymorphisms of genes controlling homocysteine/folate metabolism and cognitive function. *Neuroreport* 11 (5):1133–36.

Barry, K. E., L. J. Fleming, and V. Siskind. 1983. Seasonality of neural tube defects in Queensland. *Med J Aust* 2 (7):306.

Barthelemy, C., B. Garreau, I. Leddet, D. Ernouf, J. P. Muh, and G. Lelord. 1981. Behavioral and biological effects of oral magnesium, vitamin B6, and combined magnesium-B6 administration in autistic children. *Magnesium Bull* 3:150–53.

Barthelemy, C., B. Garreau, I. Leddet, D. Sauvage, J. P. Muh, G. Lelord, and E. Callaway. 1983. [Value of behavior scales and urinary homovanillic acid determinations in monitoring the combined treatment with vitamin B6 and magnesium of children displaying autistic behavior]. *Neuropsychiatr Enfance Adolesc* 31 (5–6):289–301.

Bassuk, A. G., and Z. Kibar. 2009. Genetic basis of neural tube defects. *Semin Pediatr Neurol* 16 (3):101–10.

Bennett, G. D., J. Vanwaes, K. Moser, T. Chaudoin, L. Starr, and T. H. Rosenquist. 2006. Failure of homocysteine to induce neural tube defects in a mouse model. *Birth Defects Res B Dev Reprod Toxicol* 77 (2):89–94.

Bhate, V., S. Deshpande, D. Bhat, N. Joshi, R. Ladkat, S. Watve, C. Fall, C. A. de Jager, H. Refsum, and C. Yajnik. 2008. Vitamin B12 status of pregnant Indian women and cognitive function in their 9-year-old children. *Food Nutr Bull* 29 (4):249–54.

Bird, E. D., E. G. Spokes, J. Barnes, A. V. MacKay, L. L. Iversen, and M. Shepherd. 1977. Increased brain dopamine and reduced glutamic acid decarboxylase and choline acetyl transferase activity in schizophrenia and related psychoses. *Lancet* 2 (8049):1157–58.

Bjelland, I., G. S. Tell, S. E. Vollset, S. Konstantinova, and P. M. Ueland. 2009. Choline in anxiety and depression: the Hordaland Health Study. *Am J Clin Nutr* 90 (4):1056–60.

Black, M. M. 2003. Micronutrient deficiencies and cognitive functioning. *J Nutr* 133 (11 Suppl 2):3927S–3931S.

Black, M. M. 2008. Effects of vitamin B12 and folate deficiency on brain development in children. *Food Nutr Bull* 29 (2 Suppl):S126–31.

Blom, H. J. 2009. Folic acid, methylation and neural tube closure in humans. *Birth Defects Res A Clin Mol Teratol* 85 (4):295–302.

Blom, H. J., G. M. Shaw, M. den Heijer, and R. H. Finnell. 2006. Neural tube defects and folate: case far from closed. *Nat Rev Neurosci* 7 (9):724–31.

Borchardt, R. T., Y. S. Wu, and B. S. Wu. 1978. Affinity labeling of histamine N-methyltransferase by 2',3'-dialdehyde derivatives of S-adenosylhomocysteine and S-adenosylmethionine. Kinetics of inactivation. *Biochemistry* 17 (20):4145–52.

Boris, M., A. Goldblatt, J. Galanko, and S. J. James. 2004. Association of MTHFR gene variants with autism. *J Am Phys Surg* 9:106–8.

Botto, L. D., and P. Mastroiacovo. 1998. Exploring gene-gene interactions in the etiology of neural tube defects. *Clin Genet* 53 (6):456–59.

Botto, L. D., and Q. Yang. 2000. 5,10-Methylenetetrahydrofolate reductase gene variants and congenital anomalies: A HuGE review. *Am J Epidemiol* 151:862–77.

Boulet, S. L., Q. Yang, C. Mai, R. S. Kirby, J. S. Collins, J. M. Robbins, R. Meyer, M. A. Canfield, and J. Mulinare. 2008. Trends in the postfortification prevalence of spina bifida and anencephaly in the United States. *Birth Defects Res A Clin Mol Teratol* 82 (7):527–32.

Bower, C., and F. J. Stanley. 1989. Dietary folate as a risk factor for neural-tube defects: evidence from a case-control study in Western Australia. *Med J Aust* 150 (11):613–19.

Boyles, A. L., A. V. Billups, K. L. Deak, D. G. Siegel, L. Mehltretter, S. H. Slifer, A. G. Bassuk, J. A. Kessler, M. C. Reed, H. F. Nijhout, T. M. George, D. S. Enterline, J. R. Gilbert, and M. C. Speer. 2006. Neural tube defects and folate pathway genes: family-based association tests of gene-gene and gene-environment interactions. *Environ Health Perspect* 114 (10):1547–52.

Brandalize, A. P., E. Bandinelli, P. A. dos Santos, I. Roisenberg, and L. Schuler-Faccini. 2009. Evaluation of C677T and A1298C polymorphisms of the MTHFR gene as maternal risk factors for Down syndrome and congenital heart defects. *Am J Med Genet A* 149A (10):2080–87.

Brody, L. C., M. Conley, C. Cox, P. N. Kirke, M. P. McKeever, J. L. Mills, A. M. Molloy, V. B. O'Leary, A. Parle-McDermott, J. M. Scott, and D. A. Swanson. 2002. A polymorphism, R653Q, in the trifunctional enzyme methylenetetrahydrofolate dehydrogenase/methenyltetrahydrofolate cyclohydrolase/formyltetrahydrofolate synthetase is a maternal genetic risk factor for neural tube defects: report of the Birth Defects Research Group. *Am J Hum Genet* 71 (5):1207–15.

Brouwer, I. A., M. van Dusseldorp, C. M. Thomas, M. Duran, J. G. Hautvast, T. K. Eskes, and R. P. Steegers-Theunissen. 1999. Low-dose folic acid supplementation decreases plasma homocysteine concentrations: a randomized trial. *Am J Clin Nutr* 69 (1):99–104.

Brown, A. S., T. Bottiglieri, C. A. Schaefer, C. P. Quesenberry, Jr., L. Liu, M. Bresnahan, and E. S. Susser. 2007. Elevated prenatal homocysteine levels as a risk factor for schizophrenia. *Arch Gen Psychiatry* 64 (1):31–39.

Brown, S. E., M. F. Fraga, I. C. Weaver, M. Berdasco, and M. Szyf. 2007. Variations in DNA methylation patterns during the cell cycle of HeLa cells. *Epigenetics* 2 (1):54–65.

Brown, W. T., I. L. Cohen, G. S. Fisch, E. G. Wolf-Schein, V. A. Jenkins, M. N. Malik, and E. C. Jenkins. 1986. High dose folic acid treatment of fragile (X) males. *Am J Med Genet* 23 (1–2):263–71.

Campbell, D. B., J. S. Sutcliffe, P. J. Ebert, R. Militerni, C. Bravaccio, S. Trillo, M. Elia, C. Schneider, R. Melmed, R. Sacco, A. M. Persico, and P. Levitt. 2006. A genetic variant that disrupts MET transcription is associated with autism. *Proc Natl Acad Sci U S A* 103 (45):16834–39.

Carrel, A. L., S. Huber, D. B. Allen, and K. V. Voelkerding. 1999. Assessment of SNRPN expression as a molecular tool in the diagnosis of Prader-Willi syndrome. *Mol Diagn* 4 (1):5–10.

Centers for Disease Control and Prevention. 1992. Recommendations for the use of folic acid to reduce the number of cases of spina bifida and other neural tube defects. *MMWR Morb Mortal Wkly Rep* 41 (RR-14):1–7.

Chiang, P. K. 1981. Conversion of 3T3-L1 fibroblasts to fat cells by an inhibitor of methylation: effect of 3-deazaadenosine. *Science* 211 (4487):1164–66.

Chiang, P. K., R. K. Gordon, J. Tal, G. C. Zeng, B. P. Doctor, K. Pardhasaradhi, and P. P. McCann. 1996. S-adenosylmethionine and methylation. *Faseb J* 10 (4):471–80.

Chiang, P. K., Y. S. Im, and G. L. Cantoni. 1980. Phospholipids biosynthesis by methylations and choline incorporation: effect of 3-deazaadenosine. *Biochem Biophys Res Commun* 94 (1):174–81.

Chiuve, S. E., E. L. Giovannucci, S. E. Hankinson, S. H. Zeisel, L. W. Dougherty, W. C. Willett, and E. B. Rimm. 2007. The association between betaine and choline intakes and the plasma concentrations of homocysteine in women. *Am J Clin Nutr* 86 (4):1073–81.

Christensen, B., L. Arbour, P. Tran, D. Leclerc, N. Sabbaghian, R. Platt, B. M. Gilfix, D. S. Rosenblatt, R. A. Gravel, P. Forbes, and R. Rozen. 1999. Genetic polymorphisms in methylenetetrahydrofolate reductase and methionine synthase, folate levels in red blood cells, and risk of neural tube defects. *Am J Med Genet* 84 (2):151–57.

Coelho, C. N., and N. W. Klein. 1990. Methionine and neural tube closure in cultured rat embryos: morphological and biochemical analyses. *Teratology* 42 (4):437–51.

Coppede, F., G. Marini, S. Bargagna, L. Stuppia, F. Minichilli, I. Fontana, R. Colognato, G. Astrea, G. Palka, and L. Migliore. 2006. Folate gene polymorphisms and the risk of Down syndrome pregnancies in young Italian women. *Am J Med Genet A* 140 (10):1083–91.

Cox, R., C. Prescott, and C. C. Irving. 1977. The effect of S-adenosylhomocysteine on DNA methylation in isolated rat liver nuclei. *Biochim Biophys Acta* 474 (4):493–99.

Craciunescu, C. N., C. D. Albright, M. H. Mar, J. Song, and S. H. Zeisel. 2003. Choline availability during embryonic development alters progenitor cell mitosis in developing mouse hippocampus. *J Nutr* 133 (11):3614–18.

Craciunescu, C. N., E. C. Brown, M. H. Mar, C. D. Albright, M. R. Nadeau, and S. H. Zeisel. 2004. Folic acid deficiency during late gestation decreases progenitor cell proliferation and increases apoptosis in fetal mouse brain. *J Nutr* 134 (1):162–66.

Cunha, A. L., M. H. Hirata, C. A. Kim, E. M. Guerra-Shinohara, K. Nonoyama, and R. D. Hirata. 2002. Metabolic effects of C677T and A1298C mutations at the MTHFR gene in Brazilian children with neural tube defects. *Clin Chim Acta* 318 (1–2):139–43.

Currenti, S. A. 2009. Understanding and determining the etiology of autism. *Cell Mol Neurobiol* (e-pub).

Cuskelly, G. J., H. McNulty, and J. M. Scott. 1999. Fortification with low amounts of folic acid makes a significant difference in folate status in young women: implications for the prevention of neural tube defects. *Am J Clin Nutr* 70 (2):234–39.

Czeizel, A. E., and I. Dudas. 1992. Prevention of the first occurrence of neural-tube defects by periconceptional vitamin supplementation. [Comment]. *N Engl J Med* 327 (26):1832–35.

D'Antuono, M., D. Merlo, and M. Avoli. 2003. Involvement of cholinergic and gabaergic systems in the fragile X knockout mice. *Neuroscience* 119 (1):9–13.

da Costa, K. A., C. E. Gaffney, L. M. Fischer, and S. H. Zeisel. 2005. Choline deficiency in mice and humans is associated with increased plasma homocysteine concentration after a methionine load. *Am J Clin Nutr* 81 (2):440–44.

Daly, S., J. L. Mills, A. M. Molloy, M. Conley, Y. J. Lee, P. N. Kirke, D. G. Weir, and J. M. Scott. 1997. Minimum effective dose of folic acid for food fortification to prevent neural-tube defects. *Lancet* 350 (9092):1666–69.

Daly, S., J. L. Mills, A. M. Molloy, M. Conley, J. McPartlin, Y. J. Lee, P. B. Young, P. N. Kirke, D. G. Weir, and J. M. Scott. 2002. Low-dose folic acid lowers plasma homocysteine levels in women of child-bearing age. *Q J Med* 95 (11):733–40.

Davis, K. L., L. E. Hollister, and P. A. Berger. 1979. Choline chloride in schizophrenia. *Am J Psychiatry* 136 (12):1581–84.

De Cabo, S. F., J. Santos, and J. Fernandez-Piqueras. 1995. Molecular and cytological evidence of S-adenosyl-L-homocysteine as an innocuous undermethylating agent in vivo. *Cytogenet Cell Genet* 71 (2):187–92.

Deguchi, T., and J. Barchas. 1971. Inhibition of transmethylations of biogenic amines by S-adenosylhomocysteine. Enhancement of transmethylation by adenosylhomocysteinase. *J Biol Chem* 246 (10):3175–81.

De Marco, P., M. G. Calevo, A. Moroni, L. Arata, E. Merello, A. Cama, R. H. Finnell, L. Andreussi, and V. Capra. 2001. Polymorphisms in genes involved in folate metabolism as risk factors for NTDs. *Eur J Pediatr Surg* 11 (Suppl 1):S14–17.

Dobo, M., and A. E. Czeizel. 1998. Long-term somatic and mental development of children after periconceptional multivitamin supplementation. *Eur J Pediatr* 157 (9):719–23.

Dolinoy, D. C., D. Huang, and R. L. Jirtle. 2007. Maternal nutrient supplementation counteracts bisphenol A-induced DNA hypomethylation in early development. *Proc Natl Acad Sci U S A* 104 (32):13056–61.

Drouva, S. V., E. LaPlante, P. Leblanc, J. J. Bechet, H. Clauser, and C. Kordon. 1986. Estradiol activates methylating enzyme(s) involved in the conversion of phosphatidylethanolamine to phosphatidylcholine in rat pituitary membranes. *Endocrinology* 119 (6):2611–22.

Economides, D. L., J. Ferguson, I. Z. Mackenzie, J. Darley, Ware, II, and M. Holmes-Siedle. 1992. Folate and vitamin B12 concentrations in maternal and fetal blood, and amniotic fluid in second trimester pregnancies complicated by neural tube defects. *Br J Obstet Gynaecol* 99 (1):23–25.

Egan, M. F., T. E. Goldberg, B. S. Kolachana, J. H. Callicott, C. M. Mazzanti, R. E. Straub, D. Goldman, and D. R. Weinberger. 2001. Effect of COMT Val108/158 Met genotype on frontal lobe function and risk for schizophrenia. *Proc Natl Acad Sci U S A* 98 (12):6917–22.

Elwood, J. M. 1975. Seasonal variation in anencephalus in Canada. *Br J Prev Soc Med* 29 (1):22–26.

Fedrick, J. 1976. Anencephalus in Scotland 1961–72. *Br J Prev Soc Med* 30 (2):132–37.

Feix, A., W. C. Winkelmayer, C. Eberle, G. Sunder-Plassmann, and M. Fodinger. 2004. Methionine synthase reductase MTRR 66A > G has no effect on total homocysteine, folate, and vitamin B12 concentrations in renal transplant patients. *Atherosclerosis* 174 (1):43–48.

Feng, L. G., Z. W. Song, F. Xin, and J. Hu. 2009. Association of plasma homocysteine and methylenetetrahydrofolate reductase C677T gene variant with schizophrenia: A Chinese Han population-based case-control study. *Psychiatry Res* 168 (3):205–8.

Findling, R. L., K. Maxwell, L. Scotese-Wojtila, J. Huang, T. Yamashita, and M. Wiznitzer. 1997. High-dose pyridoxine and magnesium administration in children with autistic disorder: an absence of salutary effects in a double-blind, placebo-controlled study. *J Autism Dev Disord* 27 (4):467–78.

Finnell, R. H., A. Gould, and O. Spiegelstein. 2003. Pathobiology and genetics of neural tube defects. *Epilepsia* 44 (Suppl 3):14–23.

Fintelman-Rodrigues, N., J. C. Correa, J. M. Santos, M. M. Pimentel, and C. B. Santos-Reboucas. 2009. Investigation of CBS, MTR, RFC-1 and TC polymorphisms as maternal risk factors for Down syndrome. *Dis Markers* 26 (4):155–61.

Fisch, G. S., I. L. Cohen, A. C. Gross, V. Jenkins, E. C. Jenkins, and W. T. Brown. 1988. Folic acid treatment of fragile X males: a further study. *Am J Med Genet* 30 (1–2):393–99.

Fisher, M. C., S. H. Zeisel, M. H. Mar, and T. W. Sadler. 2001. Inhibitors of choline uptake and metabolism cause developmental abnormalities in neurulating mouse embryos. *Teratology* 64 (2):114–22.

Fisher, M. C., S. H. Zeisel, M. H. Mar, and T. W. Sadler. 2002. Perturbations in choline metabolism cause neural tube defects in mouse embryos in vitro. *FASEB J* 16 (6):619–21.

Fraser, F. C., M. Frecker, and P. Allderdice. 1986. Seasonal variation of neural tube defects in Newfoundland and elsewhere. *Teratology* 33 (3):299–303.

Friez, M. J., J. R. Jones, K. Clarkson, H. Lubs, D. Abuelo, J. A. Bier, S. Pai, R. Simensen, C. Williams, P. F. Giampietro, C. E. Schwartz, and R. E. Stevenson. 2006. Recurrent infections, hypotonia, and mental retardation caused by duplication of MECP2 and adjacent region in Xq28. *Pediatrics* 118 (6):e1687–95.

Friso, S., S. W. Choi, D. Girelli, J. B. Mason, G. G. Dolnikowski, P. J. Bagley, O. Olivieri, P. F. Jacques, I. H. Rosenberg, R. Corrocher, and J. Selhub. 2002. A common mutation in the 5,10-methylenetetrahydrofolate reductase gene affects genomic DNA methylation through an interaction with folate status. *Proc Natl Acad Sci U S A* 99 (8):5606–11.

Frosst, P., H. J. Blom, R. Milos, P. Goyette, C. A. Sheppard, R. G. Matthews, G. J. Boers, M. den Heijer, L. A. Kluijtmans, L. P. van den Heuvel, and et al. 1995. A candidate genetic risk factor for vascular disease: a common mutation in methylenetetrahydrofolate reductase. *Nat Genet* 10 (1):111–13.

Froster-Iskenius, U., K. Bodeker, T. Oepen, R. Matthes, U. Piper, and E. Schwinger. 1986. Folic acid treatment in males and females with fragile-(X)-syndrome. *Am J Med Genet* 23 (1–2):273–89.

Gaber, K. R., M. K. Farag, S. E. Soliman, H. T. El-Bassyouni, and G. El-Kamah. 2007. Maternal vitamin B12 and the risk of fetal neural tube defects in Egyptian patients. *Clin Lab* 53 (1–2):69–75.

Garcia-Castro, I., J. M. Mato, G. Vasanthakumar, W. P. Wiesmann, E. Schiffmann, and P. K. Chiang. 1983. Paradoxical effects of adenosine on neutrophil chemotaxis. *J Biol Chem* 258 (7):4345–49.

Gardiki-Kouidou, P., and M. J. Seller. 1988. Amniotic fluid folate, vitamin B12 and transcobalamins in neural tube defects. *Clin Genet* 33 (6):441–48.

Garner, S. C., M. H. Mar, and S. H. Zeisel. 1995. Choline distribution and metabolism in pregnant rats and fetuses are influenced by the choline content of the maternal diet. *J Nutr* 125 (11):2851–58.

Gaughan, D. J., L. A. Kluijtmans, S. Barbaux, D. McMaster, I. S. Young, J. W. Yarnell, A. Evans, and A. S. Whitehead. 2001. The methionine synthase reductase (MTRR) A66G polymorphism is a novel genetic determinant of plasma homocysteine concentrations. *Atherosclerosis* 157 (2):451–56.

Gillberg, C., J. Wahlstrom, R. Johansson, M. Tornblom, and K. Albertsson-Wikland. 1986. Folic acid as an adjunct in the treatment of children with the autism fragile-X syndrome (AFRAX). *Dev Med Child Neurol* 28 (5):624–27.

Glaze, D. G., A. K. Percy, K. J. Motil, J. B. Lane, J. S. Isaacs, R. J. Schultz, J. O. Barrish, J. L. Neul, W. E. O'Brien, and E. O. Smith. 2009. A study of the treatment of Rett syndrome with folate and betaine. *J Child Neurol* 24 (5):551–56.

Glenn, M. J., E. D. Kirby, E. M. Gibson, S. J. Wong-Goodrich, T. J. Mellott, J. K. Blusztajn, and C. L. Williams. 2008. Age-related declines in exploratory behavior and markers of hippocampal plasticity are attenuated by prenatal choline supplementation in rats. *Brain Res* 1237:110–23.

Glick, J. M., S. Ross, and P. S. Leboy. 1975. S-adenosylhomocysteine inhibition of three purified tRNA methyltransferases from rat liver. *Nucleic Acids Res* 2 (10):1639–51.

Godfrey, P. S., B. K. Toone, M. W. Carney, T. G. Flynn, T. Bottiglieri, M. Laundy, I. Chanarin, and E. H. Reynolds. 1990. Enhancement of recovery from psychiatric illness by methylfolate. *Lancet* 336 (8712):392–95.

Goff, D. C., T. Bottiglieri, E. Arning, V. Shih, O. Freudenreich, A. E. Evins, D. C. Henderson, L. Baer, and J. Coyle. 2004. Folate, homocysteine, and negative symptoms in schizophrenia. *Am J Psychiatry* 161 (9):1705–8.

Goin-Kochel, R. P., A. E. Porter, S. U. Peters, M. Shinawi, T. Sahoo, and A. L. Beaudet. 2009. The MTHFR 677C→T polymorphism and behaviors in children with autism: exploratory genotype-phenotype correlations. *Autism Res* 2 (2):98–108.

Grayson, D. R., X. Jia, Y. Chen, R. P. Sharma, C. P. Mitchell, A. Guidotti, and E. Costa. 2005. Reelin promoter hypermethylation in schizophrenia. *Proc Natl Acad Sci U S A* 102 (26):9341–46.

Greenblatt, J. M., L. C. Huffman, and A. L. Reiss. 1994. Folic acid in neurodevelopment and child psychiatry. *Prog Neuropsychopharmacol Biol Psychiatry* 18 (4):647–60.

Grillo, L. B., G. L. Acacio, R. Barini, W. Pinto, Jr., and C. S. Bertuzzo. 2002. [Mutations in the methylene-tetrahydrofolate reductase gene and Down syndrome]. *Cad Saude Publica* 18 (6):1795–97.

Groenen, P. M., I. A. van Rooij, P. G. Peer, R. H. Gooskens, G. A. Zielhuis, and R. P. Steegers-Theunissen. 2004. Marginal maternal vitamin B12 status increases the risk of offspring with spina bifida. *Am J Obstet Gynecol* 191 (1):11–17.

Gueant, J. L., G. Anello, P. Bosco, R. M. Gueant-Rodriguez, A. Romano, C. Barone, P. Gerard, and C. Romano. 2005. Homocysteine and related genetic polymorphisms in Down's syndrome IQ. *J Neurol Neurosurg Psychiatry* 76 (5):706–9.

Guibert, S, T Forne, and M Weber. 2009. Dynamic regulation of DNA methylation during mammalian development. *Epigenomics* 1 (1):81–98.

Guo-Ross, S. X., S. Clark, D. A. Montoya, K. H. Jones, J. Obernier, A. K. Shetty, A. M. White, J. K. Blusztajn, W. A. Wilson, and H. S. Swartzwelder. 2002. Prenatal choline supplementation protects against postnatal neurotoxicity. *J Neurosci* 22 (1):RC195.

Guo-Ross, S. X., K. H. Jones, A. K. Shetty, W. A. Wilson, and H. S. Swartzwelder. 2003. Prenatal dietary choline availability alters postnatal neurotoxic vulnerability in the adult rat. *Neurosci Lett* 341 (2):161–63.

Gustavson, K. H., K. Dahlbom, A. Flood, G. Holmgren, H. K. Blomquist, and G. Sanner. 1985. Effect of folic acid treatment in the fragile X syndrome. *Clin Genet* 27 (5):463–67.

Hagerman, R. J., A. W. Jackson, A. Levitas, M. Braden, P. McBogg, M. Kemper, L. McGavran, R. Berry, I. Matus, and P. J. Hagerman. 1986. Oral folic acid versus placebo in the treatment of males with the fragile X syndrome. *Am J Med Genet* 23 (1–2):241–62.

Hagerman, R. J., M. Y. Ono, and P. J. Hagerman. 2005. Recent advances in fragile X: a model for autism and neurodegeneration. *Curr Opin Psychiatry* 18 (5):490–96.

Haggarty, P., D. M. Campbell, S. Duthie, K. Andrews, G. Hoad, C. Piyathilake, I. Fraser, and G. McNeill. 2008. Folic acid use in pregnancy and embryo selection. *Bjog* 115 (7):851–56.

Hakami, N., P. E. Neiman, G. P. Canellos, and J. Lazerson. 1971. Neonatal megaloblastic anemia due to inherited transcobalamin II deficiency in two siblings. *N Engl J Med* 285 (21):1163–70.

Hamon, C. G., J. A. Blair, and P. A. Barford. 1986. The effect of tetrahydrofolate on tetrahydrobiopterin metabolism. *J Ment Defic Res* 30 (Pt 2):179–83.

Hardan, A. Y., N. J. Minshew, N. M. Melhem, S. Srihari, B. Jo, R. Bansal, M. S. Keshavan, and J. A. Stanley. 2008. An MRI and proton spectroscopy study of the thalamus in children with autism. *Psychiatry Res* 163 (2):97–105.

Heijmans, B. T., E. W. Tobi, A. D. Stein, H. Putter, G. J. Blauw, E. S. Susser, P. E. Slagboom, and L. H. Lumey. 2008. Persistent epigenetic differences associated with prenatal exposure to famine in humans. *Proc Natl Acad Sci U S A* 105 (44):17046–49.

Henshaw, S. K. 1998. Unintended pregnancy in the United States. *Fam Plann Perspect* 30 (1):24–29, 46.

Hertz-Picciotto, I., and L. Delwiche. 2009. The rise in autism and the role of age at diagnosis. *Epidemiology* 20 (1):84–90.

Hirata, F., and J. Axelrod. 1980. Phospholipid methylation and biological signal transmission. *Science* 209 (4461):1082–90.

Hobbs, C. A., S. L. Sherman, P. Yi, S. E. Hopkins, C. P. Torfs, R. J. Hine, M. Pogribna, R. Rozen, and S. J. James. 2000. Polymorphisms in genes involved in folate metabolism as maternal risk factors for Down syndrome. *Am J Hum Genet* 67 (3):623–30.

Hoffman, D. R., W. E. Cornatzer, and J. A. Duerre. 1979. Relationship between tissue levels of S-adenosylmethionine, S-adenylhomocysteine, and transmethylation reactions. *Can J Biochem* 57 (1):56–65.

Hoffman, D. R., J. A. Haning, and W. E. Cornatzer. 1981. Microsomal phosphatidylethanolamine methyltransferase: inhibition by S-adenosylhomocysteine. *Lipids* 16 (8):561–67.

Hoffman, D. R., D. W. Marion, W. E. Cornatzer, and J. A. Duerre. 1980. S-Adenosylmethionine and S-adenosylhomocystein metabolism in isolated rat liver. Effects of L-methionine, L-homocystein, and adenosine. *J Biol Chem* 255 (22):10822–27.

Hogart, A., R. P. Nagarajan, K. A. Patzel, D. H. Yasui, and J. M. Lasalle. 2007. 15q11-13 GABAA receptor genes are normally biallelically expressed in brain yet are subject to epigenetic dysregulation in autism-spectrum disorders. *Hum Mol Genet* 16 (6):691–703.

Holliday, R., and G. W. Grigg. 1993. DNA methylation and mutation. *Mutat Res* 285 (1):61–67.

Holmes, G. L., Y. Yang, Z. Liu, J. M. Cermak, M. R. Sarkisian, C. E. Stafstrom, J. C. Neill, and J. K. Blusztajn. 2002. Seizure-induced memory impairment is reduced by choline supplementation before or after status epilepticus. *Epilepsy Res* 48 (1–2):3–13.

Honein, M. A., L. J. Paulozzi, T. J. Mathews, J. D. Erickson, and L. Y. Wong. 2001. Impact of folic acid fortification of the US food supply on the occurrence of neural tube defects. *JAMA* 285 (23):2981–86.

Huang, H. S., and S. Akbarian. 2007. GAD1 mRNA expression and DNA methylation in prefrontal cortex of subjects with schizophrenia. *PLoS One* 2 (8):e809.

Hublitz, P., M. Albert, and A. H. Peters. 2009. Mechanisms of transcriptional repression by histone lysine methylation. *Int J Dev Biol* 53 (2–3):335–54.

Jacques, P. F., A. G. Bostom, J. Selhub, S. Rich, R. C. Ellison, J. H. Eckfeldt, R. A. Gravel, and R. Rozen. 2003. Effects of polymorphisms of methionine synthase and methionine synthase reductase on total plasma homocysteine in the NHLBI Family Heart Study. *Atherosclerosis* 166 (1):49–55.

Jaenisch, R. 1997. DNA methylation and imprinting: why bother? *Trends Genet* 13 (8):323–29.

James, S. J., P. Cutler, S. Melnyk, S. Jernigan, L. Janak, D. W. Gaylor, and J. A. Neubrander. 2004. Metabolic biomarkers of increased oxidative stress and impaired methylation capacity in children with autism. *Am J Clin Nutr* 80 (6):1611–17.

James, S. J., S. Melnyk, G. Fuchs, T. Reid, S. Jernigan, O. Pavliv, A. Hubanks, and D. W. Gaylor. 2009. Efficacy of methylcobalamin and folinic acid treatment on glutathione redox status in children with autism. *Am J Clin Nutr* 89 (1):425–30.

James, S. J., S. Melnyk, S. Jernigan, M. A. Cleves, C. H. Halsted, D. H. Wong, P. Cutler, K. Bock, M. Boris, J. J. Bradstreet, S. M. Baker, and D. W. Gaylor. 2006. Metabolic endophenotype and related genotypes are associated with oxidative stress in children with autism. *Am J Med Genet B Neuropsychiatr Genet* 141 (8):947–56.

James, S. J., S. Melnyk, S. Jernigan, A. Hubanks, S. Rose, and D. W. Gaylor. 2008. Abnormal transmethylation/transsulfuration metabolism and DNA hypomethylation among parents of children with autism. *J Autism Dev Disord* 38 (10):1976.

James, S. J., S. Melnyk, S. Jernigan, O. Pavliv, T. Trusty, S. Lehman, L. Seidel, D. W. Gaylor, and M. A. Cleves. 2010. A functional polymorphism in the reduced folate carrier gene and DNA hypomethylation in mothers of children with autism. *Am J Med Genet B Neuropsychiatr Genet.* Published online: 12 May 2010, DOI: 10.1002/ajmg.b.31094.

James, S. J., M. Pogribna, I. P. Pogribny, S. Melnyk, R. J. Hine, J. B. Gibson, P. Yi, D. L. Tafoya, D. H. Swenson, V. L. Wilson, and D. W. Gaylor. 1999. Abnormal folate metabolism and mutation in the methylenetetrahydrofolate reductase gene may be maternal risk factors for Down syndrome. *Am J Clin Nutr* 70 (4):495–501.

Jeltsch, A. 2002. Beyond Watson and Crick: DNA methylation and molecular enzymology of DNA methyltransferases. *Chembiochem* 3 (4):274–93.

Johnson, W. G., E. S. Stenroos, J. R. Spychala, S. Chatkupt, S. X. Ming, and S. Buyske. 2004. New 19 bp deletion polymorphism in intron-1 of dihydrofolate reductase (DHFR): a risk factor for spina bifida acting in mothers during pregnancy? *Am J Med Genet A* 124 (4):339–45.

Jonas, C., T. Etienne, C. Barthelemy, J. Jouve, and N. Mariotte. 1984. [Clinical and biochemical value of Magnesium + vitamin B6 combination in the treatment of residual autism in adults]. *Therapie* 39 (6):661–69.

Jones, J. P., W. H. Meck, C. L. Williams, W. A. Wilson, and H. S. Swartzwelder. 1999. Choline availability to the developing rat fetus alters adult hippocampal long-term potentiation. *Brain Res Dev Brain Res* 118 (1–2):159–67.

Jones, P. A., and M. L. Gonzalgo. 1997. Altered DNA methylation and genome instability: a new pathway to cancer? *Proc Natl Acad Sci U S A* 94 (6):2103–5.

Jonsson, E. G., K. Larsson, M. Vares, T. Hansen, A. G. Wang, S. Djurovic, K. S. Ronningen, O. A. Andreassen, I. Agartz, T. Werge, L. Terenius, and H. Hall. 2008. Two methylenetetrahydrofolate reductase gene (MTHFR) polymorphisms, schizophrenia and bipolar disorder: an association study. *Am J Med Genet B Neuropsychiatr Genet* 147B (6):976–82.

Julvez, J., J. Fortuny, M. Mendez, M. Torrent, N. Ribas-Fito, and J. Sunyer. 2009. Maternal use of folic acid supplements during pregnancy and four-year-old neurodevelopment in a population-based birth cohort. *Paediatr Perinat Epidemiol* 23 (3):199–206.

Kalmbach, R. D., S. F. Choumenkovitch, A. P. Troen, P. F. Jacques, R. D'Agostino, and J. Selhub. 2008. A 19-base pair deletion polymorphism in dihydrofolate reductase is associated with increased unmetabolized folic acid in plasma and decreased red blood cell folate. *J Nutr* 138 (12):2323–27.

Kamen, B. A., and A. K. Smith. 2004. A review of folate receptor alpha cycling and 5-methyltetrahydrofolate accumulation with an emphasis on cell models in vitro. *Adv Drug Deliv Rev* 56 (8):1085–97.

Karson, C. N., M. F. Casanova, J. E. Kleinman, and W. S. Griffin. 1993. Choline acetyltransferase in schizophrenia. *Am J Psychiatry* 150 (3):454–59.

Karson, C. N., R. E. Mrak, M. M. Husain, and W. S. Griffin. 1996. Decreased mesopontine choline acetyltransferase levels in schizophrenia. Correlations with cognitive functions. *Mol Chem Neuropathol* 29 (2–3):181–91.

Kemperman, R. F., M. Veurink, T. van der Wal, H. Knegtering, R. Bruggeman, M. R. Fokkema, I. P. Kema, J. Korf, and F. A. Muskiet. 2006. Low essential fatty acid and B-vitamin status in a subgroup of patients with schizophrenia and its response to dietary supplementation. *Prostaglandins Leukot Essent Fatty Acids* 74 (2):75–85.

Kempisty, B., A. Mostowska, I. Gorska, M. Luczak, P. Czerski, A. Szczepankiewicz, J. Hauser, and P. P. Jagodzinski. 2006. Association of 677C>T polymorphism of methylenetetrahydrofolate reductase (MTHFR) gene with bipolar disorder and schizophrenia. *Neurosci Lett* 400 (3):267–71.

Kesler, S. R., A. A. Lightbody, and A. L. Reiss. 2009. Cholinergic dysfunction in fragile X syndrome and potential intervention: a preliminary 1H MRS study. *Am J Med Genet A* 149A (3):403–7.

Kim, Y. I., J. W. Miller, K. A. da Costa, M. Nadeau, D. Smith, J. Selhub, S. H. Zeisel, and J. B. Mason. 1994. Severe folate deficiency causes secondary depletion of choline and phosphocholine in rat liver. *J Nutr* 124 (11):2197–2203.

Kirke, P. N., A. M. Molloy, L. E. Daly, H. Burke, D. G. Weir, and J. M. Scott. 1993. Maternal plasma folate and vitamin B12 are independent risk factors for neural tube defects. *Q J Med* 86 (11):703–8.

Kluijtmans, L. A., I. S. Young, C. A. Boreham, L. Murray, D. McMaster, H. McNulty, J. J. Strain, J. McPartlin, J. M. Scott, and A. S. Whitehead. 2003. Genetic and nutritional factors contributing to hyperhomocysteinemia in young adults. *Blood* 101 (7):2483–88.

Krishna, A. P., and T. Ramakrishna. 1984. Effect of pyridoxine deficiency on the exploratory behavior of rats. *Int J Psychophysiol* 2 (1):39–43.

Leeming, R. J., and M. Lucock. 2009. Autism: is there a folate connection? *J Inherit Metab Dis* (in press).

Lejeune, J., M. O. Rethore, M. C. de Blois, and A. Ravel. 1984. [Trial of folic acid treatment in fragile X syndrome]. *Ann Genet* 27 (4):230–32.

Lelord, G., J. P. Muh, C. Barthelemy, J. Martineau, B. Garreau, and E. Callaway. 1981. Effects of pyridoxine and magnesium on autistic symptoms—initial observations. *J Autism Dev Disord* 11 (2):219–30.

Leonard, E. J., A. Skeel, P. K. Chiang, and G. L. Cantoni. 1978. The action of the adenosylhomocysteine hydrolase inhibitor, 3-deazaadenosine, on phagocytic function of mouse macrophages and human monocytes. *Biochem Biophys Res Commun* 84 (1):102–9.

Lerner, V., J. Bergman, N. Statsenko, and C. Miodownik. 2004. Vitamin B6 treatment in acute neuroleptic-induced akathisia: a randomized, double-blind, placebo-controlled study. *J Clin Psychiatry* 65 (11):1550–54.

Lerner, V., C. Miodownik, A. Kaptsan, Y. Bersudsky, I. Libov, B. A. Sela, and E. Witztum. 2007. Vitamin B6 treatment for tardive dyskinesia: a randomized, double-blind, placebo-controlled, crossover study. *J Clin Psychiatry* 68 (11):1648–54.

Lerner, V., C. Miodownik, A. Kaptsan, H. Cohen, U. Loewenthal, and M. Kotler. 2002. Vitamin B6 as add-on treatment in chronic schizophrenic and schizoaffective patients: a double-blind, placebo-controlled study. *J Clin Psychiatry* 63 (1):54–58.

Levine, J., Z. Stahl, B. A. Sela, V. Ruderman, O. Shumaico, I. Babushkin, Y. Osher, Y. Bersudsky, and R. H. Belmaker. 2006. Homocysteine-reducing strategies improve symptoms in chronic schizophrenic patients with hyperhomocysteinemia. *Biol Psychiatry* 60 (3):265–69.

Lewis, D. A., and P. Levitt. 2002. Schizophrenia as a disorder of neurodevelopment. *Annu Rev Neurosci* 25:409–32.

Lister, R., M. Pelizzola, R. H. Dowen, R. D. Hawkins, G. Hon, J. Tonti-Filippini, J. R. Nery, L. Lee, Z. Ye, Q. M. Ngo, L. Edsall, J. Antosiewicz-Bourget, R. Stewart, V. Ruotti, A. H. Millar, J. A. Thomson, B. Ren, and J. R. Ecker. 2009. Human DNA methylomes at base resolution show widespread epigenomic differences. *Nature* (e-pub).

Lovblad, K., G. Ramelli, L. Remonda, A. C. Nirkko, C. Ozdoba, and G. Schroth. 1997. Retardation of myelination due to dietary vitamin B12 deficiency: cranial MRI findings. *Pediatr Radiol* 27 (2):155–58.

Madison, L. S., T. E. Wells, T. E. Fristo, and C. G. Benesch. 1986. A controlled study of folic acid treatment in three fragile X syndrome males. *J Dev Behav Pediatr* 7 (4):253–56.

Main, P. A., M. T. Angley, P. Thomas, C. E. O'Doherty, and M. Fenech. 2010. Folate and methionine metabolism in autism: a systematic review. *Am J Clin Nutr* 91 (6):1598–620.

Martineau, J., C. Barthelemy, B. Garreau, and G. Lelord. 1985. Vitamin B6, magnesium, and combined B6-Mg: therapeutic effects in childhood autism. *Biol Psychiatry* 20 (5):467–78.

Martineau, J., C. Barthelemy, and G. Lelord. 1986. Long-term effects of combined vitamin B6-magnesium administration in an autistic child. *Biol Psychiatry* 21 (5–6):511–18.

Mavros, M., V. Chirita, O. Popescu, and B. Ferencz. 2008. [The genetic polymorphism of MTHFR gene in schizophrenia]. *Rev Med Chir Soc Med Nat Iasi* 112 (1):76–82.

Mazer, C., J. Muneyyirci, K. Taheny, N. Raio, A. Borella, and P. Whitaker-Azmitia. 1997. Serotonin depletion during synaptogenesis leads to decreased synaptic density and learning deficits in the adult rat: a possible model of neurodevelopmental disorders with cognitive deficits. *Brain Res* 760 (1–2):68–73.

McCullough, A. L., A. Kirksey, T. D. Wachs, G. P. McCabe, N. S. Bassily, Z. Bishry, O. M. Galal, G. G. Harrison, and N. W. Jerome. 1990. Vitamin B-6 status of Egyptian mothers: relation to infant behavior and maternal-infant interactions. *Am J Clin Nutr* 51 (6):1067–74.

McNulty, H., G. J. Cuskelly, and M. Ward. 2000. Response of red blood cell folate to intervention: implications for folate recommendations for the prevention of neural tube defects. *Am J Clin Nutr* 71 (5 Suppl):1308S–11S.

Meck, W. H., R. A. Smith, and C. L. Williams. 1988. Pre- and postnatal choline supplementation produces long-term facilitation of spatial memory. *Dev Psychobiol* 21 (4):339–53.

Meck, W. H., and C. L. Williams. 1997a. Characterization of the facilitative effects of perinatal choline supplementation on timing and temporal memory. *Neuroreport* 8 (13):2831–35.

Meck, W. H., and C. L. Williams. 1997b. Perinatal choline supplementation increases the threshold for chunking in spatial memory. *Neuroreport* 8 (14):3053–59.

Meck, W. H., and C. L. Williams. 1997c. Simultaneous temporal processing is sensitive to prenatal choline availability in mature and aged rats. *Neuroreport* 8 (14):3045–51.

Meck, W. H., and C. L. Williams. 1999. Choline supplementation during prenatal development reduces proactive interference in spatial memory. *Brain Res Dev Brain Res* 118 (1–2):51–59.

Meck, W. H., and C. L. Williams. 2003. Metabolic imprinting of choline by its availability during gestation: implications for memory and attentional processing across the lifespan. *Neurosci Biobehav Rev* 27 (4):385–99.

Meins, M., J. Lehmann, F. Gerresheim, J. Herchenbach, M. Hagedorn, K. Hameister, and J. T. Epplen. 2005. Submicroscopic duplication in Xq28 causes increased expression of the MECP2 gene in a boy with severe mental retardation and features of Rett syndrome. *J Med Genet* 42 (2):e12.

Mills, J. L., J. M. McPartlin, P. N. Kirke, Y. J. Lee, M. R. Conley, D. G. Weir, and J. M. Scott. 1995. Homocysteine metabolism in pregnancies complicated by neural-tube defects. [Comment]. *Lancet* 345 (8943):149–51.

Mills, J. L., J. M. Scott, P. N. Kirke, J. M. McPartlin, M. R. Conley, D. G. Weir, A. M. Molloy, and Y. J. Lee. 1996. Homocysteine and neural tube defects. *J Nutr* 126 (3):756S–760S.

Milunsky, A., H. Jick, S. S. Jick, C. L. Bruell, D. S. MacLaughlin, K. J. Rothman, and W. Willett. 1989. Multivitamin/folic acid supplementation in early pregnancy reduces the prevalence of neural tube defects. *JAMA* 262 (20):2847–52.

Miodownik, C., V. Lerner, N. Statsenko, T. Dwolatzky, B. Nemets, E. Berzak, and J. Bergman. 2006. Vitamin B6 versus mianserin and placebo in acute neuroleptic-induced akathisia: a randomized, double-blind, controlled study. *Clin Neuropharmacol* 29 (2):68–72.

Miodownik, C., V. Lerner, T. Vishne, B. A. Sela, and J. Levine. 2007. High-dose vitamin B6 decreases homocysteine serum levels in patients with schizophrenia and schizoaffective disorders: a preliminary study. *Clin Neuropharmacol* 30 (1):13–17.

Miodownik, C., A. Meoded, I. Libov, Y. Bersudsky, B. A. Sela, and V. Lerner. 2008. Pyridoxal plasma level in schizophrenic and schizoaffective patients with and without tardive dyskinesia. *Clin Neuropharmacol* 31 (4):197–203.

Moephuli, S. R., N. W. Klein, M. T. Baldwin, and H. M. Krider. 1997. Effects of methionine on the cytoplasmic distribution of actin and tubulin during neural tube closure in rat embryos. *Proc Natl Acad Sci U S A* 94 (2):543–48.

Molloy, A. M., P. Kirke, I. Hillary, D. G. Weir, and J. M. Scott. 1985. Maternal serum folate and vitamin B12 concentrations in pregnancies associated with neural tube defects. *Arch Dis Child* 60 (7):660-–6.

Molloy, A. M., P. N. Kirke, J. F. Troendle, H. Burke, M. Sutton, L. C. Brody, J. M. Scott, and J. L. Mills. 2009. Maternal vitamin B12 status and risk of neural tube defects in a population with high neural tube defect prevalence and no folic acid fortification. *Pediatrics* 123 (3):917–23.

Molloy, A. M., J. L. Mills, J. McPartlin, P. N. Kirke, J. M. Scott, and S. Daly. 2002. Maternal and fetal plasma homocysteine concentrations at birth: the influence of folate, vitamin B12, and the 5,10-methylenetetrahydrofolate reductase 677C→T variant. *Am J Obstet Gynecol* 186 (3):499–503.

Molloy, A. M., B. Orsi, D. G. Kennedy, S. Kennedy, D. G. Weir, and J. M. Scott. 1992. The relationship between the activity of methionine synthase and the ratio of S-adenosylmethionine to S-adenosylhomocysteine in the brain and other tissues of the pig. *Biochem Pharmacol* 44 (7):1349–55.

Montoya, D. A., A. M. White, C. L. Williams, J. K. Blusztajn, W. H. Meck, and H. S. Swartzwelder. 2000. Prenatal choline exposure alters hippocampal responsiveness to cholinergic stimulation in adulthood. *Brain Res Dev Brain Res* 123 (1):25–32.

Moretti, P., S. U. Peters, D. Del Gaudio, T. Sahoo, K. Hyland, T. Bottiglieri, R. J. Hopkin, E. Peach, S. H. Min, D. Goldman, B. Roa, C. A. Bacino, and F. Scaglia. 2008. Brief report: autistic symptoms, developmental regression, mental retardation, epilepsy, and dyskinesias in CNS folate deficiency. *J Autism Dev Disord* 38 (6):1170–77.

Moretti, P., T. Sahoo, K. Hyland, T. Bottiglieri, S. Peters, D. del Gaudio, B. Roa, S. Curry, H. Zhu, R. H. Finnell, J. L. Neul, V. T. Ramaekers, N. Blau, C. A. Bacino, G. Miller, and F. Scaglia. 2005. Cerebral folate deficiency with developmental delay, autism, and response to folinic acid. *Neurology* 64 (6):1088–90.

Moretti, R., P. Torre, R. M. Antonello, T. Cattaruzza, G. Cazzato, and A. Bava. 2004. Vitamin B12 and folate depletion in cognition: a review. *Neurol India* 52 (3):310–18.

Morgan, H. D., H. G. Sutherland, D. I. Martin, and E. Whitelaw. 1999. Epigenetic inheritance at the agouti locus in the mouse. *Nat Genet* 23 (3):314–18.

MRC Vitamin Study Research Group. 1991. Prevention of neural tube defects: results of the Medical Research Council Vitamin Study. *Lancet* 338 (8760):131–37.

Mulinare, J., J. F. Cordero, J. D. Erickson, and R. J. Berry. 1988. Periconceptional use of multivitamins and the occurrence of neural tube defects. *JAMA* 260 (21):3141–45.

Muntjewerff, J. W., H. Gellekink, M. den Heijer, M. L. Hoogendoorn, R. S. Kahn, R. J. Sinke, and H. J. Blom. 2008. Polymorphisms in catechol-O-methyltransferase and methylenetetrahydrofolate reductase in relation to the risk of schizophrenia. *Eur Neuropsychopharmacol* 18 (2):99–106.

Muntjewerff, J. W., M. L. Hoogendoorn, M. F. Aukes, R. S. Kahn, R. J. Sinke, H. J. Blom, and M. den Heijer. 2007. No evidence for a preferential transmission of the methylenetetrahydrofolate reductase 677T allele in families with schizophrenia offspring. *Am J Med Genet B Neuropsychiatr Genet* 144B (7):891–94.

Muntjewerff, J. W., M. L. Hoogendoorn, R. S. Kahn, R. J. Sinke, M. Den Heijer, L. A. Kluijtmans, and H. J. Blom. 2005. Hyperhomocysteinemia, methylenetetrahydrofolate reductase 677TT genotype, and the risk for schizophrenia: a Dutch population based case-control study. *Am J Med Genet B Neuropsychiatr Genet* 135B (1):69–72.

Muntjewerff, J. W., R. S. Kahn, H. J. Blom, and M. den Heijer. 2006. Homocysteine, methylenetetrahydrofolate reductase and risk of schizophrenia: a meta-analysis. *Mol Psychiatry* 11 (2):143–49.

Muntjewerff, J. W., N. van der Put, T. Eskes, B. Ellenbroek, E. Steegers, H. Blom, and F. Zitman. 2003. Homocysteine metabolism and B-vitamins in schizophrenic patients: low plasma folate as a possible independent risk factor for schizophrenia. *Psychiatry Res* 121 (1):1–9.

Nag, N., and J. E. Berger-Sweeney. 2007. Postnatal dietary choline supplementation alters behavior in a mouse model of Rett syndrome. *Neurobiol Dis* 26 (2):473–80.

Nag, N., T. J. Mellott, and J. E. Berger-Sweeney. 2008. Effects of postnatal dietary choline supplementation on motor regional brain volume and growth factor expression in a mouse model of Rett syndrome. *Brain Genes Nutr* 1237:101–9.

Nagarajan, R. P., K. A. Patzel, M. Martin, D. H. Yasui, S. E. Swanberg, I. Hertz-Picciotto, R. L. Hansen, J. van de Water, I. N. Pessah, R. Jiang, W. P. Robinson, and J. M. Lasalle. 2008. Mecp2 promoter methylation and X chromosome inactivation in autism. *Autism Res* 1 (3):169–78.

Naidich, M. J., and S. U. Ho. 2005. Case 87: Subacute combined degeneration. *Radiology* 237 (1):101–5.

Nguyen, A., T. A. Rauch, G. P. Pfeifer, and V. W. Hu. 2010. Global methylation profiling of lymphoblastoid cell lines reveals epigenetic contributions to autism spectrum disorders and a novel autism candidate gene, RORA, whose protein product is reduced in autistic brain. *FASEB J* 24 (8):3036–51.

Niculescu, M. D., Y. Yamamuro, and S. H. Zeisel. 2004. Choline availability modulates human neuroblastoma cell proliferation and alters the methylation of the promoter region of the cyclin-dependent kinase inhibitor 3 gene. *J Neurochem* 89 (5):1252–59.

Noga, A. A., and D. E. Vance. 2003. A gender-specific role for phosphatidylethanolamine N-methyltransferase-derived phosphatidylcholine in the regulation of plasma high density and very low density lipoproteins in mice. *J Biol Chem* 278 (24):21851–59.

Oakley, G. P., Jr. 1999. Folic acid fortification. *N Engl J Med* 341 (12):922–23; author reply 924.

Okano, M., D. W. Bell, D. A. Haber, and E. Li. 1999. DNA methyltransferases Dnmt3a and Dnmt3b are essential for de novo methylation and mammalian development. *Cell* 99 (3):247–57.

Okochi, T., M. Ikeda, T. Kishi, K. Kawashima, Y. Kinoshita, T. Kitajima, Y. Yamanouchi, M. Tomita, T. Inada, N. Ozaki, and N. Iwata. 2009. Meta-analysis of association between genetic variants in COMT and schizophrenia: an update. *Schizophr Res* 110 (1–3):140–48.

O'Leary, V. B., A. Parle-McDermott, A. M. Molloy, P. N. Kirke, Z. Johnson, M. Conley, J. M. Scott, and J. L. Mills. 2002. MTRR and MTHFR polymorphism: link to Down syndrome? *Am J Med Genet* 107 (2):151–55.

Olteanu, H., T. Munson, and R. Banerjee. 2002. Differences in the efficiency of reductive activation of methionine synthase and exogenous electron acceptors between the common polymorphic variants of human methionine synthase reductase. *Biochemistry* 41 (45):13378–85.

Olteanu, H., K. R. Wolthers, A. W. Munro, N. S. Scrutton, and R. Banerjee. 2004. Kinetic and thermodynamic characterization of the common polymorphic variants of human methionine synthase reductase. *Biochemistry* 43 (7):1988–97.

Oostra, B. A., and R. Willemsen. 2009. FMR1: a gene with three faces. *Biochim Biophys Acta* 1790 (6):467–77.

Ormazabal, A., R. Artuch, M. A. Vilaseca, A. Aracil, and M. Pineda. 2005. Cerebrospinal fluid concentrations of folate, biogenic amines and pterins in Rett syndrome: treatment with folinic acid. *Neuropediatrics* 36 (6):380–85.

Ou, C. Y., R. E. Stevenson, V. K. Brown, C. E. Schwartz, W. P. Allen, M. J. Khoury, R. Rozen, G. P. Oakley Jr., and M. J. Adams Jr. 1996. 5,10 Methylenetetrahydrofolate reductase genetic polymorphism as a risk factor for neural tube defects. *Am J Med Genet* 63 (4):610–14.

Pardridge, W. M. 1986. Blood–brain transport of nutrients. Introduction. *Fed Proc* 45 (7):2047–49.

Pasca, S. P., B. Nemes, L. Vlase, C. E. Gagyi, E. Dronca, A. C. Miu, and M. Dronca. 2006. High levels of homocysteine and low serum paraoxonase 1 arylesterase activity in children with autism. *Life Sci* 78 (19):2244–48.

Patterson, D. 2008. Folate metabolism and the risk of Down syndrome. *Downs Syndr Res Pract* 12 (2):93–97.

Pei, L., H. Zhu, A. Ren, Z. Li, L. Hao, and R. H. Finnell. 2005. Reduced folate carrier gene is a risk factor for neural tube defects in a Chinese population. *Birth Defects Res A Clin Mol Teratol* 73 (6):430–33.

Petronijevic, N. D., N. V. Radonjic, M. D. Ivkovic, D. Marinkovic, V. D. Piperski, B. M. Duricic, and V. R. Paunovic. 2008. Plasma homocysteine levels in young male patients in the exacerbation and remission phase of schizophrenia. *Prog Neuropsychopharmacol Biol Psychiatry* 32 (8):1921–26.

Pfeiffer, S. I., J. Norton, L. Nelson, and S. Shott. 1995. Efficacy of vitamin B6 and magnesium in the treatment of autism: a methodology review and summary of outcomes. *J Autism Dev Disord* 25 (5):481–93.

Philibert, R., T. Gunter, N. Hollenbeck, W. J. Adams, P. Bohle, H. Packer, and H. Sandhu. 2006. No association of the C677T methylenetetrahydrofolate reductase polymorphism with schizophrenia. *Psychiatr Genet* 16 (5):221–23.

Portnoi, M. F. 2009. Microduplication 22q11.2: a new chromosomal syndrome. *Eur J Med Genet* 52 (2–3):88–93.

Posey, D. L., M. J. Khoury, J. Mulinare, M. J. Adams Jr., and C. Y. Ou. 1996. Is mutated MTHFR a risk factor for neural tube defects? *Lancet* 347 (9002):686–87.

Procter, A. 1991. Enhancement of recovery from psychiatric illness by methylfolate. *Br J Psychiatry* 159:271–72.

Pufulete, M., R. Al-Ghnaniem, A. Khushal, P. Appleby, N. Harris, S. Gout, P. W. Emery, and T. A. Sanders. 2005. Effect of folic acid supplementation on genomic DNA methylation in patients with colorectal adenoma. *Gut* 54 (5):648–53.

Pugh, C. S., and R. T. Borchardt. 1982. Effects of S-adenosylhomocysteine analogues on vaccinia viral messenger ribonucleic acid synthesis and methylation. *Biochemistry* 21 (7):1535–41.

Pyapali, G. K., D. A. Turner, C. L. Williams, W. H. Meck, and H. S. Swartzwelder. 1998. Prenatal dietary choline supplementation decreases the threshold for induction of long-term potentiation in young adult rats. *J Neurophysiol* 79 (4):1790–96.

Ramaekers, V. T., and N. Blau. 2004. Cerebral folate deficiency. *Dev Med Child Neurol* 46 (12):843–51.

Ramaekers, V. T., N. Blau, J. M. Sequeira, M. C. Nassogne, and E. V. Quadros. 2007. Folate receptor autoimmunity and cerebral folate deficiency in low-functioning autism with neurological deficits. *Neuropediatrics* 38 (6):276–81.

Ramaekers, V. T., S. I. Hansen, J. Holm, T. Opladen, J. Senderek, M. Hausler, G. Heimann, B. Fowler, R. Maiwald, and N. Blau. 2003. Reduced folate transport to the CNS in female Rett patients. *Neurology* 61 (4):506–15.

Ramsahoye, B. H., D. Biniszkiewicz, F. Lyko, V. Clark, A. P. Bird, and R. Jaenisch. 2000. Non-CpG methylation is prevalent in embryonic stem cells and may be mediated by DNA methyltransferase 3a. *Proc Natl Acad Sci U S A* 97 (10):5237–42.

Rapoport, J. L., A. M. Addington, S. Frangou, and M. R. Psych. 2005. The neurodevelopmental model of schizophrenia: update 2005. *Mol Psychiatry* 10 (5):434–49.

Ray, J. G., P. R. Wyatt, M. D. Thompson, M. J. Vermeulen, C. Meier, P. Y. Wong, S. A. Farrell, and D. E. Cole. 2007. Vitamin B12 and the risk of neural tube defects in a folic-acid-fortified population. *Epidemiology* 18 (3):362–66.

Regland, B., B. V. Johansson, and C. G. Gottfries. 1994. Homocysteinemia and schizophrenia as a case of methylation deficiency. *J Neural Transm Gen Sect* 98 (2):143–52.

Reik, W., W. Dean, and J. Walter. 2001. Epigenetic reprogramming in mammalian development. *Science* 293 (5532):1089–93.

Relton, C. L., C. S. Wilding, M. S. Pearce, A. J. Laffling, P. A. Jonas, S. A. Lynch, E. J. Tawn, and J. Burn. 2004. Gene-gene interaction in folate-related genes and risk of neural tube defects in a UK population. *J Med Genet* 41 (4):256–60.

Richardson, B. 2003. DNA methylation and autoimmune disease. *Clin Immunol* 109 (1):72–79.

Richardson, B. C. 2002. Role of DNA methylation in the regulation of cell function: autoimmunity, aging and cancer. *J Nutr* 132 (8 Suppl):2401S–2405S.

Richter, B., K. Stegmann, B. Roper, I. Boddeker, E. T. Ngo, and M. C. Koch. 2001. Interaction of folate and homocysteine pathway genotypes evaluated in susceptibility to neural tube defects (NTD) in a German population. *J Hum Genet* 46 (3):105–9.

Riley, E. P., and C. L. McGee. 2005. Fetal alcohol spectrum disorders: an overview with emphasis on changes in brain and behavior. *Exp Biol Med (Maywood)* 230 (6):357–65.

Rimland, B. 1973. High dosage levels of certain vitamins in the treatment of children with severe mental disorders. In *Orthomolecular psychiatry,* ed. D. Hawkins and L. Pauling. San Francisco: Freeman.

Rimland, B. 1974. An orthomolecular study of psychotic children. *Orthomolecular Psychiatry* 3:371–77.

Rimland, B. 1988. Controversies in the treatment of autistic children: vitamin and drug therapy. *J Child Neurol* 3 Suppl:S68–72.

Rimland, B., E. Callaway, and P. Dreyfus. 1978. The effect of high doses of vitamin B6 on autistic children: a double-blind crossover study. *Am J Psychiatry* 135 (4):472–75.

Robertson, K. D., and A. P. Wolffe. 2000. DNA methylation in health and disease. *Nat Rev Genet* 1 (1):11–19.

Roffman, J. L., R. L. Gollub, V. D. Calhoun, T. H. Wassink, A. P. Weiss, B. C. Ho, T. White, V. P. Clark, J. Fries, N. C. Andreasen, D. C. Goff, and D. S. Manoach. 2008. MTHFR 677C → T genotype disrupts prefrontal function in schizophrenia through an interaction with COMT 158Val → Met. *Proc Natl Acad Sci U S A* 105 (45):17573–78.

Roffman, J. L., A. P. Weiss, T. Deckersbach, O. Freudenreich, D. C. Henderson, S. Purcell, D. H. Wong, C. H. Halsted, and D. C. Goff. 2007. Effects of the methylenetetrahydrofolate reductase (MTHFR) C677T polymorphism on executive function in schizophrenia. *Schizophr Res* 92 (1–3):181–88.

Roffman, J. L., A. P. Weiss, T. Deckersbach, O. Freudenreich, D. C. Henderson, D. H. Wong, C. H. Halsted, and D. C. Goff. 2008. Interactive effects of COMT Val108/158Met and MTHFR C677T on executive function in schizophrenia. *Am J Med Genet B Neuropsychiatr Genet* 147B (6):990–95.

Roffman, J. L., A. P. Weiss, S. Purcell, C. A. Caffalette, O. Freudenreich, D. C. Henderson, T. Bottiglieri, D. H. Wong, C. H. Halsted, and D. C. Goff. 2008. Contribution of methylenetetrahydrofolate reductase (MTHFR) polymorphisms to negative symptoms in schizophrenia. *Biol Psychiatry* 63 (1):42–48.

Rogers, E. J. 2008. Has enhanced folate status during pregnancy altered natural selection and possibly autism prevalence? A closer look at a possible link. *Med Hypotheses* doi:10.1016/j.mehy.2008.04.013.

Rosenblatt, D. S., E. A. Duschenes, F. V. Hellstrom, M. S. Golick, M. J. Vekemans, S. F. Zeesman, and E. Andermann. 1985. Folic acid blinded trial in identical twins with fragile X syndrome. *Am J Hum Genet* 37 (3):543–52.

Rosenquist, T. H., S. A. Ratashak, and J. Selhub. 1996. Homocysteine induces congenital defects of the heart and neural tube: effect of folic acid. *Proc Natl Acad Sci U S A* 93 (26):15227–32.

Roth, T. L., F. D. Lubin, M. Sodhi, and J. E. Kleinman. 2009. Epigenetic mechanisms in schizophrenia. *Biochim Biophys Acta* 1790 (9):869–77.

Rothenberg, S. P., M. P. da Costa, J. M. Sequeira, J. Cracco, J. L. Roberts, J. Weedon, and E. V. Quadros. 2004. Autoantibodies against folate receptors in women with a pregnancy complicated by a neural-tube defect. *N Engl J Med* 350 (2):134–42.

Roza, S. J., T. van Batenburg-Eddes, E. A. Steegers, V. W. Jaddoe, J. P. Mackenbach, A. Hofman, F. C. Verhulst, and H. Tiemeier. 2009. Maternal folic acid supplement use in early pregnancy and child behavioural problems: the Generation R Study. *Br J Nutr:*1–8.

Rueda, J. R., J. Ballesteros, and M. I. Tejada. 2009. Systematic review of pharmacological treatments in fragile X syndrome. *BMC Neurol* 9 (1):53.

Samaco, R. C., A. Hogart, and J. M. LaSalle. 2005. Epigenetic overlap in autism-spectrum neurodevelopmental disorders: MECP2 deficiency causes reduced expression of UBE3A and GABRB3. *Hum Mol Genet* 14 (4):483–92.

Sandahl, B. 1977. Seasonal incidence of some congenital malformations in the central nervous system in Sweden, 1965–1972. *Acta Paediatr Scand* 66 (1):65–72.

Santos, F., B. Hendrich, W. Reik, and W. Dean. 2002. Dynamic reprogramming of DNA methylation in the early mouse embryo. *Dev Biol* 241 (1):172–82.

Sazci, A., E. Ergul, Y. Guzelhan, G. Kaya, and I. Kara. 2003. Methylenetetrahydrofolate reductase gene polymorphisms in patients with schizophrenia. *Brain Res Mol Brain Res* 117 (1):104–7.

Scala, I., B. Granese, M. Sellitto, S. Salome, A. Sammartino, A. Pepe, P. Mastroiacovo, G. Sebastio, and G. Andria. 2006. Analysis of seven maternal polymorphisms of genes involved in homocysteine/folate metabolism and risk of Down syndrome offspring. *Genet Med* 8 (7):409–16.

Schanen, N. C. 2006. Epigenetics of autism spectrum disorders. *Hum Mol Genet* 15 (spec no 2):R138–50.

Schatz, R. A., T. E. Wilens, and O. Z. Sellinger. 1981. Decreased transmethylation of biogenic amines after in vivo elevation of brain S-adenosyl-l-homocysteine. *J Neurochem* 36 (5):1739–48.

Schlotz, W., A. Jones, D. I. Phillips, C. R. Gale, S. M. Robinson, and K. M. Godfrey. 2010. Lower maternal folate status in early pregnancy is associated with childhood hyperactivity and peer problems in offspring. *J Child Psychol Psychiatry* 51 (5):594–602.

Scott, J. M. 1992. Folate-vitamin B12 interrelationships in the central nervous system. *Proc Nutr Soc* 51 (2):219–24.

Scott, J. M. 1999. Folate and vitamin B12. *Proc Nutr Soc* 58 (2):441–48.

Scott, J. M., J. J. Dinn, P. Wilson, and D. G. Weir. 1981. Pathogenesis of subacute combined degeneration: a result of methyl group deficiency. *Lancet* 2 (8242):334–37.

Scott, J. M., A. M. Molloy, D. G. Kennedy, S. Kennedy, and D. G. Weir. 1994. Effects of the disruption of transmethylation in the central nervous system: an animal model. *Acta Neurol Scand Suppl* 154:27–31.

Seshadri, S., and T. Gopaldas. 1989. Impact of iron supplementation on cognitive functions in preschool and school-aged children: the Indian experience. *Am J Clin Nutr* 50 (3 Suppl):675–84; discussion 685–86.

Seshadri, S., K. Hirode, P. Naik, and S. Malhotra. 1982. Behavioural responses of young anaemic Indian children to iron-folic acid supplements. *Br J Nutr* 48 (2):233–40.

Shaw, G. M., S. L. Carmichael, W. Yang, S. Selvin, and D. M. Schaffer. 2004. Periconceptional dietary intake of choline and betaine and neural tube defects in offspring. *Am J Epidemiol* 160 (2):102–9.

Shaw, G. M., R. H. Finnell, H. J. Blom, S. L. Carmichael, S. E. Vollset, W. Yang, and P. M. Ueland. 2009. Choline and risk of neural tube defects in a folate-fortified population. *Epidemiology* 20 (5):714–19.

Shaw, G. M., E. J. Lammer, H. Zhu, M. W. Baker, E. Neri, and R. H. Finnell. 2002. Maternal periconceptional vitamin use, genetic variation of infant reduced folate carrier (A80G), and risk of spina bifida. *Am J Med Genet* 108 (1):1–6.

Shaw, G. M., R. Rozen, R. H. Finnell, C. R. Wasserman, and E. J. Lammer. 1998. Maternal vitamin use, genetic variation of infant methylenetetrahydrofolate reductase, and risk for spina bifida. *Am J Epidemiol* 148 (1):30–37.

Shaw, G. M., D. Schaffer, E. M. Velie, K. Morland, and J. A. Harris. 1995. Periconceptional vitamin use, dietary folate, and the occurrence of neural tube defects. *Epidemiology* 6 (3):219–26.

Signore, C., P. M. Ueland, J. Troendle, and J. L. Mills. 2008. Choline concentrations in human maternal and cord blood and intelligence at 5 y of age. *Am J Clin Nutr* 87 (4):896–902.

Smithells, R. W., C. Ankers, M. E. Carver, D. Lennon, C. J. Schorah, and S. Sheppard. 1977. Maternal nutrition in early pregnancy. *Br J Nutr* 38 (3):497–506.

Smithells, R. W., S. Sheppard, and C. J. Schorah. 1976. Vitamin deficiencies and neural tube defects. *Arch Dis Child* 51:944–50.

Smythies, J. R., C. G. Gottfries, and B. Regland. 1997. Disturbances of one-carbon metabolism in neuropsychiatric disorders: a review. *Biol Psychiatry* 41 (2):230–33.

Speer, M. C., J. Nye, D. McLone, G. Worley, E. C. Melvin, K. D. Viles, A. Franklin, C. Drake, J. Mackey, and T. M. George. 1999. Possible interaction of genotypes at cystathionine beta-synthase and methylenetetrahydrofolate reductase (MTHFR) in neural tube defects. NTD Collaborative Group. *Clin Genet* 56 (2):142–44.

St. Clair, D., M. Xu, P. Wang, Y. Yu, Y. Fang, F. Zhang, X. Zheng, N. Gu, G. Feng, P. Sham, and L. He. 2005. Rates of adult schizophrenia following prenatal exposure to the Chinese famine of 1959–1961. *JAMA* 294 (5):557–62.

Stadler, F., G. Kolb, L. Rubusch, S. P. Baker, E. G. Jones, and S. Akbarian. 2005. Histone methylation at gene promoters is associated with developmental regulation and region-specific expression of ionotropic and metabotropic glutamate receptors in human brain. *J Neurochem* 94 (2):324–36.

Stanislawska-Sachadyn, A., K. S. Brown, L. E. Mitchell, J. V. Woodside, I. S. Young, J. M. Scott, L. Murray, C. A. Boreham, H. McNulty, J. J. Strain, and A. S. Whitehead. 2008. An insertion/deletion polymorphism of the dihydrofolate reductase (DHFR) gene is associated with serum and red blood cell folate concentrations in women. *Hum Genet* 123 (3):289–95.

Steegers-Theunissen, R. P., G. H. Boers, F. J. Trijbels, and T. K. Eskes. 1991. Neural-tube defects and derangement of homocysteine metabolism. *N Engl J Med* 324 (3):199–200.

Steegers-Theunissen, R. P., G. H. Boers, F. J. Trijbels, J. D. Finkelstein, H. J. Blom, C. M. Thomas, G. F. Borm, M. G. Wouters, and T. K. Eskes. 1994. Maternal hyperhomocysteinemia: a risk factor for neural-tube defects? *Metabolism* 43 (12):1475–80.

Steegers-Theunissen, R. P., S. A. Obermann-Borst, D. Kremer, J. Lindemans, C. Siebel, E. A. Steegers, P. E. Slagboom, and B. T. Heijmans. 2009. Periconceptional maternal folic acid use of 400 microg per day is related to increased methylation of the IGF2 gene in the very young child. *PLoS One* 4 (11):e7845.

Stern, L. L., J. B. Mason, J. Selhub, and S. W. Choi. 2000. Genomic DNA hypomethylation, a characteristic of most cancers, is present in peripheral leukocytes of individuals who are homozygous for the C677T polymorphism in the methylenetetrahydrofolate reductase gene. *Cancer Epidemiol Biomarkers Prev* 9 (8):849–53.

Strom, C. M., R. M. Brusca, and W. J. Pizzi. 1992. Double-blind, placebo-controlled crossover study of folinic acid (Leucovorin for the treatment of fragile X syndrome. *Am J Med Genet* 44 (5):676–82.

Susser, E., R. Neugebauer, and H.W. Hoek. 1996. Schizofrenia after prenatal famine. *Arch Gen Psychiatry* 53 (1):25–31.

Tallur, K. K., D. A. Johnson, J. M. Kirk, P. A. Sandercock, and R. A. Minns. 2005. Folate-induced reversal of leukoencephalopathy and intellectual decline in methylene-tetrahydrofolate reductase deficiency: variable response in siblings. *Dev Med Child Neurol* 47 (1):53–56.

Tamura, T., R. L. Goldenberg, V. R. Chapman, K. E. Johnston, S. L. Ramey, and K. G. Nelson. 2005. Folate status of mothers during pregnancy and mental and psychomotor development of their children at five years of age. *Pediatrics* 116 (3):703–8.

Tamura, T., and E. L. Stokstad. 1973. The availability of food folate in man. *Br J Haematol* 25 (4):513–32.

Taparia, S., J. Gelineau-van Waes, T. H. Rosenquist, and R. H. Finnell. 2007. Importance of folate-homocysteine homeostasis during early embryonic development. *Clin Chem Lab Med* 45 (12):1717–27.

Temudo, T., M. Rios, C. Prior, I. Carrilho, M. Santos, P. Maciel, J. Sequeiros, M. Fonseca, J. Monteiro, P. Cabral, J. P. Vieira, A. Ormazabal, and R. Artuch. 2009. Evaluation of CSF neurotransmitters and folate in 25 patients with Rett disorder and effects of treatment. *Brain Dev* 31 (1):46–51.

Thatcher, K. N., S. Peddada, D. H. Yasui, and J. M. Lasalle. 2005. Homologous pairing of 15q11-13 imprinted domains in brain is developmentally regulated but deficient in Rett and autism samples. *Hum Mol Genet* 14 (6):785–97.

Thomas, J. D., E. J. Abou, and H. D. Dominguez. 2009. Prenatal choline supplementation mitigates the adverse effects of prenatal alcohol exposure on development in rats. *Neurotoxicol Teratol* 31 (5):303–11.

Thomas, J. D., J. S. Biane, K. A. O'Bryan, T. M. O'Neill, and H. D. Dominguez. 2007. Choline supplementation following third-trimester-equivalent alcohol exposure attenuates behavioral alterations in rats. *Behav Neurosci* 121 (1):120–30.

Thomas, J. D., M. Garrison, and T. M. O'Neill. 2004. Perinatal choline supplementation attenuates behavioral alterations associated with neonatal alcohol exposure in rats. *Neurotoxicol Teratol* 26 (1):35–45.

Thomas, J. D., M. H. La Fiette, V. R. Quinn, and E. P. Riley. 2000. Neonatal choline supplementation ameliorates the effects of prenatal alcohol exposure on a discrimination learning task in rats. *Neurotoxicol Teratol* 22 (5):703–11.

Thompson, M. D., D. E. Cole, and J. G. Ray. 2009. Vitamin B-12 and neural tube defects: the Canadian experience. *Am J Clin Nutr* 89 (2):697S–701S.

Tobi, E. W., L. H. Lumey, R. P. Talens, D. Kremer, H. Putter, A. D. Stein, P. E. Slagboom, and B. T. Heijmans. 2009. DNA methylation differences after exposure to prenatal famine are common and timing- and sex-specific. *Hum Mol Genet* 18 (21):4046–53.

Todd, R. D. 1992. Neural development is regulated by classical neurotransmitters: dopamine D2 receptor stimulation enhances neurite outgrowth. *Biol Psychiatry* 31 (8):794–807.

Trimble, M. R., J. A. Corbett, and D. Donaldson. 1980. Folic acid and mental symptoms in children with epilepsy. *J Neurol Neurosurg Psychiatry* 43 (11):1030–34.

Tucker, K. L. 2001. Methylated cytosine and the brain: a new base for neuroscience. *Neuron* 30 (3):649–52.

Turner, A. J. 1977. Commentary: The roles of folate and pteridine derivatives in neurotransmitter metabolism. *Biochem Pharmacol* 26 (11):1009–14.

Ubbink, J. B. 1995. Is an elevated circulating maternal homocysteine concentration a risk factor for neural tube defects? *Nutr Rev* 53 (6):173–75.

van den Veyver, I. B. 2002. Genetic effects of methylation diets. *Annu Rev Nutr* 22:255–82.

van der Linden, I. J., L. A. Afman, S. G. Heil, and H. J. Blom. 2006. Genetic variation in genes of folate metabolism and neural-tube defect risk. *Proc Nutr Soc* 65 (2):204–15.

van der Put, N. M., and H. J. Blom. 2000. Neural tube defects and a disturbed folate dependent homocysteine metabolism. *Eur J Obstet Gynecol Reprod Biol* 92 (1):57–61.

van der Put, N. M., F. Gabreels, E. M. Stevens, J. A. Smeitink, F. J. Trijbels, T. K. Eskes, L. P. van den Heuvel, and H. J. Blom. 1998. A second common mutation in the methylenetetrahydrofolate reductase gene: an additional risk factor for neural-tube defects? *Am J Hum Genet* 62 (5):1044–51.

van der Put, N. M., R. P. Steegers-Theunissen, P. Frosst, F. J. Trijbels, T. K. Eskes, L. P. van den Heuvel, E. C. Mariman, M. den Heyer, R. Rozen, and H. J. Blom. 1995. Mutated methylenetetrahydrofolate reductase as a risk factor for spina bifida. *Lancet* 346 (8982):1070–71.

van der Put, N. M., C. M. Thomas, T. K. Eskes, F. J. Trijbels, R. P. Steegers-Theunissen, E. C. Mariman, A. De Graaf-Hess, J. A. Smeitink, and H. J. Blom. 1997. Altered folate and vitamin B12 metabolism in families with spina bifida offspring. *QJM* 90 (8):505–10.

van der Put, N. M., H. W. van Straaten, F. J. Trijbels, and H. J. Blom. 2001. Folate, homocysteine and neural tube defects: an overview. *Exp Biol Med (Maywood)* 226 (4):243–70.

Vares, M., P. Saetre, H. Deng, G. Cai, X. Liu, T. Hansen, H. B. Rasmussen, T. Werge, I. Melle, S. Djurovic, O. A. Andreassen, I. Agartz, H. Hall, L. Terenius, and E. G. Jonsson. 2009. Association between methylenetetrahydrofolate reductase (MTHFR) C677T polymorphism and age of onset in schizophrenia. *Am J Med Genet B Neuropsychiatr Genet* (e-pub).

Vilella, E., C. Virgos, M. Murphy, L. Martorell, J. Valero, J. M. Simo, J. Joven, J. Fernandez-Ballart, and A. Labad. 2005. Further evidence that hyperhomocysteinemia and methylenetetrahydrofolate reductase C677T and A1289C polymorphisms are not risk factors for schizophrenia. *Prog Neuropsychopharmacol Biol Psychiatry* 29 (7):1169–74.

Virgos, C., L. Martorell, J. M. Simo, J. Valero, L. Figuera, J. Joven, A. Labad, and E. Vilella. 1999. Plasma homocysteine and the methylenetetrahydrofolate reductase C677T gene variant: lack of association with schizophrenia. *Neuroreport* 10 (10):2035–38.

Volcik, K. A., G. M. Shaw, E. J. Lammer, H. Zhu, and R. H. Finnell. 2003. Evaluation of infant methylenetetrahydrofolate reductase genotype, maternal vitamin use, and risk of high versus low level spina bifida defects. *Birth Defects Res A Clin Mol Teratol* 67 (3):154–57.

Volcik, K. A., G. M. Shaw, H. Zhu, E. J. Lammer, C. Laurent, and R. H. Finnell. 2003. Associations between polymorphisms within the thymidylate synthase gene and spina bifida. *Birth Defects Res A Clin Mol Teratol* 67 (11):924–28.

Wakefield, A. J., S. H. Murch, A. Anthony, J. Linnell, D. M. Casson, M. Malik, M. Berelowitz, A. P. Dhillon, M. A. Thomson, P. Harvey, A. Valentine, S. E. Davies, and J. A. Walker-Smith. 1998. Ileal-lymphoid-nodular hyperplasia, non-specific colitis, and pervasive developmental disorder in children. *Lancet* 351 (9103):637–41.

Wald, N. J., A. D. Hackshaw, R. Stone, and N. A. Sourial. 1996. Blood folic acid and vitamin B12 in relation to neural tube defects. *Br J Obstet Gynaecol* 103 (4):319–24.

Ward, B. C., N. H. Kolodny, N. Nag, and J. E. Berger-Sweeney. 2009. Neurochemical changes in a mouse model of Rett syndrome: changes over time and in response to perinatal choline nutritional supplementation. *J Neurochem* 108 (2):361–71.

Waterland, R. A., and R. L. Jirtle. 2003. Transposable elements: targets for early nutritional effects on epigenetic gene regulation. *Mol Cell Biol* 23 (15):5293–5300.

Waterland, R. A., and R. L. Jirtle. 2004. Early nutrition, epigenetic changes at transposons and imprinted genes, and enhanced susceptibility to adult chronic diseases. *Nutrition* 20 (1):63–68.

Wehby, G. L., and J. C. Murray. 2008. The effects of prenatal use of folic acid and other dietary supplements on early child development. *Matern Child Health J* 12 (2):180–7.

Whitehead, A. S., P. Gallagher, J. L. Mills, P. N. Kirke, H. Burke, A. M. Molloy, D. G. Weir, D. C. Shields, and J. M. Scott. 1995. A genetic defect in 5,10 methylenetetrahydrofolate reductase in neural tube defects. *Q J Med* 88 (11):763–66.

Wighton, M. C., J. I. Manson, I. Speed, E. Robertson, and E. Chapman. 1979. Brain damage in infancy and dietary vitamin B12 deficiency. *Med J Aust* 2 (1):1–3.

Williams, C. L., W. H. Meck, D. D. Heyer, and R. Loy. 1998. Hypertrophy of basal forebrain neurons and enhanced visuospatial memory in perinatally choline-supplemented rats. *Brain Res* 794 (2):225–38.

Wilson, A., R. Platt, Q. Wu, D. Leclerc, B. Christensen, H. Yang, R. A. Gravel, and R. Rozen. 1999. A common variant in methionine synthase reductase combined with low cobalamin (vitamin B12) increases risk for spina bifida. *Mol Genet Metab* 67 (4):317–23.

Wong-Goodrich, S. J., M. J. Glenn, T. J. Mellott, J. K. Blusztajn, W. H. Meck, and C. L. Williams. 2008. Spatial memory and hippocampal plasticity are differentially sensitive to the availability of choline in adulthood as a function of choline supply in utero. *Brain Res* 1237:153–66.

Yates, Z., and M. Lucock. 2003. Interaction between common folate polymorphisms and B-vitamin nutritional status modulates homocysteine and risk for a thrombotic event. *Mol Genet Metab* 79 (3):201–13.

Yi, P., S. Melnyk, M. Pogribna, I. P. Pogribny, R. J. Hine, and S. J. James. 2000. Increase in plasma homocysteine associated with parallel increases in plasma S-adenosylhomocysteine and lymphocyte DNA hypomethylation. *J Biol Chem* 275 (38):29318–23.

Yu, L., T. Li, Z. Robertson, J. Dean, N. F. Gu, G. Y. Feng, P. Yates, M. Sinclair, C. Crombie, D. A. Collier, N. Walker, L. He, and D. St. Clair. 2004. No association between polymorphisms of methylenetetrahydrofolate reductase gene and schizophrenia in both Chinese and Scottish populations. *Mol Psychiatry* 9 (12):1063–65.

Zeisel, S. H. 2006. The fetal origins of memory: the role of dietary choline in optimal brain development. *J Pediatr* 149 (5 Suppl):S131–36.

Zeisel, S. H. 2009. Importance of methyl donors during reproduction. *Am J Clin Nutr* 89 (2):673S–677S.

Zeisel, S. H., and M. D. Niculescu. 2006. Perinatal choline influences brain structure and function. *Nutr Rev* 64 (4):197–203.

Zhang, T., R. Xin, X. Gu, F. Wang, L. Pei, L. Lin, G. Chen, J. Wu, and X. Zheng. 2009. Maternal serum vitamin B12, folate and homocysteine and the risk of neural tube defects in the offspring in a high-risk area of China. *Public Health Nutr* 12 (5):680–86.

Zintzaras, E. 2006. C677T and A1298C methylenetetrahydrofolate reductase gene polymorphisms in schizophrenia, bipolar disorder and depression: a meta-analysis of genetic association studies. *Psychiatr Genet* 16 (3):105–15.

Zittoun, J. 1993. Anemias due to disorder of folate, vitamin B_{12} and transcobalamin metabolism. *La Revue du praticien* 43 (11):1358–63.

14 Dietary Factors and the Emerging Role of Epigenetics in Neurodegenerative Diseases

Lucia Migliore and Fabio Coppedè

CONTENTS

14.1 Nutrition, Evolution, and Diseases 363
14.2 Nutrition and the Epigenome: Folate Metabolism and DNA Methylation ... 364
14.3 Oxidative Stress and Neurodegeneration: The Link between Oxidative Stress, DNA Repair, and DNA Methylation 366
14.4 Epigenetic Studies in AD (Including Cellular and Animal Models) 368
14.5 Epigenetic Studies in Neurodegenerative Diseases Other than AD 369
14.6 Dietary Factors in the Treatment and Prevention of Neurodegenerative Diseases: Possible Epigenetic Implications 369
14.7 Conclusions 372
References 374

14.1 NUTRITION, EVOLUTION, AND DISEASES

There is evidence of a strong relationship between human dietary habits and changes of the human genome, since the beginning of the evolution of our species. Today, population growth, globalization, and economic pressure powerfully affect diets worldwide. It is generally accepted that socioeconomic status influences dietary habits as well as health-related outcomes in various parts of the world (Vlismas et al. 2009). The modern environment encourages a sedentary lifestyle and provides easy access to processed food, which leads to a reduction of energy expenditure and increased caloric intake (Y. S. Lee 2009). Indeed, dietary habits affect several human diseases, including among others, obesity, cardiovascular disease, diabetes, cancer, and age-related neurodegenerative diseases (Joseph et al. 2009). Within this context, recent studies suggest that the reduction of caloric intake and the consumption of diets rich in antioxidants and anti-inflammatory components, such as those found in fruits, nuts, vegetables, and spices, may lower age-related cognitive declines and the risk of developing neurodegenerative disease (Joseph et al. 2009). Studies in rodents (see Sections 14.3 and 14.4) have demonstrated that early-life exposure to neurotoxic compounds during brain development or dietary modifications can modify the epigenome with consequences on the levels of expression of Alzheimer's disease (AD)-related genes later in life (Fuso et al. 2009; Zawia et al. 2009). These studies have been paralleled by others performed in human neuroblastoma cell cultures and demonstrate that the deprivation of B vitamins from the media resulted in epigenetic modifications and altered expression of AD-related genes (Fuso et al. 2005, 2007). Overall, there is increasing indication that environmental and particularly dietary factors could affect neurodegeneration by modifying the epigenome. The aim of this chapter is to review the possible epigenetic effects of dietary factors and their relevance to neurodegeneration and neuroprotection.

14.2 NUTRITION AND THE EPIGENOME: FOLATE METABOLISM AND DNA METHYLATION

Folates are essential nutrients required for one-carbon biosynthetic and epigenetic processes. They are derived entirely from dietary sources, mainly from the consumption of green vegetables, fruits, cereals, and meat. Folic acid is the synthetic form added to foods and found in dietary supplements. After intestinal absorption, folate metabolism requires reduction and methylation into the liver to form 5-methyltetrahydrofolate (5-methylTHF), release into the blood, and cellular uptake; then it can be used for the synthesis of DNA and RNA precursors or for the conversion of homocysteine (Hcy) to methionine, which is then used to form the main DNA methylating agent S-adenosylmethionine (SAM). Folic acid is converted to a natural biological form of the vitamin as it passes through the intestinal wall, with enzymatic reduction and methylation resulting in the circulating form of the vitamin, 5-methylTHF (Bailey and Gregory 1999; Coppedè 2009).

Folate does not cross biological membranes by diffusion alone, but requires several transport systems to enter the cells, the best characterized being the reduced folate carrier (RFC1). Methylenetetrahydrofolate reductase (MTHFR) is the first enzyme in the DNA methylation pathway since it reduces 5,10-methylentetrahydrofolate (5,10-MTHF) to 5-methylTHF. Subsequently, methionine synthase (MTR) transfers a methyl group from 5-methylTHF to homocysteine (Hcy), forming methionine and tetrahydrofolate (THF). Methionine is then converted to SAM in a reaction catalyzed by methionine adenosyltransferase (MAT). Most of the SAM generated is used in transmethylation reactions, whereby SAM is converted to S-adenosyl homocysteine (SAH) by transferring the methyl group to diverse biological acceptors, including proteins and DNA. Cobalamin (or vitamin B_{12}) is a cofactor of MTR, and methionine synthase reductase (MTRR) is required for the maintenance of MTR in its active state. If not converted into methionine, Hcy can be condensed with serine to form cystathionine in a reaction catalyzed by cystathionine β-synthase (CBS), which requires vitamin B_6 as a cofactor. Cystathionine can then be utilized to form the antioxidant compound glutathione (GSH). Indeed, SAM allosterically regulates

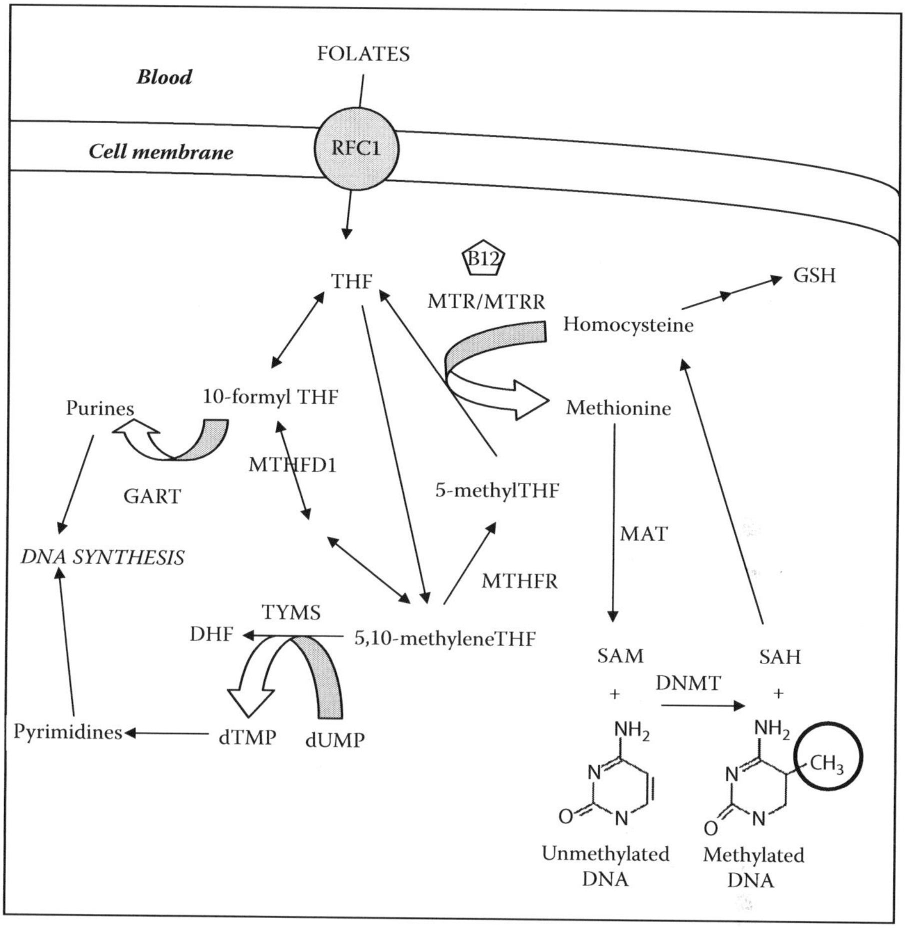

FIGURE 14.1 Simplified overview of the human folate metabolic pathway. *Enzymes:* DNMT: DNA methyltransferase; GART: phosphoribosylglycineamide transformylase; MAT: methionine adenosyltransferase; MTHFD1: methylenetetrahydrofolate dehydrogenase; MTHFR: methylenetetrahydrofolate reductase; MTR: methionine synthase; MTRR: methionine synthase reductase; RFC1: reduced folate carrier; TYMS: thymidilate synthase. *Metabolites:* DHF: dihydrofolate; GSH: glutathione; THF: tetrahydrofolate; dTMP: deoxythymidine monophosphate; dUMP: deoxyuridine monophosphate; SAH: S-adenosylhomocysteine SAM: S-adenosylmethionine. *Cofactors:* B_{12}: vitamin B_{12}.

the activity of CBS (Coppedè 2009). A diagram illustrating folate metabolism is shown in Figure 14.1.

A deficiency in cellular folates results in aberrant DNA methylation, point mutations, chromosome breakage, and increased frequency of micronuclei (MN), defective chromosome recombination, and aneuploidy (Fenech, 2001). Impaired folate metabolism, resulting from the presence of common functional polymorphisms of genes encoding for metabolic enzymes, has been associated with several human diseases including cardiovascular diseases (Smulders and Stehouwer 2005; Trabetti 2008), various kinds of cancer

(Bolufer et al. 2006; Kono and Chen 2005; Lin et al. 2007; Skibola et al. 2004), neural tube defects (Molloy et al. 2009; van der Put et al. 2001), Down syndrome (Coppedè 2009), and neurodegenerative diseases (Anello et al. 2004; Bi et al. 2009). Moreover, recent studies in cells and animal models indicate that low folate levels induce epigenetic modifications, possibly affecting the expression of several genes (Linhart et al. 2009; Pogribny et al. 2008; Wasson et al. 2006). However, studies attempting to demonstrate whether or not folate deficiency from the diet induces global DNA hypomethylation have yielded conflicting or opposite results, depending on the cells or the organs tested (Maloney et al. 2007; Pogribny et al. 2008; Wasson et al. 2006).

There is indication that plasma Hcy levels are increased in AD patients and folate levels decreased. The levels of vitamins B_{12} and B_6 have been studied less frequently than folate and Hcy, but there is some indication that also vitamin B_{12} levels might be decreased in AD patients (Van Dam and Van Gool 2009). We have recently genotyped 118 AD patients and 105 controls for the presence of the *MTHFR* 677C>T polymorphism. No significant difference in *MTHFR* 677T allele and genotype frequencies was observed between the two groups. We also measured blood levels of folate, Hcy, and vitamin B_{12} in a subgroup of AD patients and controls, observing significantly increased Hcy levels in AD patients, lower folate levels in AD subjects, and vitamin B_{12} levels that were similar between the two groups. An inverse correlation between blood Hcy and folate values was found in AD subjects, and the *MTHFR* 677TT genotype was associated with higher plasma Hcy levels (Coppedè, Ricci et al. 2009). Hyperhomocysteinemia has been repeatedly reported in Parkinson's disease (PD) patients; the increase, however, seems mostly related to the methylated catabolism of l-Dopa, the main pharmacological treatment of PD. Therefore hyperhomocysteinemia may not be specific to this movement disorder, the condition being, in fact, rather the result of the combinations of different factors, including the pharmacological treatment (Martignoni et al. 2007). Elevated plasma Hcy levels have been observed in amyotrophic lateral sclerosis (ALS) patients (Zoccolella et al. 2008), and studies in ALS animal models have shown that folic acid protects motor neurons against increased Hcy, inflammation, and apoptosis (Zhang et al. 2008). Cerebral folate deficiency (CFD) is associated with low levels of 5-methylTHF in the cerebrospinal fluid (CSF) with normal folate levels in the plasma and red blood cells. The onset of symptoms caused by the deficiency of folates in the brain is at around 4 to 6 months of age. This is followed by delayed development, with deceleration of head growth, hypotonia, and ataxia, followed in one-third of children by dyskinesias (choreo-athetosis, hemiballismus), spasticity, speech difficulties, and epilepsy (Gordon 2009). The possible contribution of folate deficiency to epigenetic modifications in patients with neurodegenerative diseases will be discussed in the next sections.

14.3 OXIDATIVE STRESS AND NEURODEGENERATION: THE LINK BETWEEN OXIDATIVE STRESS, DNA REPAIR, AND DNA METHYLATION

Oxidative stress describes a condition in which cellular antioxidant defenses are insufficient to keep the levels of reactive oxygen species (ROS) below a toxic threshold. Increasing evidence suggests crucial implications for oxidative stress in several steps of the pathogenesis of many neurodegenerative diseases, and the current opinion is that it could have a causative role, instead of being an epiphenomenon of the pathological processes. Significant biological changes related to a condition of oxidative stress have been found not only in brain tissues but also in peripheral tissues of individuals affected by AD, mild cognitive impairment (MCI; which is a pre-dementia phase preceding AD), PD, ALS, and Huntington's disease (HD) among others (Lovell et al., 1999, 2007; Migliore, Fontana, Colognato et al. 2005; Migliore, Fontana, Trippi et al. 2005). Overall, there is substantial evidence indicating that a condition of increased oxidative damage to lipids, proteins, and nucleic acids characterizes neurodegenerative diseases (Mancuso et al., 2006). We have recently reviewed environmental factors that could contribute to a condition of increased oxidative stress in neurodegenerative diseases (Migliore and Coppedè 2009a, 2009b): among them metals, pesticides, air pollutants, and the inflammation resulting from head traumas could play an important role in increasing oxidative damage, whereas the consumption of fruits, fish, and vegetables, rich in antioxidant compounds, could partially counteract human exposure to oxidant compounds (see Section 14.6). Glutathione is one of most important endogenous antioxidant compounds deriving from the trans-sulfuration pathway of Hcy. There is some indication that GSH levels are lower in AD brain regions and blood cells; moreover, the GSH content is significantly lower in the substantia nigra of PD patients (Ballatori et al. 2009). We measured oxidative DNA damage and glutathione S-transferase (GST) activity in plasma from PD patients and controls. Oxidative damage to purine bases was higher in PD patients, and the GST enzymatic activity in PD patients was lower than in healthy controls (Migliore et al. 2001).

Increasing evidence suggests that neurodegenerative diseases are also characterized by deficiencies of proteins involved in the repair of oxidative DNA damage (Cardozo-Pelaez et al. 2005; Coppedè et al. 2007, 2010; Coppedè, Migheli, Lo Gerfo 2009; Coppedè, Migheli, Ceravolo 2009, 2010; Coppedè and Migliore 2009, 2010; Fukae et al. 2005; Jarem et al. 2009; Wong et al. 2008). Recently Zawia et al. (2009) proposed a model linking epigenetic modifications, oxidative DNA damage, DNA repair, and AD. One of the most studied epigenetic modifications is the change of methylation patterns of CpG-rich regions in the promoter of specific genes, resulting in gene silencing (hypermethylation) or overexpression (hypomethylation). The authors observed that environmental influences occurring during brain development of rats, such as exposure to the xenobiotic metal lead (Pb), inhibit DNA-methyltransferases, thus resulting in hypomethylation of the promoters of genes associated with AD, such as *APP* (the gene encoding the amyloid precursor protein) and *BACE1* (the β-secretase gene that cleaves the amyloid precursor protein APP). This early life imprint was sustained and triggered later in life to increase the levels of APP and amyloid-beta (Aβ). These latent effects were accompanied by an increase of cerebral oxoguanine (8-oxo-G) levels, indicating that epigenetic imprinting in early life influenced the expression of AD-related genes and promoted DNA damage and AD pathogenesis. Whereas AD-related genes were overexpressed late in life, others were repressed, suggesting that the early life perturbations resulted in hypomethylation of some genes as well as hypermethylation of others (Basha et al. 2005; Bolin et al. 2006; Wu et al. 2008). However, hypermethylated genes are more susceptible to Aβ-induced oxidative DNA damage since methylation of cytosines at CpG dinucleotides restricts the ability of oxoguanine DNA glycosylase (OGG1) to repair an adjacent 8-oxo-G. Therefore the authors concluded that although the conditions leading to early life hypo- or hypermethylation of specific genes are not yet fully understood, these changes can have an impact on gene expression and imprint susceptibility to oxidative DNA damage in the aged brain (Zawia et al. 2009). Kruman and colleagues (2002) observed that when maintained on a folic acid-deficient diet, *APP* mutant transgenic mice exhibited increased oxidative DNA damage and hippocampal neurodegeneration. The authors suggested that folic acid deficiency could impair DNA repair in neurons, which sensitizes them to oxidative DNA damage induced by Aβ.

14.4 EPIGENETIC STUDIES IN AD (INCLUDING CELLULAR AND ANIMAL MODELS)

Alongside the studies performed by Zawia and colleagues on AD animal models described in the previous section, several studies performed on neuroblastoma cells suggest that the manipulation of environmental factors can epigenetically modify the expression of AD-related genes and proteins. Particularly, the levels of methylation of CpG islands in the promoters of the *APP* and the *PSEN*1 (Presenilin 1, the core of the γ-secretase activity that cleaves APP) genes were analyzed on human neuroblastoma SK-N-SH or SK-N-BE cell lines, and it was observed that under conditions of folate and vitamin B_{12} deprivation from the media, the status of methylation of the promoter of the *PSEN1* gene underwent a variation, with a subsequent deregulation of the production of presenilin1, BACE, and APP proteins (Fuso et al. 2005). Both γ-secretase and β-secretase are required during the amyloidogenic cleavage of APP, leading to the formation of Aβ fragments. Therefore this study confirmed that some of the genes responsible for the production of Aβ fragments in AD can be regulated through an epigenetic mechanism depending on the cellular availability of folate and B_{12} vitamins, and involving the production of SAM and the status of methylation of CpG islands in the DNA. Subsequently, the authors observed, on the same experimental model, that Hcy accumulation induced through vitamin B deprivation could impair the "methylation potential" of the cells with subsequent presenilin 1, BACE, and amyloid-beta up-regulation (Fuso et al. 2007). The same group recently observed that B-vitamin (folate, vitamin B_{12}, and vitamin B_6) deprivation induces hyperhomocysteinemia and brain SAH, depletes brain SAM, and enhances PSEN1 and BACE expression and Aβ deposition in mice (Fuso et al. 2008). Similarly, they observed that B-vitamin deprivation induces *PSEN1* epigenetic changes in mice (Fuso et al. 2009).

Recent studies on murine cerebral endothelial cells have demonstrated that Aβ reduces global DNA methylation while increasing DNA methylation of the gene-encoding neprilysin (NEP), one of the enzymes responsible for Aβ degradation, thus suppressing the NEP expression in mRNA and protein levels. These results indicate that Aβ induces epigenetic effects, suggesting that DNA methylation might be part of a vicious cycle involving the reduction in NEP expression along with a resultant increase in Aβ accumulation, which in turn induces global DNA hypomethylation (Chen et al. 2009). Studies performed in postmortem brain samples and lymphocytes of late-onset AD patients revealed a notably age-specific epigenetic drift, supporting a potential role of epigenetic effects in the development of the disease. Particularly, some genes that participate in Aβ processing (*PSEN1*, *APOE*) and methylation homeostasis (*MTHFR, DNMT1*) showed a significant interindividual epigenetic variability, which could contribute to AD pathology (Wang et al. 2008).

14.5 EPIGENETIC STUDIES IN NEURODEGENERATIVE DISEASES OTHER THAN AD

Studies about the possible role of epigenetics in other neurodegenerative diseases are still in their infancy. Few data exists on the aberrant DNA methylation patterns and histone modification profiles of DNA sequences of genes that have a fundamental role in neurodegenerative diseases other than AD, such as PD, HD, and ALS. To explain the variable phenotypic expressivity (the age of onset, severity, and/or penetrance of the pathological phenotype), disturbances in methylation levels have been involved. Moreover, observations on whether a particular disease is more commonly inherited from an individual's mother or father are suggestive of an involvement of imprinted genes (Chao et al. 2009). DNA methylation is dynamically regulated in the human cerebral cortex throughout the life span, involves differentiated neurons, and affects a substantial portion of genes predominantly by an age-related increase (Siegmund et al. 2007). Sporadic ALS (SALS) results from the death of motor neurons in the brain and spinal cord. It has been proposed that epigenetic silencing of genes vital for motor neuron function could underlie SALS. However, the promoter of genes thought to be implicated in SALS—*SOD1* and *VEGF*, or two common human isoforms of the metallothionein family—has not been found with inappropriate methylation levels (Oates and Pamphlett 2007; Morahan et al. 2007). In PD patients the TNF-alpha promoter DNA from substantia nigra was found significantly less methylated in comparison to DNA from cortex; however, although there was a tendency for hypomethylation in PD, the analysis revealed no particular pattern in PD patients compared to controls (Pieper et al. 2008).

Some of the DNA methylations and histone modifications observed in neurodegenerative disorders such as AD, PD, and HD, and in other neurological disorders such as multiple sclerosis, epilepsy, and ALS (see the recent review by Urdinguio et al. 2009) might prove to be useful as early biomarker indicators of disease, while they might also be targeted with epigenetic drugs such as DNA-demethylating drugs and histone demethylase inhibitors (already clinically approved for the treatment of subtypes of leukaemia and lymphoma) (Urdinguio et al. 2009).

14.6 DIETARY FACTORS IN THE TREATMENT AND PREVENTION OF NEURODEGENERATIVE DISEASES: POSSIBLE EPIGENETIC IMPLICATIONS

The findings on the early involvement of oxidative stress in many neurodegenerative diseases (Keller et al. 2005; Migliore, Fontana, Trippi et al. 2005; Moreira et al. 2007; Chang et al. 2008) have led to the idea that lifestyle factors, and especially the diet, may counteract oxidative damage and could be of help in aging and neurodegeneration. In general the potential therapeutical treatments are currently divided in two different categories: vitamin and nonvitamin antioxidants, mainly including the phytochemicals.

Studies in vitro or in animal models have indicated that many compounds can decrease neurodegeneration, excitotoxicity, oxidative stress, apoptosis, protein aggregation, and disturbance of Ca2+ homeostasis, and compensate the energy impairment. Zhao (2009) reviewed the studies on the protecting effects of natural antioxidants on in vitro models of neurodegenerative diseases. The protective effect of green tea polyphenols, such as epigallocatechin-3-gallate (EGCG), on neurons against apoptosis has been shown in cellular and animal PD models. Also nicotine and genistein, the most active component of soy isoflavone, have been found to protect against Aβ-induced apoptosis of hippocampal neurons in transgenic mouse AD models (Zhao 2009). The potential therapeutic efficacy of creatine, coenzyme Q10, idebenone, synthetic triterpenoids, and mitochondria- targeted antioxidants and peptides (SS-31) in in vitro studies and in animal models of PD, HD, ALS, and AD has been reviewed by Chaturvedi and Beal (2008).

It is well known that many vitamins directly scavenge ROS but can also up-regulate the antioxidant capacity of the oxidative defense system of the body. Among the antioxidant treatments, those using vitamin E, vitamin E analogs, and vitamin C have been evaluated over several years. Oral supplementation of vitamin C and vitamin E alone and in combination have been shown to decrease oxidative DNA damage in animal studies in vivo, in vitro, and in situ. A diet fortified with antioxidants (vitamin E, vitamin C, alpha-lipoic acid, l-carnitine) in combination with a program of behavioral enrichment was capable of reducing the levels of oxidative damage and increasing the activity and expression of key endogenous antioxidant enzymes in the aging canine brain (Opii et al. 2008). Results of a prospective observational study ($n = 4740$) suggested that the combined use of 400 IU of vitamin E daily and 500 mg of vitamin C daily for at least 3 years was associated with the reduction of AD prevalence and incidence. However, subsequent meta-analyses reached the conclusions that in the absence of prospective, randomized, controlled clinical trials documenting benefits that outweigh recently documented morbidity and mortality risks, vitamin E supplements should not be recommended for primary or secondary prevention of AD (Boothby and Doering 2005). In 2005, over 187 studies on specific antioxidants in the prevention of AD were evaluated (Frank and Gupta 2005). Among nonvitamin agents taken into account there were aged garlic extract, curcumin, melatonin, resveratrol, Ginkgo biloba extract, and green tea. However, the conclusions were that while the clinical value of antioxidants for the prevention of AD is often ambiguous, some can be recommended based upon epidemiological evidence, known benefits for prevention of other diseases, and benign nature of the substance; but for drawing more conclusive information about their usefulness, long-term prospective studies are indeed recommended (Frank and Gupta 2005).

Curcumin and ferulic acid, two powerful antioxidants, the first from the curry spice turmeric and the second a major constituent of fruits and vegetables, have emerged as strong inducers of the heat shock response. Food supplementation with curcumin and ferulic acid is considered a nutritional approach to reduce oxidative damage and amyloid pathology in AD (Calabrese et al. 2006).

Recent experimental evidence suggests protective and trophic effects of ginseng in the memory function of AD. Ginseng extract has been reported to have antioxidant potential and scavenge superoxide radicals (S. T. Lee et al. 2008).

Several epidemiological studies indicate that moderate consumption of red wine is associated with a lower incidence of dementia and AD. Red wine is enriched in antioxidant polyphenols with potential neuroprotective activities, and in vivo data have clearly demonstrated the neuroprotective properties of the naturally occurring polyphenol resveratrol in rodent models for stress and diseases (Vingtdeux et al. 2008).

Higher intake of food rich in antioxidants such as fruit and vegetables confers protection against the development of ALS even if no statistically significant dose-response relationship was observed between intake of beta-carotene, vitamin C, and vitamin E and the risk of ALS (Okamoto et al. 2009).

Indeed, many intervention trials with antioxidants are presently considered inconclusive in demonstrating benefits in humans. The main observation supporting these findings is that the administration of antioxidants in subjects who already had extensive pathology may be too late. Moreover, the doses used may not be the most effective: there is limited evidence to date that lower doses and/or mixtures of antioxidants might have more benefit than higher doses of single agents (Halliwell 2009).

Since diet is a major source of antioxidants, attention has been devoted also to various antioxidants supplied to human body through diet, both vegetarian as well as nonvegetarian. Although epidemiological data on diet and AD have been conflicting (Luchsinger et al. 2007), higher adherence to the Mediterranean diet (rich in fruits, vegetables, legumes, cereals, and fish) is associated with a trend for reduced risk for developing MCI and with reduced risk for MCI conversion to AD (Scarmeas et al. 2009). The main source of fat of the Mediterranean diet is olive oil, which has been shown to be effective against oxidative stress-associated diseases and also with aging. Besides its richness in monounsaturated fatty acid, the oleic acid, olive oil is a good source of phytochemicals with antioxidant properties, such as polyphenolic compounds, squalene, and alpha-tocopherol. Moreover, in the context of the Mediterranean diet, the benefits associated with the consumption of several functional components may be intensified by certain forms of food preparation (Ortega et al. 2006; Covas 2008). In a recent meta-analysis on more than 1.5 million people, a higher adherence to the Mediterranean diet is associated with a significantly reduced risk of incidence and mortality from all causes and from cardiovascular, neoplastic, and neurodegenerative diseases (Sofi et al. 2009). The Mediterranean diet is also rich in polyunsaturated fatty acids, of which the principal sources are fatty fish (salmon, tuna, and mackerel). Epidemiological evidence supports a possible role of fatty acid intake in maintaining adequate cognitive functioning and possibly for the prevention and management of cognitive decline and dementia, but not when the AD process has already taken over (Solfrizzi et al. 2006).

Despite the huge and increasing amount of data on food polyphenols that have potential antiaging and brain-protective activities, data partially confirmed by the results of some epidemiological studies, the research in this field is still incomplete, and questions about bioavailability, biotransformation, synergism with other dietary factors, mechanisms of the antioxidant activity, and risks inherent to their possible pro-oxidant activities are still unanswered (Rossi et al. 2008). Indeed, only a few studies were focused on the ability of phytonutrients from dietary supplementation to cross the blood–brain barrier (BBB). Polyphenolic compounds present in blueberry-supplemented rat diet can be detected in various brain regions important for learning and memory (Andres-Lacueva et al. 2005). Green tea catechins are brain permeable (Mandel et al. 2006) and, after oral ingestion, can be found in the rat brain as in vivo metabolite, 3'-O-methyl epicatechin (Abd El Monhsen et al. 2002). Experimental studies so far have reported beneficial effect of curcumin in ameliorating the increased BBB permeability in hypoxic conditions (Kaur and Ling 2008); however, the bioavailability of curcumin is very low, since the drug is rapidly metabolized by conjugation (Kelloff et al. 1996).

Recent findings suggest that several phytochemicals exhibit biphasic dose responses on cells with low doses, activating signaling pathways that result in increased expression of genes encoding survival proteins, as in the case of the Keap1/Nrf2/ARE pathway activated by curcumin and NAD/NADH-sirtuin-1 activated by resveratrol. Consistently, the neuroprotective roles of dietary antioxidants including curcumin, acetyl-L-carnitine, and carnosine have been demonstrated through the activation of these redox-sensitive intracellular pathways (Calabrese et al. 2008). Interesting findings provide evidence that one mechanism of cancer chemoprevention by sulforaphane, an isothiocyanate found in cruciferous vegetables, such as broccoli and Brussels sprouts, is via epigenetic changes associated with inhibition of histone deacetylase activity. Moreover, other diet components such as butyrate, biotin, lipoic acid, garlic organosulfur compounds, and metabolites of vitamin E also have structural features compatible with histone deacetylase activity inhibition (Dashwood and Ho 2007). It was also suggested that polyphenolic compounds found in fruits such as blueberries may exert their beneficial effects through signal transduction and neuronal communication (Lau et al. 2007).

The best-known epigenetic marker is DNA methylation, and diet is a major aspect of the environment that may influence DNA methylation patterns thus providing an important common link between a variety of disease conditions and nutrition. Several reports have been published in recent years indicating that phytochemicals may reactivate genes silenced by aberrant methylation. Dietary phytochemicals, particularly catechol-containing polyphenols, were shown to inhibit DNA methyltransferases and reactivate epigenetically silenced genes (Paluszczak et al. 2008). On the other hand, food-derived compounds are able to induce epigenetic changes in intestinal mucosa during colorectal tumorigenesis (Nyström and Mutanen 2009). Furthermore, recent work in cell cultures and animal models suggests that the beneficial effects of resveratrol intake against the neurodegenerative process in AD could imply the gene expression regulation in the mechanism of Aβ clearance (Vingtdeux et al. 2008). Figure 14.2 summarizes possible implications of dietary factors to epigenetic modifications of genes relevant to neurodegeneration.

14.7 CONCLUSIONS

The evolution of the human diet over the past 10,000 years from a Paleolithic diet to our current modern pattern of intake has resulted in profound changes in feeding behavior. Shifts have occurred from diets high in fruits, vegetables, lean meats,

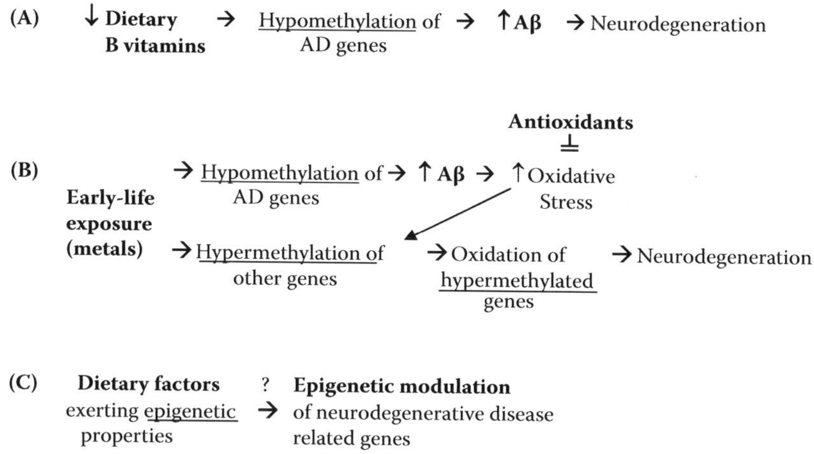

FIGURE 14.2 Possible implications of dietary factors to epigenetic modifications of genes relevant to neurodegeneration. A. Studies in rodents and in human neuroblastoma cells have demonstrated that dietary B vitamin deprivation causes epigenetic modifications of AD-related genes with subsequent increased production of the Aβ peptide (see Section 14.4 for details). B. Studies in rodents suggest that early-life exposure to environmental agents can result, later in life, in hypomethylation of AD-related genes with increased production of the Aβ peptide, as well as in hypermethylation of other genes. Hypermethylated genes are less accessible to DNA repair proteins and more prone to be damaged by Aβ-induced oxidative species. Within this context, dietary antioxidants could counteract the production of oxidative species and delay the onset of neurodegeneration (see Section 14.3 for details). C. In the near future, dietary factors with known epigenetic properties could be used to modulate the expression of neurodegenerative disease-related genes (see Section 14.6 for details).

and seafood to processed foods high in sodium and hydrogenated fats and low in fiber (Jew et al. 2009). Moreover, in recent years the diet has been recognized as a potential source of hazardous chemicals (i.e., heterocyclic aromatic amines, nitroso compounds, food additives, pesticide residuals). Both these dietary changes and the presence of hazardous chemicals in food have adversely affected dietary parameters known to be related to health, resulting in an increase in obesity as well as chronic age-related diseases such as cardiovascular disease, diabetes, and cancer.

We can now reconsider diet as a potential source of protective compounds, provided that our diet habits change by accepting dietary patterns rich in fruits and vegetables including omega-3 fatty acids, polyphenols, fiber, and plant sterols. Interestingly, one of the brain structures associated with learning and memory, as well as mood, is the hippocampus. The hippocampus is the region undergoing selective neurodegeneration in AD, but it is also one of the two structures in the adult brain where the formation of newborn neurons, or neurogenesis, persists. The level of neurogenesis in the adult hippocampus has been linked directly to cognition and mood. Therefore modulation of adult hippocampal neurogenesis by diet emerges as a possible mechanism by which nutrition impacts on mental health. Moreover, adult hippocampal neurogenesis is also subject to epigenetic regulation, and both DNA methylation and histone acetylation are important. For example, the histone deacetylase inhibitor, valproic acid, induces neuronal differentiation of adult hippocampal progenitors most likely through the induction of neurogenic transcription factors. Adult hippocampal neurogenesis responds to neurodegenerative diseases such as AD and PD. Several studies have reported decreased adult hippocampal neurogenesis in AD mouse models, and mouse models of PD show a decrease in the survival rate of newborn hippocampal neurons. Moreover, the integration of newborn neurons is disrupted by central nervous system (CNS) inflammation (reviewed in Stangl and Thuret 2009). Studies in rodents have demonstrated that caloric restriction, omega 3 fatty acids, flavonoids, blueberry, and low concentrations of curcumin increased adult hippocampal neurogenesis. On the contrary, folate deficiency, increased Hcy levels, zinc deficiency, vitamin A deficiency, and high-fat diets decreased or inhibited adult hippocampal neurogenesis (Stangl and Thuret 2009).

One of the most exciting areas for future research remains the study of transgenerational effects of the human exposure to environmental factors. The studies described in Section 14.3 (Basha et al. 2005; Bolin et al. 2006; Wu et al. 2008) suggest that early-life exposure of rodents to metals, happening during fetal brain development, results in epigenetic modifications of several genes that are not triggered until late in life. Further studies are required to clarify if something similar happens in neurons and to understand if this process is not reversible or if it can be reversed or attenuated by dietary interventions later in life. The contribution of dietary factors during brain development needs to be clarified as well.

In conclusion, the discovery of epigenetic properties of several foods, coupled with their antioxidant and neuroprotective ones, should lead us to think that we must pay attention to our diet from the beginning of our life, given that a wrong dietary habit could potentially affect not only our own health but even that of our offspring.

REFERENCES

Abd El Mohsen, M. M., G. Kuhnle, A. R. Rechner, H. Schroeter, S. Rose, P. Jenner, and C. A. Rice-Evans. 2002. Uptake and metabolism of epicatechin and its access to brain after oral ingestion. *Free Rad Biol Med* 33:1693–1702.

Andres-Lacueva, C., B. Shukitt-Hale, R. L. Galli, O. Jauregui, R. M. Lamuela-Raventos, and J. A. Joseph. 2005. Anthocyanins in aged blueberry-fed rats are found centrally and may enhance memory. *Nutr Neurosci* 8:111–20.

Anello, G., R. M. Guéant-Rodríguez, P. Bosco, J. L. Guéant, A. Romano, B. Namour, R. Spada, F. Caraci, G. Pourié, J. L. Daval, and R. Ferri. 2004. Homocysteine and methylenetetrahydrofolate reductase polymorphism in Alzheimer's disease. *Neuroreport* 15:859–61.

Bailey, L. B., and J. F. Gregory III. 1999. Folate metabolism and requirements. *J Nut* 129:779–82.

Ballatori, N., S. M. Krance, S. Notenboom, S. Shi, K. Tieu, and C. L. Hammond. 2009. Glutathione dysregulation and the etiology and progression of human diseases. *Biol Chem* 390:191–214.

Basha, M. R., W. Wei, S. A. Bakheet, N. Benitez, H. K. Siddiqi, Y. W. Ge, D. K. Lahiri, and N. H. Zawia. 2005. The fetal basis of amyloidogenesis: exposure to lead and latent overexpression of amyloid precursor protein and beta-amyloid in the aging brain. *J Neurosci* 25:823–29.

Bi, X. H., H. L. Zhao, Z. X. Zhang, and J. W. Zhang. 2009. Association of RFC1 A80G and MTHFR C677T polymorphisms with Alzheimer's disease. *Neurobiol Aging* 30:1601–7.

Bolin, C. M., R. Basha, D. Cox, N. H. Zawia, B. Maloney, D. K. Lahiri, and F. Cardozo-Pelaez. 2006. Exposure to lead and the developmental origin of oxidative DNA damage in the aging brain. *FASEB J* 20:788–90.

Bolufer, P., E. Barragan, M. Collado, J. Cervera, J. A. López, and M. A. Sanz. 2006. Influence of genetic polymorphisms on the risk of developing leukemia and on disease progression. *Leuk Res* 30:1471–91.

Boothby, L. A., and P. L. Doering. 2005. Vitamin C and vitamin E for Alzheimer's disease. *Ann Pharmacother* 39:2073–80.

Calabrese, V., C. Cornelius, C. Mancuso, G. Pennisi, S. Calafato, F. Bellia, T. E. Bates, A. M. Giuffrida Stella, T. Schapira, A. T. Dinkova Kostova, and E. Rizzarelli. 2008. Cellular stress response: a novel target for chemoprevention and nutritional neuroprotection in aging, neurodegenerative disorders and longevity. *Neurochem Res* 33:2444–71.

Calabrese, V., E. Guagliano, M. Sapienza, C. Mancuso, D. A. Butterfield, and A. M. Stella. 2006. Redox regulation of cellular stress response in neurodegenerative disorders. *Ital J Biochem* 55:263–82.

Cardozo-Pelaez, F., D. P. Cox, and C. Bolin. 2005. Lack of the DNA repair enzyme OGG1 sensitizes dopamine neurons to manganese toxicity during development. *Gene Expr* 12: 315–23.

Chang, Y., Q. Kong, X. Shan, G. Tian, H. Ilieva, D.W. Cleveland, J. D. Rothstein, D. R Borchelt, P. C. Wong, and C. L. Lin. 2008. Messenger RNA oxidation occurs early in disease pathogenesis and promotes motor neuron degeneration in ALS. *PLoS ONE* 3:e2849.

Chao, M. G., S. V. Ramagopalan, and B. M. Herrera. 2009. Epigenetics in multiple sclerosis susceptibility: difference in transgenerational risk localizes to the major histocompatibility complex. *Hum Mol Genet* 18:261–66.

Chaturvedi, R. K., and M. F. Beal. 2008. Mitochondrial approaches for neuroprotection. *Ann N Y Acad Sci* 1147:395–412.

Chen, K. L., S. S. Wang, Y. Y. Yang, R. Y. Yuan, R. M. Chen, and C. J. Hu. 2009. The epigenetic effects of amyloid-beta(1-40) on global DNA and neprilysin genes in murine cerebral endothelial cells. *Biochem Biophys Res Commun* 378: 57–61.

Coppedè, F. 2009. The complex relationship between folate/homocysteine metabolism and risk of Down syndrome. *Mutat Res* 682:54–70.

Coppedè, F., A. Lo Gerfo, C. Carlesi, S. Piazza, M. Mancuso, L. Pasquali, L. Murri, L. Migliore, and G. Siciliano. 2010. Lack of association between the APEX1 Asp148Glu polymorphism and sporadic amyotrophic lateral sclerosis. *Neurobiol Aging* 31:353–55

Coppedè, F., M. Mancuso, A. Lo Gerfo, C. Carlesi, S. Piazza, A. Rocchi, L. Petrozzi, C. Nesti, D. Micheli, A. Bacci, L. Migliore, L. Murri, and G. Siciliano. 2007. Association of the hOGG1 Ser326Cys polymorphism with sporadic amyotrophic lateral sclerosis. *Neurosci Lett* 420:163–68.

Coppedè, F., F. Migheli, R. Ceravolo, E. Bregant, A. Rocchi, L. Petrozzi, E. Unti, R. Lonigro, G. Siciliano, and L. Migliore. 2009c. The hOGG1 Ser326Cys polymorphism and Huntington's disease. *Toxicology* (in press).

Coppedè, F., F. Migheli, A. Lo Gerfo, M. R. Fabbrizi, C. Carlesi, M. Mancuso, S. Corti, N. Mezzina, R. Del Bo, G. P. Comi, G. Siciliano, and L. Migliore. 2009b. Association study between XRCC1 gene polymorphisms and sporadic amyotrophic lateral sclerosis. *Amyotroph Lateral Scler* (in press).

Coppedè, F., and L. Migliore. 2009. DNA damage and repair in Alzheimer's disease. *Curr Alzheimer Res* 6:36–47.

———. 2010. DNA repair in premature aging disorders and neurodegeneration. *Curr Aging Sci* 3:3–19.

Coppedè, F., G. Ricci, E. Caldarazzo Jenco, G. Siciliano, and L. Migliore. 2009a. A complex relationship between the methylenetetrahydrofolate reductase 677C>T polymorphism, blood homocysteine, and folate values in patients with Alzheimer's disease. Paper presented at the annual meeting of the European Federation of Neurological Societies, Florence.

Covas, M. I. 2008. Bioactive effects of olive oil phenolic compounds in humans: reduction of heart disease factors and oxidative damage. *Inflammopharmacology* 16:216–18.

Dashwood, R. H., and E. Ho. 2007. Dietary histone deacetylase inhibitors: from cells to mice to man. *Semin Cancer Biol* 17:363–69.

Fenech, M. 2001. The role of folic acid and vitamin B12 in genomic stability of human cells. *Mutat Res* 475: 57–67.

Frank, B., and S. Gupta. 2005. A review of antioxidants and Alzheimer's disease. *Ann Clin Psychiatry* 17:269–86.

Fukae, J., M. Takanashi, S. Kubo, K. Nishioka, Y. Nakabeppu, H. Mori, Y. Mizuno, and N. Hattori. 2005. Expression of 8-oxoguanine DNA glycosylase (OGG1) in Parkinson's disease and related neurodegenerative disorders. *Acta Neuropathol* 109:256–62.

Fuso, A., R. A. Cavallaro, A. Zampelli, F. D'Anselmi, P. Piscopo, A. Confaloni, and S. Scarpa. 2007. Gamma-secretase is differentially modulated by alterations of homocysteine cycle in neuroblastoma and glioblastoma cells. *J. Alzheimer's Dis* 11:275–90.

Fuso, A., V. Nicolia, R.A. Cavallaro, L. Ricceri, F. D'Anselmi, P. Coluccia, G. Calamandrei, and S. Scarpa. 2008. B-vitamin deprivation induces hyperhomocysteinemia and brain S-adenosylhomocysteine, depletes brain S-adenosylmethionine, and enhances PS1 and BACE expression and amyloid-beta deposition in mice. *Mol Cell Neurosci* 37:731–46.

Fuso, A., V. Nicolia, A. Pasqualato, M. T. Fiorenza, R. A. Cavallaro, and S. Scarpa. 2009. Changes in Presenilin 1 gene methylation pattern in diet-induced B vitamin deficiency. *Neurobiol Aging* (in press).

Fuso, A., L. Seminara, R. A. Cavallaro, F. D'Anselmi, and S. Scarpa. 2005. S-adenosylmethionine/homocysteine cycle alterations modify DNA methylation status with consequent deregulation of PS1 and BACE and beta-amyloid production. *Mol Cell Neurosci* 28:195–204.

Gordon, N. 2009. Cerebral folate deficiency. *Dev Med Child Neurol* 51:180–82.

Halliwell, B. 2009. The wanderings of a free radical. *Free Radic Biol Med* 46:531–42.

Jarem, D. A., N. R Wilson, and S. Delaney. 2009. Structure-dependent DNA damage and repair in a trinucleotide repeat sequence. *Biochemistry.* 48:6655–63.

Jew, S., S. S. AbuMweis, and P. J. Jones. 2009. Evolution of the human diet: linking our ancestral diet to modern functional foods as a means of chronic disease prevention. *J Med Food* 12:925–34.

Joseph, J., G. Cole, E. Head, and D. Ingram. 2009. Nutrition, brain aging, and neurodegeneration. *J Neurosci* 29:12795–801.

Kaur, C., and E. A. Ling. 2008. Blood brain barrier in hypoxic-ischemic conditions. *Curr Neurovasc Res* 5:71–81.

Keller, J. N., F. A. Schmitt, S. W. Scheff, Q. Ding, Q. Chen, D. A. Butterfield, and W. R. Markesbery. 2005. Evidence of increased oxidative damage in subjects with mild cognitive impairment. *Neurology* 64:1152–56.

Kelloff, G. J., J. A. Crowell, E. T. Hawk, and V. E. Steele. 1996. Strategy and planning for chemopreventive drug development: clinical development plans II. *J Cell Biochem Suppl.* 26:54–71.

Kono, S., and K. Chen. 2005. Genetic polymorphisms of methylenetetrahydrofolate reductase and colorectal cancer and adenoma. *Cancer Sci* 96:535–42.

Kruman, I. I., T. S. Kumaravel, A. Lohani, W. A. Pedersen, R. G. Cutler, Y. Kruman, N. Haughey, J. Lee, M. Evans, and M. P. Mattson. 2002. Folic acid deficiency and homocysteine impair DNA repair in hippocampal neurons and sensitize them to amyloid toxicity in experimental models of Alzheimer's disease. *J Neurosci* 22:1752–62.

Lau, F. C., B. Shukitt-Hale, and J. A. Joseph. 2007. Nutritional intervention in brain aging: reducing the effects of inflammation and oxidative stress. *Subcell Biochem* 42:299–318.

Lee, S. T., K. Chu, J. Sim, J. H. Heo, and M. Kim. 2008. Panax ginseng enhances cognitive performance in Alzheimer disease. *Alzheimer Dis Assoc Disord* 22:222–26.

Lee, Y. S. 2009. The role of genes in the current obesity epidemic. *Ann Acad Med Singapore* 38:45–3.

Lin, D., H. Li, W. Tan, X. Miao, and L. Wang. 2007. Genetic polymorphisms in folate-metabolizing enzymes and risk of gastroesophageal cancers: a potential nutrient-gene interaction in cancer development. *Forum Nutr* 60:140–45.

Linhart, H. G., A. Troen, G. W. Bell, E. Cantu, W. H. Chao, E. Moran, E. Steine, T. He, and R. Jaenisch. 2009. Folate deficiency induces genomic uracil misincorporation and hypomethylation but does not increase DNA point mutations. *Gastroenterology* 136:227–235.e3.

Lovell, M. A., S. P. Gabbita, and W. R. Markesbery. 1999. Increased DNA oxidation and decreased levels of repair products in Alzheimer's disease ventricular CSF. *J Neurochem.* 72:771–76.

Lovell, M. A., and W. R. Markesbery. 2007. Oxidative DNA damage in mild cognitive impairment and late-stage Alzheimer's disease. *Nucleic Acids Res* 35:7497–504.

Luchsinger, J. A., J. M. Noble, and N. Scarmeas. 2007. Diet and Alzheimer's disease. *Curr Neurol Neurosci Rep* 7:366–72.

Maloney, C. A., S. M. Hay, and W. D. Rees. 2007. Folate deficiency during pregnancy impacts on methyl metabolism without affecting global DNA methylation in the rat fetus. *Br J Nutr* 97:1090–98.

Mancuso, M., F. Coppedè, L. Migliore, G. Siciliano, and L. Murri. 2006. Mitochondrial dysfunction, oxidative stress and neurodegeneration. *J Alzheimers Dis* 10:59–73.

Mandel, S., T. Amit, L. Reznichenko, O. Weinreb, and M. B. Youdim. 2006. Green tea catechins as brain-permeable, natural iron chelators-antioxidants for the treatment of neurodegenerative disorders. *Mol Nutr Food Res* 50:229–34.

Martignoni, E., C. Tassorelli, G. Nappi, R. Zangaglia, C. Pacchetti, and F. Blandini. 2007. Homocysteine and Parkinson's disease: a dangerous liaison? *J Neurol Sci* 257:31–37.

Migliore, L., and F. Coppedè. 2009a. Environmental-induced oxidative stress in neurodegenerative disorders and aging. *Mutat Res* 674:73–84.

———. 2009b. Genetics, environmental factors and the emerging role of epigenetics in neurodegenerative diseases. *Mutat Res* 667:82–97.

Migliore, L., I. Fontana, R. Colognato, F. Coppedè, G. Siciliano, and L. Murri. 2005. Searching for the role and the most suitable biomarkers of oxidative stress in Alzheimer's disease and in other neurodegenerative diseases. *Neurobiol Aging* 26:587–95.

Migliore, L, I. Fontana, F. Trippi, R. Colognato, F. Coppedè, G. Tognoni, B. Nucciarone, and G. Siciliano. 2005. Oxidative DNA damage in peripheral leukocytes of mild cognitive impairment and AD patients. *Neurobiol Aging* 26:567–73.

Migliore, L., R. Scarpato, F. Coppedè, L. Petrozzi, U. Bonuccelli, and V. Rodilla. 2001. Chromosome and oxidative damage biomarkers in lymphocytes of Parkinson's disease patients. *Int J Hyg Environ Health* 204:61–66.

Molloy, A. M., L. C. Brody, J. L. Mills, J. M. Scott, and P. N. Kirke. 2009. The search for genetic polymorphisms in the homocysteine/folate pathway that contribute to the etiology of human neural tube defects. *Birth Defects Res A Clin Mol Teratol* 85:285–94.

Morahan, J. M., B. Yu, R. J. Trent, and R. Pamphlett. 2007. Are metallothionein genes silenced in ALS? *Toxicol Lett* 168:83–87.

Moreira, P. I., A. Nunomura, K. Honda, G. Aliev, G. Casadesus, X. Zhu, M. A. Smith, and G. Perry. 2007. The key role of oxidative stress in Alzheimer's disease. In *Oxidative Stress and Neurodegenerative Disorders,* ed. G. Ali Qureshi and S. Hassan Parvez. Amsterdam: Elsevier, 451–66.

Nyström, M., and M. Mutanen. 2009. Diet and epigenetics in colon cancer. *World J Gastroenterol* 15:257–63.

Oates, N., and R. Pamphlett. 2007. An epigenetic analysis of SOD1 and VEGF in ALS. *Amyotroph Lateral Scler* 8:83–8.

Okamoto, K., T. Kihira, G. Kobashi, M. Washio, S. Sasaki, T. Yokoyama, Y. Miyake, N. Sakamoto, Y. Inaba, and M. Nagai. 2009. Fruit and vegetable intake and risk of amyotrophic lateral sclerosis in Japan. *Neuroepidemiology* 32:251–56.

Opii, W. O., G. Joshi, E. Head, N. W. Milgram, B. A. Muggenburg, J. B. Klein, W. M. Pierce, C. W. Cotman, and D. A. Butterfield. 2008. Proteomic identification of brain proteins in the canine model of human aging following a long-term treatment with antioxidants and a program of behavioral enrichment: relevance to Alzheimer's disease. *Neurobiol Aging* 29:51–70.

Ortega, R. M. 2006. Importance of functional foods in the Mediterranean diet. *Public Health Nutr* 9:1136–40.

Paluszczak, J., V. Krajka-Kúzniak, and W. Baer-Dubowska. 2009. The effect of dietary polyphenols on the epigenetic regulation of gene expression in MCF7 breast cancer cells. *Toxicol Lett* (in press).

Pieper, H. C., B. O. Evert, O. Kaut, P. F. Riederer, A. Waha, and U. Wullner. 2008. Different methylation of the TNF-alpha promoter in cortex and substantia nigra: implications for selective neuronal vulnerability. *Neurobiol Dis* 32:521–27.

Pogribny, I. P., A. R. Karpf, S. R. James, S. Melnyk,, T. Han, and V. P. Tryndyak. 2008. Epigenetic alterations in the brains of Fisher 344 rats induced by long-term administration of folate/methyl-deficient diet. *Brain Res* 1237:25–34.

Rossi, L., S. Mazzitelli, M. Arciello, C. R. Capo, and G. Rotilio. 2008. Benefits from dietary polyphenols for brain aging and Alzheimer's disease. *Neurochem Res* 33:2390–2400.

Scarmeas, N., Y. Stern, R. Mayeux, J. Manly, N. Schupf, and J. A. Luchsinger. 2009. Mediterranean diet and mild cognitive impairment. *Arch Neurol* 66:216–25.

Siegmund, K. D., C. M. Connor, M. Campan, T. I. Long, D. J. Weisenberger, D. Biniszkiewicz, R. Jaenisch, P. W. Laird, and S. Akbarian. 2007. DNAmethylation in the human cerebral cortex is dynamically regulated throughout the life span and involves differentiated neurons. *PLoS ONE* 2:e895.

Skibola C. F., M. S. Forrest, F. Coppedè, L. Agana, A. Hubbard, M. T. Smith, P. M. Bracci, and E. A. Holly. 2004. Polymorphisms and haplotypes in folate-metabolizing genes and risk of non-Hodgkin lymphoma. *Blood* 104:2155–62.

Smulders, Y. M., and C. D. Stehouwer. 2005. Folate metabolism and cardiovascular disease. *Semin Vasc Med* 5:87–97.

Sofi, F., R. Abbate, G. F. Gensini, and A. Casini. 2009. Evidences on the relationship between Mediterranean diet and health status. *Recenti Prog Med* 100:127–31.

Solfrizzi, V., A. M. Colacicco, A. D'Introno, C. Capurso, F. Torres, C. Rizzo, A. Capurso, and F. Panza. 2006. Dietary intake of unsaturated fatty acids and age-related cognitive decline: an 8.5-year follow-up of the Italian Longitudinal Study on Aging. *Neurobiol Aging* 27:1694–1704.

Stangl, D., and S. Thuret. 2009. Impact of diet on adult hippocampal neurogenesis. *Genes Nutr* 4:271–82.

Trabetti, E. 2008. Homocysteine, MTHFR gene polymorphisms, and cardio-cerebrovascular risk. *J Appl Genet* 49:267–82.

Urdinguio, R. G., J. V. Sanchez-Mut, and M. Esteller. 2009. Epigenetic mechanisms in neurological diseases: genes, syndromes, and therapies. *Lancet Neurol* 8:1056–72.

van Dam, F., and W. A. van Gool. 2009. Hyperhomocysteinemia and Alzheimer's disease: a systematic review. *Arch Gerontol Geriatr* 48:425–30.

van der Put, N. M., H. W. van Straaten, F. J. Trijbels, and H. J. Blom. 2001. Folate, homocysteine and neural tube defects: an overview. *Exp Biol Med* 226:243–70.

Vingtdeux, V., U. Dreses-Werringloer, H. Zhao, P. Davies, and P. Marambaud. 2008. Therapeutic potential of resveratrol in Alzheimer's disease. *BMC Neurosci* 9:S6.

Vlismas, K., V. Stavrinos, and D. B. Panagiotakos. 2009. Socio-economic status, dietary habits and health-related outcomes in various parts of the world: a review. *Cent Eur J Public Health* 17:55–63.

Wang, S. C., B. Oelze, and A. Schumacher. 2008. Age-specific epigenetic drift in late-onset Alzheimer's disease. *PLoS One* 3:e2698.

Wasson, G. R., A. P. McGlynn, H. McNulty, and S. L. O'Reilly. 2006. Global DNA and p53 region-specific hypomethylation in human colonic cells is induced by folate depletion and reversed by folate supplementation. *J Nutr* 136:2748–53.

Wong, A. W., G. P. McCallum, W. Jeng, and P. G. Wells. 2008. Oxoguanine glycosylase 1 protects against methamphetamine-enhanced fetal brain oxidative DNA damage and neurodevelopmental deficits. *J Neurosci* 28:9047–54.

Wu, J., M. R. Basha, B. Brock, D. P. Cox, F. Cardozo-Pelaez, C. A. McPherson, J. Harry, D. C. Rice, B. Maloney, D. Chen, D. K. Lahiri, and N. H. Zawia. 2008. Alzheimer's disease (AD)-like pathology in aged monkeys after infantile exposure to environmental metal lead (Pb): evidence for a developmental origin and environmental link for AD. *J Neurosci* 28:3–9.

Zawia, N. H., D. K. Lahiri, and F. Cardozo-Pelaez. 2009. Epigenetics, oxidative stress, and Alzheimer disease. *Free Radic Biol Med* 46:1241–49.

Zhang, X., S. Chen, L. Li, Q. Wang, and W. Le. 2008. Folic acid protects motor neurons against the increased homocysteine, inflammation and apoptosis in SOD1 G93A transgenic mice. *Neuropharmacology* 54:1112–19.

Zhao, B. 2009. Natural antioxidants protect neurons in Alzheimer's disease and Parkinson's disease. *Neurochem Res* 34:630–38.

Zoccolella, S., I. L. Simone, P. Lamberti, V. Samarelli, R. Tortelli, L. Serlenga, and G. Logroscino. 2008. Elevated plasma homocysteine levels have been observed in patients with amyotrophic lateral sclerosis. *Neurology* 70:222–25.

Index

1-C, *see* Maternal one-carbon (1-C)-supplemented diet
293 cells, 91
2OG, *see* 2-Oxoglutarate oxygenase fold (2OG)
677 TT genotype, 324–325

A

A1298C polymorphism, 57
A549 cells, 200
ABCA1, 84–85
  gene methylation, 129
  prenatal famine and, 306
Abdominal obesity, 262–263, *see also* Visceral obesity
  in metabolic syndrome, 259–260
Aβ, *see* Amyloid-beta (Aβ)
ACCORD, 215, 216
Acetylation, 78, *see also* Histone modification
  effect on histones, 81
  of H3-lysine, 3
  of histones, 22, 110
  unfolds chromatin, 219
Acetylcholine synthesis
  choline role in, 145
  in schizophrenics, 333
Acetyl-CoA, 111
  in metabolic syndrome, 261
  in schizophrenics, 333
Acetyl-L-carnitine neuroprotective, 372
Acetylserotonin methyltransferase inhibited by SAH, 339
Action in Diabetes and Vascular Disease: Preterax and Diamicron MR Controlled Evaluation (ADVANCE) trial, 215
Action to Control Cardiovascular Risk in Diabetes (ACCORD) trial, 215, 216
Activation, *see also* Transcriptional activation
  of mobile repetitive DNA elements, 113
  of proto-oncogenes, 113
Activator protein-1 (AP-1), 190–191
  in chronic respiratory disease, 191
  inactivated by EGCG, 200
  inhibited by resveratrol, 200
Active transcription promoter pathways, 14
Acute lymphoblastic leukemia (ALL), 249–250
Acute myeloid leukemia, 167
Acute promyelocytic leukemias, 81, 82
Acyl-CoA oxidase generational transmission, 243
1-Acylglycerol-3-phosphate defects, 263
AD, *see* Alzheimer's disease (AD)
ADAM10, 165, 167
ADAM17, 165
Adenocarcinomas, *see* Colorectal cancer
Adenoma, *see* Colorectal cancer
Adenomatous polyposis coli gene, 89
Adhesion molecules in COPD, 187
Adipocyte methylation in insulin resistance study, 6
Adipokine secretion in metabolic syndrome, 260
Adiponectin
  and breast cancer risk, 247
  defects in, 263
Adolescent females and folate deficiency, 144–145
ADP-ribosylation of histones, 110
β-2 Adrenergic receptor defects, 263
Adult obesity and metabolic imprinting, ix
Adult-onset disease hypothesis, 85
ADVANCE, 215
Advanced glycation end products (AGEs) in diabetic complications, 221
Aflatoxin increases risk of HCC with hepatitis, 144
African Americans
  hypertensive vascular disease in, 5
  vascular disease prevalence, 5
Age effects of folate supplementation, 47
Age-related disorders
  DNA methylation in, 14
  histone acetylation in, 14
AGEs, 221
Aging
  as accumulation of epigenetic damage, 304
  and calorie restriction, 16
  and diabetes, 268
  DNA methylation patterns and, 63–64
  epigenetic alterations accumulate, 307
  epigenetic changes in specific genes, 304
  gene effects on metabolism, 20
  gene silencing in, 20
  and homocysteine levels, 296
  humans and lower mammals differ, 21
  methylation in liver, 64
  oxidative brain damage, 368
  silenced genes in, 21
  Sir2 protein effects, 20
AGO1, 83
agouti gene, 60–61
Agouti mouse
  affected by dietary methyl donors, 15

Avy, 85–86
Avy locus, 299
Bisphenol A effects, 61
coat color effects and disease risk, 85–86
dietary effects on DNA methylation, 85
effects of maternal diet, 220–221
methyl donor actions, 336–337
methyl group donor supplementation, 39–40
phenotype changes due to methylation in utero, 273–274
pre-natal folate status effects, 60–61
pseudoagouti, 299
AIDS patients and folate deficiency, 144–145
Air pollution
exacerbates COPD, 186
and neurodegenerative diseases, 367
Airway inflammation, 186
Akt signaling pathway, 164–165, 248
ALA, 302
Ala222Val polymorphism and neural tube defects, 324–325
Alanine aminotransferase, *see* Glutamic pyruvate transaminase 2 (*GPT2*)
Alcohol exposure and HCC development, 97
Alcohol intake
in colorectal cancer patients, 56–57
and CpG methylation, 268
effect on MTHFR polymorphisms, 57
folate deficiency and, 144–145
HCC major cause, 144
and reduced effects of heart disease, 22
Alcoholism and B vitamins, 300
Alipoprotein E, 271
ALL, 249–250
Allele-specific gene expression, 3
All-trans-retinoic acid in epigenetic health, 301
Allyl mercaptan (AM)
effects of, 23
more potent that DADS or SAC, 91
Alpha-linolenic acid (ALA), 302
Alpha-lipoic acid, for neurodegenerative disease treatment, 370
Alpha-tocopherols
in epigenetic health, 302
in olive oil, 371
ALS, *see* Amyotrophic lateral sclerosis (ALS)
Alu methylation state, 3
Alzheimer's disease (AD), *see also* Neurodegenerative disease
antioxidants for treatment, 370
cognitive decline in, 5–6
curcumin for treatment, 370
and DNA repair, 367
and epigenetic drift, 304
epigenetic studies, 368–369
ferulic acid for treatment, 370
genotyping data, 6
hippocampal neurogenesis in, 374
homocysteine levels in patients, 366
lead exposure and, 367
lower GSH levels in, 367
Mediterranean diet effects, 371
methylation state, 6
and oxidative stress, 367
possible treatments, 370
preventive actions, 364
PUFAs ineffective once decline starts, 371
red wine and incidence, 371
AM (Allyl mercaptan), *see* Allyl mercaptan (AM)
Amino acids
metabolism, 271
modifications in histones, 231
role in signaling, 132
supplementation with, 133
Amyloid-beta (Aβ)
and AD, 367
and B vitamin deficiency, 368
and DNA methylation state, 368
formation of, 368
induces apoptosis, 370
induces epigenetic effects, 368
induction of DNA damage, 368
neprilysin (NEP), 368
processing genes, 369
resveratrol action in clearance of, 372
upregulation, 368
Amyloidogenic cleavage, 368
Amyotrophic lateral sclerosis (ALS), *see also* Neurodegenerative disease
antioxidant foods for, 371
homocysteine levels in patients, 366
modifications in, 369
and oxidative stress, 367
possible treatments, 370
Anacardic acid activity, 277
Analysis of clinical samples for epigenetic effects, 4
Anemia and folic acid in cognitive functioning, 335
Anencephaly, 274
Aneuploidy, 34
Angelman syndrome (AS), 339–340
Animal models not identical to humans, 65
Annurca apple, 277
polyphenol source, 89
Annurca polyphenol extract (APE), 89
activity of, 277
Anorexigenic neurons, 272
Anthocyanins, 192
Antiangiogenic agent, 169
Antiapoptotic agent, 165
Antiapoptotic targets, 162
Antiarrhythmic agents
flavonoid interference with, 201

polyphenol interaction, 202
Antibiotic polyphenol interaction, 202
Anticipation
definition, 306
nutritional exposure and, 308
Anticoagulant flavanal interaction, 202
Anticonvulsants
flavonoid interference with, 201
interference with folate metabolism, 300
Antifungal agent polyphenol interaction, 202
Antihistamine interfered with by flavonoids, 201
Antihypertensive polyphenol interaction, 202
Antimicrobial agents deplete folate, 300
Anti-miR-155 therapy, 168
Anti-miR-181b as HCC inhibitor, 168
Antineoplastic agents deplete folate, 300
Antioxidant capacity, 187
Antioxidant dietary compounds could counter oxidants, 367
Antioxidant *Ligustrum lucidum*, 276
Antioxidant response elements (ARE), 198, 199
Antioxidants
many trials inconclusive, 371
for neurodegenerative disease treatment, 370
Antiplatelet drug flavanal interaction, 202
Antiproliferative effect of diallyl disulfide, 91
Antipsychotic medication and MEK1 gene, 303
Antisense orientation, 33
Anti-sense RNA, 154
Anxiolytic flavonoid interference, 201
AOM (Azoxymethane), 41
AP-1, *see* Activator protein-1 (AP-1)
aP2, 261
in diabetes, 267
APC gene, 56
AP-citrate lyase, 220
APC^{min} mice, 91
treatment with SFN, 114
Apc-specific DNA methylation, 41
APE, *see* Annurca polyphenol extract (APE)
API-5, *see* Apoptosis inhibitor-5 (API-5)
Apididymal adipocytes, 242
Apigenin beneficial in COPD, 199
APOA1-C3-A4 gene cluster, 272
ApoB degradation in dyslipidemia, 263
APOC11, 271
APOE gene in Aβ processing, 369
Apoptosis
in aging, 20
butyrate effect on, 90
in cancer disease progression, 77
effect of curcumin, 97
EGCG induces, 195
indole-3-carbinol effects on, 98
MCP30 trigger, 115
p53 effects, 196
in prostate cancer cells, 115
reduction of, 370
regulated by transcription factors, 195
regulation by miRNA, 83–84
SIRT1 effect on, 195
sulforaphane effect on, 91
Apoptosis following DNA damage, 20
Apoptosis inhibitor-5 (API-5) in HCC, 165
Appetite regulation in the brain, 86
APP gene
and AD, 367
and CpG methylation, 368
and neurodegenerative diseases, 368
Arabidopsis thaliana miRNA, 155
ARE, 198
ARE/EpRE, 199
Arginine
acetylation of in histone, 3
maternal supplementation, 133
residue methylation, 112
supplementation in livestock, 133
Argonaute-1 (AGO1), 83
Argonaute 2, 156
Argonaute proteins, 156
Artemisia dracunculus L., 276
Arterial oxygen saturation improved by polyphenols, 199
AS (Angelman syndrome), 339–340
Ascl2/Mash2 gene, 234
Ascorbic acid, *see* Vitamin C
ASD, *see* Autism spectrum disorders (ASD)
ASGR (Asialoglycoprotein receptor), 170
Asialoglycoprotein receptor (ASGR), 170
Asparagus as folate source, 36, 318
Aspergillus flavus, 144
Aspergillus fumigatus and respiratory disease, 191
Aspirin/Folate Polyp Prevention Study, 51
Asthma, 7
abnormal inflammation and, 186
epigenetic factors and environment, 189–191
exposures to allergens, 189–190
HAT activity in, 191
HDAC activity in, 191
inflammatory response in, 187
maternal smoking and, 191
steroid resistance in, 192
Astrocytes, 294
AT1b angiotensin receptor (AT1b), 267
Ataxia in CFD, 366
ATF2/c-Jun in chronic respiratory disease, 191
Atherogenic state, 260
Atherosclerosis
choline deficiency as cause, 145
LDL action, 263
and maternal diet, 131
metabolic imprinting and, ix
methylation role in disease processes, 5

phenotype switching in, 5
Athletic performance affected by intrauterine growth restriction, 130
ATP
production impairment, 266–267
synthesis into SAM, 323
Autism spectrum disorders (ASD), 328ff
brain development factors, 339
B vitamin treatment for, 330–331
candidate genes, 340–341
chromosome 15q11-13, 340
chromosome regions linked, 340
co-occurrence with other disorders, 339–340
DNA hypomethylation in, 341
epigenetic basis, 6
folate levels in, 329
folate metabolism efficiency and, 330
folic acid supplementation and protein expression, 328
folic acid supplementation may increase incidence, 328
fragile X syndrome, 332
gender factors in, 340
genetic variants and, 333
genomic imprinting regions, 340
neurotransmitter formation and, 329
Rett syndrome, 331–332
SAH elevated in, 341
treatment options, 333
vitamin B_{12} levels in, 329
Autoimmunity and age-dependent methylation changes, 19
Autophagy affected by SIRT1, 195
Autosomal diseases, 274
Avoidance behavior learning, 335–336
Avy locus, *see* Agouti mouse
Axin2 gene and maternal high-fat diet, 236
Axin fused gene, 60–61
epiallele (AxinFu) and maternal supplementation, 87
Ayurveda, 21
5-Aza-2′-deoxycytidine (5-aza-2dC) (Dacogen), 169
demethylating agent, 89
as DNMT inhibitor, 34
5-AzaC, 170
Azacytidine reverses cytosine methylation, 3
5-Azacytidine (Vidaza), 147, 169
to identify tumor suppressor miRNA, 169
Azoxymethane (AOM) as colon carcinogen, 41

B

BACE, 368
BACE1 gene and AD, 367
Baicalein in epigenetic health, 301
Baker's yeast as folate source, 318
BALF (Bronchoalveolar lavage fluid), 187
Barker hypothesis, 245, 304
Bax-induced apoptosis, 20
Bcl-2, 262
Bcl-w in HCC, 165
B cell function, 14
BCL2 inactivation, 3
BCL2 protein, 97
FMD diet effect on, 162
BCL6 in carcinogenesis, 168
BCL6 protein, 97
FMD diet effect on, 162
BD, *see* Bipolar disorders (BPD or BD)
BDNF (Brain-derived neurotrophic factor), 302
Beans as folic acid source, 16
Beckwith-Wiedemann syndrome (BWS), 274
gene imprinting and, 4
Behavioral enrichment for neurodegenerative disease treatment, 370
Berries as resveratrol source, 22
Beta-carotene as antioxidant food component, 371
Beta-cell development, 240
effect of maternal dietary restriction, 238
in insulin receptor, 262
and maternal diet, 241–242
maternal diet and, 241
Betaine
choline metabolite, 321
dependent methylation pathways, 87–88
in homocysteine/methionine conversion, 323
hypomethylation effects, 273
in maternal diet, 336
and NTD risks, 327–328
prenatal supplementation, 87
supplementation for Rett syndrome, 331
supplementation in agouti mouse, 85–86
Betain-homocysteine methyltransferase (BHMT), 323
BHMT rs3733890 variant, 325
BHMT rs3733890 variant and NTD risks, 325–326
Bevacizumab, 169
BHMT (Betain-homocysteine methyltransferase), 323
BHMT rs3733890 variant, 325
NTD risks, 325–326
Biallelic C to T mutation, 145
Bilberry as resveratrol source, 22
Bile salt regulation, 271
Bile salts as nutrient sensor, 271
Bioactive food components, *see also* specific nutrients
for ALS, 371
cancer prevention and, 1
for COPD symptoms, 199, 200
and diet habits, 373

effect on histone-modifying enzymes, 95
enzyme effects of, 270
food sources of, 78
genes controlling, 8
influence on cancer disease progression, 77
modify epigenetic mechanisms, 380
plant sources of, 24
research areas for, 8–9
roles of, 277
Biomarkers
for inflammatory airway diseases, 189
inflammatory markers for COPD, 187
lacking for liver carcinoma, 143–144
for lung disease, 188–189
for neurodegenerative diseases, 369
Biomass fuel burning exacerbates COPD, 186
Biotin
in chromatin modification, 269
and hepatic cancer, 95
may inhibit histone deacetylase, 372
modification of lysine residues by, 116
Biotin deficiency link to cancer, 116
Biotinidase, 116
Biotinylation
effect on histones, 81
enzymes for, 116
factors influencing, 116
histone modification, 3
of histones, 110
implicit cancer link, 116
reversible process, 95
Bipolar disorders (BPD or BD), 6–7, 287, *see also* Major psychoses
brain development factors, 339
characteristics of, 287
contraindication for SAM, 301
DNA methylation in, 291
epigenetic brain factors, 288
overlap with SCZ, 288
skewed X chromosome inactivation in females, 294
valproic acid adjuvant for treatment, 23
Birth defects and maternal obesity, 236
Birth process risks for obese mothers, 130–131
Birth weight, *see also* Maternal nutrition
as ALL risk factor, 249–250
diabetes risk and, 240–241
disease correlation with, 129
and later obesity, 272
Bisphenol A (BPA)
folic acid counteracts effects, 61
maternal exposure to, 86
Bitter melon, *see also Momordica charantia*, MCP30 source, 115
Blood mononuclear cells, 53
Blueberry
absorption of phytonutrients from, 372
and hippocampal neurogenesis, 374
polyphenol effects, 372
as resveratrol source, 22
Bmi-1
EGCG effect on, 96
overexpression, 96
Bottle-fed infants vs. breast-fed, 238
Boundary elements, 219
BPD, *see* Bipolar disorders (BPD or BD)
BPH-1, 91
BRAC1 silenced in cancer, 148
Brain
affected by methyl group donor deficiency, 39
atrophy, 319
disease-related tissues, 292
folate storage, 319
methylome, 293
tissue methylation studies, 294
fetal tissue and NTDs, 328
Brain-derived neurotrophic factor (BDNF)
DHA effect on, 302
DNA methylation in, 303
Brain development, *see also* Cognitive functioning
choline in, 321, 334
dopamine in, 339
folic acid in, 334
serotonin in, 339
Breads (fortified) as folic acid source, 16
Breast cancer, *see also* Mammary cancer
demethylation in, 112
epigenetic factors in, 248
estrogen exposure in utero, 248
estrogen-induced, 113
folate intake and promoter methylation, 59
and folate status, 36
folate status gene interaction, 59
gene expression in, 88
hypomethylation and, 14
lysine sites altered in, 111–112
metastatic liver carcinoma and, 143
miR-21 activity in, 98
miRNA expression in, 84
predictive factors, 246
susceptibility and fetal environment, 246
Breast-fed infants vs. bottle-fed, 238
Breast-feeding, 245
Broccoli, 277
effect of SFN in sprouts, 91–92
as folate source, 36
folate source, 318
SFN in, 114
4′-Bromoflavone as chemopreventive agent, 198
Bronchoalveolar lavage fluid (BALF) of smokers, 187
Brown adipose tissue, 133
Brussels sprouts as folate source, 36, 318

Butein anti-cancer properties, 115
Butterfly orchid tree as resveratrol source, 22
Butyrate, *see also* Dietary fiber ferment
  in cancer chemoprevention, 194
  effect on histone, 269
  HDAC inhibition by, 115
  HDAC inhibitor, 90
  inhibits HDAC enzymes, 90
  may inhibit histone deacetylase, 372
  processes affected by, 90
  similar to S-allylmercaptocysteine, 195
B vitamins, *see also* specific vitamins
  and behavioral outcomes, 336
  and cognitive functioning, 334–335
  deficiency
    Alzheimer's disease gene expression, 364
    effects of, 368
  and DNA methylation in the colon, 47–48
  as methyl donors, 327
  in neurodevelopment, 335
  supplementation, 300
BWS (Beckwith-Wiedemann syndrome), *see* Beckwith-Wiedemann syndrome (BWS)
BxPC-3 cells, 97

C

C1298A polymorphism, 322
C19MC miRNA cluster, 156
C57BL/6J mice, 97
C57BL6 mice, 163ff
C677T (Ala222Val) polymorphism in neural tube defects, 324–325
C677T genotype, 49, 50
C677T polymorphism, 322
Ca2+ homeostasis, 370
CACNA1G gene, 57
Caco2 cell line, 48, 91
Caenorhabditis elegans
  miRNA, 155
  model, 94
  resveratrol extends life span of, 23
Caffeic acid phenethyl ester (CAPE), 198
  induces HO-1, 199
Calcium channel antagonists flavonoid interaction, 201
CALERIE (Comprehensive Assessment of Long-Term Effects of Reducing Intake of Energy), 16, *see also* Calorie restriction; Dietary restriction
CALM1 gene, 304
Calorie restriction, *see also* Comprehensive Assessment of Long-Term Effects of Reducing Intake of Energy (CALERIE); Dietary restriction
  decreases cancer susceptibility, 94
  effects of, 16
  and hippocampal neurogenesis, 374
  for life extension, 23–24
  metabolic changes from, 270
  type III HDAC activity increased by, 94
cAMP response element binding protein (CRB), 198
Cancer, *see also* Carcinogenesis; specific cancers
  abrupt genetic changes, 64–65
  age-dependent methylation changes and, 19
  Bisphenol A increases risk, 61
  dietary HDAC inhibitors, 114–115
  disease process, 77
    hypermethylation in, 80
    PcG proteins, 82
    polycomb complexes and, 83
  disease progression
    epigenetic factors in, 112–113
    markers for, 113
  DNA methylation and, 33–36
  DNA methylation low in tumor cells, 14
  early nutrition and, 274
  folate and, 36–38
  hypomethylation and, 14
  IGF2 imprinting loss, 4
  inactivating mutations in, 35
  Ligustrum lucidum for treatment of, 276
  loss of normal DNA methylation, 16
  and maternal diet, 245–250
  Mediterranean diet effects, 24
  metabolic imprinting and, ix
  methylation in, 112
  miR expression and, 143
  p53 inactivating mutations in, 35
  predisposition and nutrition, ix
  progressive process, 109
  resveratrol as antiproliferative, 22
  risk in agouti mouse, 85–86
  risk modulation, 77
Cancer cells
  altered lysine methylation in, 112
  DNA methylation in, 264
  effects of EGCG, 195
  histone lysine deacetylation, 380
  histone marking, 83
  hypermethylated genes, 83
  imbalance between acetylation processes, 380
CAPE (Caffeic acid phenethyl ester), 198, 199
"Cap 'n collar" transcription factors, 198
Carbohydrate metabolism regulation and maternal diet, 243
Carbohydrate-responsive element binding protein (ChREBP), 271
Carbonylation, 289
Carcinogen bioactivation in cancer disease progression, 77
Carcinogenesis, *see also* Cancer; specific cancers

acquired somatic mutations, 35
butyrate effect on, 90
hypermutability of methylated cytosine, 35
hypomethylation of tumor suppressor genes, 147
methylation in, 35
transcription silencing and, 35
two-hit theory, 147–148
Cardiac troponin gene, 268
Cardioprotective wine activity, 22
Cardiovascular disease (CVD), *see also* Heart disease; Hypertension; Ischemic heart disease; Vascular disease
birth weight correlation, 129
CD36 gene defect, 263
as complication of diabetes, 215–216
effect of early nutrition, 274, 304
effect of prenatal maternal diet, 221
folate pathway in, 7
and impaired folate metabolism, 365–366
Mediterranean diet effects, 24, 371
metabolic imprinting and, ix
in metabolic syndrome, 260, 268
metabolic syndrome risk factor, 260
and polyphenolic food constituents, 16
Cardiovascular neuronal disorders, 143
Carnosine as neuroprotective, 372
β-Carotene as plasma antioxidant, 187
Cashew nut shell liquid, 277
Catechins
beneficial in COPD, 199
modulate NF-κG activation, 192
Catecholamines, 323
B_{12} vital for, 319
SAM in synthesis of, 338
synthesis of, 338
Catechol-O-methyltransferase (COMT), 292, 323
Catechol-O-methyltransferase (COMT) gene inhibited by SAH, 339
β-Catenin, 236
Cauliflower, 114
Caveolin-1, 248
CBS (Cystathionine β-synthase), *see* Cystathionine β-synthase (CBS)
CC genotype, 50
CD36 gene defect, 263–264
CD4+ cells increased in asthma, 187
Cd68, 235
CDAA, *see* Choline-deficient L-amino acid-defined (CDAA)
CDAA diet, 167
CDH13 gene, 59
CDH14 gene, 304
CDK, *see* Cyclin-dependent kinase (CDK)
CDK6
in carcinogenesis, 168
in HCC, 165
CDKN1A promoter, 90
CDKN1B/p27, 165
CDKN1C/p57, 165
CDKN2A gene, 57
CDKN3, 321
Cdlm1c gene, 234
C/EBPa silenced in cancer, 148
C/EBPβ
in female offspring of maternal LP diet, 243
in HCC, 168
Cell adhesion molecules in diabetic complications, 221
Cell cycle arrest
butyrate effect on, 90
SFN effect on, 91
Cell cycle in aging, 20
Cell death mediation, 20
SIRT1 effect on, 196
Cell differentiation, 83–84
Cell-growth arrest, 195
β-Cell inhibition, 267
Cell membrane signaling and choline, 145
Cell proliferation
estrogen receptors and, 248
regulation by miRNA, 83–84
Cell-to-cell connection regulated by miRNA, 97
Cellular heredity, 264
Cellular memory
methylation in, 291
of nutritional exposure, 275
Cellular proliferation, 271
in cancer disease progression, 77
Cellular signaling in cancer disease progression, 77
Cellular stress response 1 gene, 89
Cellular vulnerability, 93
Central nervous system inflammation, 374
Centromere abnormalities, 113
Cereals
as folate source, 36
as folic acid source, 16
Cerebral folate deficiency (CFD), 366
Cerebral oxoguanine (8-oxo-G), 367
Cerebrovascular disease and cognitive decline, 5–6
Cervical cancer
folate deficiency and, 144–145
and folate status, 36
hypomethylation and, 14
Cervical dysplasia, 57
folate status and, 59
CFD (Cerebral folate deficiency), 366
c-fos, 249
CG dinucleotide in epigenetic apparatus, 79
CGI, *see* CpG islands (CGI)
5,10-CH_2-THF, 322
5-CH_3-THF in methyl trap, 322–323

Chemical agents for gene activation, 3
Chemical carcinogens potentiated by folate deficiency, 145
Chemotaxis, 339
Chemotherapeutics polyphenol interaction, 202
Chicken as methionine source, 16
Chicken miRNA, 155
Child Behavior checklist, 336
Children's Psychiatric Rating Scale (CPRS), 331
Chinese Famine
- prenatal nutrition and psychosis, 305
- and schizophrenia risk, 332

ChIP, *see* Chromatin immunoprecipitation (CHIP) assays
CHO-K1 cell line, 48
Cholesterol metabolism in aging, 20
Choline
- animal model diets to induce deficiency, 146–147
- avoidance behavior learning and, 335–336
- and behavioral outcomes, 336
- betain as metabolite, 321
- in cognitive functioning, 334
- deficiency, 96
  - conditions caused by, 145
- demand during pregnancy, 321
- dependent methylation pathways, 87–88
- effect of restriction on fetal development, 235
- in folate cycle, 235
- food sources of, 16, 321
- functions of, 145
- and histone modification, 268
- in maternal diet, 336
- methyl group donor, 38–39
- neural tube closure and, 321
- and NTD risks, 327
- perinatal supplementation, 321
- period of developmental sensitivity, 321
- prenatal supplementation, 87
- relationship with folate, 323
- required nutrient, 145
- risk modification by, 116
- supplementation for Rett syndrome, 331
- supplementation in agouti mouse, 85–86

Choline acetyltransferase in schizophrenics, 333
Choline chloride as schizophrenia treatment, 333
Choline deficiency and DNA hypomethylation, 15
Choline-deficient L-amino acid-defined (CDAA), 163ff
Cholinergic dysfunction in fragile X syndrome, 332
Choline sufficient and amin acid defined (CSAA) diet, 164
ChREBP (Carbohydrate-responsive element binding protein), 271
Chromatic compactation and relaxation, 2
Chromatic structure, 80
Chromatin
- in epigenetic apparatus, 79
- formation of, 218
- inactive conformation, 33
- in megaloblasts, 319
- structure, 289

Chromatin-associated proteins altered by diet, 15
Chromatin decondensation, 113
Chromatin factors modified by bioactive food components, 380
Chromatin immunoprecipitation (CHIP) assays, 83
Chromatin modifiers, 98
Chromatin organization changes, 379
Chromatin packaging affected by histones, 110
Chromatin regulation
- methylation action in, 110
- by proteins, 82

Chromatin remodeling, ix, 78
- ATP-dependent, 82
- compaction, 271
- environmental effects integrated during, 379
- in gene silencing, 20
- leptin and, 272
- miRNA action in, 98
- only a part of modifications, 79
- SIR proteins in, 19
- SIRT1 effect on, 195
- in starvation state, 271

Chromatin structure, 230
- maternal LP diet effects, 243

Chromatin unfolding, 81
Chromosomal instability, 34
- caused by transposon activation, 3

Chromosomal mutations in miR genes, 168
Chromosome 13 (13q14) chronic lymphocytic leukemia link, 157
Chromosome condensation affected by histone modification, 81
Chromosome segregation defects, 113
Chronic airway obstruction disease, *see* COPD
Chronic cough, 199–200
Chronic inflammation in physiology of in COPD, 186
Chronic lymphocytic leukemia (CLL) link, 157
Chronic non-neuronal disease, 5–6
Chronic obstructive pulmonary disease (COPD)
- altered activity in lungs, 194
- chronic cough, 199–200
- effects of, 186
- exacerbating factors, 186
- glutathione decreased in lungs, 187
- HDAC reduced activity, 194
- inflammatory response in, 187
- Nrf-RelA imbalance, 198
- polyphenol benefits, 199–201

restoration of glucocorticoid sensitivity, 194
SIRT1 levels reduced in, 198
SIRT1 modulation for intervention, 198
steroid resistance in, 192
symptoms, 186
Chronic stress as risk factor for major psychosis, 288–289
Cigarette smoking, *see* Smoking
CIMP, *see* CpG island methylation phenotype (CIMP)
Circadian regulator, 94
Cirrhosis, 144
Cisplatin for HCC treatment, 169
9-Cis retinol dehydrogenase, 94
Citrullination, 289
Citrulline, 133
Citrus fruit as folate source, 36
C-Jun N-terminal kinase, 261
Clinical Global Impression Scale, 331
Clinical trials
folate human intervention trials, 52–53
folate status, 49ff
folic acid supplementation, 51
CLL, *see* Chronic lymphocytic leukemia (CLL) link
Clozapine, 303
c-myc
identified by RLGS-M, 153
mRNA expression, 90
PEITC effect on, 92
Cobalamin, *see* Vitamin B_{12}
Code types, 2
Coenzyme Q10 for neurodegenerative disease treatment, 370
Cognitive decline, 5–6
dietary effects on, 363–364
neuropathology, 5
PUFAs for prevention, 371
Cognitive functioning
choline role in, 334
PUFAs for maintenance, 371
Collagen expression, 267
Colon Cancer Family Registry Folate Study, 8
Colonic adenomas, *see* Colorectal cancer
Colorectal cancer
Adenoma
folate status and colorectal cancer, 36
folate supplementation and, 36
adenomatous polyposis coli gene, 89
Annurca apple protective, 89
bioactive food components in tumorigenesis, 372
Bmi-1 overexpressed in, 96
clinical trials, 50–51
demethylation in, 112
effect of folate on DMH as carcinogen, 47
folate clinical trials, 51
folate deficiency and, 41, 144–145
folate depletion and aging, 64
folate exposure *in utero* and, 63
folate metabolic pathway and, 36
folate pathway in, 7
and folate status, 36
folate status and colorectal cancer, 36
folate supplementation clinical trials, 50–51
gene expression in, 88
homocysteine levels in patients, 53
hypomethylation and, 14
inactivating mutations in, 35
lysine sites altered in, 111–112
metastatic liver carcinoma and, 143
methyl group donor deficiency and, 39
miR-181 elevation, 167
miRNA expression in, 84
p53 inactivating mutations in, 35
polyphenols protective, 89
SAM-to-SAH ratios, 53
selenium in prevention of, 89–90
Compound epigenetic mosaic, 87
Comprehensive Assessment of Long-Term Effects of Reducing Intake of Energy (CALERIE), 16
COMT, *see* Catechol-O-methyltransferase (COMT) gene
COMT Val158Met genetic variants as schizophrenia risk factor, 333
COMT variant in autism, 329–330
Concanavalin, 271
Condensation suppresses transcription, 81
Constitutive heterochromatin, 113
DNA methylation in, 19
Contraceptives interfere with folate metabolism, 300
COPD, *see* Chronic obstructive pulmonary disease (COPD)
Copper
effect on HAT, 94
in maternal diet, 250
Corn lilly as resveratrol source, 22
Coronary heart disease, *see* Heart disease
Corticosol inactivation, 239
Corticosteroid resistance, *see also* Glucocorticoid resistance
Corticosteroids
failure to ameliorate lung inflammation, 192
for inflammatory airway diseases, 192
Corticosterone levels, 239, 262
COX-2 expression suppressed by curcumin, 200
CpG, *see also* Cytosine-guanine (CpG) dinucleotide
DNMT1, 3a, 3b, 33
methyltransferases, 33
CpG253 methylation, 190–191
CpG dinucleotide, 32

CpG island methylation phenotype (CIMP), 56, 57
  in colorectal cancer cells, 89
CpG islands (CGI), 33
  DNA methylation of, 78
  DNA methylation of miRNAs in, 168
  exon 1, 169
  gene silencing in, 35
  hypermethylation and lung health, 188
  intron 1, 169
  methylation and APP, 368
  methylation and tumor development, 56
  methylation of, 36
  in promoter regions, 80
  SOX10 in schizophrenia, 292
  unmethylated, 80
CpG methylation, 7, 16, 19, 32
  effect of dietary restriction on, 235
  and folate status, 44
  most widely studied, 79
CpG sites
  germline mutation hot spots, 35
  methylation state of, 339–340
CpG suppression, 32
CPRS, *see* Children's Psychiatric Rating Scale (CPRS)
CRABP1 gene, 57
Cranberry as resveratrol source, 22
CRB, *see* cAMP response element binding protein (CRB)
C-reactive protein (CRP) in metabolic syndrome, 260
Creatine
  in epigenetic health, 301
  for neurodegenerative disease treatment, 370
  synthesis from SAM, 133
CREB, 198
  in chronic respiratory disease, 191
CREB-binding protein, 219
CRP, *see* C-reactive protein (CRP)
Cruciferous vegetables, *see also* Isothiocyanates
  effects of, 23
  HDAC inhibitors in, 194
  indole-3-carbinol in, 98
  inhibit HDAC enzymes, 90
  SFN-Cys source, 195
  source of selenium, 115
  source of sulforaphane, 115
  sulforaphane in, 114
CSAA diet, *see* Choline sufficient and amin acid defined (CSAA) diet
CT genotype, 50
CTNNA2 gene, 304
Curcuma longa L., 302
  source of curcumin, 97
Curcumin, 302
  activates Nrf2, 199
  activity of, 277
  bioavailability of, 372
  chemopreventive agent, 198
  effect on HAT, 94
  effect on hepatoma, 94
  effects of, 97
  in epigenetic health, 301
  and glucocorticoid function, 192–193
  hepatic β-glucuronidation of, 202
  and hippocampal neurogenesis, 374
  IL-8 suppressor, 200
  induces heat shock response, 370
  induces HO-1, 199
  inhibits NF-κB activation, 200
  interaction with piperin, 202
  modulates NF-κG activation, 192
  for neurodegenerative disease treatment, 370
  neuroprotective, 372
  reduces brain membrane permeability, 372
  reverses modifications of HDAC2, 200
  safety of, 202
  suppresses COX-2, 200
  suppresses neutrophil recruitment, 200
  tissues concentrated in, 202
  treatment strategy for inflammatory disease, 200
CVD, *see* Cardiovascular disease (CVD)
Cyanocobalamin, 319
Cyclin D1, 165, 248
Cyclin decrease in prostate cancer cells, 93
Cyclin-dependent kinase (CDK), 35–36
  inhibitors, 165
Cyclin G1, 165
Cyclin kinase inhibitor, 195
Cyclooxygenase-2 promoters, 221
CYLD, 93
CYP34A, *see* Cytochrome P450 (CYP 3A4)
Cystathionine β-synthase (CBS), 323, 364
  844ins68 variant, 326, 335
  SAH regulation of, 364–365
  variants and NTD risk, 326
Cystathionine in folate cycle, 364
Cysteine
  cadmium interaction, 89
  synthesis from SAM, 133
  zinc interaction, 89
Cytochrome P450 (CYP 3A4), 201
Cytokines in diabetic complications, 221
Cytosine
  deamination to uracil, 35
  hypermutability of methylated, 35
  methylation in cardiac function, 268
Cytosine-guanine (CpG) dinucleotide, 32
Cytosine methylation, 32
  reversal by chemical treatment, 3
Cytosine residue methylation site, 110
Cytosine-to-thymine transitions, 35

D

D1, *see* Cyclin D1
D2 receptor gene (DRD2), 291–292
Dacogen, *see* 5-Aza-2′-deoxycytidine (5-aza-2dC) (Dacogen)
DADS (Diallyl disulfide), *see* Diallyl disulfide (DADS)
Danio rerio miRNA, 155
DAPK1 gene, 152
 silenced in cancer, 148
Darakchasava, 21
DCCT (Diabetes Control Complications Trial), 215
DDIT4 protein, 165
Deacetylate histones, 8
Deacetylation suppresses transcription, 81
Deadly quartet, 260
Deamination, 289
Decitabine, *see* 5-Aza-2′-deoxycytidine (5-aza-2dC) (Dacogen)
Deerberry as resveratrol source, 22
Degraded RNA, 154
del, *see* 3′-UTR 6 bp deletion (del)
Dementia
 neuropathology, 5
 red wine and incidence, 371
Demethylase MBD2, 33
Demethylation in cancer disease progression, 35
Demyelination, 339
5′-Deoxyadenosylcobalamin, 320
5-Deoxy-azycytidine, 147
Deoxycytidine reverses cytosine methylation, 3
2-Deoxyglucose and cancer risk reduction, 94
Derepression of silenced tumor suppressor genes, 92–93
DES, *see* Diethylstilbestrol (DES)
Detoxification
 in aging, 20
 genes, 198
Developmental origins of health and disease, 268
Dexamethasone, 192
 activity increased by theophylline, 194
 gestational exposure, 243–244
DGCR8 (RNA binding protein Pasha), 156
DHA (Docosahexaenoic acid), 302
DHFR (Dihydrofolate reductase), *see* Dihydrofolate reductase (DHFR)
Diabetes, *see also* Insulin resistance
 Artemisia dracunculus L. for treatment of, 276
 birth weight as predictive, 241
 complications, 215–216
  glucose as gene activator in, 216
  vascular inflammation in, 221
 damage leading to, 266
 effect of early nutrition, 274, 304
 effect of maternal nutrition, 239–245
 epigenetic factors in, 266–267
 and HCC, 144
 heritability of, 264–265
 insulin resistance and, 261–262
 Ligustrum lucidum for treatment of, 276
 maternal high-fat diet and, 273
 metabolic imprinting and, ix
 and metabolic syndrome, 259–260
 methyl metabolism and, 276
 miR expression and, 143
 mitochondrial oxidative phosphorylation, 261
 mortality and inheritance, 274
 and NASH, 165
 oxidative phosphorylation, 262
 PEPCK overexpression and, 242
 prediabetes, 245
 predictive factors, 240
 predisposing factors for, 268
 predisposition to, 242
 protective phenotype, 261–262
 risk in agouti mouse, 85–86
 transgenerational, 245
 type 1 inheritance, 274
 type 2 diabetes mellitus (T2DM or T2D), 259–260
  and epigenetic regulation in utero, 6
 types of, 239–240
Diabetes Control Complications Trial (DCCT), 215
Diallyl disulfide (DADS), *see also* Garlic
 in *Alliums*, 114
 in cancer chemoprevention, 194
 cell cycle arrest by, 91
 converted to S-allylmercaptocysteine, 195
 effect on histone, 269
 effects of, 23
 global HDAC inhibitor, 195
 inhibits HDAC enzymes, 90
Diamicron MR, 215
Dicer enzyme, 156
Dicer null mice, 156–157
Diesel exposure and respiratory disease, 191
Dietary agents need drug-style study, 117
Dietary components
 cancer preventive, 78
 cancer susceptibility and, 14
 could counter oxidants, 367
 in disease progression, 7ff
 and gene activation, 88
 HDAC/HAT activity modulation, 24
 histone acetylation modulation, 24
 in major psychoses, 298ff
 methylation influence of, 14
 of methyl group donors, 39
 modifies miRNA expression, 98
Dietary effects

of ethnic food, 308
timing of, 84ff
Dietary fiber ferment, *see also* Butyrate, inhibits HDAC enzymes, 90
Dietary habits, historical changes in, 363–364
Dietary HDAC inhibitors, 114–115
Dietary historical changes, 372–373
Dietary inadequacy leading to hepatocarcinogenesis, 96
Dietary methyl group donors
agouti mouse affected by, 15
deficiency and hepatocarcinogenesis, 96
effect on histone lysine methylation, 116
Dietary polyphenols, *see* Polyphenols; Sirtuins
Dietary restriction, *see also* Calorie restriction; Comprehensive Assessment of Long-Term Effects of Reducing Intake of Energy (CALERIE)
and chromatin remodeling, 271
effects of, 20–21
during gestation, 273
SIRT1 role, 22
Dietary signature, ix, 271
removal of paternal imprints, ix
Dietary supplementation epigenetic effects, 87
Diet as epigenetic modulator, 268ff
in major psychoses, 289
Diethylstilbestrol (DES)
mechanisms of effects, 249
multigenerational effects, 246, 249
Differentiation
butyrate effect on, 90
SIRT1 effect on, 195, 196
Diffuse large B cell lymphomas, 168
Digoxin polyphenol interaction, 202
Dihydrofolate, 322
Dihydrofolate reductase (DHFR), 322
19 bp del variant, 325
folic acid as inhibitor, 51
variant, 329
in autism, 330
Dihydroxybergamottin in grapefruit juice, 201
Dimethylhydrazine (DMH), 41
effect of folate on action as colon carcinogen, 47
Disease predisposition
and calorie restriction, 16
and maternal diet, 236
Dkk1 gene, 236
DMH, *see* Dimethylhydrazine (DMH)
DNA base transition mutations, 147
DNA damage
methyl group donor deficiency and, 39
in neurodegenerative diseases, 367
SIRT1 influences repair, 20
DNA-demethylating drugs, 369
DNA demethylation
carcinogenesis and, 34
nutrient influence on, 78
role in DNA methylation, 80
DNA/helicase/ATPase-containing complex, 82
DNA hypermethylation
in gene silencing, 14
in liver, 87–88
site-specific, 19
DNA hypomethylation, *see also* Hypomethylation
agents for, 147
in ASD, 341
and folate deficiency, 366
and folate status, 48
in lupus, 14–15
promotes active transcription, 14
tumor development and, 34
DNAJA2, 94
DNA methylation, ix
aberrant, 80
aberrant in ASD, 341
affected by dietary components, 88
age-associated decline, 19
alteration by diet, 15
in asthma, 190
in BD patients, 291
bioactive food components modify, 380
blocking during replication, 80
in brain tissues, 294
cancer and, 33–36
in cancer cells, 264
in cardiac physiology, 268
clinical trials of folic acid supplementation, 51
component of epigenetics, 2
dietary effects on, 85
dietary methyl group donor deficiency, 39
in DNA damage and repair, 7
effect on histones, 81
folate affect on p*53*, 44
folate deficiency and, 42–43
site and gene specificity, 49
and folate levels, 336–337
and folate metabolism, 364–366
folate supplementation *in utero*
brain effects, 61
liver effects, 61
folic acid compensates for low capacity, 327
folic acid supplementation and, 51
in gene silencing, 20
gene silencing, 149
in gene switching, 16
green tea balances status, 22
histone modification and, 23
inhibits binding of transcription factors, 80
irreversible changes in, 39
levels and folate status, 40

in liver, 273
loss of normal patterns in disease progression, 16
and lung disease, 189
mechanisms of, 128
mediation of, 33
mitotic inheritance, 80
in neurobehavioral disorder, 290–291
and NTD risks, 328
only a part of modifications, 79
p53 methylation affected by folate, 44
perinatal stages, 337–338
prenatal famine exposure effects, 84–85
principal epigenetic mechanism, 218
principal epigenetic modification, 14
restoring normal patterns, 23
in SCZ patients, 291–292
sever folate deficiency and, 41
SIRT1 role, 22
timing of dietary effects, 84ff
timing of folate deficiency and, 87–88
DNA methylation patterns, 78
DNA methyl binding protein, 80
DNA methyltransferase
aberrant induced by methyl deficiency, 96
in lupus, 14–15
maintain function, 148
nutrient influence on, 78
types of, 110
DNA mutations in cancer disease progression, 80
DNA promoter regions, 2
DNA silencing, 110, *see also* Gene silencing
DNMT
cadmium inhibition, 89
catalyst, 79
compensatory upregulation, 45
cytosine-to-thymine transitions, 35
dietary components interfere with activity, 88
expression reduction by APE, 89
folate deficiency and, 48
folate deficiency and inhibition, 89
gene deletion, 35
inhibition, 276
5-aza-2′-deoxycytidine, 34
reexpression of silenced genes, 35
by selenium, 89
targeting of, 80
up-regulation, 39
zinc inhibition, 89
DNMT1, 33
action of, 338
individual variability of, 369
maintenance methylation by, 291
maintenance methyltransferase, 79, 110
DNMT1, 3a, 3b, 33
in human cancers, 35–36
in major psychoses, 298
Dnmt1 gene mutation, 34, 35, 235
and maternal diet, 235
DNMT3A
de novo methylation, 79
methyltransferase, 110
DNMT3a, 3b, 33
in early embryo, 337–338
methylation by, 291
DNMT3B
de novo methylation, 79
methyltransferase, 110
DNMT3L coordinates methyltransferases, 110
Docosahexaenoic acid (DHA), 302
Dopamine, 323
in brain development, 339
Dot1a-Af9 pathway, 267
Down syndrome, 333–334
Doxorubicin, 168
for HCC treatment, 169
DRD2, 291–292
Drosophila forkhead transcription factor (FOXO), 195–196
in calorie restriction, 270
Drosophila melanogaster
lifespan extended by resveratrol, 23
miRNA, 155
models, 94
Drug use
effect on lung health, 188
risk factor for major psychosis, 288–289
Dutch Hunger Winter, 84–85, 272
Dutch Hunger Winter Study, 238
DNA methylation effects, 306
generational transmission of low birth weight, 243
and IGF2 methylation, 337
prenatal nutrition and psychosis, 305
and schizophrenia risk, 332
Dysbindin, 292
Dyskinesia in CFD, 366
Dyslipidemia
in diabetic complications, 221
hyperinsulinemia and, 262
LPL gene mutations, 263
in metabolic syndrome, 259–260, 263
and PPAR, 273

E

E2F3 protein, 97
FMD diet effect on, 162
in HCC, 165
E-cadherin gene, 59
E-CAD silenced in cancer, 148
ECM accumulation, 222
ECM (Extracellular matrix (ECM)-related gene expression), 216

ECM-modifying genes, 221
EDIC (Epidemiology of Diabetes Interventions and Complications trial), 215
EED gene, 304
EGCG
activates detoxifying enzymes, 199
activity of, 277
beneficial in COPD, 199
decaffeinated extracts, 201
effect on Bmi-1, 96
effect on P-glycoprotein, 202
effects of, 195
modulates NF-κB/AP-1 activity, 200
reduces skin cancer cell survival, 96
safety of, 202–203
Eggs as choline source, 16, 321
Eicosapentaenoic acid (EPA), 302
ELAM-1 (Endothelial leukocyte adhesion molecule 1), 187
Embryogenesis, 33
Embryonic kidney 293 cells, 91
Embryonic neurodevelopment, 338–339
Embryonic stem cell differentiation, 95–96
Embryonic stem (ES) cells, 290
Emrl and maternal diet, 235
Endocrine signaling, 195
Endocrine status *in utero*
amino acids as signaling molecules, 132
epigenetic effects, 125
Endogenous retroviruses, *see also* Retroviruses, in human DNA, 32
Endoplasmic reticulum X-box binding protein-1, 261
Endothelial leukocyte adhesion molecule 1 (ELAM-1), 187
Endothelium-dependent relaxation, 129
and maternal or nursing diet, 131
Energy metabolism and epigenetic modifications, 220
Environmental factors
affect genome, 1
age-related methylation and, 64
in asthma, 189–190
cause altered gene expression, 1
in neurodegenerative diseases, 367
triggers, 128
types of, 1, 2 (fig-)
Environmental tobacco smoke (ETS), 189, *see also* Smoking
role in asthma, 190
Enzymatic modification of histones, 2
Enzyme induction of carcinogenesis, 98
Enzymes, *see also* specific enzymes
in histone modifications, 221–223
methyl transferases, 3
transcriptional repressors, 148
Eosinophils increased in asthma, 187
EPA (Eicosapentaenoic acid), 302
EpCAM+ HCC cells, 167
Epidemiology of Diabetes Interventions and Complications (EDIC) trial, 215
Epidermal growth factor and early mammary tumors, 248
Epigallocatechin 3-gallate (EGCG)
in DNA methylation, 302
DNMT inhibition by, 276
HDAC2 and, 194
for neurodegenerative disease treatment, 370
restores methylation patterns, 88
Epigenetic abnormality reversibility, 109
Epigenetic changes
gradual and progressive, 64–65
in major psychoses, 291–295
occur throughout life, 15–16
transmission to future generations, 235
Epigenetic drift, 304
age-dependent, 304
age-specific, 368–369
Epigenetic drugs, 369
Epigenetic dysregulation, 112–113
Epigenetic effect type, 127–128
Epigenetic information as dynamic, 1
Epigenetic inheritance and nutrition, ix
Epigenetic instability, 292
Epigenetic marks
as biomarkers, 98
and life experiences, 5–6
silenced heterochromatin signature, 113
Epigenetic modifications
in asthma, 190
dietary components influence, 270
DNA methylation, 14
fetal cell plasticity enables, 128
and impaired folate metabolism, 365–366
mechanisms for, 219–220
regulators of processes, 17, 19
schematic, 17, 379
specific disease associations, 18–19
tune gene expression, 266
types of, 110
Epigenetic modulation
by diet, 268ff
in metabolic syndrome, 268ff
Epigenetic programming, 265
Epigenetic regulation, ix
by diet, 269
by dietary components, 78
and disease susceptibility, 4–7
disturbed during disease processes, 1
genomic imprinting, ix
and health, 4–7
nature-nurture gap, 13
of normal cells, 109
SIRT1 role, 196

Epigenetics
action of, 264
basic explanation, 216–220
components of, 2
definition, 1, 14, 32, 110, 125, 216, 230, 288
effect on lung health, 188
Lamarckian theory as origin, 264
studies of neurodegenerative disease, 369
term coined, 13
term derivation, 127–128
Epigenetic therapy, 275–276
Epigenome
determines use of genetic information, 1
maps, 5
Epigenomic analysis development efforts, 5
Epigenomics Roadmap Program, 4–7
Epigenomics vascular disease approach, 5
Epilepsy
in CFD, 366
modifications in, 369
Epithelial-mesenchymal transition, 97
ERα gene, 59
ERα methylation, 57
ERα receptors, 248
ERβ receptors, 248
ERK pathway signaling, 14–15
ES (Embryonic stem cells), 290
Esophageal cancer
folate deficiency and, 144–145
and folate status, 36
gene expression in, 88
hypomethylation and, 14
MGMT gene and folate effect, 59
ESR1 (Estrogen receptor 1), 98
Essential fatty acids, 302
Estrogen exposure in utero, 248
Estrogen-induced mammary gland carcinogenesis, 113
Estrogen metabolism, 98
Estrogen receptor 1, 98
Estrogen receptor-alpha, 165
Estrogen receptors
epigenetic modification of, 248
as nutrient sensors, 271
Estrogens as carcinogens, 246, 247
Ethnic density as protective factor for psychoses, 308
Ethnic food as protective factor for psychoses, 308
Ethnic variation in disease processes, 5
ETS (Environmental tobacco smoke), 189, 190
Eucalyptus as resveratrol source, 22
Euchromatin, 79, 80
structure of, 218–219
Euglycemia in metabolic syndrome, 260–261
Excitotoxicity reduction, 370
Executive function processes, 296
genetic variants and, 333
Exportin 5/RAn GTPase, 156
Extracellular matrix (ECM)-related gene expression, 216
EZH2, 83

F

F_2-isoprostanes in COPD process, 187
FAD (Flavin adenine dinucleotide), 322
Famine effects, 84–85, *see also* Chinese Famine; Dutch Hunger Winter
neural tube defects, 323
Farnesoid X receptor (FXR), 271
Fate-accelerated aging, 22
Fat metabolism affected by PPAR, 273
Fatty acid-binding proteins, 261
Fatty acids in metabolic syndrome, 260–261
Fatty fish in Mediterranean diet, 371
Fatty liver disease, 93–94
protective phenotype, 261–262
Fava beans as genistein source, 117
Ferulic acid induces heat shock response, 370
Fetal development and methyl group levels, 235
Fetal hyperglycemia, 245
Fetal liver specific miRNA, 157
Fetal macrosomia, 244
Fetal origins of adult onset disease, 94
Fetal programming, 125ff, 126–128, 230, 268
definition, 126, 127
and maternal diet, 236–239
FEV_1 (Forced expiratory volume), 186, 199
Fibrinogen in metabolic syndrome, 260
Fibronectin, 165, 222
Finger domains, 110
Fish as methionine source, 16
Flavanals, 192
effect on platelet aggregation, 202
Flavin adenine dinucleotide (FAD), 322
Flavones, 192
beneficial in COPD, 199
Flavonoids, *see also* Resveratrol
drug metabolism interference by, 201
effects on asthmatics, 201
and hippocampal neurogenesis, 374
reduce iron absorption, 202
types of, 192
Flavonols, 192
beneficial in COPD, 199
Flurouracil for HCC treatment, 169
Flushing (livestock), 126
FMD (Folate-methionine-choline deficient diet), *see* Folate-methionine-choline deficient (FMD) diet
FMR1 and fragile X syndrome, 332, 340

FOCM (Folic acid one-carbon metabolism), *see* Folic acid one-carbon metabolism (FOCM)
Folate, 47, *see also* Folic acid; Folinic acid
  absorption of, 322
  in AD patients, 366
  age effects on action of, 47
  autism and levels of, 329
  and Aβ formation, 368
  cancer and, 36–38
  cancer risk and, 36
  can compensate for vitamin B_{12} deficiency, 319
  cofactor for methyl groups, 235
  and cognitive functioning, 335
  colorectal cancer and, 37
  colorectal cancer and exposure *in utero*, 63
  cycle, 364
  deficiency and timing and degree of effect, 40–41
  digestion of, 300
  digestive absorption of, 318
  disruption of functions of, 38
  DNA methylation and, 38ff
  and DNA methylation status, 336–337
  DNMT activity inhibition, 89
  effect on biomarkers for colorectal cancer, 36
  effect on DMH as colon carcinogen, 47
  effect on MTHFR polymorphisms, 57
  environmental factors and, 47–48
  food sources of, 36
  gene variants in metabolism, 37
  and histone modification, 268
  and homocysteine, 296
  homocysteine levels reduced by, 327
  impaired metabolism and disease, 365–366
  intestinal microflora synthesis, 45
  in maternal diet, 250
  metabolic conversion of, 364
  metabolism, 322
  as methyl donor, 319
  methyl donor actions, 322
  methyl group donor, 38–39
  and MTHFR 677 genotype, 337
  natural form of, 300
  natural form of vitamin B_9, 318
  neural tube defect preventive, 319
  pathway genes in autism, 329–330
  protects from high homocysteine levels, 327
  relationship with choline, 323
  and Rett syndrome, 331
  role in the body, 319
  selenium and, 47
  smoking protective, 59
  storage locations in the body, 319
  synthesis in rats, 145–146
  transport mechanisms, 364
  transport of, 322
  treatment option, 300
Folate binding protein gene (Folbp1+/-), 61
Folate deficiency
  cancer susceptibility, 144–145
  cerebral, 330
  clinical trials, 50
  colon cancer risk, 145
  and colorectal cancer, 41
  compensated for by choline, 323
  conditions caused by, 144–145
  DNA hypomethylation and, 48
  DNA hypomethylation induced by, 15
  DNA methylation and, 42–43
  and DNA methylation in the colon, 47–48
  effects of, 40, 319
  effects on colon cells, 48
  effects site and gene specific, 49
  extent and durational effects, 45–46, 47
  HCC and, 144–145
  hypermethylation of brain cells, 299
  intervention for, 159
  megaloblastic anemia, 319
  neural tube defects from, 323
  SAM depletion caused by, 147
  severe, 41
  timing and effects, 87–88
Folate-dependent gene variants in autism, 328
Folate depletion
  and colorectal carcinogenesis, 38
  and DNA hypomethylation, 38
  and DNA methylation, 64
Folate metabolism, 325
  and DNA methylation, 364–366
  and Down syndrome, 334
  efficiency and autism, 330
Folate metabolites, 57
Folate-methionine-choline deficient (FMD) diet, 145–146
  mechanisms for liver damage by, 147
  microRNAs dysregulated by, 158
  miRNA dysregulated, 157ff
  protein expression effects, 150
Folate pathway disease association, 7
Folate receptors, 322
  antibodies, 330
  autoantibodies, 326–327
Folate status
  and alcohol in colorectal cancer patients, 56–57
  clinical trials, 49ff
  and DNA methylation observational studies, 54–55
  gene-specific DNA methylation, 56, 58–59
  gene-specific DNA methylation observational studies, 58–59
  and global DNA methylation status, 56

methylation state in adenoma patients, 57
methyltransferase expression, 57
MTHFR enzyme activity and, 56
Folate supplementation, 48–49
and adenoma risk, 36
in adult humans, 88
and cancer risk, 37
clinical trials, 49ff
and DNA methylation in observational study, 56
gender differences, 49
hyperhomocysteinemia reversed by, 88
hypomethylation reversed by, 88
may exacerbate B12 deficiency, 300
Folbp1+/– protein binding gene, 61
Folic acid, *see also* Folate; Folinic acid
absorption and metabolism, 324
for ALS treatment, 366
as autism treatment, 330–331
and behavioral outcomes, 336
in cognitive functioning, 334
compensates for lower DNA methylation capacity, 327
deficiency, 96
and Aβ induced damage, 368
and oxidative DNA damage, 368
effect of supplementation on homocysteine levels, 299–300
in folate cycle, 235
food sources of, 16
hypomethylation effects, 273
and IGF2 methylation, 337
in maternal diet, 336
methylation effects of, 14
and neural tube defects, 276
prenatal supplementation, 87
protective of mtDNA, 145
as psychiatric treatment, 333
risk modification by, 116
synthetic folate form, 36
synthetic form of folate, 318
treatment for fragile X syndrome, 332
Folic acid one-carbon metabolism (FOCM), 7, *see also* One-carbon metabolic pathway
Folic acid supplementation
in agouti mouse, 85–86
autism levels, 328
clinical trial, 51
counteracts Bisphenol A effects, 61
effect on prospective major psychosis rates, 308
in food products, 324
insufficient to avert autism, 328
may increase autism occurrence, 328
neural tube defect preventive, 324
persistent effects, 51
during pregnancy, 60–63, 273
Folinic acid, *see also* Folate; Folic acid
as autism treatment, 330–331
for Rett syndrome, 331
supplementation during pregnancy, 61
treatment for CSF abnormalities, 330
FOLR2 gene, 322
Folypoly-γ-glutamate carboxypeptidase, 322
Food sources of bioactive food components, 78, *see also* Bioactive food components
Forced expiratory volume (FEV_1), 186, 199
Forkhead transcription factor, 20
FOS oncogene, 165
FOXD3 gene, 152
FOXO1/3, 197
FOXO3/4, 198
FOXO3a, 93
FOXO (Drosophila forkhead transcription factor), *see* Drosophila forkhead transcription factor (FOXO)
FoxP1, 169
Fragile X syndrome (FXS), 332
causes of, 340
co-occurrence with ASD, 339–340
Free radicals
liver damage caused by, 147
resveratrol reduction of, 22
Fruit intake, *see* Bioactive food components
FTO, 267
Furanocoumarins in grapefruit juice, 201
FXR (Farnesoid X receptor), 271
FXS, *see* Fragile X syndrome (FXS)

G

G1, 165
G1958A (rs2236225) variant, 325
G1/S transition, 165
G_2/M phase, 91
G9a histone, 323
downregulation, 112, 113
GABAergic
inhibition in ASD, 340
neurons, 341
neurotransmission in major psychosis, 292–293
GABRB3 in ASD, 340
GAD1 in brain tissue, 295
GAD67
in brain tissues, 295
in schizophrenia, 341
GADD43, 197
Gain of function modification, 81
Gallus gallus miRNA, 155
Gamma-tocopherols in epigenetic health, 302
Garcinia indica fruit, 277
Garcinol activity, 277

Garlic, *see also* Diallyl disulfide (DADS)
diallyl disulfide in, 114
effects of, 23
extract for neurodegenerative disease treatment, 370
HDAC inhibitors in, 114
inhibits HDAC enzymes, 90
as methionine source, 16
organosulfur compounds, 372
organosulfur compounds in, 91
Gastrointestinal polyps, 91
GCN5 *N*-acetyltransferase (GNAT), 111
GCPII (Glutamate carboxypeptidase II) gene, 322, 326
GDD45, 196
Gel electrophoresis technique, 150
Gender effects on gene methylation, 84–85
Gender factors, 242–243
in $B_{6\,response}$, 301
in DNA methylation in SCZ patients, 291
Dutch Hunger Winter, 337
effect of low-protein maternal diet, 235
in glucose metabolism, 242
glucose uptake, 242
in GLUT4 expression, 242–243
in inheritance, 274–275
inherited nutrition related effects, 274–275
insulin resistance, 242–243
in longevity inheritance, 274
MECP2 and ASD, 340
PEMT activity, 321
in perinatal nutrition and gene methylation, 306
in schizophrenic brain tissue, 295
vinclozolin exposure, 273
X chromosome vulnerability in males, 340
Gene activation, 13
and calorie restriction, 20–21
by chemical agents, 3
Gene disruption caused by transposon activation, 3
Gene-environment interactions during development, 21, 379
Gene expression
influences on, 20
modifications and, 79
Gene imprinting
component of epigenetics, 2
definition, 3
diseases regulated epigenetically by, 3–4
Gene regulation mechanisms, 127–128
nutrient effects, 270–272
Gene silencing, 13
and calorie restriction, 20–21
by dietary components, 270
and histone biotinylation, 116
histone lysine acetylation in, 111
methylation in, 339
methylation state and, 273
miRNAs in, 98
pathways for, 14
PcG proteins, 82
in SALS, *see* Sporadic ALS (SALS)
types of, 98
Genes plus environment in psychoses, 288
Gene switching mechanism, 79
Gene therapy for hepatocarcinoma, 170
Genetic anticipation, 303
Genetic manipulation for life extension, 23–24
Genetic risk factors, 115–116
Genetics, definition, 32
Gene transcription, 195
Genistein, *see also* Soybeans
action opposed by other soybean components, 93
activation of aberrantly silenced suppressor genes, 93
chemopreventive agent, 117
in DNA methylation, 302
DNA methylation and, 86
effects on offspring, 249
EGCG source, 88
in epigenetic health, 302
hypermethylation in agouti mice, 299
induced expression of tumor suppression genes, 93
in maternal diet, 247
for neurodegenerative disease treatment, 370
offspring obesity protection, 274
possible breast cancer reduction by, 117
possible prostate cancer reduction by, 117
supplementation in agouti mouse, 86
Genome stability, 113
Genomewide association studies (GWAS), 6, 8
Genomic imprinting, ix, 230
definition, 130
loss of normal DNA methylation, 16
methylation action in, 110
methylation in, 339
regions near ASD loci, 340
and sex of transmitting parent, 274
Genomic instability, 14
Genomic stability, 113
Genotoxic carcinogens, 113
Genotypes and diet-related disease, ix
Germline transmission, 220–221
Ginkgo biloba extract, 370
Ginseng
effects of, 371
for memory function in AD, 371
Glaucoma, 6
Glial cell cancer, 98
Glioblastomas, 84
Gliomas, 167

Glomerulosclerosis, 129–130
Glucagon promoter, 267
Glucocorticoid receptor (GR), 194
 in diabetes, 266
 effect of folic acid, 276
 effect of maternal low-protein diet, 273
 generational transmission, 243
 and hypertension, 273
Glucocorticoid resistance, *see also* Corticosteroid resistance
 in COPD patients, 194
 in inflammatory lung disease, 192
Glucocorticoids
 effects of administration during pregnancy, 239
 gestational exposure, 243–244
 restoration of sensitivity in COPD, 194
Glucose
 effect on histone acetylation, 220
 gene-activator in diabetic complications, 216
 influence on transcription process, 219
 tolerance affected by prenatal maternal diet, 221
Glucose control, 215–216
Glucose homeostasis, 156–157
Glucose intolerance
 and insulin resistance, 262
 progression to diabetes, 262
Glucose metabolism
 maternal LP diet and, 242–243
 in metabolic syndrome, 259–260
Glucose tolerance
 11-hydroxysteroid dehydrogenase 1 in, 262
 and birth size, 241
 effect of maternal diet, 241–242
 in metabolic syndrome, 259–260
Glucose transporter 2 (Glut2), 223
Glucose transporter 4 (GLUT4), 242–243
 in diabetes, 267
Glucose uptake and maternal diet, 242
Glut2 (Glucose transporter 2), 223
GLUT4 (Glucose transporter 4), *see* Glucose transporter 4 (GLUT4)
Glutamate carboxypeptidase II (GCPII) gene, 322, 326
Glutamates, *see* Folate
Glutamergic neurotransmission in major psychosis, 292–293
Glutamic pyruvate transaminase, 94
Glutamic pyruvate transaminase 2 (*GPT2*), 94
Glutamine maternal supplementation, 133
Glutamine transaminase K in selenium metabolism, 92
Glutathione gene (GSTP1)
 methylated in prostate cancer, 3
 methylation level as screening tool, 3
 silenced in cancer, 148
Glutathione (GSH), 364
 antioxidant, 367
 levels, 199
 levels in neurodegenerative diseases, 367
 synthesis, 200
 thiol antioxidant, 187
Glutathione S-transferase B class gene, 91
Glutathione-S-transferase gene (GSTP1), 115
Glutathione S-transferase (GST), 367
Glutathione-S-transferase M1 (GSTM1) in autism, 329–330
glutathione S-transferase-pi, 88
Glycemic variability, 221–222
Glycine deficiency effects, 133
Glycine-N-methylase, 276
Glycine-N-methyl transferase, 276
Glycolytic glucokinase levels, 242
Glycoprotein for B_{12} absorption, 319
Glycosaminoglycans, 263
GM-CSF in inflammatory lung disease, 192
GNASAS, 84–85
 gene methylation, 129
  prenatal famine and, 306
GNAT (GCN5 *N*-acetyltransferase), 111
Gnetum as resveratrol source, 22
GPT2 (Glutamic pyruvate transaminase 2), 94
Grains as folate source, 36
Grandparental effect, 274
Grapefruit juice
 contraindicated during various treatments, 201
 drug interference by, 201
Grape juice, 21
Green tea, 277
 adverse effects of, 201
 catechins brain permeable, 372
 catechin source, 192
 decaffeinated extracts, 201
 effects of, 22
 EGCG source, 88
 for neurodegenerative disease treatment, 370
Gr gene, 61
GR (Glucocorticoid receptor), *see* Glucocorticoid receptor (GR)
Grv10 gene, 234
GSH, *see* Glutathione (GSH)
GST (Glutathione S-transferase), 367
GSTM1 (Glutathione-S-transferase M1) in autism, 329–330
GSTP1 (Glutathione gene), *see* Glutathione gene (GSTP1)
GSTP1 (Pi-class glutathione-S-transferase), *see* Pi-class glutathione-S-transferase (GSTP1)
GWAS, 6, 8

H

H19 gene, 88, 233, 234
H1 histone, 271
H2A-H2B dimers, 80
H3-H4 tetramer, 80
H3K12, 94
H3K14, 272
H3K20, 112
H3K27, 112, 220
H3K27me2, 295
H3K27me3, 95, 96, 272, 295, 339
H3K36, 220
H3K4, 112, 221, 222
H3K4me, 220, 222
H3K4me2, 223
H3K4me3, 223, 295, 339
H3K79, 220
H3K9, 89, 112, 113
H3-K9, 196
H3K9, 220
H3K9me1/2, 272
H3K9me3, 223
H3-lysine, 3
H3 methylation, 82, 269
H4K12, 272
H4K16, 196, 272
H4K20, 112, 113, 196, 220
H4K8, 272
Haplotypes in neurobehavioral disorders, 303
HAT/HDAC imbalance, 192–193
HAT (Histone acetyltransferase), *see* Histone acetyltransferase (HAT)
HBV/HCV-induced HCC, 144
H-cadherin gene, 48
HCC (Hepatocarcinogenesis), *see* Hepatocarcinogenesis (HCC)
HCEC cell line, 48
HCT116 cell line, 48
HCT116 colorectal cancer cells, 91
HCV RNA in HCC cells, 166–167
Hcy (Homocysteine), *see* Homocysteine (Hcy)
HDAC10, 111
HDAC11, 111
HDAC1-3, 111
HDAC1 mRNA, 93–94
HDAC2
  curcumin reverses modifications, 200
  epigenetic effects on lung health, 192
  modulation by polyphenols, 192–195
  reduced activity in COPD patients, 194
HDAC4, 169
HDAC4-7, 111
HDAC8, 91
HDAC8 inhibition by MSP, 92, 111
HDAC9, 111
HDAC/HAT imbalance, 192–193
HDAC (Histone deacetylase), *see* Histone deacetylase (HDAC)
HDAC SIR2, 94
HD (Huntington's disease), *see* Huntington's disease (HD)
HDL (High density lipoprotein), *see* High density lipoprotein (HDL)
Head trauma and neurodegenerative diseases, 367
Health Professionals Follow-Up Study, 56
Heart defects and maternal obesity, 236
Heart disease, *see* Cardiovascular disease (CVD)
Heart-specific RNA, 154
Heat shock response, 370
Hematopoietic cells, 319
Hematopoietic differentiation, 167
Heme oxygenase-1 (HO-1), 199
Hemiballismus in CFD, 366
Hep3B cells, 94
Hepatic β-glucuronidation of curcumin, 202
Hepatic gluconeogenic phosphoenolpyruvate carboxykinase (PEPCK), 240, 242
  generational transmission, 243
Hepatic phosphatase and tensin homolog (PTEN), 97
  genistein effect on, 93
Hepatic steatosis, 262
Hepatitis C infection, 144
Hepatitis in Dicer null mice, 156–157
Hepatocarcinogenesis (HCC), 96
  alcohol consumption and, 144
  diabetes and, 144
  FMD diet-induced, 157ff
  HBV/HCV-induced, 144
  hepatitis C infection and, 144
  methyl-deficiency induced, 270
  in mice, 146
  microRNA influence on, 142ff
  microRNAs altered in, 164–165
  miR-122 in, 161
  miR-155 in, 168
  miR-181 in, 167–168
  in rats, 145–146
  risk increased with NASH, 165
  therapeutic intervention limited, 169
Hepatocellular carcinoma
  miRNA alterations in, 157
  prevalence of, 143
Hepatocyte injury and NASH, 165
Hepatocyte regeneration, 161
Hepatoma, 94
Heterochromatin, 79, 80
  constitutive, 113
  conversion to euchromatin, 218–219
  DNA methylation in, 19
  role of, 219
Heterochromatin proteins, 273

Heterozygosity loss, 34
HF, *see* Maternal high-fat (HF) diet
HGH, *see* Human growth hormone (HGH)
HGH in antiaging treatments, 24–25
High density lipoprotein (HDL) levels
  action of, 263
  in dyslipidemia, 263
  in metabolic syndrome, 260
High-fat diet
  and hippocampal neurogenesis, 374
  rate of spontaneous HCC, 97
High-fat maternal diet, 93–94
High methyl donor diet
  effect on lung health, 188
  respiratory symptoms and, 191
Hippocampal function, 334
  in AD, 373
Hippocampal neurogenesis, 374
Hippocampal progenitor cells, 321
Histamine N-methyltransferase inhibited by SAH, 339
Histone
  in epigenetic apparatus, 79
  neutralization; *see also* H3-lysine
Histone-3-Lysine-4 methylation, 220
Histone acetylation
  affected in vivo, 91
  butyrate effect on, 90
  controllers of, 19
  enhanced by catalase, 94
  enhanced by superoxide dismutase, 94
  increased by genistein, 93
  lunasin inhibition, 93
  NF-κB role, 192
  plant sources of bioactive compounds, 24
  suppressed by hydrogen peroxide, 94
Histone acetylation promotes active transcription, 14
Histone acetyltransferase (HAT), 19
  abnormal activity and cancer, 81
  activity, in COPD patients, 194
  altered activity in cancer cells, 111–112
  in diabetes, 266
  effect on lung health, 192
  enzyme family, 81
  families of, 111
  function of, 219–220
  functions of, 111
  HAT/HDAC imbalance, 192–193
  interaction with NF-κB, 221
  modulation by polyphenols, 193
  MOZ/YBF2/SAS2/TIP60 (MYST) family, 111
  plant sources of bioactive compounds, 24
  role in cardiac muscle, 268
Histone biotinylation
  cancer link, 116
  and gene silencing, 116
Histone carbonylation and calorie restriction, 94
Histone code, 81, 219
Histone deacetylase (HDAC), 19, *see also* Sirtuins
  aberrant targeting of, 81–82
  abnormal activity and cancer, 81
  action of, 80
  altered activity in cancer cells, 111–112
  in cardiac hypertrophy, 268
  classes of, 111
  class III, 20
  deacetylation by, 23
  effect on lung health, 192
  enzyme family, 81
  functions of, 219–220
  inhibitors, 380
    as chemotherapeutic agents, 82
    Istodax, 114
    treatment for diabetic complications, 224
    Zolinza, 114
  life preservation effect, 22
  no effect on activity by SFN, 195
  plant sources of bioactive compounds, 24
  in prefrontal cortex of SCZ patients, 295
  and sirtuins, 203
  three types of, 81
  type III activity increased by calorie restriction, 94
Histone deacetylase inhibitors, 303
Histone deacetylation
  and lung disease, 189
  SIR proteins in, 19–20
Histone demethylase and metabolic syndromes, 8
Histone demethylase inhibitors, 369
Histone demethylase (Jhdm2a), 220
Histone G9a, 323
Histone H1, 271
Histone H2A, 82
Histone H3, 82
  diallyl disulfide effect on acetylation, 91
  effect of SFN in sprouts, 91–92
  selenium effect on, 89
Histone H3 lysine 27 (H3K27me3) trimethylation, 83
Histone H4
  diallyl disulfide effect on acetylation, 91
  effect of SFN in sprouts, 91–92
  increased acetylation of, 91
Histone hyperacetylation and cell cycle arrest, 91
Histone hypoacetylation in gene silencing, 14
Histone lysine acetylation and methylation, 380
Histone lysine acetylation state altered in cancer cells, 111–112
Histone lysine methylation, 116
Histone marking, 83
Histone methylation

action of, 220
in developing embryo, 339
and lung disease, 189
not permanent, 221
and one-carbon metabolic pathway, 115–116
and schizophrenia risk, 341
Histone methyltransferase and hypoglycemic memory, 216
Histone methyltransferase (HMTase), 221, 222
inhibited by SAH, 339
Histone modification, 78, 230
acetylation best characterized, 14
affects DNA-based events, 81
of amino terminal portion, 2
by bioactive food components, 90ff
in brain tissues, 294–295
cancer risk modification, 81
cause chromatin structural changes, 81
in chronic diseases, 269
component of epigenetics, 2
by dietary components, 268
DNA methylation and, 23
enzyme interactions and, 110
enzymes, 98
epigenetic mechanisms, 110
gene expression modification, 81
in gene silencing, 20
in gene switching, 16
in inflammatory lung disease, 191
and lung disease, 189
nutrient influence on, 78
only a part of modifications, 79
preceding carcinogenesis, 113
processes, 22
in proinflammatory genes, 269
types of, 110
Histone modifications by bioactive food components, 380
Histone modifying enzymes, 222
Histone modulation by grapes, 22
Histones
acetylated in response to cigarette smoke, 193
acetylation state, 3
amino acid modifications in, 231
biotinylation sites on, 116
in cancer disease progression, 110–112
in chromatin, 218
in COPD patients, 193
definition, 110
enzymatic modification of, 2
and euchromatin states, 3
genistein effect on phosphorylation, 117
global loss of, 112
and heterochromatin states, 3
NF-κB alteration of, 187
phosphorylation targeted by genistein, 117
and SAM, 298
SIRT1 effect on, 196
Histones H3 and H4, 92
Histone system, aberrant, 96
HIV protease inhibitors, 201
^{hm}C, *see* 5-Hydroxymethylcytosine (^{hm}C)
HMG-CoA reductase inhibitors, 201
hMLH1 gene, 56, 57
HMLH1 promoters, 89
hMLH2 gene, 57
hMSH2 gene, 57
HMTase, *see* Histone methyltransferase (HMTase)
HMT functioning, 112
normalization of, 380
4-HNE, *see* 4-Hydroxy-2-nonenal (4-HNE)
HO-1 (Heme oxygenase), 199
Holocarboxylase synthetase, 116
Homeodomain-only protein (Hop), 268
Homocysteine (Hcy)
accumulation effects, 368
in AD patients, 366
B_{12} vital for conversion to methionine, 319
in cerebrospinal fluid, 296
and choline deficiency, 323
and cognitive deficits, 296
and cognitive functioning, 335
in colorectal cancer patients, 53
dietary effects on levels, 299–300
in DNA damage and repair, 7
and Down syndrome, 334
and executive dysfunction, 296
in folate cycle, 364
and folate deficiency, 323
folate in conversion to methionine, 322
folate/methyl deficiency and, 299
folate protection from high levels of, 327
folate status and, 53
and global DNA methylation status, 298
and hippocampal neurogenesis, 374
increased by folate deficiency, 15
levels and folate status, 37–38
levels associated with autism, 329
lowering levels for treatment of major psychoses, 301
in major psychoses, 296
maternal levels
and Down syndrome, 334
and schizophrenia risk, 332
maternally regulated, 330
and methionine metabolism, 273
and methylcobalamin, 322
and NTD risks, 326, 327
relationship with choline, 323
relationship with folate, 323
remethylation pathway, 323
remethylation to methionine, 38
risk factor for major psychosis, 288–289

SNP polymorphisms and, 324–325
as teratogen, 327
treatment option, 300
Hop (Homeodomain-only protein), 268
Hormone levels in utero, 248–249
Hormone precursors, 132
Hormone receptors as nutrient sensors, 271
Hoxa-10 and -11 expression, 249
Hoxa1 gene, 95
HP1-like protein, 271
HPA (Hypothalamic-pituitary-adrenal axis), 304
HSD11B2, 267
11β-HSD2, 267
HSL, 267
HT29 cells, 91, 92
Htr2c expression, 340
Human growth hormone (HGH) in antiaging treatments, 24–25
Human mutL homolog 1, 88
Huntington's disease (HD), *see also* Neurodegenerative disease
and oxidative stress, 367
possible treatments, 370
Hydrocephaly and maternal obesity, 236
4-Hydroxy-2-nonenal (4-HNE), 187
8-Hydroxydeoxy guanine (8-OHdG), 147
5-Hydroxymethylcytosine (^{hm}C), 290
11-hydroxysteroid dehydrogenase 1, 262
11 β-Hydroxysteroid dehydrogenase 2 (11β-HSD2), 267
Hyperactivity and folate in early pregnancy, 336
Hyperemesis gravidarum and fetal nutrition, 128–129
Hyperglycemia
damage leading to diabetes, 266
in diabetic complications, 221
effect on proatherogenic genes, 270
effect on proinflammatory genes, 223
epigenetic factors in, 7
and maternal diet, 241
and reduced methylation, 223
Hyperglycemic memory, 216
Hyperhomocysteinemia, 61
clinical trials, 51
DNA methylation reduced in, 88
fetal survival rates and, 328
folic acid supplementation during pregnancy, 61
in major psychoses, 296
in Parkinson's disease patients, 366
vitamin treatment, 299–300
Hyperinsulinemia
CD36 gene defect, 263–264
epigenetic factors in, 7
IGF-1 treatment of undernourished offspring, 238
in metabolic syndrome, 260–261
Hyperleptinemia, 7
Hyperlipidemic diseases
damage leading to diabetes, 266
and HCC, 144
treatment with grape components, 21
Hypermethylation process, 2
in aging cells, 20
in gene silencing, 20
inactivates tumor suppressor genes, 2
silencing of TSGs, 147
Hypermutability of methylated cytosine, 35
Hyperphagia
effect of leptin on, 239
effect of maternal dietary restriction, 238
Hypertension, *see also* Cardiovascular disease (CVD)
birth weight effect, 127
in diabetes, 267
effect of methyl donor restriction, 235
fetal insults and development, 267
and GR receptor, 273
heritability of, 264–265
11-hydroxysteroid dehydrogenase 1 in, 262
and insulin resistance, 263–264
maternal diet and, 133
maternal high-fat diet and, 273
and maternal protein restriction, 129
metabolic imprinting and, ix
in metabolic syndrome, 259–260, 263–264
pathenogenesis, 267
Hypertensive vascular disease in African Americans, 5
Hypertriacylglyceroemia, 262
Hypoglycemia
deleterious long-term effects, 216
epigenetic influence though transient, 222
epigenetic modification by, 216
Hypoinsulinemia and maternal diet, 241
Hypomethylation, *see also* DNA hypomethylation
in aging cells, 20
carcinogenesis and global, 34
chromatin stability affected by, 147
global and cancer effects, 14
methyl group deficiency and, 96
Hypothalamic-pituitary-adrenal (HPA) axis, 304
Hypotonia, 366
Hypoxia as risk factor for major psychosis, 288–289

I

I3C (Indole-3-carbinol), 98
IAP (Intracisternal-A particle), 86
Ibuprofen, 303
ICAM-1 (Intercellular adhesion molecule), 187
ICRs, *see* Imprinting control regions (ICRs)

ID4 gene, 152
  silenced in cancer, 148
Idebenone for neurodegenerative disease treatment, 370
Identical twins discordance in bipolar disorders, 6–7
Iff1R, 167
Ifg2/H19 gene cluster, 233, 234
Ifgpl gene silencing, 21
IFN (Interferon), 190
IgE levels in respiratory disease, 191
IGF-1
  and breast cancer risk, 247
  and early mammary tumors, 248
  reduced in offspring by eclampsia, 15
IGF1R in HCC, 165
IGF-1 treatment of undernourished offspring, 238
IGF2, *see* Insulinlike growth factor II (IGF2)
IGF2 gene, 233, 234
  and maternal diet, 234
  maternal folate and methylation state, 62–63
  methylation following prenatal famine, 129
  promoter methylation of, 57
IGF2 (Insulin-like growth factor II), *see* Insulin-like growth factor II (IGF2)
IGF2 methylation
  Dutch Hunger Winter and, 306
  malnutrition and, 306
Igf2P0, 233–234
IGT, 262
IκBα kinase, 200
IL10, 84–85
IL10 gene methylation, 129
  prenatal famine and, 306
IL-4
  in chronic respiratory disease, 191
  in inflammatory lung disease, 190
IL-6, 223
IL-8
  in COPD process, 187
  in inflammatory lung disease, 192
  suppressed by curcumin, 200
  suppression of, 194
Ile22Met variant, 325
Immunosenescence, 14
Immunosuppressant flavonoid interference, 201
Impaired glucose tolerance complications, 215–216
Imprinted genes, 33
  placental, 234
Imprinting control regions (ICRs), ix, 4, 233, 340
Imprinting loss, 34
Inattention behavior and folate in early pregnancy, 336
Indole-3-carbinol (I3C), 98
Indoor cooking exacerbates COPD, 186
Infection exacerbates COPD, 186
Inflammation
  miR expression and, 143
  and neurodegenerative diseases, 367
  regulated by transcription factors, 195
Inflammatory cytokines, 168
  up-regulation in diabetes, 221
Inflammatory process proteins maternal diet and, 129–130
Inflammatory response in respiratory disorders, 187
Inheritance of acquired characters, 264
INK4a, 189
Insl/2, 223
INS gene, 266
INSIGF gene methylation, 129
  prenatal famine and, 306
Insulin 1/2 (Insl/2), 223
Insulin gene coding, 266
Insulin/glucose ratio, 7
Insulin-like growth factor II (IGF2), 84–85, 242, *see also* IGF
Insulinlike growth factor II (IGF2)
  cancer link, 274
  DMR methylation, 337
  loss of imprinting, 4
Insulin metabolism in metabolic syndrome, 259–260
Insulin receptor deletion, 262
Insulin receptor substrate-1 (IRS1), 261–262
Insulin resistance, *see also* Diabetes; Metabolic syndrome
  adipocyte methylation study, 6
  basis of, 261–262
  birth weight correlation, 129
  and CVD, 260
  effect of methyl donor restriction, 235
  effects of maternal diet, 242
  and glucose intolerance, 262
  and HCC, 144
  and hypertension, 263
  lack vasodilatory effect of insulin, 263
  lipoatrophy, 262
  LPL gene mutations, 263
  and maternal diet, 131, 133
    and type 2 diabetes development, 240
  maternal high-fat diet and, 273
  maternal low-protein diet effects, 243–244
  in metabolic syndrome, 259–262
  metabolic syndrome risk factor, 260
  mitochondrial oxidative phosphorylation, 261–262
  and NAFLD, 165
  and NASH, 165
  possible neonatal intervention, 128
  protective phenotype, 261–262
  resveratrol protective, 22–23

Insulin sensitivity in metabolic syndrome, 260–261
Intercellular adhesion molecule (ICAM-1), 187
Interferon-γ promoter, 189–190
Interferon (IFN) in lung disease, 190
Interindividual variance, 301, 304, 308
Interleukin, *see* specific IL designations
Internal initiation, 32–33
Intestinal microflora synthesize folate, 45
Intracisternal-A particle (IAP), 86
Intrahepatic metastasis, 167
Intrauterine growth restriction (IUGR), 130, 231–232, *see also* Maternal nutrition
  and beta-cell development, 241
  diabetes risk and, 241
  effect on glucose metabolism, 242
  and maternal diet, 237–238
Intrinsic factor for B_{12} absorption, 319
Intronic miR, 156
Invasion inhibition, 90
Iron
  deficiency in pregnant women, 239
  in maternal diet, 237, 250
IRS1 (Insulin receptor substrate-1), 261–262
Ischemic heart disease, 129, *see also* Cardiovascular disease (CVD)
  maternal diet and, 125
β-Islet cells, 266
Isoflavones, 192
Isothiocyanates
  in cruciferous vegetables, 92
  in epigenetic health, 302
  HDAC inhibitors, 92
  inhibits HDAC enzymes, 90
Istodax, 114
ITGAL (CD11a) promoter, 14
IUGR (Intrauterine growth restriction), *see* Intrauterine growth restriction (IUGR)

J

Jackfruit as resveratrol source, 22
JB6 cells, 200
Jhdm2a (Histone demethylase), 220
Jumonji histone demethylases (JHDM), 112

K

K27me3, *see* Lysine 27 (K27me3)
K27 methyltransferase EZH2, 95
K4, 230
K4K5, 273
K9, 231
Kaempferol beneficial in COPD, 199
Kale as sulforaphane source, 114
KAT, *see* Histone acetyltransferase (HAT)
KB cell line, 48
KDM1, 112
Keap1, 199
Keap1/Nrf2/ARE pathway, 372
α-Keto acid metabolites, 92
Ketogenesis, 271
Kidney beans as methionine source, 16
Kidney disease and diabetes, 267
Kidneys
  folate in, 322
  folate storage in, 319
Kinked tail in agouti mice, 60–61
Kiwis as omega-3 fatty acid source, 302
KMT, *see* Lysine methyltransferase (HMT)
Knudson's two-hit theory of carcinogenesis, 147–148
Ko-jo-kon, 21, 22
Kroc Collaborative Study Group, 216
Ku70-Bax complex, 197
Ku70 deacetylation, 20
KVLQT1 gene, 268

L

Lactoferrin gene, 249
Lamarckian theory, 264
  transgenerational responses resemble, 275
Lamin A/C defects, 263
Lariant intermediates, 156
Laser capture microdissection, 6
Latent Early-life Associated Regulation (LEARn) model, 304
L-carnitine for neurodegenerative disease treatment, 370
LDL (Low density lipoprotein), *see* Low density lipoprotein (LDL)
LDL-TAG, 263
l-Dopa in Parkinson's disease, 366
Lead exposure, 367
Leafy vegetables as folic acid source, 16
LEARn model, 304
Legacy effect in diabetes, 215
Legumes as folate source, 36
LEP, 84–85
LEP gene methylation, 129
  prenatal famine and, 306
Leptin
  and breast cancer risk, 247
  in breast-fed infants, 238
  effect in adults, 272
  effects in neonate, 239
  metabolism affected by maternal diet, 243–244
  during neonatal period, 272
  resistance, 239
  role of, 239
Leptin treatment
  effects of, 239

of undernourished offspring, 238
Let-7, 157
oncogene inactivation by, 3
Let-7a, 164
let-7a3, 168
let-7 family, 157
Lettuce as choline source, 16
Leukemia
acute lymphoblastic, 249–250
lysine sites altered in, 111–112
Leukemia and HDA, 81
Leukemia cells, copper effect on, 94
Leukemia risk and folate status, 36
Leukemia zinc finger, 81–82
Lewy Body disease, 5–6
LFA-1 overexpression, 14
Life experience and disease process, 5–6
Life span duration, 19
Lifestyle and lung health, 188
Ligustrum lucidum, 276
Lima beans as folate source, 36
LINE-1 gene, 59
LINE-1 (Long interspersed nucleotide elements), 51
LINE methylation state, 3
Lingonberry as resveratrol source, 22
Lipid metabolism, 165
Lipid oxidation and HDL levels, 263
Lipid peroxidation products in COPD process, 186–187
Lipid transport and choline, 145
Lipoatrophy, 262
Lipoic acid, 372
Lipopolysaccharide (LPS), 187, 200
Lipoproteinlipase (LPL), 263
Lipoprotein synthesis, 145
Liver
choline source, 321
DNA hypermethylation, 87–88
as folate source, 36
folate source, 318
folate storage in, 319
Liver cancer
CDAA diet and, 163
as choline source, 16
demethylation in, 113
in Dicer null mice, 156–157
early process, 95
fetal tissue affected by maternal diet, 93–94
FMD diet effect on, 161, 162
folate deficiency and, 144–145
hypomethylation and, 14
induction of, 113
metastatic, 143
methionine prevents malignant changes, 147
methyl group donor deficiency, 39
miR-155 in, 168
miR-21 activity in, 98
and NAFLD, 165
progression in animal model, 146
survival rates, 169
tumor formation induced by methyl group deficiency, 96
tumor suppressor miRNA silenced in, 169
Liver cells
dietary effects on, 271–272
effect of maternal low-protein diet, 273
folate and aging, 64
proliferation affected by FMD diet, 161, 162
starvation effects, 271
Liver disease
FMD diet and, 147
HCC development similar to methyl-deficiency HCC, 97
and maternal diet, 131
nonalcoholic fatty liver disease, 97
Liver fibrosis
CDAA diet and, 163
FMD diet effect on, 161
Liver-specific miRNA, 157
Liver X receptor (LXR), 271
Livestock nutrition management, 125, 126, 129–130
LnCap
selenium effects, 89
SFN effect on, 91
LOI (Loss of imprinting), 4
Longevity
grandparental effect, 274
Mediterranean diet effects, 24
Long interspersed nucleotide elements (LINE-1), 3, 51, 59
Long noncoding RNA, 154
Long terminal repeat retrotransposons, 116
Loss of function modification, 81
Loss of imprinting (LOI), 4
Lou Gehrig's disease, *see* Amyotrophic lateral sclerosis (ALS)
Low density lipoprotein (LDL), 260
atherogenic character, 263
modification of, 263
LP, *see* Maternal low-protein (LP) diet
LPL (Lipoproteinlipase), 263
LPS (Lipopolysaccharide), 187, 200
LSD1 (Lysine-specific demethylase 1), *see* Lysine-specific demethylase 1 (LSD1)
Lunasin, 93
Lung cancer
chemotherapeutic strategy, 198
demethylation in, 112
dietary polyphenol as chemopreventive, 200
and folate status, 36
hypomethylation and, 14
lysine sites altered in, 111–112

miR-21 activity in, 98
miRNA expression in, 84
non-small-cell and folate, 59
p16 methylation in, 189
Lung disease
early environmental exposure and, 188
epigenetic factors in, 188
may begin very early in life, 188
polyphenol regulation of inflammation, 192–193
Lupus erythematosus (SLE), 14
Luteolin, 199
LXR (Liver X receptor), 271
Lycopene, 302
Lymphoblastoid cells, 291
Lymphocyte DNA methylation, 61, 63
folate status and, 49
Lymphocytes, 64
Lymphoma
genetics, 34–35
miRNA expression in, 84
Lysine, *see also* H3-lysine; Histones
acetylation
in cancer disease progression, 111–112
cancer treatment by correction of aberrant, 114
acetylation of in histone, 3
Lysine 27 (K27me3), 65
Lysine ε-amino group, 111
methylation by SAM, 112
Lysine (K)4, 230
Lysine K9, 231
Lysine methyltransferase (HMT), 112
specificity of, 221
Lysine residues
modification by biotin, 116
site of enzymatic modifications, 2
Lysine sites altered in cancer cells, 111–112
Lysine-specific demethylase 1 (LSD1), 112, 221, 223

M

Mackerel as source of PUFAs, 371
Maclura pomifera, 277
Macrophage phagocytosis, 339
Macrophages in COPD process, 186
MafA, 223
Magnesium in autism treatment, 331
Maintenance methylase (DNMT1), 33, 291
Major psychoses, 287, *see also* Bipolar disorders (BPD or BD); Schizophrenia (SCZ)
brain methylome differences in, 293
clinical parameters in, 296
dysbindin risk factor, 292
environmental factors and, 296
epigenetic alteration in, 291–295
epigenetic paradigm of, 304
epigenetic patterns in disease-related tissues, 292
folate treatment for, 300
glutamergic neurotransmission in, 292–293
malnutrition and, 304–306
methylation changes associated with, *see* Major psychoses
one-carbon metabolism in, 296–298
prefrontal cortex function in, 296
and prenatal folic acid, 308
risk factors for, 288–289
SAH in, 298
SAM in, 298
skewed X chromosome inactivation in females, 294
transgenerational effects, 303–304
mal1, 261
Malnutrition
and major psychoses, 304–306
vitamin treatment, 300
Mammary cancer, *see* Breast cancer
Manic depressive disorder, *see* Bipolar disorders (BPD or BD)
MAPK (Mitogen-activated protein-kinase) pathway, 303
Maspin in HCC, 165
Mast cells increased in asthma, 187
Maternal anemia, 239
Maternal calorie-restricted diet, 232
Maternal diet and offspring cancer risk, 247
Maternal dietary restriction timing, 238
Maternal disomy, 340
Maternal genetic influences in neurodevelopment, 340
Maternal high-fat (HF) diet, 232
increases E2, 248
offspring disease susceptibility, 273
placental effects, 236
Maternal hyperglycemia, 244
Maternal infection as risk factor for major psychosis, 288–289
Maternal iron deficiency, 239
Maternal leptin intake and breast cancer risk, 248
Maternal low-protein (LP) diet, 232
corticosol inactivation, 239
effect on GR, 273
effects of, 235, 241–243
glucose metabolism effects, 242, 243–244
offspring breast cancer risk, 248–249
offspring disease susceptibility, 273
offspring gender differences, 242–243
thrifty phenotype hypothesis, 238
Maternally imprinted genes, 233
Maternal methyl donor deficiency, 270
Maternal modifier genes remove paternal imprints, ix

Maternal nutrition, *see also* Neural tube defects (NTD); specific nutrients
- and ALL risk, 249–250
- amino acid supplementation, 133
- animal experiments in, 233
- arginine supplementation, 133
- and beta-cell development, 241
- birth weight effects, 128–129
- and cancers, 245–250
- choline and NTD risks, 327
- cognitive functioning and, 334–335
- fetal obesity, 131
- fetal origins of adult onset disease, 94
- fetal skeletal muscle development, 131
- folate absorption and metabolism, 324
- folate intake effects, 60–63
- folate intake studies, 62–63
- folate metabolism, 325–326
- folate supplementation timing, 335
- folic acid supplementation, 324
- high-fat diet effects, 131, 221
- histone acetylation affected by, 93–94
- hyperemesis gravidarum, 128–129
- impaired vascular function and, 129
- importance, 230
- influence of imbalance, ix
- insulin metabolism and, 131
- intrauterine growth restriction, 130
- and IUGR, 237
- and lean mass, 245
- leptin intake, 248
- and leukemia risk, 250
- liver effects, 131
- low-protein diet, 235
- maternal heritability, 299
- micronutrients, 250
- myogenesis effects, 131
- neural tube defects and, 323
- in neurodevelopment, 318
- obesity during pregnancy, 126
- offspring breast cancer risk, 248
- placental angiogenesis effects, 131
- postnatal health connection, 125
- prenatal famine exposure, 84–85
- protein intake, 241
- protein restriction effects, 129
- risk factor for major psychosis, 288–289
- and schizophrenia risk, 332, 341
- specific diets, 232
- and type 2 diabetes, 239–245
- undernutrition, 125, 126

Maternal obesity and diabetes risk, 244

Maternal one-carbon (1-C)-supplemented diet, 232

Maternal overnutrition, 130–131

Maternal prediabetes, 245

Maternal stress as risk factor for major psychosis, 288–289

MAT (Methionine adenosyltransferase), 364

Matrix metalloproteinases, 261

MBD, *see* Methyl binding domain proteins (MBD); Methyl-CpG binding proteins (MBD)

MBD1 and ASD, 340

MBD2
- action of, 80
- DNA methyl binding protein, 80

MBD2 demethylase, 33

MBD2 expression, in human cancers, 35–36

MBD3, *see* Methyl-CpG binding protein domain 3 (MBD3)

MBPs, *see* Methylated binding proteins (MBPs)

^{m}C, 290

Mcl-1, 165

MCI, *see* Mild cognitive impairment (MCI)

Mcp-1 (Monocyte chemoattractant protein-1), 223

MCP30, *see also* Bitter melon; *Momordica charantia*
- as cancer preventive, 115
- HDAC-1 inhibitor, 115
- inhibits HDAC activity, 92

MeCP2, 340
- action of, 80

MeCP2 deficiency in ASD, 340

MECP2 gene, 331
- in Rett syndrome, 340
- Rett syndrome and, 331–332

Medical Research Council (MRC) Vitamin Study Research Group, 324

Medications may deplete B vitamins, 300

Mediterranean diet
- effect on AD, 371
- effects of, 24

MEG3 gene methylation, 129
- prenatal famine and, 306

Megaloblastic anemia caused by folate deficiency, 319

MEGF, 84–85

Meiotically heritable alterations, 127–128

MEK1 gene correlated with antipsychotic use, 303

Melatonin for neurodegenerative disease treatment, 370

Memory capacity and choline, 321

MESA (Multi-Ethnic Study of Atherosclerosis), 5

Messenger RNA (mRNA), 3

Mest/Peg1 gene, 234

Metabolic compromise, 236

Metabolic diseases
- and polyphenolic food constituents, 16
- possible neonatal intervention, 128

Metabolic memory, 216

in diabetes, 215
long-term effects of hypoglycemia, 216
Metabolic syndrome, 7–8
birth size and risk, 240–241
birth weight correlation, 129
characteristics of, 259–260
epigenetic basis for, 268
epigenetic factors in, 265–289
and HCC, 144
heritability of, 264–265
11-hydroxysteroid dehydrogenase 1 in, 262
hypertension as risk factor, 260
IUGR and, 237–238
lipoatrophy, 262
predisposing factors for, 268
proinflammatory state, 260
resistance to, 262
single-gene defects in, 263
Metabolism and leptin, 239
Metabolites necessary for homeostasis, 132
Metalloproteinases, 261
Metallothionein family, 369
Metals
latent activity of, 374
and neurodegenerative diseases, 367
Metastable epialleles, 87
Metastasis
butyrate effect on, 90
in HCC, 167
MET gene, 169
in ASD, 341
Methionine
B_{12} vital for conversion from homocysteine, 319
deficiency, 96
effect of restriction on fetal development, 235
effect on MTHFR polymorphisms, 57
effect on SAM:SAH ratio, 300
in folate cycle, 235
folate in conversion from homocysteine, 322
food sources of, 16
in HCC, 165
and histone modification, 268
may exacerbate schizophrenia, 300
methyl group donor, 38–39
and NTD risks, 327–328
prenatal supplementation, 87
prevents malignant liver changes, 147
risk modification by, 116
sources of, 300
supplementation risks, 300–301
synthesis into SAM, 323
synthetase deactivated by nitrous oxide, 320–321
Methionine adenosyltransferase (MAT), 323, 364
Methionine deficiency, 15
Methionine metabolism, 7
Methionine synthase defect, 322–323
Methionine synthase reductase (MTRR), 322, *see also* MTRR
in folate cycle, 364
Methionine synthetase, 322
Methylated binding proteins (MBPs), 3
Methylation, 78, *see also* Histone modification
histone modification, 3
of histones, 22, 110
Methylation analysis (RLGS-M), 150–151
Methylation capacity, 323
Methylation changes with aging, 17, 19
Methylation index, 298
Methylation levels as diagnostic for prostate cancer, 3
Methylation patterns
cancer types vary in, 49
and lung health, 188
and maternal diet, 234
in preimplantation embryos, 327
Methylation potential of cells, 368
Methylation profile
in asthma, 7
in bipolar disorders, 6–7
in retinal cells, 6
Methylation state
in AD patients, 367
atherosclerosis and, 5
during embryogenesis, 233
equilibrium needed, 3
and lung disease, 189
metabolism pathways and, 273
and phenotype switching, 5
of repetitive gene regions, 3
transcription silencing and, 35
Methyl binding domain proteins (MBD), 149, *see also* MBD
Methylcobalamin and homocysteine, 322
Methyl-CpG-binding protein 2 (MeCP2)
gender factors in, 340
interaction of, 110
in Rett syndrome, 340
Methyl-CpG binding protein domain 3 (MBD3), 232
Methyl-CpG binding proteins (MBD), 33
5-Methylcytosines, 32
deamination to thymine, 35
formed by methylation, 110
Methylcytosine-to-thymine transition mutations, 32
Methyl deficient induced HCC, 96–97
Methyl donor precursor deficiency, 15
Methyl donor restriction, 235
Methyl donors in agouti mouse, 15
Methylenetetrahydrofolate dehydrogenase (MTHFD), 325
5,10-Methylenetetrahydrofolate (5,10-MTHF), 57

correlation with genotypes, 59
reduction of, 364
Methylene tetrahydrofolate (MTHFR) reductase gene, 145
Methylenetetrahydrofolate reductase (MTHFR), 36, *see also* MTHFR
deficiency mimics Rett syndrome, 332
folate metabolism and, 38
folate transport by, 364
genotype polymorphism in clinical trials, 49, 50
individual variability of, 369
polymorphisms and carcinogenesis, 37–38
role schematic, 37
Methylenetetrahydrofolate reductase (MTHFR) enzyme, 322
Methylfolate as psychiatric treatment, 333
Methyl group dietary deficiency, 96
effect on histone, 269
factors in, 235
hepatocarcinogenesis and, 96
hepatocarcinogenetic, 269
induced in rats, 145–146
in major psychoses, 298ff
Methyl group dietary sources, 115–116, 132
choline, 145
Methyl group donors, 38–39
dietary deficiency of, 39
5-Methylhydrofolate activity, 338
Methylmalonic acid (MMA)
and cognitive functioning, 335
effects of elevated levels, 320
indicator of B_{12} deficiency, 329
treatment option, 300
Methylmalonyl coenzyme A mutase (MUT), 320
Methylome in brain tissue, 293
β-Methylselenopyruvate (MSP), 92
6-Methylsulfinylhexyl isothiocyanate, 23
5-Methyltetrahydrofolate-homocysteine methyltransferase (MTR), 322
Methyltetrahydrofolate in plasma, 300
Methyltetrahydrofolate supplementation, 51
5-MethylTHF
and CFD, 366
production of, 364
5-Methyl THF monoglutamate, 318–319
MethylTHF to THF conversion, 319
Methyltransferase actions, 323
Methyl transferases, 3
Methyl trap, 322–323
MGMT gene and folate effect in esophageal cancer, 59
Micronodular cirrhosis, 147
Micronuclei (MN), 365
MicroRNA (miRNA or miR), 2, *see also* miR; Noncoding RNA
affected by CDAA diet, 164
coordinates methyltransferases, 110
database of, 155
definition, 3
discovery history, 154
in endocrine function, 267
in fat metabolism, 267
functions of, 156–157
genes epigenetically modified in cancer cells, 98
human identified types, 155
influence on disease development, 143
intronic, 156
largest cluster, 156
liver specific, 157
liver specific expression, 157
modifications of, 98
most DNA devoted to coding for, 154
profile, 2
protein regulation by, 156
regulation in asthma, 190
regulatory functions of, 97
species types in database, 155
therapeutic potential, 143
therapy for hepatocarcinoma, 170
transcriptional control by, 267
Migration
and major psychoses, 308
risk factor for major psychosis, 288–289
Mild cognitive impairment (MCI)
and AD, 367
Mediterranean diet effects, 371
and oxidative stress, 367
MINT-1, -2, -3, 57
miR, *see* MicroRNA (miRNA or miR)
miR-1
heart-specific, 154
tumor suppressor, 169
miR-101 in HCC, 165
miR-1-1 in HCC cell lines, 169
miR-122, 156–157, 165
affected by CDAA diet, 164
conserved across species, 155
in HCC, 165
in hepatocellular carcinoma, 270
in human HCC, 161
in liver and HCC, 166–167
liver specific, 154, 157
in NASH patients, 166
miR-124a in carcinogenesis, 168
miR-126, 84
liver specific, 157
miR-127
in carcinogenesis, 168
FMD diet effect on, 161
miR-127 gene, 97
miR-130a, indole-3-carbinol effects on, 98
miR-130 in tumors, 159

miR-133 heart-specific, 154
miR-142 affected by CDAA diet, 164
miR-143, 84
in diabetes, 267
in HCC, 165
miR-145, 84
miR-146b, indole-3-carbinol effects on, 98
miR-146 gene, 97
miR-148 liver specific, 157
miR-150 affected by CDAA diet, 164
miR-155
affected by CDAA diet, 164
CDAA diet and, 168
in HCC, 168
in hepatocarcinogenesis, 270
NASH and, 168
miR-155 gene, 97
miR-15 chronic lymphocytic leukemia link, 157
miR-15/miR-16 oncogene inactivation, 3
miR-16
affected by CDAA diet, 164
chronic lymphocytic leukemia link, 157
FMD diet effect on, 162
in NASH patients, 166
miR-16a gene, 97
miR-16a liver specific, 157
miR-17-92 family liver specific, 157
miR-17 family affected by CDAA diet, 164
miR-181
affected by CDAA diet, 164
in HCC, 167–168
miR-181b/d and CDAA diet, 167
miR-181b in NASH patients, 166
miR-182 gene, 97
miR-18a in HCC, 165
miR-192
in diabetic kidney disease, 267
liver specific, 157
miR-194 liver specific, 157
miR-195
affected by CDAA diet, 164
in HCC, 165
miR-199/199* in NASH patients, 166
miR-199a* in HCC, 165
miR-200 affected by CDAA diet, 164
miR-200b gene, 97
miR-203 in pancreatic cancer, 168
miR-21
affected by CDAA diet, 164
indole-3-carbinol effects on, 98
in NASH patients, 166
preneoplastic transformation, 159
in solid tumors, 98
in specific cancers, 98
miR-210 in pancreatic cancer, 168
miR-214
affected by CDAA diet, 164
in NASH patients, 166
miR-21 gene, 97
miR-221
affected by CDAA diet, 164
liver specific, 157
in NASH patients, 166
miR-221/222
affected by CDAA diet, 164
in HCC, 165
in NASH patients, 166
miR-221/222 gene, 97
miR-222
affected by CDAA diet, 164
in NASH patients, 166
in pancreatic cancer, 168
miR-223
affected by CDAA diet, 164
in HCC, 165
liver specific, 157
miR-224 in HCC, 165
miR-23a,b affected by FMD diet, 157
miR-23a in NASH patients, 166
miR-23b in NASH patients, 166
miR-23 preneoplastic transformation, 159
miR-24 in NASH patients, 166
miR-26a liver specific, 157
miR-27b liver specific, 157
miR-30d in diabetes, 267
miR-30d liver specific, 157
miR-31 gene, 97
miR-320, 83
miR-342 affected by CDAA diet, 164
miR-346 affected by CDAA diet, 164
miR-34a
affected by CDAA diet, 164
FMD diet effect on, 161
in NASH patients, 166
miR-34a gene, 97
miR-375, 156–157
in diabetes, 267
miR-377, indole-3-carbinol effects on, 98
miR-383 affected by CDAA diet, 164
miR-483 fetal liver specific, 157
miR-487a affected by CDAA diet, 164
miR-92a fetal liver specific, 157
miR-96 affected by CDAA diet, 164
miR-99b
affected by CDAA diet, 164
in NASH patients, 166
miR-induced gene-silencing complex (miRISC), 156
miRISC, *see* miR-induced gene-silencing complex (miRISC)
miR-miR-143 liver specific, 157
miRNA, *see* MicroRNA (miRNA or miR)
miRNA-199a* curcumin effect on, 97
miRNA-22 curcumin effect on, 97–98

miRNA profiling, 84
diagnostic use of, 3
miRNAs
in cancer pathogenesis, 83–84
chronic lymphocytic leukemia link, 157
curcumin effect on, 97
diagnostic use of, 3
dysregulated in NASH patients, 166
functions of, 3
specificity of, 3
miR signature as biomarker, 157
Mismatch repair genes
in CpG islands, 35
MLH1, 56
Mitochondrial DNA (mtDNA) modulated by folate, 145
Mitochondrial function effects of prenatal nutrition, 131
Mitochondrial oxidative phosphorylation in insulin resistance, 261–262
Mitochondrial role in vascular function, 131
Mitochondria-targeted antioxidants for neurodegenerative disease treatment, 370
Mitogen-activated protein kinase 3 gene modulated by diallyl disulfide, 91
Mitogen-activated protein-kinase (MAPK) pathway, 303
Mitosis and SIR proteins, 19
Mitotically heritable alterations, 127–128
Mitotic inheritance, 80
Mitotic recombination, 34
Mlh1, 34
MLH1 gene, 56
promoter methylation of, 57
MMA, *see* Methylmalonic acid (MMA)
MN, *see* Micronuclei (MN)
MnSOD, 196
Momordica charantia
apoptosis induction by seed component, 92
HDAC-1 inhibitor source, 115
Monocyte chemoattractant protein-1 (Mcp-1), 222
Monocyte gene expression in atherosclerosis, 5
Monoglutamates in folate absorption, 318
Monozygotic (MZ) twins
epigenetic differences in, 85
epigenetic instability in, 292
methylation and histone differences, 289
methylation differences in major psychoses, 291–292
skewed X chromosome inactivation in females, 294
Monozygotic twins, *see* Twins
Morbidity birth weight correlation, 129
Morris hepatoma 7777 cells, 91
Movement disorders in schizophrenia, 332
MOZ/YBF2/SAS2/TIP60 (MYST) family, 111
MRC (Medical Research Council Vitamin Study Research Group), 324
mRNA c-myc expression, butyrate effect on, 90
mRNA degradation, 83
mRNA methyltransferase inhibited by SAH, 339
MSC (Se-methyl-selenocysteine), 92
MSP (β-Methylselenopyruvate), 92
mtDNA, *see* Mitochondrial DNA (mtDNA)
5,10-MTHF, *see* 5,10-Methylenetetrahydrofolate (5,10-MTHF)
MTHFD, *see* Methylenetetrahydrofolate dehydrogenase (MTHFD)
MTHFR, *see* Methylenetetrahydrofolate reductase (MTHFR)
MTHFR 677 C/C genotype, 337
MTHFR 677T allele, 334
MTHFR 677TT genotype, 324–325, 328
MTHFR 677 T/T genotype, 337
MTHFR 677TT genotype
in AD patients, 366
DNA methylation, 55–56
folate status and hypermethylation, 59
MTHFR C677T polymorphism
and autism, 330
folate intake and methylation in adenomas, 57
MTHFR C677T variant, 324–325
in mother or offspring, 326
NTD risks, 326, 327
schizophrenia risk increased with, 333
MTHFR polymorphisms, 324–325
A1298C, 57
alcohol effects, 57
and autism, 329
in cognitive functioning, 335
and methylation level, 53, 55
MTHFR 677C/T, 366
TT genotypes, 53, 55
MTHFR TT genotype, 56
mTOR pathway, 165
MTR, *see* 5-Methyltetrahydrofolate-homocysteine methyltransferase (MTR)
MTR A2756G (rs1805087) variant, 325
MTRR, *see* Methionine synthase reductase (MTRR)
MTRR 66 GG genotype, 326
MTRR A66G (Ile22Met), 325
MTRR A66G variant, 326
MTRR variant and autism, 329
MTR variants, 326
in cognitive functioning, 335
NTD risks, 326
Mulberry as resveratrol source, 22
Multi-Ethnic Study of Atherosclerosis (MESA), 5
Multikinase inhibitor, 169
Multiple sclerosis, modifications in, 369

Muscle damage and choline deficiency, 145
Muscular degeneration, 6
Mutation increases, 34
mutL homolog 1, 88
MUT (Methylmalonyl coenzyme A mutase), 320
Mycolytic thiols, 186
Myelin abnormalities, 320
  and cognitive functioning, 335
Myelodysplastic syndrome, 170
Myeloid leukemia, 167
Myelopathy, 320–321
Myocardial infarction, *see also* Cardiovascular disease (CVD)
  in diabetics, 215–216
  HDAC inhibitors in treatment, 224
Myocyte enhancer factor-2, 268
Myotrophin, 267
MYST, *see* MOZ/YBF2/SAS2/TIP60 (MYST) family
MZ, *see* Monozygotic (MZ) twins

N

N-acetyl aspartate levels in Rett syndrome, 331
N-acetylglucosaminylation, 263
NAD+
  catalytic mechanism, 111
  cleavage, 19–20
  dependence of sirtuins, 195
  depletion, 194
  oxidative stress and, 196
NAD+-dependent sirtuins, 220
  in calorie restriction, 270
Nad/NADH-sirtuin-1, 372
NADPH (Nicotinamide adenine dinucleotide phosphate), 322
NAFLD (Non-alcoholic fatty liver disease), 165
NANOG expression, 276
Naringenin
  drug interference by, 201
  effect on P-glycoprotein, 202
NASH, *see* Nonalcoholic steatohepatitis (NASH)
NASH-induced hepatocarcinogenesis, 163ff
NASH patients
  cirrhosis and HCC in, 165
  disease associations, 165
  miRNAs dysregulated, 166
Nasopharyngeal carcinoma cells, 48
National Institutes of Health (NIH) Epigenomics Roadmap Program, 4–7
Nature-nurture gap, 13
NCM356 cell line, 48
NCM460 cell line, 48
NcRNA (Nonprotein coding RNA), 154
NDUFB6, 270
NEFA (Nonessential fatty acid flux), 262, 263
Neil 1, 262
Neonatal cold resistance, 133
Neonatal nutrition, ix
Neoplasia patients
  CpG hypermethylation, 57
  folate levels, 57
Neoplastic disease, 371
Neprilysin (NEP), 368
nestin gene, 95
Netherlands Cohort Study on Diet and Cancer, 56
Neural tube defects (NTD), 323–328, *see also* Spina bifida
  B vitamins and decreased risk, 327
  choline role in, 321
  folate pathway in, 7
  global DNA methylation and, 328
  homocysteine effects, 327
  interaction of gene variants, 326
  and maternal obesity, 236
  metabolism efficiency and, 326
  prevented by adequate folate, 319
  protein myelination in, 339
  receptor autoantibodies, 326–327
  risk levels, 325
  variation in findings, 325
  vitamin B_{12} status, 327
Neuroblastoma risk and folate status, 36
Neurodegenerative disease
  deficiencies in DNA repair, 367
  dietary factors in treatments, 369–372
  and impaired folate metabolism, 365–366
  Mediterranean diet effects, 371
  and polyphenolic food constituents, 16
  reduction of, 370
  resveratrol action in, 372
  studies of, 369
Neurodevelopment
  choline role in, 321
  disorders of, 339–340
  epigenetic factors in, 339–340
  SAM activity in, 338
Neuroendocrine function role of leptin, 239
NEUROG1 gene, 57
Neurological disorders, vitamin B_{12} deficiency, 319
Neurologic disorders
  age-dependent methylation changes and, 19
  choline deficiency as cause, 145
Neuronal maturation, 331
Neuropathy from pernicious anemia, 319
Neurophils in COPD process, 186
Neuroprotective *Ligustrum lucidum*, 276
Neurotransmission
  choline role in, 145
  in major psychosis, 292–293
  and SAM, 298
Neurotransmission in major psychosis, 292–293
Neurotransmitters

B_{12} vital for, 319
PLP metabolism in, 329
synthesis of, 339
Neutrophil recruitment suppressed by curcumin, 200
NF-E2, 198
NF-κB (Nuclear factor kappa B), *see* Nuclear factor kappa B (NF-κB)
NF-κB/RelA antagonist, 198
Nicotinamide, 49
Nicotinamide adenine dinucleotide (NAD), 81
Nicotinamide adenine dinucleotide phosphate (NADPH), 322
Nicotinamide adenine nucleotide (NAD+)-dependent catalytic mechanism, 111
Nicotine, 370
NIH 3T3 cells
E1A-transfected fibroblast cells, 93
folate deficiency and, 48
NIMH Global Obsessive Compulsive Scale, 331
Nitric oxide release affected by prenatal and early nutrition, 131
Nitrous oxide, myelopathy caused by, 320–321
N-methyl-D-aspartate (NMDA) receptors, 292–293
Nonalcoholic fatty liver disease (NAFLD), 97
choline deficiency as cause, 145
emerging major health problem, 165
and maternal diet, 131
Nonalcoholic steatohepatitis (NASH), 97, *see also* NASH
obesity as risk factor, 144
Noncoding RNA (ncRNA), 79, 154
affected by methyl deficiency, 270
bioactive food components modify, 380
dietary modulation of, 98
in gene expression, 83
majority of mammalian genome, 142
Nonessential fatty acid (NEFA) flux, 262, 263
Nongenotoxic carcinogens, 113
Non-neuronal disease, 5–6
Non-small-cell lung cancer
Bmi-1 overexpressed in, 96
methylation in promoter regions, 188–189
Nonsteroidal anti-inflammatory drugs (NSAIDs) flavanal interaction, 202
Normoglycemia, 215
NOTCH1 protein, 97
FMD diet effect on, 162
Not I site methylation analysis, 151–152
Npas2 gene, 94
Nrf2, 198–199
activation by resveratrol, 200
Nrf2-Keapl-ARE activators, 199
Nrf-RelA imbalance, 198
NSAIDs, *see* Nonsteroidal anti-inflammatory drugs (NSAIDs)
NTD, *see* Neural tube defects (NTD)
Nuclear β-catenin, 236
Nuclear factor-erythroid derived 2 (NF-E2) binding sites, 198
Nuclear factor kappa B (NF-κB)
in acetylation of histones, 192
activated in diabetes, 221
antagonist, 198
in calorie restriction, 270
expression persists, 221–222
HDAC regulates activity, 23
in inflammatory reaction, 187
inhibited by resveratrol, 200
inhibited by SIRT1, 198
interaction with HATs, 221
in metabolic syndrome, 261
p65 recruitment, 222
polyphenol inhibition of acetylation, 193–194
SIRT1-mediated repression, 23
suppressed by curcumin, 200
Nuclear organization, 81
Nucleosome remodeling and deacetylase corepressor complex (NuRD), 82
Nucleosomes, 230
structure, 2, 80
tighter packaging, 33
Nur77, 243
NuRD (Nucleosome remodeling and deacetylase corepressor complex), 82
Nurse's Health Study, 56
Nursing diet, 131
Nutrient effects in regulation, 132, 133
Nutrient metabolism and epigenetic modifications, 220
Nutrient sensors in gene regulation, 270–272
Nutrients in gene regulation, 270–272
effects and timing of, 272–274
Nutrigenomics, definition, ix
Nutritional deficiency
degree of, 305
in major psychoses, 298ff
timing of, 305
Nutritional epigenetics, 8–9
Nutritional exposure and anticipation in major psychoses, 308
Nutritional Prevention of Cancer trial, 89
selenium-enriched yeast, 92
Nutrition as risk factor for major psychosis, 289
Nutrition in utero, 15
Nuts as methionine source, 16
NZ-01, 276

O

O^6-methylguanine methyltransferase, 88
O^6-MGMT gene, 56, 57
O-acetyl-ADP-ribose formation, 19–20

O-acyltransferase, 263
Obesity
early nutrition and, 274
epigenetic factors and overfeeding, 7–8
epigenetic factors in, 85
and epigenetic modifications, 220
fetal programming for, 236–239
and genistein supplementation, 86
and HCC, 144
11-hydroxysteroid dehydrogenase 1 in, 262
hyperinsulinemia and, 262
IGF-1 treatment of undernourished offspring, 238
increased incidence of metabolic syndrome, 262
and insulin resistance, 261–262
maternal high-fat diet and, 273
maternal status during gestation, 130–131
mitochondrial oxidative phosphorylation, 261
and NASH, 165
and polyphenolic food constituents, 16
during pregnancy, 126
pregnancy risks, 236
proinflammatory state, 260
protective phenotype, 261–262
resveratrol protective, 22–23
risk in agouti mouse, 85–86
Observational studies of folate status and DNA methylation, 53ff
Occupational dust exposure exacerbates COPD, 186
OGG1 (Oxoguanine DNA glycosylase), 368
8-OHdG (8-Hydroxydeoxy guanine), 147
Oleic acid in olive oil, 371
Oligodendrocytes, 294
Olive oil
for oxidative stress, 371
polyphenols in, 371
Omega-3 fatty acids
and hippocampal neurogenesis, 374
for schizophrenia treatment, 302
3′-O-methyl epicatechin, 372
Oncogenes
activation by demethylation, 2
hypomethylated in disease states, 2
Oncogenic targets, 162
Oncomir in solid tumors, 98
One-carbon metabolic pathway, 295–296, *see also* Folic acid one-carbon metabolism (FOCM)
and Down syndrome, 334
epigenetic reprogramming of, 273
folate in, 319
genetic variants as schizophrenia risk factors, 333
and histone methylation, 115–116
variants in autism, 329–330
Onion as quercetin source, 192
Oral contraceptives interfere with folate metabolism interference with folate metabolism, 300
Oranges as folate source, 318
Orexigenic neurons, 272
Organ meats as folate source, 36
Organosulfur compounds, *see also* Diallyl disulfide (DADS); Garlic
alter HDAC activity, 23
alter histone acetylation, 23
HDAC inhibitors, 91
Oropharynx malignancy and folate status, 36
Orotic acid, 272
Ovarian cancer
demethylation in, 112
and folate status, 36
Overfeeding, *see* Obesity
Oxidative stress
cause of COPD, 186
chemically induced, 198
in COPD process, 186
damage leading to diabetes, 266
in diabetic complications, 221
and glucose intolerance, 262
and neurodegeneration, 366–368
p53 accumulation, 196
polyphenol extract for, 199
reduction of, 370
8-Oxo-G (Cerebral oxoguanine), 367
2-Oxoglutarate oxygenase fold (2OG), 267
Oxoguanine DNA glycosylase (OGG1), 368
Oxysterols as nutrient sensor, 271

P

p110β, 240, 242
p14(ARF) promoters, 89
p14 gene, 56
p15^{INK4A} hypermethylation, 59
p15 silenced, 148
p16
and lung disease, 189
methylation state of, 189
p161INK4a, 88
p16 *(CDKN2A)*, 248–249
p16 cell cycle regulator, 56
p16 gene, 56, 57
genistein induced expression of, 93
methylation and folate in aging subjects, 64
promoter methylation, 57
promoter methylation of, 59
p16(INK4a) promoters, 89
p16 methylation, 39
p16 silenced in cancer, 148
p21 gene
genistein induced expression of, 93

PEITC effect on, 92
silencing of, 82
and tumorogenesis, 35–36
p21 mRNA expression, 91
p21 promoter and expression, 91
p21 protein in colorectal cancer cells, 90
$p21^{WAF}$, 195
P21WAF1 luciferase reporter, 92
p21WAF1 promoter, 91
$p27^{kip1}$, 197
p300, 219
in diabetes, 266
p300/CBT acetyltransferase, 111
p300 coactivator, 268
p53
acetylation status of, 196
in calorie restriction, 270
deacetylated by SIRT1, 195
dependent apoptosis, 198
effect of resveratrol, 197
genistein effect on, 93
SIRT1-mediated repression, 23
p53 deacetylase, 20
p53-dependent apoptosis, 36
p53 gene
in aging, 20–21
folate deficiency and DNA hypomethylation, 48
p53 hypomethylation, 39, 41
p53 methylation affected by folate, 44
p53 region methylated CpG sites, 35
p53 transcription factor, 91
p65 gene, 221–222
pAkt, 248
Paleolithic diet, 372
Pancreas
development and maternal diet, 241–242
folate storage, 319
Pancreatic and duodenal homeobox 1 (Pdx1), 223, 242
Pancreatic cancer
curcumin effect on, 98
curcumin effect on cells, 97
demethylation in, 112
and folate status, 36
miR-181 elevation, 167
miR-21 activity in, 98
miRNAs elevated expression, 168
PaO_2, 199
Parasitic DNA, 32–33
Parasitic sequences, 34
Parental germline gamete gender influence, ix
Parent-of-origin effect, 274
in neurodegenerative diseases, 369
Parkinson's disease (PD), *see also* Neurodegenerative disease
EGCG for treatment, 370
GST activity in, 367
hyperhomocysteinemia in patients, 366
hypomethylation in, 369
lower GSH levels in, 367
oxidative damage in, 367
possible treatments, 370
TNF-alpha promoter, 369
PARP1, *see* Poly (ADP-ribose) polymerase (PARP1)
Partridgeberry as resveratrol source, 22
Parvalbumin-immunoreactive neurons, 294
PASR (Polymerase-associated small RNA), 154
Paternal age
offspring's increased risk of schizophrenia, 303–304
risk factor for major psychosis, 288–289
Paternal disomy, 340
Paternal genes, 233
Paternal genetic influences in neurodevelopment, 340
Paternally imprinted genes, 233
Pax2 regulator, 267
PC-3, 91
PcG, *see* Polycomb group (PcG) of proteins
PcG complex 2 binding, 65
PcG complex cooperation, 96
PcG genes in cancers, 83
PD, *see* Parkinson's disease (PD)
PDCD4 in HCC, 165
PDD (Pervasive developmental disorder), 359
Pdx1 (Pancreatic and duodenal homeobox 1), 223, 262
Peanuts
as choline source, 16
as resveratrol source, 22
Peas as folic acid source, 16
Pedersen hypothesis, 245
Peg10 gene, 234
Peg3, 235
and maternal diet, 235
Peg3 gene, 234
PEITC (Phenethyl isothiocyanate), 92
PEMT, *see* Phosphatidylethanolamine N-methyltransferase (PEMT)
PEPCK (Hepatic gluconeogenic phosphoenolpyrubate carboxykinase), *see* Hepatic gluconeogenic phosphoenolpyrubate carboxykinase (PEPCK)
Peptides (SS-31) for neurodegenerative disease treatment, 370
Peptidylprolyl isomerase E-like gene (PPIEL), 291
Pericentromeric satellites, 304
Perinatal nutrition
critical period for, 337–338
gene methylation effects, 306
influence of, ix

Peripheral blood mononuclear cells, 53
Peripheral circadian regulator, 94
Pernicious anemia, 319
 neurological effects of, 319
 vitamin B_{12} deficiency, 319
Peroxisome proliferator-activated receptor-alpha (PPAR-alpha), 220
Pervasive developmental disorder (PDD), 329
Pesticides and neurodegenerative diseases, 367
PGC-1α deacetylation, 23
P-glycoproteins, 202
Pharmacologic studies needed for dietary agents, 117
Phase 2 detoxification genes, 198
PHD finger domains, 110
Phenethyl isothiocyanate (PEITC), 92
Phenotype affected by maternal methyl group supplementation, 39–40
Phenotype switching in atherosclerosis, 5
Phlda2 gene, 234
Phosphatidylcholine, formation of, 321
Phosphatidylethanolamine methyltransferase inhibited by SAH, 339
Phosphatidylethanolamine N-methyltransferase (PEMT), 321
Phosphatidylinositol (PI3-kinase), 242
Phosphocreatine levels in brain, 301
Phospholipid metabolism and liver damage, 147
Phospholipids synthesis
 B_{12} vital for, 319
 choline role in, 145
 and SAM, 298
Phosphorylation, 78, *see also* Histone modification
 effect on histones, 81
 histone modification, 3
 of histones, 22, 110
 and lung disease, 189
Phytoalexin, 197
Phytochemicals
 biphasic dose responses, 372
 in cruciferous vegetables, 98
 may reactivate silenced genes, 372
 for neurodegenerative disease treatment, 370
PI3-kinase, *see* Phosphatidylinositol (PI3-kinase)
Piceatannol, 202
 in blueberries, 115
Pi-class glutathione-S-transferase (GSTP1), 89
Piperin interaction with curcumin, 202
piRNAs coordinate methyltransferases, 110
Piwi-interacting RNA, 154
PKC zeta, 242
Placenta affected by high-fat diet, 236
Placental development, 235
Placental imprinted genes, 233–234
Placental insufficiency, 231–232
Placental Wnt pathway, 236
Plaque growth and HDL levels, 263
Plasma antioxidants, 187
Plasma cholesterol regulation, 156–157
PLP, *see* Pyridoxal 5 phosphate (PLP)
Pluripotency, 82
PMA-stimulated JB6 cells, 200
PNN, *see* Preneoplastic nodules (PNN)
Poaceae as resveratrol source, 22
Pol I activity, 272
Pollutants effect on lung health, 188
POLR3D, 83
Poly (ADP-ribose) polymerase (PARP1), 194
Polyamines synthesized from SAM, 133
Polycomb group (PcG) of proteins, 82–83, *see also* PcG
Polycomb protein group, 3
Polycomb repressive complex 1 bioactive food components modify, 380
Polycomb repressive complexes, 95–96
Polyglutamates, *see* Folate
Polygonum cuspidatum, 21, 22
Polymerase-associated small RNA (PASR), 154
Polyphenol E, 202–203
Polyphenolic food constituents, 16
Polyphenols, *see also* Catechins; Epigallocatechin 3-gallate; Flavonoids; Resveratrol
 activate Sirt1, 94
 adverse effects of, 201–202
 Annurca apple, 89
 with anti-cancer properties, 115
 anti-inflammatory, 199
 bioavailability of, 202–203
 chemopreventive agents, 198
 as chemopreventive for lung disorders, 196, 200
 for chronic lung disease, 191ff
 in combination with steroids, 194
 drug metabolism interference by, 201
 effect on P-glycoprotein, 202
 effects of, 199
 effects of digestion, 203
 flavonoids, 192
 HDAC modulation, 192–195
 for improved lung function, 199–201
 increase HDAC2 activity, 194
 inflammation modulation, 186
 inflammatory response mediation by, 194
 neuronal communication effects, 372
 for oxidative stress, 186
 redox signaling modulation, 186
 in red wine, 371
 for reversing epigenetic changes in lung disorders, 199
 signal transduction effects, 372
 SIRT1 interaction, 196
 sirtuin activation by, 22

Polyps reduced by treatment, 34
Polyunsaturated fatty acids (PUFA)
in Mediterranean diet, 371
for schizophrenia treatment, 302
Pomegranate extracts in epigenetic health, 301
Pomiferin, 277
Pon2, 235
Portulaca oleracea as omega-3 fatty acid source, 302
Postmenopausal women, *see also* Aging
folate depleted diets in, 49
methylation state of, 64
Posttranslational histone modification, 78
PPAR, 240
and dyslipidemia, 273
effect of maternal low-protein diet, 273
fat metabolism effects, 273
PPAR-1 defects, 262
PPARα, 271, 276
PPAR-alpha, *see* Peroxisome proliferator-activated receptor-alpha (PPAR-alpha)
PPAR coactivator-1 in metabolic syndrome, 261
PPARγ
in diabetes, 266
generational transmission, 243
PPAR-γ 2 in diabetes, 267
PPARγ coactivator-1
in diabetes, 266
genetic variant, 262
in metabolic syndrome, 261
PPARGC1A, 266–267
Ppary gene, 61
PPIEL (Peptidylprolyl isomerase E-like gene), 291
Prader-Willi syndrome (PWS), 274
co-occurrence with ASD, 339–340
genetic defects in, 340
loci involved, 340
pRb, 248–249
Prediabetes, 244, 245, 262
Preepimutation in schizophrenia, 295
Prefrontal cortex
executive function processes in, 296
GAD1 in, 295
in major psychoses, 292
SAM levels, 298
Pregnancy, *see also* Folic acid; maternal entries
critical window for folate/B vitamin effects, 337
dietary exposures, 60–63
folate deficiency in, 144–145
maternal folate intake studies, 62–63
in obese women, 130–131
preimplantation embryos reprogrammed, 327
Prenatal famine exposure, 84–85
Preneoplastic nodules (PNN), 152
FMD diet effect on, 157, 159
Presenilin 1 (PSEN1), 368
Preterax, 215
Primary miRs (pri-miR), 156
Prohormone convertase, 7
Proinflammatory genes in diabetic complications, 223
Proline isomerization, 289
Promoter CpG islands, *see* CpG islands
Promoter hypermethylation and aging, 63–64
Promoter methylation and colorectal carcinomas, 57
Promyelocytic leukemia
protein, 81–82
zinc finger, 81–82
Promyelocytic leukemias
HDAC and, 81
NuRD complex and, 82
Proopiomelanocortin gene, 7
Prostate cancer
cyclin decrease in cells, 93
and folate status, 36
gene expression in, 88
hypomethylation and, 14
MCP30 therapy, 115
metastatic liver carcinoma and, 143
selenium treatment of, 115
Prostate cancer cells, *see also* specific cell types, PEITC effect on, 92
Prostate epithelial cells, 91
Prostatic intreepithelial neoplasia, 92
Proteases in COPD process, 186
Protein aggregation reduction, 370
Protein deacetylation, 19–20
Protein expression affected by folic acid supplementation *in utero*, 61
Protein kinase B, 242
Protein kinase C, 261
Protein kinase C-mediated signal transduction, 147
Protein tyrosine phosphatase receptor-type O (PTPRO), 152–153
Proto-oncogene
activation, 113
hypomethylation, 39
upgregulation of, 34
PSEN1, *see* Presenilin 1 (PSEN1)
PSEN1 gene in Aβ processing, 369
Pseudagouti mice, 299
pSrc, 248
Psychological distress contribution to disease process, 5–6
Psychoses, *see* Major psychoses
PTEN, in HCC, 164–165
Pten gene silencing, 21
Pteroyl-L-glutamic acid, *see* Folic acid
Pteroylmonoglutamic acid, 300, *see also* Folate

PTPRO (Protein tyrosine phosphatase receptor-type O), 152–153
PUFA, *see* Polyunsaturated fatty acids (PUFA)
Purine synthesis in DNA damage and repair, 7
Purkinje cells, 290
Purslane as omega-3 fatty acid source, 302
PWS, *see* Prader-Willi syndrome (PWS)
Pyridoxal 5 phosphate (PLP) in autism, 329
Pyridoxal kinase activity in autism, 329
Pyridoxal phosphate, *see* Vitamin B_6
Pyridoxine, *see* Vitamin B_6

Q

Quercetin
  activates Sirt1, 94
  adverse effects of, 201
  beneficial in COPD, 199
  drug interference by, 201
  effect on P-glycoprotein, 202
  in epigenetic health, 301
  modulates NF-κG activation, 192
  sources of, 115

R

RA, *see* Retinoic acid (RA)
RAR, *see* Retinoic acid receptor (RAR)
RAR-B_2 gene, 59
Rare-cutting restriction sites, 151–152
ras, 248
RAS, *see* Renin-angiotensin (RAS) system
RASSF1A gene, 56
RB silenced, 148
Rdh12, 94
Reactive oxygen species (ROS)
  in COPD process, 186
  and neurodegeneration, 366–367
  vitamin scavenging of, 370
Rearrangement promotion, 34
Recombination affected by histone modification, 81
Red cells folate storage, 319
Redox-sensitive intracellular pathways, 372
Reduced folate carrier (RFC1) enzyme, 322, 364
Red wine
  antioxidant polyphenols in, 371
  cardioprotective effect, 21, 22
  and dementia incidence, 371
  resveratrol source, 192
Reelin (RELN) gene, 341
RelA/p65, 192
  acetylation in smokers, 194
  deacetylated by SIRT1, 195
Religious Orders Study, 6
RELN, 341
  in schizophrenia, 341
Renal dysfunction
  effect of maternal diet, 127
  maternal diet and, 129–130
Renin-angiotensin (RAS) system, 267
  in carcinogenesis, 168
  inactivation, 3
Repair affected by histone modification, 81
Repetitive gene regions methylation state, 3
Replication affected by histone modification, 81
Repression, *see* Transcriptional repression
Restriction Landmark Genomic Scanning (RLGS), 150
  methylation analysis (RLGS-M), 150
Restriction length polymorphisms (RFLPs), 88
Resveratrol, 21–25, *see also* Polyphenols
  activates Nrf2, 199
  activates Sirt1, 94
  for AD, 372
  aging retardation by, 23
  antiproliferative agent for cancer treatment, 21–22
  bioavailability of, 202
  caloric restriction mimic, 23
  cardioprotective, 22
  cellular protection by, 198
  chemopreventive agent, 198
  dietary restriction mimetic, 22–23
  in drug design, 23
  effects of, 115
  effects on SIRT1, 197
  induces HO-1, 199
  inhibits AP-1 activation, 200
  inhibits inflammatory cytokine release in COPD, 200
  inhibits NF-κB activation, 200
  insulin resistance protective, 22–23
  lung anti-inflammatory, 200
  modulates NF-κG activation, 192
  for neurodegenerative disease treatment, 370
  and neurodegenerative processes, 372
  obesity protective, 22–23
  piceatannol, 202
  plant sources of, 22
  reduces effects of free radicals, 22
  Sir2 proteins stimulated by, 22
  sources of, 115
  for stress, 371
  stress response by plants, 22
Retinoic acid (RA), 95
  in epigenetic health, 301
Retinoic acid receptor beta, 88
Retinoic acid receptor (RAR)
  action inhibited, 81–82
  chromosomal changes in, 81–82
  as nutrient sensor, 271
Retinoic X receptor (RXR) as nutrient sensor, 271

Retinopathy and hyperglycemia memory, 216
Retrotransposons, 116
Retroviruses in human DNA, 32
Rett syndrome (RTT), 331–332
  co-occurrence with ASD, 339–340
  genetic defects in, 340
  mutations in MECP2, 340
REV-ERBalpha, 166–167
Reverse transcriptase-PCR analysis, 88
RFC1, *see* Reduced folate carrier (RFC1) enzyme
RFLPs, *see* Restriction length polymorphisms (RFLPs)
Rheum as resveratrol source, 22
Rhus verniciflua as source of butein, 115
Ribosylation, 289
RING finger domains, 110
Risk factors modifiable, 115–116
RIZ1, downregulation, 113
RIZ1 downregulation, 112
RKO colorectal cancer cells, 89
RLGS, *see* Restriction Landmark Genomic Scanning
RLGS-M, *see* Restriction Landmark Genomic Scanning
RNA
  noncoding, 79, 380
  types of microRNA, 154
RNA binding protein Pasha (DGCR8), 156
RNA interference, 32
RNAi protein, 83
RNA polymerase III (RNase III), 156
RNase III, *see* RNA polymerase III (RNase III)
ROAA gene in ASD, 341
ROS quenching, 194
RsaI RFLP, 88
Rtl1 gene, 234
RTT, *see* Rett syndrome (RTT)
RUNX3 gene, 57
RUNX3 gene methylation and folate, 59
Rush Memory and Aging Project, 6
RXR, as nutrient sensor, *see* Retinoic X receptor (RXR)

S

Saccharomyces cerevisiae life extended by resveratrol, 23
S-adenosylhomocysteine hydrolase (SAHH), 323
S-adenosylhomocysteine (SAH)
  accumulation, 327
  age effects on action of, 47
  changes with age, 64
  elevated in ASD, 341
  in folate cycle, 364
  folate deficiency and levels, 48
  folate effects in brain, 44–45
  folate effects in colorectum, 44–45
  and folate/methyl deficient diet, 299
  in folate normal subjects, 53
  and folate pathway, 38
  and folate status, 41
  increased by folate deficiency, 15
  levels and folate status, 40
  in major psychoses, 298
  methylation inhibitor, 323
  methyltransferase inhibition in brain development, 339
  and NTD risks, 327
  regulation CBS, 364–365
  selenium influence on levels, 47
S-adenosylmethionine (SAM)
  age effects on action of, 47
  as antidepressant, 301
  availability influenced, 268
  B_{12} vital for production of, 319
  and Aβ formation, 368
  changes with age, 64
  decreased by folate deficiency, 15
  depletion caused by folate deficiency, 147
  effect on phosphocreatine in brain, 301
  in folate cycle, 364
  folate deficiency and levels, 48
  folate effects in brain, 44–45
  folate effects in colorectum, 44–45
  folate in production of, 235
  and folate/methyl deficient diet, 299
  in folate normal subjects, 53
  and folate status, 41
  in folic acid metabolism, 270
  levels and folate status, 40
  methyl donor, 33
  methyl source for lysine methylation, 112
  nutrient influence on, 78
  in one-carbon metabolism, 298
  pathway to, 38
  polyamines synthesized from, 133
  regulation by methyl sources, 132, 133
  synthesis mechanism, 133
  for tetrahydrobiopterin synthesis, 338
  universal methyl donor, 79, 110, 115–116
SAHA, 194
SAHH (S-adenosylhomocysteine hydrolase), 323
SAH (S-adenosylhomocysteine), *see* S-adenosylhomocysteine (SAH)
S-allylmercaptocysteine (SAC)
  cell cycle arrest by, 91
  effects of, 23
  produce in vivo from DADS, 195
  similar to butyrate, 195
Salmon as source of PUFAs, 371
SALS (Sporadic ALS), 369
SAM (S-adenosylmethionine), *see* S-adenosylmethionine (SAM)
SAM:SAH ratio, 298

and B vitamin deficiency, 368
in colorectal cancer, 53
and folate/methyl deficient diet, 299
methionine pushes to SAH, 300
Schizophrenia (SCZ), 277, *see also* Major psychoses
brain development factors, 339
characteristics of, 287
and Chinese famine, 305
clozapine action, 303
DNA methylation in, 291
and Dutch Hunger Winter, 305
epigenetic brain factors, 288
epigenetic changes in, 6
epigenetic mechanisms in, 341
folic acid supplementation, 299–300
gender differences in DNA methylation, 291
genetic risk factors, 333
H3K27me2 in, 295
histone 3 in, 295
histone methylation modifications in, 341
histone status in brain tissue, 295
homocysteine effects of maternal levels and risk, 332
and homocysteine levels, 296, 298
increase with paternal age, 303–304
major psychosis, 287
and maternal folic acid deficiency, 308
MEK1 gene and antipsychotic use, 303
methylation pathologies in, 333
methylfolate treatment, 333
neurodevelopmental disorder, 332
NMDA abnormalities in, 293
overlap with BD, 288
reduced BDNF levels in, 302
reduced omega-3 levels in, 302
risk and short birth interval, 308
SOX10 status, 292
valproic acid adjuvant for treatment, 23
vitamin B_6 supplementation, 299–300, 301
Scots pine, as resveratrol source, 22
SCZ, *see* Schizophrenia (SCZ)
S-D rats, 238
γ-Secretase, 368
β-Secretase gene and AD, 367
β-Secretase in amyloidogenic cleavage of APP, 368
Seed sequence, 156
SELECT trial, 92
Selenite, 115
Selenium
in color cancer prevention, 89–90
co-occurrence with sulforaphane, 115
decreases HDAC activity, 115
DNMT inhibitor, 89
enriched yeat as prostate cancer protective, 92
in epigenetic health, 302
folate and, 47
metabolism and HDAC expression, 92
metabolites, 92
as plasma antioxidant, 187
selenite treatment, 89
speciation, 92
for treatment of prostate cancer, 115
Selenomethionine, *see* Selenium
Se-methyl-selenocysteine (MSC), 92
Senescence
cellular stress-induced, 20
p53 effects, 196
SIRT1 effect on, 195
Serine
acetylation of in histone, 3
deficiency effects, 133
phosphorylation of insulin receptor, 261–262
Serine hydroxymethyltransferase (SHMT), 322
Serotonin deficit effects, 339
Serotonin specific reuptake inhibitors, 201
Serotonin synthesis, 338
SET domain-containing protein (Set7/9), 216, 222, 223
Sex-determining region Y-box containing gene 10 (SOX10), 292
SFN, *see* Sulforaphane (SFN)
SFN-Cys, *see* SFN-cysteine (SFN-Cys)
SFN-cysteine (SFN-Cys), 195
Sfrp5 gene, 236
SHMT, *see* Serine hydroxymethyltransferase (SHMT)
Short birth interval
risks and folic acid deficiency, 308
and schizophrenia risk, 308
Silencing
definition; *see also* Gene silencing, 32
methylation in, 112
of second allele, 233
Silent information regulator (SIR) proteins, 19
Silver-Russell syndrome (SRS), 4
Single-nucleotide polymorphism
folate metabolism, 324–325
in neurobehavioral disorders, 303
SIR, *see* Silent information regulator (SIR) proteins
Sir2
effect on life span, 20
inactivates p53, 20–21
SIR-2 enzymes, 19–20, 81
SIR2 family, 111
Sir2 proteins
life preservation effect, 22
resveratrol stimulates, 22
siRNA, *see* Small interfering RNA (siRNA)
SIRT1
activation, 22

in apoptosis control, 20
apoptosis effects of, 196
in calorie restriction, 270
degradation by cigarette smoke, 381
expression, 94
inhibition decreases stress resistance, 198
inhibits NF-κB, 198
localization sites, 196
oxidative stress and, 196
regulation of lung inflammation, 381
role in cancer modulation, 196
role in dietary restriction; *see also* Calorie restriction
role in lung disorders, 196
sensitizes tumor cells to TNF, 198
SIRT1-dependent mechanism, 23
SIRT1-mediated repression
of NFκB, 23
of p53, 23
Sirtuins (SIRT or SIRT1), 20, *see also* Histone deacetylase (HDAC)
activation of, 115
anti-aging effects of, 23
cancer protective in calorie restriction, 94
class III HDACs, 195
delay cancer development, 115
and HDACs, 203
increase life span, 115
inhibitors to target cancer cells, 198
NAD+-dependent, 195
natural compounds affect activity, 196–197
polyphenol activation of, 22
SAHA does not inhibit, 195
SIRT1-7, 111
stimulated by resveratrol, 22
trichostatin A does not inhibit, 195
Skewed X chromosome inactivation, 294
SLC19A1, *see* Solute carrier family 19, member 1 (SLC19A1)
SLC19A1 A80G (rs1051266) variant, 325
and autism, 330
NTD risks, 325
SLC19A1 variant, 329
Slc38a4 gene, 234
SLE, *see* Lupus erythematosus (SLE)
Smad2, 3 and 4, 167
Small interfering RNA (siRNA), 154, 223
in diabetic complications, 222
for HCC gene therapy, 170
Small nuclear riboprotein N gene (SNRPN), 340
Small nuclear RNA (snRNA), 154
Small nucleolar RNA (snoRNA), 154
Small RNAs, 83, *see also* MicroRNA (miRNA or miR)
Smoking, *see also* Environmental tobacco smoke (ETS)
COPD cause, 186
effects from early life exposure, 188
fetal effects, 237, 245
folate deficiency and, 144–145
folate intake as protective, 59
inhibition of inflammation caused by, 198
maternal and asthma incidence, 191
and methylation in lung inflammatory disease, 189
reduces HDAC2, 194
SIRT1 levels and, 196
SMS, *see* Spermine synthase (SMS) gene
Snat2, 235
snoRNA (Small nucleolar RNA), 154
SNP (rs6265-val66met), 303
snRNA (Small nuclear RNA), 154
SNRPN, *see* Small nuclear riboprotein N gene (SNRPN)
Socioeconomic environment
effect on lung health, 188
neural tube defects and, 323
SOCS1 gene, 57
SOD1 gene, 369
Sodium retention in metabolic syndrome, 260
Solute carrier family 19, member 1 (SLC19A1), 322
Somatic mutations caused by chemotherapy, 144–145
Sorfenib, 169
SOX10, *see* Sex-determining region Y-box containing gene 10 (SOX10)
Soy, *see* Genistein; Soybeans
Soybeans, *see also* Genistein
components with opposing action, 93
EGCG source, 88
as genistein source, 117
SP1 transcription factor, 98
Sp3 transcription factor, 91
Sparkleberry as resveratrol source, 22
Spasticity in CFD, 366
Species differences, 65
Specimen types for epigenetic analysis, 4
Spermine synthase (SMS) gene, 291
Spina bifida, 274, *see also* Neural tube defects
genetic risk factor, 324–325
Spinach as folate source, 318
Spinal cord degeneration, 320
Spontaneous hepatocarcinogenesis (HCC), 96
Sporadic ALS (SALS), 369
Spruce as resveratrol source, 22
Squalene in olive oil, 371
Squamous cell carcinoma, 96
Src, 248
SRF, 165
in HCC, 167
SRS, *see* Silver-Russell syndrome (SRS)
SRT1720, 198

SS-31 for neurodegenerative disease treatment, 370
Starvation effects, 271–272
Steatohepatitis, 163
Steatosis
FMD diet effect on, 161
and NAFLD, 165
Stem cell aging, 25
Stem cell defects, 25
Stem cell pluripotency, 82
Stem cell proliferation, 4
loss of imprinting of IGF2, 4
Steroid resistance in asthma and COPD, 192
STMN1, 165
STMN2 gene, 304
Stomach cancer
and folate status, 36
hypomethylation and, 14
Stress resistance
regulated by transcription factors, 195
SIRT1 and, 198
Substantia nigra, 367, 369
Succinyl coenzyme A, 320
Suckling diet, 131
Sulforaphane (SFN)
activates Nrf2, 199
activity of, 277
apoptosis induction by, 91–92
in cancer chemoprevention, 194
cancer protective action, 372
chemopreventive agent, 198
co-occurrence with selenium, 115
effect on histone, 269
effects of, 23
HDAC inhibition by, 114–115
HDAC inhibitor, 91–92
inhibits histone deacetylase, 372
no effect on HDAC activity, 195
Sumoylation
effect on histones, 81
histone modification, 3
of histones, 110
and lung disease, 189
Sunflower seeds as folic acid source, 16
Suppressor gene aberrations liver damage caused by, 147
Suv39h1/SuVa39, *see* Variegation 3-9 homologue 1 (Suv39h1/SuVa39)
Suv39h downregulation, 113
Suv4-20h2 downregulation, 112, 113
SuVa39/Suv39h1, 223
SW480 colorectal cancer cells, 89
SW48 colorectal cancer cells, 89
SW620 cell line, 48
SWI/SNF, *see* SWItch/Sucrose NonFermentable (SWI/SNF) complex
SWItch/Sucrose NonFermentable (SWI/SNF) complex, 82
Syndrome X, *see* Metabolic syndrome
System A amino acid transporter, 232
Systems biology approach, 8

T

T2D, *see* Diabetes
T2DM, *see* Diabetes
TAG, *see* Triacylglycerol (TAG) synthesis
Tardive dyskinesia and vitamin B_6 levels, 332
Taurine
in epigenetic health, 301
synthesis from SAM, 133
T cell DNA methylation and NTD risks, 327
T cell function, 14–15
T cell lymphomas, 35
TCF21 gene, 152
TCN2 776 G alleles and Down syndrome, 334
TCN2 gene, 323
variant and autism, 329
Tea, *see* Green tea
Telomere abnormalities, 113
TET1, 290
Tetrahydrobiopterin synthesis, 338
Tetrahydrofolate, 57
Tetrahydrofolate (THF), 322
in folate cycle, 364
in folic acid metabolism, 318
metabolization of, 322
TGF-β, 267
TGFβ pathway, 167
TGS, *see* Transcriptional gene silencing (TGS)
Thalamus levels of choline metabolites in autism, 329
Theophylline
in epigenetic health, 301
increased dexamethasone activity, 194
increases HDAC2 activity, 194
THF, *see* Tetrahydrofolate (THF)
Th helper cells, 190, 191
Thiols for oxidative stress, 186
Thrifty phenotype hypothesis, 238, 268, 304
and Dutch Hunger Winter, 272
Thrombosis and HDL levels, 263
Thymidylate production, 51
Thymidylate synthase (TS or TYMS) variant, 325
Thymidylate synthesis in DNA damage and repair, 7
Thyroid cancer
hypomethylation and, 14
miRNA expression in, 84
Timing of cellular vulnerability, 93, 272
in asthma, 189–190
prenatal famine and, 306

TIMP1 in HCC, 165
TIMP3, 165
TNF, *see* Tumor necrosis factor (TNF)
TNF-α
 in COPD process, 187
 in diabetic complications, 221
TNF-induced inflammatory genes, 222
Tobacco liver carcinogenicity, 144
Tofu as methionine source, 16
Toxins and lung health, 188
TRAIL-induced apoptosis in HCC, 165
Transcobalamin, 323
Transcriptional activation, 112, *see also* Activation
Transcriptional co-repressor, 33
Transcriptional gene silencing (TGS), 83
Transcriptional interference, 33
Transcriptional repression, 112
Transcriptional repressors, 148
Transcriptional silencing
 by hypermethylation, 80
 by methylated CpG islands, 33
 by miRNAs, 83
 SIR proteins in, 19
Transcription factor binding, 341
Transformation of cells, 93
Transgenerational effects
 epigenetic mechanisms, 86
 psychosis transmission, 303–304
Transgenerational effects of environmental factors, 374
Translation suppression of mRNA, 3
Translocation caused by transposon activation, 3
Transposon activation and chromosomal instability, 3
Transposon insertion, ix
Transposons in human DNA, 32
Transsulfuration cycle, 323
Triacylglycerol (TAG) synthesis, 263
Trichostatin A
 HDAC inhibitor, 90, 194
 to identify tumor suppressor miRNA, 169
Triglyceride levels in metabolic syndrome, 260
3,5,40-Trihydroxystilbene, *see* Resveratrol
Trimethylglycine, *see* Betaine
Trisomy 21, 333–334
Triterpenoids for neurodegenerative disease treatment, 370
tRNA methyltransferase inhibited by SAH, 339
Trophectoderm, 233
TS, *see* Thymidylate synthase (TS or TYMS)
TSG, *see* Tumor suppressor genes (TSG)
TT genotypes, 53, 55
 NTD risks, 325
Tumor behavior affected by bioactive food components, 77
Tumor cells
 characteristics of, 80
 H4 modification, 81
 histone methylation in, 196
Tumor development
 de novo methylation and, 39
 inactivation of, 80
 SFN suppression of, 91
Tumorigenesis
 HDAC inhibitors, 380
 imbalance in acetylation reactions, 380
Tumor necrosis factor (TNF), 198
Tumor progression in animal model, 146
Tumor promoter methylation, 57
Tumor suppressor genes (TSG)
 in CpG islands, 35
 derepression of silenced genes, 92–93
 epigenetic factors in lung health, 188
 green tea effects, 22
 identification of, 152
 inactivated by hypermethylation of promoters, 2
 inactivation of, 80
 silencing by hypomethylation, 147
 SIRT1 localization, 196
Tuna as source of PUFAs, 371
Turmeric source, 192
Turnip greens as folate source, 318
Twins, *see* Monozygotic (MZ) twins
TWIST2 gene, 152
Two-hit theory of carcinogenesis, 147–148
TYMS, *see* Thymidylate synthase (TS or TYMS)
Tyrosine kinase receptor in HCC, 165

U

U937 cells, 200
UBE3A mutations, 340
Ubiquitin, 110
Ubiquitination, 78, *see also* Histone modification
 effect on histones, 81
 of histone H2A, 82
 histone modification, 3
 of histones, 22, 110
 and lung disease, 189
Ubiquitin ligase, 340
UHRF1 protein, 110
UKPDS, *see* UK Prospective Diabetes Study (UKPDS)
UK Prospective Diabetes Study (UKPDS), 215–216
Ulcerative colitis, 36
Uncoupling protein-2, 262
Undernutrition, 125, 126
Untranslated region (UTR), 156
3′-Untranslated region (UTR), 156
Uracil
 folate deficiency and, 48

misincorporation, 48
Uracil incorporation in DNA, 7
Urban environment as risk factor for major psychosis, 288–289
Urban environment effects on major psychoses, 308
Uremia
clinical trials, 51
DNA methylation reduced in, 88
Uterine cancer, 249
5′-UTR, 156
UTR, *see* Untranslated region
3′-UTR 6 bp deletion (del) variant, 325

V

Valproic acid
adjuvant for antipsychotics, 23
effect on epigenetic process, 303
in neuronal differentiation, 374
Variegation 3-9 homologue 1 (Suv39h1/SuVa39), 216
Vascular cell adhesion molecule (VCAM-1), 187
Vascular disease, *see also* Cardiovascular disease
birth weight correlation, 129
dysfunction and maternal protein restriction, 129
Vascular epigenomics, 5
Vascular inflammation and HDL levels, 263
Vascular insulin resistance and possible neonatal intervention, 128
Vasculopathic genes, 5
Vasculoprotective genes, 5
Vasoconstriction due to fat deposition, 263
VCAM-1 (Vascular cell adhesion molecule), 187
Vegan diet and vitamin B_{12} deficiency, 319
Vegetables, *see also* Bioactive food components
as folate source, 36
as folic acid source, 16
source of bioactive food components, 78
Vegetarian diet and vitamin B_{12} deficiency, 319
VEGF, 369
Vertical transmission of genomic imprints during pregnancy, 190
Victimization as risk factor for major psychosis, 288–289
Vidaza, *see* 5-Azacytidine (Vidaza)
Vinclozolin, 273
Vinyl-carbamate carcinogen, 98
Viral hepatitis, 97
Viral infection, 143
Visceral obesity in metabolic syndrome, 260
Vitamin A
deficiency and lung health, 188
and hippocampal neurogenesis, 374
retinoic acid (RA), 95
Vitamin B_{12}, 319–321
in AD, 366
in autism, 329
and Aβ formation, 368
clinical trials, 51
and cognitive functioning, 335
deficiency, 96
and methyl trap, 322–323
deficiency effects, 319
and DNA methylation in the colon, 47–48
exacerbated by folate deficiency, 300
in folate cycle, 235, 322
and homocysteine, 296
hypomethylation effects, 273
intrinsic factor for absorption, 319
methyl group donor, 38–39
MTR cofactor, 364
and NTD risks, 326, 327
prenatal supplementation, 87
as psychiatric treatment, 333
risk modification by, 116
supplementation in agouti mouse, 85–86
for tetrahydrobiopterin synthesis, 338
transport for uptake, 323
treatment option, 300
vital functions of, 319
Vitamin B_6
in AD, 366
in autism, 329
as autism treatment, 331
avoidance behavior learning and, 335–336
and behavioral outcomes, 336
cofactor in transsulfuration cycle, 323
and DNA methylation in the colon, 47–48
effect of supplementation on homocysteine levels, 299–300
in folate cycle, 235
and homocysteine, 296
lowers homocysteine levels, 301
as psychiatric treatment, 333
for psychosis treatment, 301
risk modification by, 116
in schizophrenia, 332
in tardive dyskinesia, 332
THF cofactor, 322
Vitamin B_9, *see* Folate
Vitamin B deficiency and homocysteine accumulation, 368
Vitamin B supplementation, 300
Vitamin C
as antioxidant, 371
effect of flavonoids on, 202
and iron absorption, 239
for neurodegenerative disease treatment, 370
for oxidative stress, 186
as plasma antioxidant, 187

Vitamin D deficiency, 188, 308
Vitamin deficiency and DNA methylation in the colon, 47–48
Vitamin E
 as antioxidant, 371
 in epigenetic health, 302
 maternal deficiency and asthma, 190
 metabolites may inhibit histone deacetylase, 372
 for neurodegenerative disease treatment, 370
 for oxidative stress, 186
 as plasma antioxidant, 187
Vitamins, *see also* specific vitamins
 in maternal diet, 250
 scavenge reactive oxygen species, 370
VLDL, 263
V-maf musculoaponeurotic fibrosarcoma oncogene homolog A (MafA), 223
Volume expansion, 260

W

Western-style diet
 effect on offspring, 238
 and psychotic disorders, 303
Wheat germ as choline source, 321
White hellebore as resveratrol source, 22
Wine has cardioprotective effect, 22
Wnt3a gene, 236
Wnt inhibitory factor-1, 200
Wnt pathway, 236

X

X chromosome and mental retardation in males, 340
X chromosome inactivation, 110, 339
 gene imprinting and, 4
 loss of normal DNA methylation, 16
XCI erasure and reestablishment, 340
Xenobiotics, 198
Xenopus levis miRNA, 155

Y

Yeast as folate source, 36

Z

Zebrafish miRNA, 155
ZFHX1B protein, 97
 FMD diet effect on, 162
Zinc
 catalytic mechanism, 111
 and CpG methylation, 268
 in folate cycle, 235
 and hippocampal neurogenesis, 374
 hypomethylation effects, 273
 maternal deficiency and asthma, 190
 in maternal diet, 250
 prenatal supplementation, 87
 supplementation in agouti mouse, 85–86
Zinc finger, 81–82
Zolinza, 114